SDGs and Textiles

Series Editor

Hafeezullah Memon, International Institute of Silk, Zhejiang Sci-Tech University, Hangzhou, China

The book series "SDGs and Textiles" addresses the strategies to achieve sustainable development goals (SDGs) in the present, past, and future. It presents books about the present and future policies of textile ministries of different countries, and books related to sustainability education around different parts of the world in the textile sector. Moreover, it would welcome the conference proceeding related to SDGs and Textiles. The series would cover books comparing the sustainability and SDGs of different institutions and countries. The individual book volumes in the series are thematic. The goal of each book is to give readers a comprehensive overview of a different area of sustainability in the textile sector. As a collection, the series provides valuable resources to a broad audience in academia, the research community, industry, and anyone looking to expand their knowledge of SDGs and Textiles.

Textiles and life are together – life cannot be separated from textiles as it is the most important need for human beings after food. In 2015, the United Nations General Assembly proposed 17 interlinked global goals to be achieved by 2030. Since then, academia and industry have paid much attention to achieving these goals. Textile found its close relation with almost all of these 17 goals.

SDG 1 - No Poverty: Poverty would never be overcome by a charity only; it is essential to develop people's skills to have a better and wealthy life. Thus, the textile can be considered an excellent discipline to achieve this goal by creating jobs and small and medium businesses.

SDG 2 - Zero Hunger: Through the effective utilization of advanced application of Agrotech Textiles, it is possible to have higher crop yields and save crops from rough weather, unexpected rains, floods, insects, etc.; thus, geotextiles play an essential in achieving this goal of sustainable development.

SDG 3 - Good Health & well-being: There has been much health consciousness after Covid19, and medical textiles assist in getting good health and well-being.

SDG 4 - Learning & Education: Textile or fashion has remained a significant discipline for societies for ages, and there has always remained much to explore in this field. Textile-related universities may play a vital role by offering free access to their education resources, training and spreading information among the locals.

SDG 5 - Gender Equality: The textile sector is one of the industrial sectors that accepted gender equality long ago; in particular, the garment sector has more females than males. Thus, the textile sector has been doing gender equality. Moreover, there has been a recent trend for Gender Neutral Clothing, which need worth studying and may further assist gender equality.

SDG 6 - Clean Water & Sanitation: Textiles could be achieved through filtration, and of course, textile is one of the critical materials for filtration.

SDG 7 - Affordable & Clean Energy: With the recent advancement in material science and engineering, the textile sector has come on the front for, not only by using this clean energy during textile production but also by assisting the production of this clean energy, either in the form of wind turbines blades made of textile composites or by energy harvesting from T-Shirts, etc.

SDG 8 - Decent Work: Recently, there has been much attention that the textile workers are not paid well, labor rights are not cared about, etc.

SDG 9 - Industry and innovation: Textile Industry always follows innovation; the textile companies that do not chase innovation cannot survive in the market.

SDG 10 - Reduced Inequalities: Getting better life and well-being would help reduce inequalities in the textile industry.

SDG 11 - Sustainable Cities: Sustainable Textile Cities through Buildtech and transport textiles.

SDG 12 - Consumption and Production: Textile and garment consumption and production all come under.

SDG 13 - Climate Action: Oekotech or Ecotech Textile, waste management of textiles are upfront to achieve this goal of sustainable development.

SDG 14 - Life Below Water: Mitigating microfiber waste in rivers and oceans may come under the context of it. There has been much attention on this subject after passing the bill at the parliament level of the UK.

SDG 15 - Life on Land: Geotech or Geotextiles studies life on land.

SDG 16 - Peace, Justice, and Strong Institutions: Protective textiles are doing their best to achieve peace, justice, and strong institutions.

SDG 17 - Partnerships for the Goals: The application of textiles to achieve sustainable development goals is only an example. In all textiles sectors, combined efforts of all the goals are essential to achieve true sustainability.

Shou-Xiang Jiang · Raphael Kanyire Seidu ·
Benjamin Tawiah
Editors

Advances in Textile Materials and Processing Techniques for Sustainability

Springer

Editors
Shou-Xiang Jiang
School of Fashion and Textiles (SFT)
Research Centre of Textiles for Future Fashion
The Hong Kong Polytechnic University
Hung Hom, Hong Kong, China

Raphael Kanyire Seidu
School of Fashion and Textiles (SFT)
The Hong Kong Polytechnic University
Hung Hom, Hong Kong, China

Benjamin Tawiah
Department of Industrial Art (Textiles and Fashion)
Kwame Nkrumah University of Science and Technology
Kumasi, Ghana

ISSN 2948-1236 ISSN 2948-1244 (electronic)
SDGs and Textiles
ISBN 978-981-95-0468-8 ISBN 978-981-95-0469-5 (eBook)
https://doi.org/10.1007/978-981-95-0469-5

This Springer imprint is published by the registered company Springer Nature Singapore Pte Ltd.
The registered company address is: 152 Beach Road, #21-01/04 Gateway East, Singapore 189721, Singapore

This book is dedicated to all the contributors, academic institutions, and stakeholders who have made this book possible. Their dedication to advancing knowledge and fostering innovation in sustainable textiles is commendable and vital for the industry's evolution.

Preface

In recent years, the textile industry has faced increasing scrutiny regarding its environmental impact and sustainability practices. As one of the largest sectors globally, the production and processing of textiles consume vast amounts of resources and generate significant waste. This reality has prompted a critical need for innovative solutions that not only enhance the quality and functionality of textile materials but also minimise ecological footprints. The book, *Advances in Textile Materials and Processing Techniques for Sustainability* addresses these pressing challenges by presenting a comprehensive overview of the latest developments in textile science and technology. This book brings together contributions from leading researchers, industry experts, and practitioners who are at the forefront of sustainable textile innovation. The first section of the book delves into advanced textile materials, exploring eco-friendly fibers, biodegradable polymers, and smart textiles that contribute to sustainability. These materials are not only designed to reduce environmental impact but also to meet the evolving demands of consumers for high-performance, multifunctional textiles.

The subsequent sections focus on processing techniques that enhance sustainability throughout the textile supply chain. Topics such as waterless dyeing methods, energy-efficient manufacturing processes, and waste reduction strategies are thoroughly examined. Moreover, the book emphasises the importance of a circular economy in textiles, advocating for practices that promote recycling, reusing, and repurposing materials. It highlights successful case studies and innovative business models that demonstrate how sustainability can be woven into the very fabric of the textile industry.

This compilation is not just a resource for researchers and academics; it is also intended for industry professionals, policymakers, and students who are keen to understand the complexities of textile sustainability. The book fosters collaboration and knowledge exchange, inspiring a collective effort toward a more sustainable future for the textile sector.

The book, *Advances in Textile Materials and Processing Techniques for Sustainability* serves as a timely and essential contribution to the ongoing discourse on sustainable practices in textiles. We invite readers to explore these advancements and consider their implications for the future of the industry, as we collectively strive for a more sustainable and responsible approach to textile production and consumption. We hope this book will not only inform but also inspire action towards a more sustainable textile industry, one that respects both people and the planet.

Hung Hom, Hong Kong, China Shou-Xiang Jiang
Hung Hom, Hong Kong, China Raphael Kanyire Seidu
Kumasi, Ghana Benjamin Tawiah

Acknowledgements

The compilation of *Advances in Textile Materials and Processing Techniques for Sustainability* would not have been possible without the invaluable contributions and support of numerous individuals and organizations. First and foremost, we extend our heartfelt gratitude to the esteemed authors and researchers who have shared their expertise and insights within these pages. Their dedication to advancing knowledge in the field of sustainable textiles is both inspiring and essential for fostering innovation. Each chapter reflects a commitment to addressing the pressing challenges faced by the textile industry, and we are deeply appreciative of their efforts. We also wish to acknowledge the support of academic institutions and research organizations that have provided the foundation for this work. Their commitment to sustainability and collaboration has been instrumental in shaping the discourse around textile innovation. We particularly want to thank the Hong Kong polytechnic University and the Kwame Nkrumah University for the support and received during the compilation of this all important book.

Our sincere thanks go to the reviewers whose constructive feedback and critical insights have greatly enhanced the quality of this book. Their expertise has ensured that the content is not only rigorous but also relevant to current industry challenges and opportunities. Additionally, we would like to express our appreciation to the publishing team for their professionalism and dedication throughout the production process. Their attention to detail and commitment to excellence have been invaluable in bringing this project to fruition.

Finally, we acknowledge the support of our families and friends, whose encouragement and understanding have sustained us during the writing and editing process. Their belief in the importance of our work has been a source of motivation and strength.

As we present this book to the academic community and industry practitioners, we hope it serves as a catalyst for dialogue, collaboration, and action towards a more sustainable future in textiles. Thank you for being part of this journey.

About This Book

This book focuses on innovative materials and processing techniques in the textile industry aimed at promoting sustainability. It addresses environmental concerns, resource efficiency, and the development of eco-friendly textiles. The initial chapters introduce the concept of sustainability in textiles, discussing the environmental impact of traditional textile manufacturing. It outlines the need for sustainable practices and sets the stage for the innovations discussed in subsequent chapters. The second half of the chapter explored the various sustainable materials used in textile production, such as organic cotton, hemp, recycled fibres, etc. It emphasizes the importance of material selection in reducing the carbon footprint of textiles. It further discusses the innovative dyeing methods that minimize water usage and harmful chemicals. Techniques such as digital printing and natural dyes are discussed, highlighting their benefits for both the environment and the textile quality.

The third half of the book addresses the issue of textile waste, presenting strategies for recycling and upcycling materials. It discusses the circular economy model and its application in the textile industry. Further, energy consumption in textile manufacturing is a critical concern, hence a couple of chapters are dedicated to critically examining the technologies and practices that enhance energy efficiency, including the use of renewable energy sources and energy-efficient machinery.

Subsequently, a chapter has been dedicated to discussing the methodology of Life Cycle Assessment, providing insights into the evaluation of the environmental impact of textile products from production to disposal. Furthermore, a comprehensive discourse focusing on understanding consumer attitudes towards sustainability is presented examining the market trends, consumer preferences, and the role of education in promoting sustainable textile choices.

Finally, the emerging trends and technologies that are shaping the future of the textile industry has been discussed thoroughly focusing on smart textiles, biotechnology, and the integration of sustainability into textile business models. Further, ethical and issues of regulatory compliance in the manufacturing of sustainable textile is presented. The book *Advances in Textile Materials and Processing Techniques for Sustainability* serves as a vital resource for researchers, practitioners, and students interested in sustainable practices within the textile industry. It provides a comprehensive overview of current advancements and future directions, emphasizing the importance of sustainability in textile manufacturing

Contents

Editors and Contributors

About the Editors

Shou-Xiang Jiang is currently a Professor of fashion and textiles at The Hong Kong Polytechnic University. He teaches a broad range of subjects such as innovative textile design approach, advanced textile and fashion, and international fashion and textiles design. Prof. Jiang's research deals with innovative textile effects using physical and chemical treatment technologies. He is internationally renowned for his development of metallised textiles. Leading a team that has developed a sputtering coating technology with research funding from the HKSAR government and sponsorships from the textile industries, Prof. Kinor Jiang has designed and launched an anhydrous system that uses metallising technologies to produce functional and decorative textiles for the production line. He holds 30 granted and filed patents based on the coating technology. Prof. Jiang has published more than 200 journal papers and more than 10 books. His designs and art projects have been exhibited worldwide and received more than 30 awards.

Raphael Kanyire Seidu is a Research Assistant at the School of Fashion and Textiles at The Hong Kong Polytechnic University, Hong Kong. He holds a Ph.D. in Fashion and Textiles Design from The Hong Kong Polytechnic University, Hong Kong SAR. He was also a Visiting Research Student at Nottingham Trent University (NTU) in the UK. He has several publications in leading journals in textile, design, technology, and material science. His recent paper was adjudged the "Second-Best Paper" presented at the 12th Sustainable Education and Development Research Conference in Somanya, Ghana. His research interest focuses on retro-reflective materials, smart textiles, cloth culture, weaving, and print technologies.

Dr. Benjamin Tawiah holds a B.A. in Textiles and a Master of Engineering from Kwame Nkrumah University of Science and Technology, Ghana, and Jiangnan University, China, respectively. He also has a Postgraduate Diploma in Management Information Systems from GIMPA and a Ph.D. in Textile Technology (Chemistry) from Hong Kong Polytechnic University. Previously, he was a postdoctoral fellow at Hong Kong Polytechnic University and City University of Hong Kong. Currently, he lectures at Kwame Nkrumah University of Science and Technology, focusing on polymer nanomaterial composites, flame retardant polymer composites, and second-generation batteries for advanced grid energy storage.

Contributors

Richard Acquaye Department of Textile Design and Technology, Takoradi Technical University, Takoradi, Ghana

Edward Apau Department of Visual and Industrial Art, Sunyani Technical University, Sunyani, Ghana

Morrison Appiagyei Boadi Department of Visual and Industrial Art, Sunyani Technical University, Sunyani, Ghana

Frederick Boamah Department of Fashion Design and Textiles Studies, Kumasi Technical University, Kumasi, Ghana

Edem Kwami Buami Department of Industrial Art, Ho Technical University, Ho, Ghana

Mehedi Hasan Chaion Department of Textile Engineering, Jashore University of Science and Technology, Jashore, Bangladesh

Morolake Dairo Manchester Fashion Institute, Manchester Metropolitan University, Manchester, UK

Emmanuel Kwame Danso Department of Fashion Design and Textiles Studies, Kumasi Technical University, Kumasi, Ghana

Bernice Akweley Djangmah Department of Industrial Art, Kwame Nkrumah University of Science and Technology, Kumasi, Ghana

Joseph Nii Offei Dodoo Fashion Department, Tema Technical Institute, Accra, Ghana

Benjamin Eghan Faculty of Art, Design and Humanities, De Montfort University, Leicester, UK

Bin Fei School of Fashion and Textiles, The Hong Kong Polytechnic University, Kowloon, Hong Kong, P. R. China

George Kwame Fobiri Department of Fashion Design and Textiles Studies, Kumasi Technical University, Kumasi, Ghana

Vandana Gupta Department of Fashion & Textile Design, Parul University, Vadodara, India

Ramratan Guru Assistant Professor, School of Design, Mody University of Science and Technology, SIkar, Rajasthan, India

Sylvia Asheley Gyampoh Department of Fashion Design, Radford University College, Accra, Ghana

Shouxiang Jiang School of Fashion and Textiles, The Hong Kong Polytechnic University, Kowloon, Hong Kong SAR, China;
Research Centre of Textiles for Future Fashion, The Hong Kong Polytechnic University, Kowloon, Hong Kong SAR, China

Tahera Kalsoom Manchester Fashion Institute, Manchester Metropolitan University, Manchester, UK

Rowena Fatchu Kansanba Department of Fashion Design and Textiles Studies, Kumasi Technical University, Kumasi, Ghana

Raji Rafiu King Guangdong Provincial Key Laboratory of Electronic Functional Materials and Devices, Huizhou University, Huizhou, China

Chanchal Kumar Kundu Department of Textile Engineering, Jashore University of Science and Technology, Jashore, Bangladesh;

School of Fashion and Textiles, The Hong Kong Polytechnic University, Kowloon, Hong Kong, P. R. China

Haijin Liu School of Environment, Key Laboratory for Yellow River and Huaihe River Water Environment and Pollution Control, Ministry of Education, Henan Key Laboratory for Environmental Pollution Control, Henan Normal University, Xinxiang, PR China

Mutala Jentina Nakpan Hubei Key Laboratory of Energy Storage and Power Battery, Shiyan Key Laboratory of Quantum Information and Precision Optics, Shiyan, China

Sharon Nunoo Manchester Fashion Institute, Manchester Metropolitan University, Manchester, UK

Emmanuel Abankwah Ofori School of Automotive Engineering, Hubei University of Automotive Technology, Shiyan, China;
Hubei Key Laboratory of Energy Storage and Power Battery, Shiyan, China;
Shiyan Key Laboratory of Quantum Information and Precision Optics (QIPO), Shiyan, China

Hannah Owusu Ansah Hubei Key Laboratory of Energy Storage and Power Battery, Shiyan Key Laboratory of Quantum Information and Precision Optics, Shiyan, China;
School of Automotive Materials, Hubei University of Automotive Technology, Shiyan, China

Dzingai Pethile Department of Fibre and Polymer Materials Engineering, Faculty of Engineering, National University of Science and Technology, Bulawayo, Zimbabwe

Jyoti Rani Parul Institute of Design, Parul University, Vadodara, Gujarat, India

Phoebe Naa Afaaley Sackeyfio Department of Industrial Art (Textiles), Kwame Nkrumah University of Science and Technology, Kumasi, Ghana

Sakthivel Santhanam Assistant Professor, School of Design, Presidency University, Bangalore, India

Bismark Sarkodie Department of Industrial Art (Textiles), Kwame Nkrumah University of Science and Technology, Kumasi, Ghana;
School of Environment, Key Laboratory for Yellow River and Huaihe River Water Environment and Pollution Control, Ministry of Education, Henan Key Laboratory for Environmental Pollution Control, Henan Normal University, Xinxiang, PR China

Raphael Kanyire Seidu School of Fashion and Textiles, The Hong Kong Polytechnic University, Hung Hom, Kowloon, Hong Kong SAR, China

Kyungeun Sung Faculty of Arts, Design and Humanities, De Montfort University, Leicester, UK

Benjamin Tawiah Department of Industrial Art (Textiles and Fashion), College of Art and Built Environment, Kwame Nkrumah University of Science and Technology, PMB, Kumasi, Ghana;
School of Fashion and Textiles (SFT), The Hong Kong Polytechnic University, Hong Kong SAR, China

Chipambwa Walter Department of Clothing and Textile Technology, School of Art and Design, Chinhoyi University of Technology, Chinhoyi, Zimbabwe

Chapter 1
Foundations of Textile Sustainability: Challenges and Opportunities

Raphael Kanyire Seidu, Benjamin Tawiah, and Shouxiang Jiang

Abstract The textile industry has several production processes and distribution channels that have severe economic, social and environmental impacts. Conventional practices such as using toxic dyes, large quantities of water and synthetic fibres amongst others, have exacerbated the widespread harmful impacts. This has resulted in the industry being a major contributor to water consumption, environmental pollution and carbon footprints, hence ensuring textile sustainability is imperative. This introductory chapter delves into the foundation of textile sustainability, focusing on the core challenges in the industry influencing sustainable practices and opportunities to advance such measures. The chapter begins with key challenges of the textile industry, which has advanced in the implementation of textile sustainability. Highlighting these include the environmental impacts, consumption of resources, social and ethical challenges. Further discussion on the eco-friendly materials, processes, and circular economy opportunities for sustainability is presented. Integrating these strategies and practices, such as recycling, reuse and cleaner production technologies, offers significant transformative measures to limit pollution and waste that are toxic and harmful to the environment. The chapter uncovers the roles certifications, policy regulations, consumer behaviour and engagement play in effectively advancing sustainability in the textile industry. Key research directions for the future are discussed, serving as a significant material for consumers, manufacturers, researchers and policymakers in the field.

R. K. Seidu (✉) · B. Tawiah · S. Jiang (✉)
School of Fashion and Textiles, The Hong Kong Polytechnic University, Hung Hom, Kowloon, Hong Kong SAR, China
e-mail: raphael-kanyire.seidu@connect.polyu.hk

S. Jiang
e-mail: kinor.j@polyu.edu.hk

B. Tawiah
Department of Industrial Art, Kwame Nkrumah University of Science and Technology, Kumasi, Ghana

S. Jiang
Research Centre of Textiles for Future Fashion, The Hong Kong Polytechnic University, Kowloon, Hong Kong SAR, China

S.-X. Jiang et al. (eds.), *Advances in Textile Materials and Processing Techniques for Sustainability*, SDGs and Textiles, https://doi.org/10.1007/978-981-95-0469-5_1

Keywords Textile sustainability · Material and process solutions · Circular economy · Challenges · Certifications · Consumer behaviour

1.1 Introduction

The benefits of the textile industry, such as the provision of goods, economic growth and a source of employment, cannot be overemphasised. Most importantly, the socio-economic development of developing and developed countries is influenced by the textile industry [1]. The industry comprises several activities, from sourcing raw materials to the eventual distribution of textile products to consumers. As such, the operation in the textile industry results in substantial consumption of raw materials, water, chemicals and energy. This adversely impacts the environment (Fig. 1.1) by contributing to air, water and land pollution [2]. These environmental and social challenges linked to the consumption and production activities of the industry are critical under the current discourse on global sustainability. By effectively addressing these challenges, the textile industry can significantly transform and transition towards sustainability.

Textile sustainability plays a critical role in significantly reshaping the growth, production and competitive advantages of the textile industry by advancing technologies and materials that reduce carbon footprint and waste generation. Thus, advancement in this area drives for effective sustainable approaches that foster innovation and informed policies for the industry. It was thus not surprising that the contribution of research to this field has been on the continuous rise, as confirmed

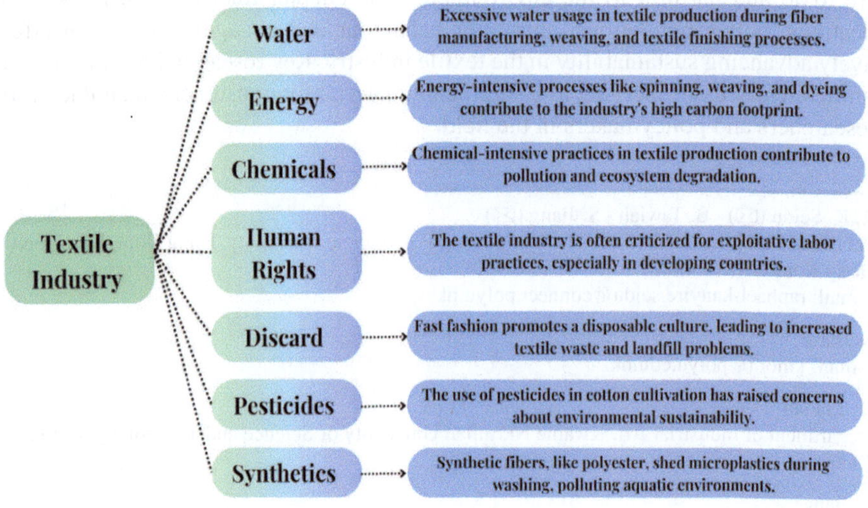

Fig. 1.1 Impact of the textile industry. Reprinted with permission from [3] Copyright 2024 Elsevier Ltd.

by the results of the Scopus database (Fig. 1.2). Herein, the appropriate research blocks of ("textile" AND "sustainability") were conducted on May 14, 2025, where the initial results of 4,829 documents comprising articles, reviews, books, chapters, conference papers, notes, amongst others, were published. The number of publications above one hundred was recorded in the year 2016, of which subsequent years have seen large research outputs. This implies the importance of advancing textile sustainability through eco-friendly materials, technologies and processes aimed at reducing environmental impacts. For example, the implementation and adoption of green manufacturing processes have emerged as a potential alternative to advance sustainability in the textile industry [4]. Furthermore, India, the United States and China, out of fifteen countries, were the top contributing countries to the subject area of textile sustainability (Fig. 1.2), where key findings would advance their efforts for a more sustainable textile industry. In fact, India, China and Bangladesh are among the top producers of textiles and garments in the world [5], hence, issues of wastewater, consumption of large quantities of chemicals and water are common [6].

The concept of textile sustainability aims at ensuring social responsibility, safe workspace conditions, efficient production process and supply chain, which results in minimising environmental impacts and responsible consumption by consumers. Textile sustainability broadly encompasses the adoption of eco-friendly or bio-based raw materials, cleaner or sustainable production processes, ethical labour practices

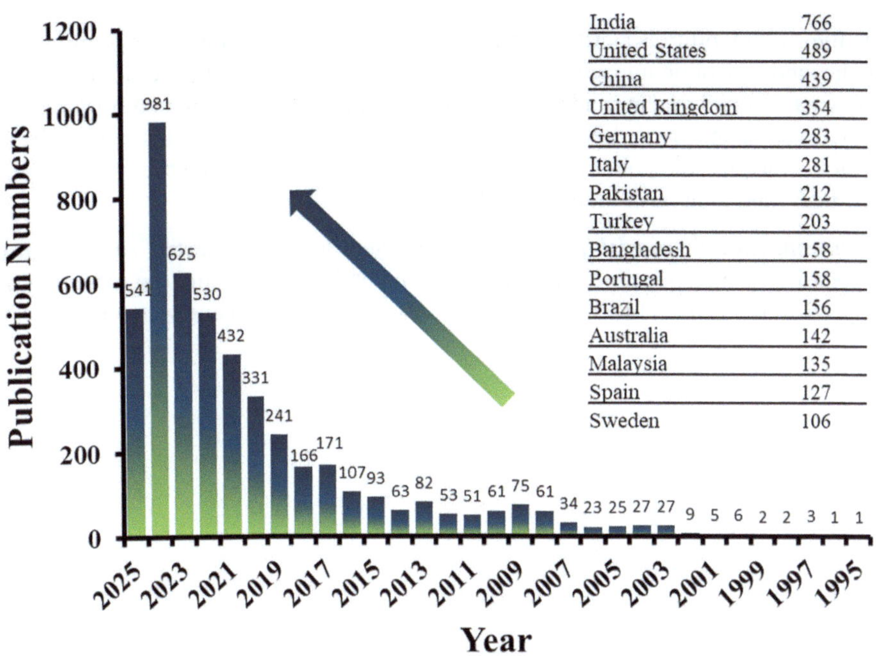

Fig. 1.2 Publication trends of textile sustainability based on Scopus data (as of May 14, 2025)

and innovative technologies, which collectively lead to social, economic, and environmental benefits. It is thus not surprising to identify key initiatives by the UNEP textile initiative [7], the EU strategy for sustainable and circular textiles [8] and the Sustainable Development Goals (SDGs), which have provided core targets to make the textile industry sustainable. Key literature such are recent studies by Li and Sindhuphak [9], Thakker and Sun [10] and a recent book by Gardetti and Muthu [11] has extensively discussed how the United Nations Sustainable Development Goals can drive the textile industry to be responsible and make the planet a safe place by 2030 and beyond. As an introductory section to this book titled "Advances in Textile Materials and Processing Techniques for Sustainability", this chapter emphasises the challenges for achieving textile sustainability, opportunities for sustainability through materials, technology, and circular economy, the role of certifications, policy regulations, consumer behaviour and engagement, implications of textile sustainability and envisioning future directions for progress in textile sustainability. These core areas in this chapter provide a brief overview of the intriguing areas that the rest of the chapters will discuss, contributing to policy development and implementation in achieving Sustainable Development Goals (SDGs) in the textile industry.

1.2 Foundations of Textile Sustainability

Textile sustainability is a critical framework aimed at mitigating the environmental, social, and economic impacts of the textile industry [12, 13]. This multidimensional approach is structured around three foundational pillars: environmental stewardship, social responsibility, and economic viability (Fig. 1.3). These pillars collectively inform strategies and policies that seek to balance resource efficiency, ethical labour practices, and long-term industry resilience.

Environmental Stewardship: Environmental stewardship in textile sustainability emphasises the responsible management of natural resources and the reduction of ecological footprints [14]. Key considerations include minimising water and energy

Fig. 1.3 Foundations of textile sustainability

consumption, reducing chemical pollution, and promoting circular economy principles such as recycling and biodegradable materials [15–17]. Advancements in sustainable fibre production, eco-friendly dyeing techniques, and waste management are central to ensuring a reduced environmental impact.

Social Responsibility: The textile industry profoundly affects global labour markets and communities, making social responsibility an indispensable component of sustainability [18]. Ethical labour practices, fair wages, and safe working conditions are essential to fostering a responsible supply chain [19]. Additionally, consumer awareness and transparency initiatives, such as certifications for ethical sourcing and worker protections, contribute to improving social standards and equity within the industry [20].

Economic Viability: For sustainability to be effectively implemented, it must be financially sustainable. Economic viability entails fostering innovation that balances profitability with sustainable practices [21]. Investment in sustainable technologies, cost-effective production methods, and business models that prioritize long-term resilience over short-term gains play a crucial role in ensuring that sustainability efforts remain economically feasible [22]. Strategic collaboration among industry stakeholders further enhances resource efficiency and market competitiveness.

The integration of these three pillars in textile manufacturing will help transition the industry towards a more sustainable future that meets both present and future demands while minimising adverse impacts on society and the environment.

1.3 Challenges for Achieving Textile Sustainability

The textile industry has contributed effectively to the employment, trade and revenue aspects of a country's economy [2, 23]. For example, the textile industry contributes effectively to the Asian economy and about 8.5% to the GDP of Pakistan [24]. In the UK, the textile and fashion industry contributed about 1.3 million jobs and £62 billion in GDP [25]. In fact, both skilled and unskilled labourers are engaged in the various sections of the textile industry, contributing to employment opportunities [26]. Aside from this, the textile industry is a major consumer of water, has labour-related issues, transparency issues and contributor to environmental pollution [2]. These impact the textile industry to fully achieve sustainability. Achieving the latter comes with social, economic and environmental challenges.

Social Challenges

The textile industry has contributed to the rate of employment and revenues of a country's economy, hence influencing its economic and societal growth [27]. However, the industry faces several social issues that negatively impact the industry's goal of transitioning towards sustainability. Key challenges or issues are working conditions, labour rights and transparency in the supply chain. Workplace conditions have been very challenging with excessive overtime or long hours, unsafe environments and low wages. Due to large orders and the mandate of the industry to meet clients' needs, workers would have to work overtime and long hours. In most cases, workers face exploitative practices while working under pressure. Working such long hours exposes workers to serious musculoskeletal disorders, such as low back pain [28], neck pain, stress on muscles, and joint pains [29], amongst others. Subsequently, the production processes at the workplace release fumes, dust particles and fine fibres, which pose some health problems in unsafe environments. These problems occur due to the non-availability of PPE supply and its use [26, 30, 31]. For example, dusty environments lead to certain respiratory disorders such as byssinosis, cough, asthma and bronchitis amongst others [32, 33]. These health-related problems have been extensively reported by Seidu et al. [26], Lillypet et al. [34], and Kabir et al. [35], hence advocating for improved safety for workers. These health issues could affect their productivity, a situation that could affect their employment status. It is thus essential to maintain a safe work environment to ensure workers' productivity and well-being. The rate of unemployment exposes both skilled and unskilled workers to unfair treatment of low wages from managers of the textile industry. Subsequently, the quest to make a living and the fear of job loss further exposed these workers to low-wage exploitation.

The issue of labour rights has been central to the textile industry, where workers in developing countries and some Asian textile-producing countries are exposed to frequent labour violations [36]. These, however, go against the guiding principles of improving and protecting the human rights of workers as stipulated by the United Nations [37]. The issue of labour rights was prominent with the 2013 Rana Plaza incident in Bangladesh, which called for strict adherence to maintaining human rights and improved working conditions. Aside from this, the issue of child labour is also a major challenge where children are engaged in certain activities of the production process and supply chain of the textile industry. For example, activities such as cotton seed production, cotton harvesting and cotton removing widely practiced in India, Uzbekistan and Egypt respectively, were reported as core activities that employed child labour [38]. The report further revealed that, poverty, demand for cheap labour and low adult wages as causative factors pushing for child labour in the textile industry. In some cases, these children are often forced and abused to work under unsafe conditions that pose serious threats to their health. The textile industry is composed of a complex supply chain that promotes its opaque nature and lack of transparency for consumers [2]. Due to this, the practice of hiding core information about material sourcing, production and operations, amongst others, leads to a lack of transparency in the textile industry. This situation makes it difficult for consumers to get the right information, hence leading to trust issues. The complex nature of the supply chain further makes it difficult to improve environmental and sourcing responsibility [2].

Economic Challenges

The textile industry faces several economic challenges that impact its ability to achieve sustainability. Aside from the economic benefits, key economic challenges such as the need for ethical labour practices, production processes, and high cost of raw materials have hindered the industry from achieving sustainability. Ensuring ethical labour and workspace practices would require paying the appropriate living wages for workers, which increases the cost of operations in the industry [39]. Subsequently, sustainable approaches have been introduced to improve production processes, reduce waste generation and the consumption of water and energy. These practices are cost-effective and require large investments from the textile industry. Even though these sustainable measures are beneficial, they put extreme financial strain on the manufacturing companies [40]. Furthermore, it will require investment from the companies to retrain workers to manage and implement new sustainable technologies. Alternatively, upgrading conventional machines in the company with new technologies is expensive to purchase and manage. This implies that firms would have to increase their budget to cover the sustainable machinery and practices to help minimise the environmental impacts.

Environmental Challenges

Aside from the exploitative and working conditions for workers, the various activities in the textile industry are a major contributor to environmental problems. These environmental challenges include pollution, high energy and water consumption, reliance on fossil fuels and waste generation, which hinder the industry from fully achieving sustainability. The manufacturing process in the textile industry utilises several materials and approaches that tend to pollute the land, air and water. Dyes from both natural and synthetic sources, together with their auxiliaries, are widely used in colouring textile materials. These synthetic dyes are widely consumed by the textile industry [41] due to their fastness, range of colours and ease of use [42]. Amongst these, azo dyes occupy about 50% of the synthetic dyes used by the textile industry [43]. The discharge of non-bonded dyes, additives and chemicals as wastewater enters the oceans, lakes and streams, which has a significant impact on aquatic organisms in water bodies [44]. The discharge of large quantities of untreated wastewater from the dyeing process leads to oxidative stress, affecting the behaviour, growth and development of aquatic organisms in water bodies [45]. For example, Congo red (about 100 mg/L) and RP2B (10 mg/L) have growth suppression [46] and inhibition (about 78%) [47] respectively on algae. The adsorption and effects of these dyes on aquatic organisms are transmitted through the food chain. These sea-related foods are consumed by humans, hence extending the reach of heavy metals and other dye chemicals to the human organs [48]. The ingestion of these poses a serious health risk to humans (Fig. 1.4), such as bladder cancer, pulmonary disorders, mammary gland cancer, and nervous system dysfunction [44], disorders of the CNS and dermatitis [48]. These effects, amongst others, are further discussed in detail in a report by Singha et al. [44].

Subsequently, the production processes in the textile industry, like weaving and spinning, release large quantities of sublimates and dust that cause air pollution at the workplace. These fine particles pose a serious health risk to workers [49]. Harmful emissions of gases like nitrogen, carbon monoxide and sulphur dioxide are produced from manufacturing machines [50]. Alternatively, gases such as carbon dioxide and methane are commonly produced from decomposed woollen textile materials that

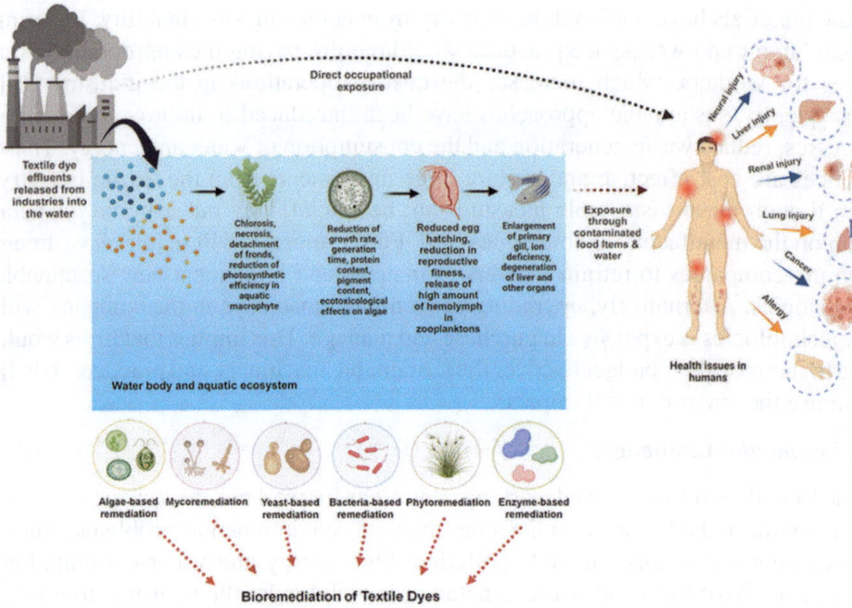

Fig. 1.4 Effects of textile dyeing process on aquatic life and human body. Reprinted with permission from [45] Copyright 2024 Elsevier Ltd.

are disposed of in landfills [51]. All these practices and others have resulted in the textile industry contributing to about 10% of the global carbon emissions [52]. Other country-specific studies have reported on the influence of the textile industry on carbon footprint or emissions from China [53, 54], Greece [55], Indonesia [56] and India [57]. For example, the raw material processing, energy utilisation and wastewater significantly contribute to the emission of carbon dioxide [58].

High energy and water consumption also remain major challenges that affect the textile industry's ability to achieve sustainability. The textile industry is characterised by several production processes, such as fibre preparation, fabric production and finishing, amongst others, that utilise varying machineries that consume high energy. Energy consumption in the textile industry comes in several forms, like natural gas, coal, LPG and oil (to produce steam) and electricity [15]. The latter is used to power the air conditioning, office, lighting, and machinery, whereas thermal energy from a stream is used in finishing, bleaching, washing and dyeing processes in the textile industry. Such high consumption of energy results in environmental impacts [59], making the issue of high energy a major concern in the textile industry. The textile industry is also a major consumer of water, with about 93 billion cubic metres of water yearly [60]. However, the consumption rate varies from one sub-sector to another due to the length and characteristics of the process, the methods applied, and the textile materials to be processed [61]. The processing of natural fibres requires large amounts of water as compared to synthetic fibres [62]. For example, about 2,700 L of water are required to produce a cotton t-shirt and about 3,781 L of water for denim jeans [60]. The more water is consumed, the larger quantities of wastewater are also produced, causing serious environmental impacts. This, coupled with the excessive water consumption, has resulted in the negative impact of textile wet processing [63].

The textile industry relies on fossil fuels for energy-intensive manufacturing processes like finishing, drying and dyeing, in manufacturing countries such as Bangladesh, Vietnam, India, China and Indonesia [64]. Alternatively, fossil fuels or petrochemicals are widely used for synthetic fibre production, with the most common fibres being polyester and nylon [65]. In fact, the production of polyester has been on the rise due to its cheap production, strength and performance. This has further increased their application for garment production, hence promoting fast fashion. The production process and end-of-life of textile materials generate significant amounts of waste, posing serious environmental threats. It is reported that about 92 million tons of textile waste is from the textile industry [66]. Herein, textile waste is categorised under post-industrial waste (from discarded packaging, automotive and medical textiles) [67], post-consumer waste (from used and discarded textiles after their end-of-life) and pre-consumer waste (from textile manufacturing processes) [66]. These waste when disposed of and burnt produce harmful gases and substances, for pollution in the environment. This practice thus calls for effective approaches to safeguard the safety of human health and animals.

1.4 Opportunities for Sustainability: Materials, Technology, Circular Economy

Sustainability in the textile industry has presented several opportunities in materials, new technologies for production processes, and circular economy strategies. The integration of these core opportunities helps the textile industry to reduce the environmental impacts, ensure cost savings and promote a more sustainable future with new market prospects. It further promotes a more ethical manufacturing approach, a socially and environmentally responsible mindset [68].

Materials

In the textile industry, the use of conventional materials like synthetic fibres for textile and fashion products has resulted in serious environmental pollution. This has driven the advancement of alternative materials with unique properties. These are non-toxic and ecologically friendly materials for textile production that provide an alternative approach to reduce the environmental footprint. As consumers increasingly prioritise the utilisation of eco-friendly materials, the textile industry is innovating and advancing the production of sustainable materials. These include natural fibres (like jute, hemp, flax, bamboo and organic cotton amongst others) that are safe, biodegradable and eco-friendly [40], bio-based fibres [69, 70], lab-grown materials and recycled polyester [2], lyocell, polylactic acid, and lenpur fibres amongst others

[71]. Further discussion of these sustainable textile materials is reported in relevant studies like Shibly and Hoque [2], Rahaman ct al. [40], Maduna and Patnaik [71] and Eyupoglu [72]. Figure 1.5a and b illustrate the widely used eco-friendly fibres and raw materials for eco-textiles, respectively. For example, organic cotton fibres are produced from non-genetically modified seeds, fertilisers and pesticides [2]. Aside from their unique properties, organic cotton fibres reduce water and soil pollution [71]. Importantly, sustainable textile materials prioritise ethical practices, biodegradability and eco-friendliness. Other core benefits include greenhouse gas reduction, improved material performance and properties, and environmental conservation [2]. Aside from these benefits, influential factors from consumers, such as economic considerations, social responsibility and awareness of the environment, are central to influencing their purchasing and utilisation of sustainable textile materials [73].

New Technologies for Production Processes

Due to the complex nature of the production and supply chain process, the textile industry is a major contributor to about 5–10% of greenhouse gas emissions globally [75], pollution, water and energy consumption. The latter is said to increase by 30% in 2030 [76]. As such, key technological innovations in the textile industry play a significant role in reducing pollution and resource consumption. These technological innovations are applied in fibre production, textile conversion, transportation and the eventual disposal of the textiles in the environment (Fig. 1.6). Sourcing and production of fibres is the beginning stage in textiles production. Herein, the environmental impacts of synthetic fibres have led to a push for more environmentally friendly fibres. Several sustainable innovations in fibre production such as recycled and bio-based synthetics, cotton and regenerated cellulosic fibres, have been advanced for reduced energy and water consumption. For example, there is a reduction of about 72% in greenhouse gas emissions and a reduction in energy of about 50–85% when using recycled PET as compared to virgin PET [77]. These advancements, as reported by Zhang et al. [75] are limited to fibre production and not disposal.

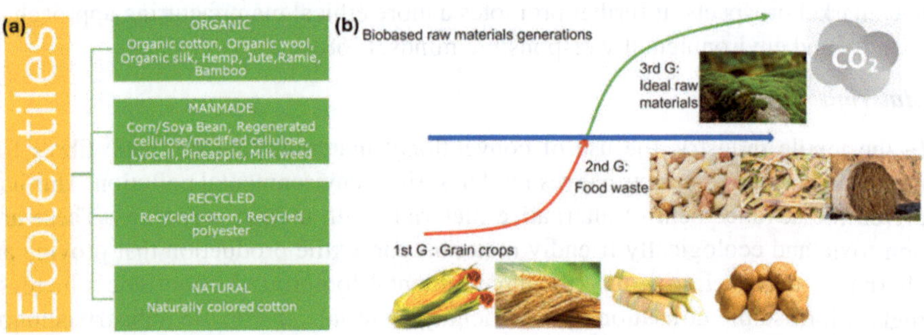

Fig. 1.5 **a** Widely used eco-friendly fibres for eco-textiles [74] (This article is an open-access article distributed under the terms and conditions of the creative commons attribution (CC BY) license), **b** Textile fibres from bio-based raw materials [69] (This is an open-access article distributed under the terms of the creative commons attribution license (CC BY))

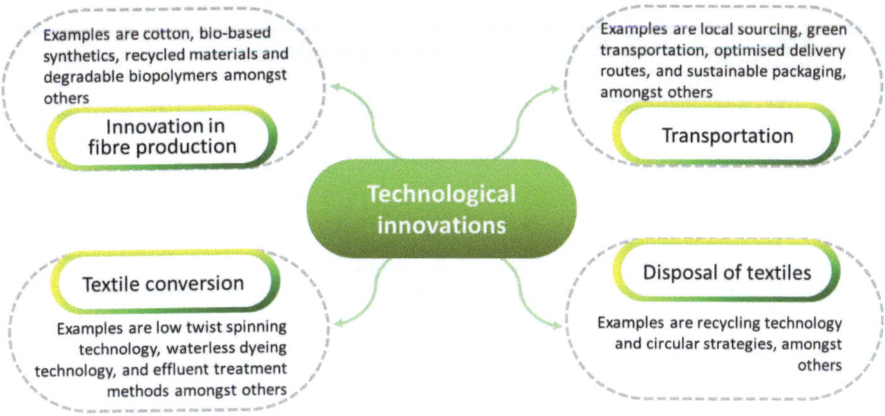

Fig. 1.6 Key technological innovations fostering sustainability in the textile industry

Alternatively, the manufacturing of textiles faces core challenges of large consumption of chemicals, energy and water, which promote environmental concerns. To address these, innovative manufacturing technologies (Fig. 1.7) have been reported by Zhang et al. [75] which pushes the industry towards sustainability. For example, "optimised nozzle geometry" of low-pressure and high volume nozzles is reported to save about 26% of energy [78]. The adoption of these technologies facilitates an effective textile production process with reduced energy, water and chemical consumption. The manufactured textile materials are further transported to distant places globally. Key innovations here are local sourcing, green transportation, optimised delivery routes, and sustainable packaging, amongst others. Long-term transportation of goods that contribute to the carbon footprint is gradually being replaced with prioritised local sourcing to limit the delivery times, ensure operational efficiency, to reduce environmental impacts [79]. These textile materials at the end-of-life are easily disposed of into the environment, causing serious threats. However, relevant innovative technologies like recycling technology and circular strategies, amongst others, have extended the life cycle of textiles, reduced waste generation and pollution in the environment.

Circular Economy Strategies

The textile industry is widely known as a major chemical, energy and water consumer, which has led to environmental pollution and global carbon emissions [80]. Significant use of hazardous chemicals for various production and finishing processes has contributed to polluting the environment [81]. Alternatively, the linear approach for textile life cycles leads to the indiscriminate incineration or disposal of used textiles, with about 12% recycled. It is, however, observed that the actual lifetime of the textiles is not fully realised before they are discarded [82]. This practice, coupled with the cheap textile products from fast fashion, has led to an increased generation of textile waste in the environment. Such negative impacts are minimised with the

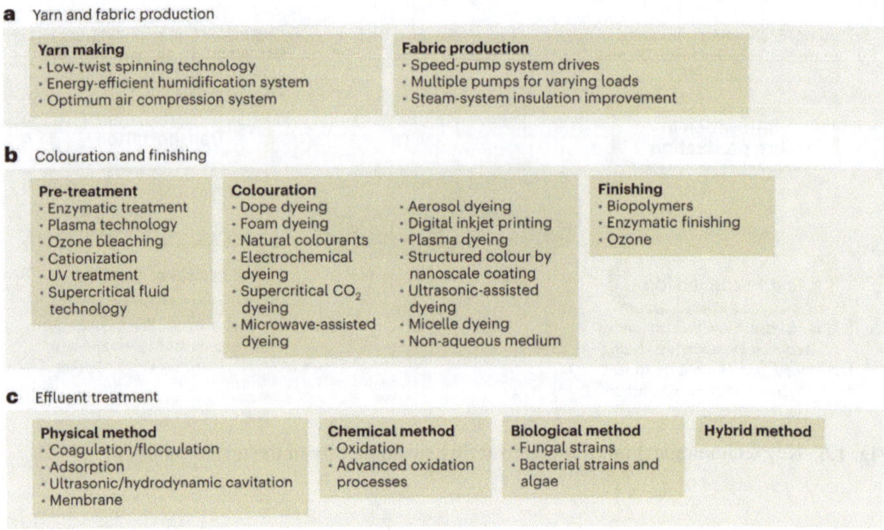

Fig. 1.7 Innovative manufacturing technologies driving textile sustainability. Reprinted with permission from [75] Copyright 2022, Springer Nature Ltd.

advancement of the circular economy (CE). This approach, as described by Geissdoerfer et al. [83] is "a regenerative system in which resource input and waste, emission, and energy leakage are minimised by slowing, closing, and narrowing material and energy loops". This approach has pushed for economically sustainable practices [84] that offer sustainable development and environmental preservation. Key benefits of CE are long-term preservation, generation of revenue [85], minimising waste generation and maximising resource efficiency [86].

The core circular economy strategies have evolved from the initial 3R approach of reduce, reuse and recycle to a 9R framework [86]. These include "refuse, rethink, reduce, reuse, repair, refurbish, remanufacture, repurchase, recycle, and recover" [80, 86]. Figure 1.8 illustrates the various areas in the supply chain of textiles where CE strategies are adopted. However, the full utilisation of these strategies is limited complexity and nature of the materials [87]. CE principles are widely applied in the supply chain of the textile industry to ensure waste from production and post-consumer use is reduced [88]. With a gradual adoption of CE strategies in Asia and Africa, Europe and the United States have, however, advanced the use of this for effective circularity [89]. Subsequently, the implementation of CE principles has been initiated by key global brands like Inditex, H&M, Gap, and Burberry in their supply chain [90]. Figure 1.9 further provides the various CE principles or strategies employed in the textile industry. The adoption of these strategies is driven by core factors such as changes in regulations [91], sustainable technological advancement [92], targets for reduced CO2 emissions and energy consumption [93], and a new material source [94].

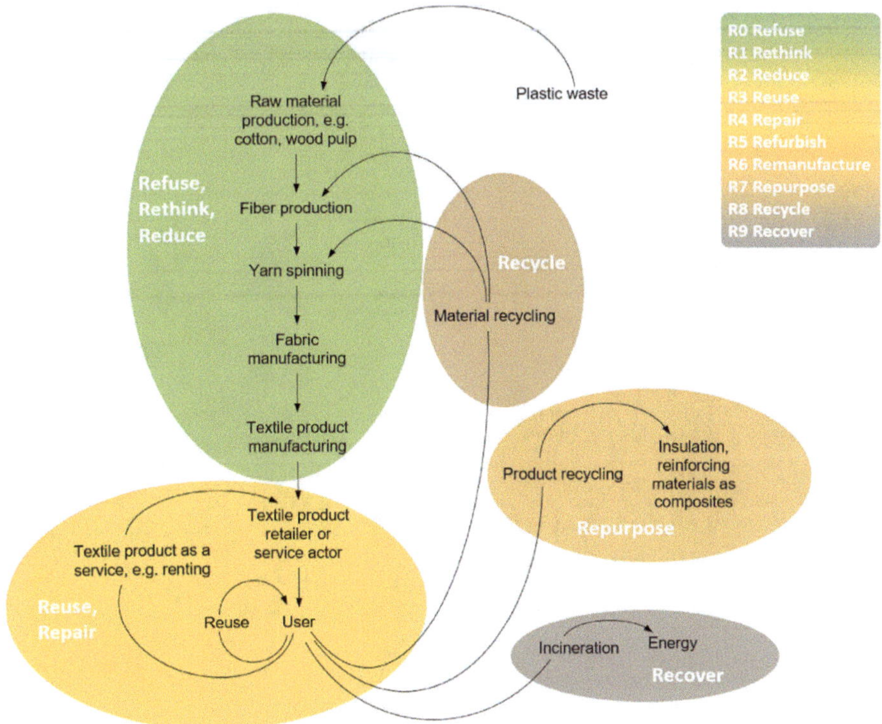

Fig. 1.8 CE strategies in the supply chain of the textile industry [80]. (This is an open-access article distributed under the terms of the creative commons CC-BY license)

Concurrently, key barriers or challenges have influenced the effective adoption and transition of the textile industry towards CE. These include the complexity of the supply chain that affects circular transparency [95], lack of policies by companies for CE and lack of education and awareness [96], unclear economic viability [97], low quality perception by end-users [98], limited technology [99], and the absence of strategic planning for adopting CE processes [100]. Aside from these barriers, several key enablers such as innovation and technology, commitment and awareness from stakeholders, shortened processes for reverse logistics, consumer participation and policy implementation, promote the transition of the textile industry towards CE [96]. Consumer behaviour further motivates them to adopt any of the CE strategies, such as recycling, repair and reuse amongst others, for post-consumer textile materials. These responsible behaviours significantly influence the consumer to purchase, use and dispose of the textile products without causing environmental impacts in the long term [101]. Key motivational factors have been outlined to influence the concept of recycling, repair and reuse among consumers, environmental and economic concerns [102, 103], education [104, 105], style and uniqueness [106]. As noted by [102] and [107], convenience and family subjective norms, respectively, further play a significant role in sustainable post-consumer behaviour like recycling and resale of

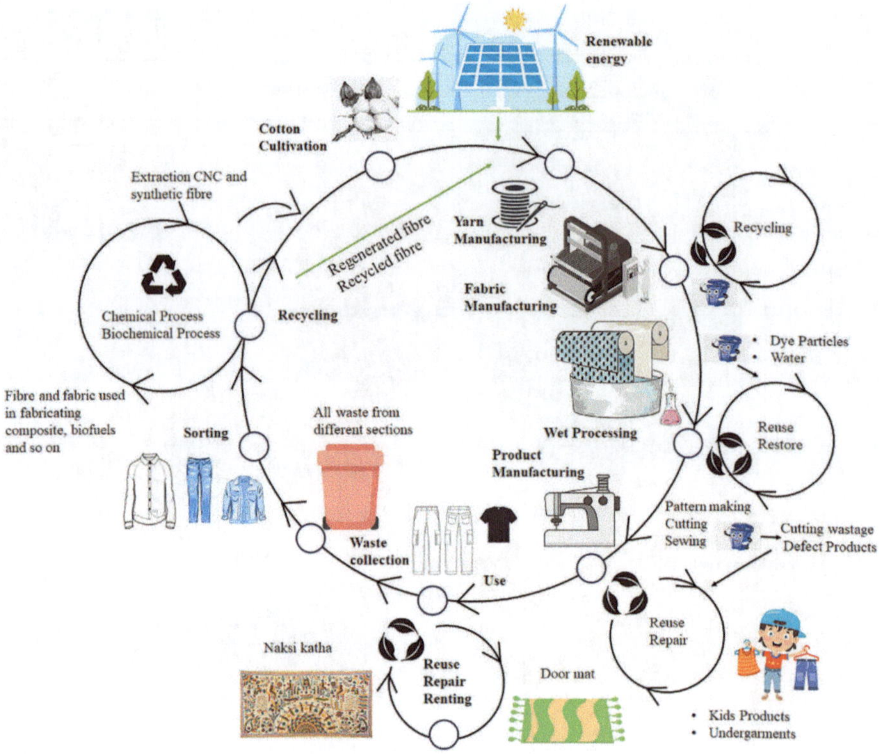

Fig. 1.9 Circular economy in the textile industry [85] (This is an open-access article distributed under the terms of the creative commons CC BY license)

textile products. Other motivational factors that influence consumers to adopt the CE strategies are further reported in a study by Kemi and Zilahy [107]. These behaviours by consumers foster a responsible and sustainable environment.

Other key opportunities for sustainability in the textile industry also include the roles played by strict policy regulations and certifications, consumer behaviour and engagement that can foster a shift towards eco-friendly materials, transparency, traceability, and trust.

1.5 Role of Certifications and Policy Regulations

The textile industry, which consumes large chemicals, water and energy, is one of the major polluting industries, causing ecological footprints and environmental impacts [82]. These impacts stress the need for effective policy regulations and certifications to reduce the negative threats from waste from production processes and the end-of-life of the textile. Policy regulations and certifications play a pivotal role in advancing textiles' sustainability by ensuring strict establishment of compliance and standards that foster effective changes.

Transparency, responsibility to the environment, and ethical practices are core themes of certifications for textile sustainability. Basic aspects such as labour standards and rights, management of chemicals and the origin of fibres are the basic requirements covered in certifications [108]. Furthermore, it is described as an essential tool which bridges the gaps between "intention and verification" [108]. Certifications come in the form of marks, symbols and logos [109] are commonly displayed on textile products for easy recognition by consumers. They provide information that all social and environmental standards are met by the products to promote a sustainability drive. They further provide information about the production process of a textile through standardised communications methods such as labels. Brands widely use the latter to improve gaps in consumer knowledge [110], purchasing decisions, and provide "unobservable product attributes or practices" [109]. This way, consumers trust such products from brands, hence their intention and purchasing decisions are influenced [111]. A clear benefit for both brands and consumers. Several certifications form critical foundations for textile sustainability in the industry to enhance decision-making and transparency. These include Global Organic Textile Standard (GOTS), OEKO-TEX® Standards, Bluesign® and Zero Discharge of Hazardous Chemicals [112]. However, when brands adopt an illegal approach of greenwashing, it tends to mislead consumers, hence affecting their trust. This further influences transparency between consumers and brands.

Policy regulations also play a central role in driving sustainable practices and solutions in the textile industry. These regulations provide frameworks that enhance sustainable practices, reduce waste and environmental impacts. For example, policy regulations can force manufacturers to establish core collection systems for used textile materials, and provide collaborative efforts and platforms [113]. Other regulatory measures by the European Union include the Registration, Evaluation, Authorisation, and Restriction of Chemicals (REACH), Ecodesign for sustainable products regulation and the EU strategy for sustainable and circular textiles [112]. However, certain textile export countries like Turkey and Bangladesh find it difficult to comply with the EU environmental regulations and rules [114]. These environmental problems are waste management, consumption of energy, water and air pollution. The issues of investment were identified as the problems faced by Turkey, whereas safety and labour rights were core problems faced in Bangladesh. To address the issues in the latter, the Accord on Fire and Building Safety in Bangladesh was established to ensure safe working conditions for workers in the textile and fashion industry [115, 116]. In Turkey, however, the Sustainability Strategy and Action Plan was developed to ensure the textile industry reduces its environmental impacts, ensures traceability and transparency to meet the European Green Deal [117]. Other key global policies and regulations are comprehensively reported by Silina et al. [118] to ensure sustainability in the textile industry.

1.6 Role of Consumer Behaviour and Engagement

Consumers have remained a major player in the marketing and sales of textile and fashion products on the market. The behaviours and choices of consumers significantly contribute to increased rates of consumption of textile products and their eventual disposal [119], hence the generation of large quantities of waste. The latter is attributed to the trendy nature and cheap prices associated with clothes [120]. This implies that their purchasing behaviours, decisions and choices potentially shape and influence the textile industry [121]. As such, the behaviour and engagement of consumers further play a critical role in textile sustainability for effective measures to limit environmental impacts. Thus, it is paramount to effectively understand consumer behaviours [122] because of their unsustainable and uncontrollable patterns [121]. By keeping track of their behaviour patterns, manufacturers and authorities can effectively find solutions to limit waste disposal. With this, consumer behaviour and engagement roles are critical for textile sustainability. This section discusses the relevant theories, factors and initiatives for consumer behaviour and engagement (Fig. 1.10).

Theories

In any consumer study, relevant frameworks are developed based on the appropriate theories to clearly understand the attitudes, motivations, decisions, preferences and engagement amongst others of consumers towards sustainable textiles on the markets. Aside from this, the use of theories influences the development of business and policy strategies to revamp consumers' behaviours and engagement. There are several theories commonly used in consumer-related studies; however, the most commonly used theories are the Theory of Planned Behaviour (TPB), the Theory of Reasoned Action (TRA). The Theory of Planned Behaviour (TPB) (Fig. 1.11) was developed by Ajzen [123], where the individual's intentions play a central factor in influencing their behaviour. These intentions are the motivational factors that drive one's behaviour by indicating their willingness and planning efforts to achieve a set goal. Furthermore, there is a higher likelihood of executing a behaviour based on an increased intention to perform the act [123]. This implies that a person's intention to act in certain behaviours can be predicted based on their attitudes, perceived behavioural control and subjective norms. Due to the uniqueness of the TPB, it is widely used by several studies to understand consumer behaviour towards sustainable textile products [124]. Furthermore, such behaviours are guided by the control beliefs, normative

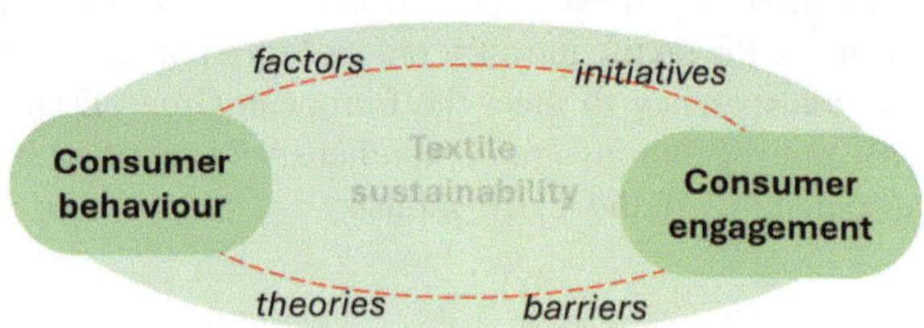

Fig. 1.10 Key components for consumer behaviour and consumer engagement

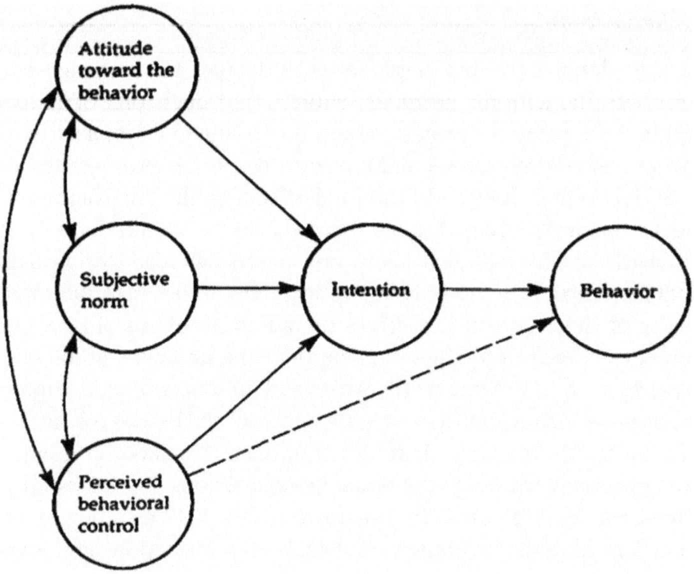

Fig. 1.11 The theory of planned behaviour (TPB). Reprinted with permission from [123] Copyright 1991 Elsevier Inc.

beliefs and behavioural beliefs of consumers [125]. The use of the three drivers, such as behavioural control, subjective norms and attitudes, the consumption decisions and intentions of consumers can be evaluated [126]. The Theory of Reasoned Action (TRA), developed by [127], examines the behavioural intention based on behaviours. Herein, behavioural intention, which consists of subjective norms and attitude [128], represents the likelihood of a person to perform an action and adopt a particular behaviour [129]. This means that consumers are more likely to adopt a certain behaviour if their attitude increases [130]. TRA is, however, limited in accounting for perceived behavioural control, which measures the capability of consumers to perform a behaviour [131]. Due to this limitation, the TPB was developed to address this.

Subsequently, other commonly used theories are the Value-Belief-Norm (VBN) Theory by Stern et al. [132], Generational Cohort Theory, Behavioral Reasoning Theory (BRT), Attitude-Behavior-Context (ABC) model, The Social Cognitive Theory (SCT), Corporate Social Responsibility Theory, Consumer Perceived Value (CPV) Theory and The Elaboration Likelihood Model as identified by Balasubramanian et al. [124] in their study. The adoption of these common theories, studies can essentially explain consumer behaviours and predict their outcomes for effective design and implementation of interventions, initiatives, incentives and structures, to promote a sustainable environment.

Consumer Behaviour

Consumer behaviour in the purchase, usage, effective maintenance and eventual disposal refers to the actions, attitudes, choices and decisions of individuals that have a serious influence on the ethical and environmental impacts of sustainable textile products. Such preferences and motivations further promote the purchase of these textiles due to their long-term minimal effects on the environment [126, 133]. These have a direct influence on the failure or success of the products on the market. The shift towards eco-friendly textiles by consumers can lead to significant support for sustainability in the textile industry [134]. This is because the environmental consciousness of these consumers drives the adoption of sustainable practices by manufacturers or brands in production, supply chain, and even at the end-of-life of the textile products to limit large textile waste and pollution-related impacts. Despite the role of consumer behaviour in driving the markets and brands towards sustainable practices in the textile industry, there are significant barriers that hinder adoption and prevent consumers from making sustainable decisions or choices on sustainable textiles. These barriers are carefully illustrated in Fig. 1.12 as reported by [135].

These barriers are clearly hinged on a lack of information, greenwashing and trust, a lack of transparency and costs [134, 136]. Information on the environmental benefits or impacts of textile sustainability for consumers is imperative to ensure they have the appropriate knowledge. The absence of such information would lead to a lack of knowledge and unawareness of consumers of the true benefits that textile sustainability could have on the ecosystem. As such, ensuring effective training of marketers, providing support systems through public awareness on relevant information on websites, leaflets and labels [137], would significantly push for the right access to information by consumers. Through advertisement, consumers are well aware and informed about the sustainable textile products, hence, a lack of these will ultimately affect their purchase behaviour. This draws the importance of knowledge and information about sustainable textile products, leading to a change in their attitudes, purchasing patterns, and willingness to pay, as widely evident in relevant studies [73, 138–140]. As such, once consumers have more experience and information about these products, they tend to develop a positive attitude, consumption pattern and behaviours [140]. The issue of greenwashing and trust is another key barrier which negatively affects consumers' behaviour towards sustainable textiles. The act of greenwashing leads to the provision and publication of false information and claims about the eco-friendliness of a textile product. Greenwashing (Fig. 1.13) is considered a "deliberate deception" [141] and intentional deception of consumers [142]. Consumers are misled into making choices based on the false or exaggerated claims made by companies in the industry. These practices by brands negatively affect the adoption and purchasing attitudes of consumers. Such an approach, adopted and engaged by brands, leads to doubts and loss of trust in the claims of sustainability [143]. It was thus not surprising to note that relevant studies [143–145] have established the negative impacts the practice of greenwashing by brands has on consumer trust and their eventual purchase decisions towards sustainable textiles. By confusing consumers via greenwashing could ultimately affect the brand's efforts to promote

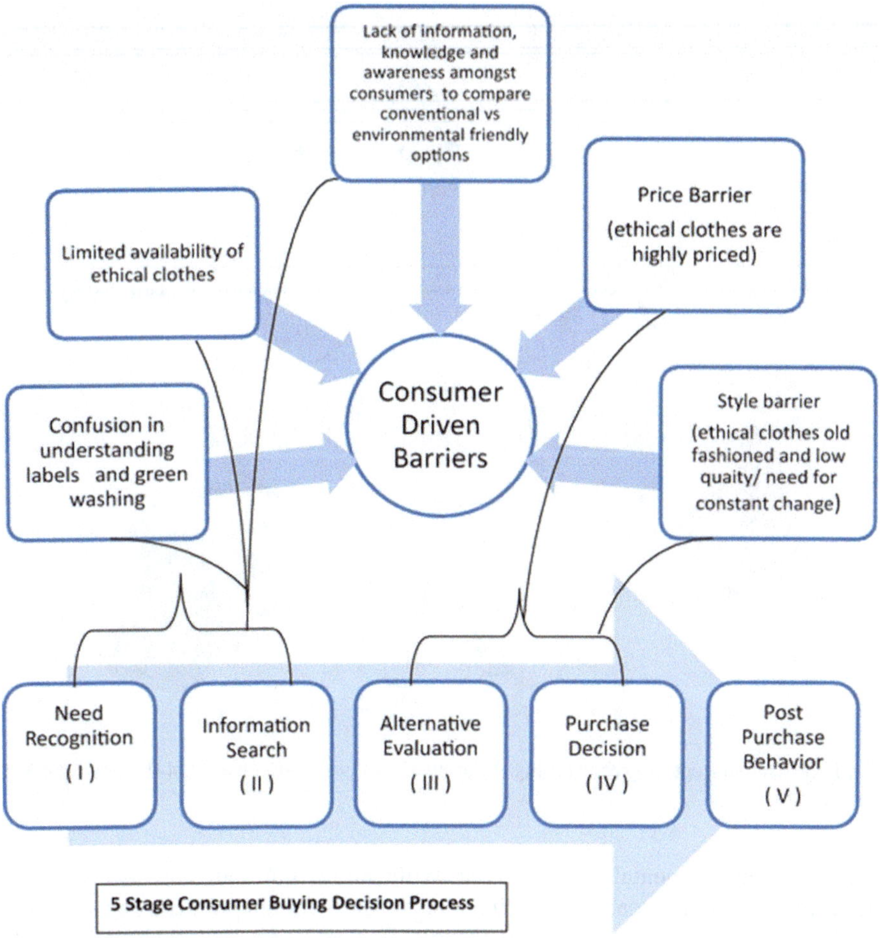

Fig. 1.12 Barriers driving for consumer purchase decisions. Reprinted with permission from [135] Copyright 2017, Springer science business media Dordrecht

textile sustainability to help limit any environmental impacts. Aside from this, there are negative impacts on the brand's image and trust [146], and the reputation and social responsibility of brands [147], which affects the consumer's behaviour for other similar products from other brands in the industry [148]. Avoiding this approach would aid in brand-consumer trust for improved purchasing behaviours.

Transparency in the source of materials used, their production methods, and their related supply chain is important to attract and build consumers' trust towards sustainable textiles. This refers to the ability of brands to openly make known their practices, operations and production information to consumers [149]. The provision of such key information for consumers further leads to the authenticity and trustworthiness of the brands. This practice by brands makes it easier for consumers to understand the

Fig. 1.13 The concept of greenwashing. Reprinted with permission from [144] Copyright 2022 Elsevier B.V.

source and environmental impacts of the textile products they are purchasing. A situation which could influence their behaviours towards certain products and brands on the market. Hence, these consumers can make informed choices and decisions. The absence of transparency is another barrier that affects consumers' behaviours. The non-disclosure of information on their production processes could lead to negative perceptions by the consumers, hence they feel reluctant to exercise their purchasing power. Consumers are demanding transparency, a situation which drives accountability by brands [149]. Furthermore, aside from these benefits, key challenges such as the cost of technology or systems to track and provide information for transparency, the complex nature of the supply chain and the lack of standardisation affect brands to fully implement transparency. With large investments in innovative technological infrastructure, brands can easily deliver on making their operation transparent for consumers. The cost and price of sustainable textile products are one of the major barriers [150, 151]. The behaviour and attitudes of consumers towards sustainable textile products change when they perceive that such products are priced higher. Such practice deters consumers from purchasing such products, even though they know the environmental benefits. This leads to a situation where these consumers would rather turn to purchase products from fast fashion [152]. However, once consumers perceive such products to be of good quality and value, they are willing to pay more [150]. As such price or cost of the products is no longer the right determinant of consumer behaviour. This implies that consumers will overlook the high price tag for sustainable textiles once the products are aligned with ethical standards, practices and have less environmental impact. The effectiveness of this is dependent on the information available on such products for consumers to make informed decisions.

Lastly, product availability is also another significant barrier. Consumers build experience with textile products they use often; hence, their choices are largely influenced by such experience. As such, consumers can easily repeat their shopping activities for particular textile products based on their past experiences and availability [153]. This practice further limited the attitudes and behaviour of consumers to purchase sustainable textiles, most especially when such products are not available on the market. This factor, coupled with high cost or price, affects the behaviours and intentions of consumers towards sustainable or eco-friendly textiles [151]. Additionally, once the consumers find it difficult to locate sustainable textile products on the market, that discourages them, even if such consumers are environmentally conscious. This gap must be filled by brands, by making available at all times their sustainable textile products on the market, a situation that will contribute effectively to the shift towards environmentally friendly products on the market.

Consumer Engagement

With the need for sustainable practices, materials and processes, consumer engagement drives meaningful changes in the textile industry towards sustainability. As the demand for sustainable textile products continues to increase, manufacturers will be forced to shift their production lines and strategies to accommodate the change in purchasing patterns of consumers. For example, they can send strong messages to manufacturers when they advocate for a transparent supply chain, purchase eco-friendly products from responsible brands in the market [121]. Consumer engagement draws on direct involvement, awareness, and behaviours in the sustainable practices and products from brands for informed decisions. However, key barriers such as limited availability of sustainable textiles, lack of trust and information; and key drivers such as adoption of digital tools, effective communication, and transparency, all influence consumer engagement [121]. In particular, the adoption of the digital product passport concept provides sufficient information for the product, which is easily accessible to consumers [154]. Several strategies, such as educational campaigns, recycling programs, digital solutions or platforms, and cloth rental services, amongst others, have been advanced to enhance and promote consumer engagement. A lack of the latter would create several blockages for the effective implementation of sustainable models in businesses [155]. One of such key engagement tools is digital solutions or technologies that can be easily used to engage consumers. Relevant engagement solutions for consumers, such as social marketing, recommender systems, traceability platforms, non-gamified and gamified applications, have been discussed extensively in a comprehensive study by Estrela Ferreira et al. [156]. In fact, these solutions promote effective tracking, available information, and transparency for informed decisions. For recommender systems, consumers are provided with personalised services and content for recommendations to online purchases based on their previous attitude, behaviour, and preferences [157]. This approach is made easier due to the large use of the internet for online shopping by consumers, hence companies have integrated various recommender systems on their websites [156]. Another approach, gamification or eco-gamification, is also adopted by companies to engage consumers in a rewarding and fun way. For example, an eco-gamification platform secured via blockchain technology was proposed by Das et al. [158] to enhance engagement with consumers in Bangladesh. The study revealed that the processes from production to recycling could be advanced in a closed loop to ensure the platform contributes effectively to sustainable practices in the industry in Bangladesh and beyond.

Aside from these, there are several initiatives by various brands in the industry to promote consumer engagement that drives long-term changes in behaviours, attitudes, preferences, and participation. Educational awareness and campaigns have been advanced in the industry to ensure effective engagement. Most especially, the #WhoMadeMyClothes campaign, which was triggered by the 2013 Rana Plaza disaster in Bangladesh, started by the Fashion Revolution, sparked a worldwide campaign by consumers to understand where and who produced their clothes [159]. This was triggered by the 2013 Rana Plaza disaster in Bangladesh. Such educational campaigns have forced brands to reinvent their processes to ensure transparency. In fact, in the latest 2023 report by the Fashion Revolution [160], it was revealed that the supplier list was disclosed by the majority of the brands (about 52%). This approach aimed at promoting transparency for consumers to know how their clothes are made. Other initiatives such as the worn wear by Patagonia, the garment collecting program by H&M, and "buy better, wear longer" by Levi's, amongst others, aim at strengthening responsible consumption and engagement by consumers. The concept of upcycling, which is widely used by some of the brand's initiatives, forms an important marketing strategy to promote effective textile sustainability in the industry [161].

1.7 Implications of Textile Sustainability

The textile industry is one of the major contributors to environmental degradation, particularly due to its massive consumption of water, chemical pollution, and waste management [75]. The dyeing process employs large quantities of water and synthetic chemicals, resulting in water scarcity, soil degradation, and toxic emissions. Furthermore, the industry is a major consumer of high energy and fossil fuels for powering machinery and fibre production. All activities that contribute to the carbon footprint of the textile industry. These negative impacts have led to a push for textile sustainability, which has economic, social, and environmental implications.

The environmental impact is heightened by the enormous amount of textile waste, with an entire truckload of garments discarded or burned every second. In response, sustainable textile practices are gaining traction. These include the use of environmentally friendly materials such as recycled fibre and organic cotton, as well as cleaner production methods that consume less water and release fewer toxic chemicals. Transitioning from a linear economy to a circular economy of durable, resellable, reusable, and recyclable products is essential. This practice promotes responsible disposal of textiles, thereby reducing textile waste. These changes not only lessen the environmental footprint of the textile industry but also contribute to a broader global sustainability initiative. This has further encouraged environmentally responsible textile companies to adopt sustainable techniques in their production processes [162]. Other environmental benefits of textile sustainability include reduced resource depletion and water conservation, which protect natural resources and preserve water supplies.

The concept of textile sustainability carries significant social implications, particularly regarding workers' rights, community well-being, and consumer attitudes. Historically, textile production has been associated with exploitative labour practices marked by low wages, hazardous working conditions, and long hours [163], especially in developing nations that serve as global manufacturing hubs. Consequently, social sustainability in the textile industry promotes equity, dignity, and environmental responsibility throughout the value chain. Sustainable initiatives within the textile sector aim to rectify these imbalances by advocating for ethical sourcing, commitment to fair trade practices, and enhanced transparency in supply chains. This includes ensuring that workers receive fair wages, operate in safe conditions, and benefit from essential labour protections. Such measures mitigate their exposure to harmful substances and chemicals, thereby safeguarding their health and well-being.

The shift towards sustainability in the textile industry has significant economic impacts on both manufacturers and consumers. These economic benefits include a new stream of jobs and revenues, resource savings, reduced production costs, and lower environmental and waste disposal costs. The activities such as collection, sorting, transportation and the actual recycling in the textile recycling process open up and create several business opportunities for individuals [24]. As a result, people can easily set up small companies and engage in employment under any of the recycling approaches, hence providing a good source of revenue. Alternatively, the adoption of biodegradable materials, sustainable practices and technologies has led to a reduction in production costs, energy and water consumption. Others, such as reduced chemical consumption, led to reduced treatment costs and concentrations in wastewater. Furthermore, textile industries that have invested in sustainability are likely to experience operational efficiency, reduce waste, and improve reputation, which can be transformed into increased customer loyalty and market competitiveness. This information on the brand's practices and operations for consumers [149] ensures their trust and trustworthiness.

1.8 Conclusions and Future Research Directions

The foundations for textile sustainability are achieved by addressing the significant challenges within the textile industry that tend to affect the adoption and implementation. By carefully addressing key challenges surrounding the environmental impacts, consumption of resources (such as water, energy and raw materials), generation of waste and the associated social issues with regards to low wages, poor working conditions and labour violations, the textile industry can significantly lay a firm foundation towards achieving sustainability. In addition to this, by leveraging opportunities for improvement and innovations, the industry can build upon good foundations that drive textile sustainability. These opportunities include adopting sustainable practices, innovative technologies and materials that lead to reduced chemicals, energy and water consumption, safe working conditions, improved social conditions and reduced environmental footprint. Through effective collaborations with consumers, policymakers and other stakeholders, the textile industry can drive meaningful changes that are built on trust, transparency, and traceability.

To further advance sustainability in the textile industry, future research directions can investigate five (5) key areas such as technological advancement in CE strategies and supply chain, consumer and market analysis studies, impact analysis of sustainable practices and approaches, effective network communication framework for social engagement and lastly the development of green processes for fibre and textile manufacturing. Firstly, technology plays a critical role in helping the textile industry transition into sustainability to reduce environmental impacts. The advancement and integration of technology into the production processes, industry workspace, circular strategies and the supply chain, amongst others, revolutionised the industry. Herein, we envisage that the integration of key innovations such as blockchain technology, AI, IoT, etc. can significantly ensure efficiency, optimisation and traceability in the textile industry. This would further reduce the consumption of water and energy, which are typical in textile conversion processes [75].

Secondly, market and consumer analysis studies are central for textile manufacturing companies and brands to understand the core factors driving consumers for sustainable materials. These in-depth analyses would drive effective certifications, policy direction and development, transitioning the industry to build trust, ensure transparency, and adopt sustainable initiatives. Such studies would result in robust conclusions on consumer purchase intentions and influencing factors for sustainable textile and clothing [164]. Subsequently, future research studies on impact analysis of sustainable practices and approaches are critical to assess and measure the effectiveness of key sustainability approaches driving digital traceability, ethical labour practices, circular economy models and eco-friendly or bio-based materials, for efficient consumption of resources (water, energy and raw materials) and reduction in the carbon footprint. Thirdly, an effective network communication framework for social engagement is imperative to improve the delivery of information from companies to consumers. Future studies could investigate effective communication tools and channels that improve industry-consumer engagement, affirming the earlier position by

Abbate et al. [164]. Lastly, we envisage that the development of green processes for fibre and textile manufacturing will contribute to the foundations of textile sustainability. Studies can focus on technological advancement for resource efficiency for energy optimisation, waste management and reduction, green processing technologies and sustainable materials, that transition the industry to sustainability, hence meeting the relevant SDGs goals.

References

1. Zhao, H., and B. Lin. 2019. Resources allocation and more efficient use of energy in China's textile industry. *Energy* 185: 111–120.
2. Shibly, M. A. H. and M. M. U. Hoque. 2025. Sustainability challenges of the textile industry, in textile industry and the environment—challenges, recent development and future, ed. C.-L. Hui, IntechOpen: Rijeka.
3. Provin, A. P., et al. 2024. Closing the cycle: Circular economy strategies for the textile industry using banana farming waste. *Journal of Cleaner Production* 470: 143352.
4. Sarker, M. S. I., and I. Bartok. 2024. Global trends of green manufacturing research in the textile industry using bibliometric analysis. *Case Studies in Chemical and Environmental Engineering* 9: 100578.
5. Harsanto, B., et al. 2023. Sustainability innovation in the textile industry: A systematic review. *Sustainability* 15 (2): 1549.
6. Rahman, M. T. and N. Khandaker. 2019. Physical properties of dewatered waste sludge from textile processing wastewater facility for construction material application. Singapore: Springer Singapore.
7. United Nations Environment Programme (UNEP). 2025. *UNEP textile initiative.* [cited 2025 May 2]. https://www.unep.org/topics/chemicals-and-pollution-action/circularity-sectors/sustainable-and-circular-textiles.
8. European Commission. 2025. *EU strategy for sustainable and circular textiles.* [cited 2025 May 4]. https://environment.ec.europa.eu/strategy/textiles-strategy_en#:~:text=The%20%E2%80%9CReset%20the%20Trend%E2%80%9D%20campaign,in%20all%20EU%20Member%20States.
9. Li, C., and A. Sindhuphak. 2024. Establishing sustainable education for fashion and textile design according to the SDG. *Journal of Lifestyle and SDGs Review* 4 (2): e01637.
10. Thakker, A. M., and D. Sun. 2023. Sustainable development goals for textiles and fashion. *Environmental Science and Pollution Research* 30 (46): 101989–102009.
11. Miguel Angel, G. and M. Subramanian Senthilkannan. 2019. The UN sustainable development goals for the textile and fashion industry. 1 ed. Textile science and clothing technology. Springer Singapore, 75.
12. Broega, A., C. Jordão, and S. Martins. 2017. Textile sustainability: reuse of clean waste from the textile and apparel industry. In *IOP conference series: materials science and engineering.* IOP Publishing.
13. Patwary, S. 2020. Clothing and textile sustainability: Current state of environmental challenges and the ways forward. *Textile and Leather Review* 3 (3): 158–173.
14. Lee, K. E. 2017. Environmental sustainability in the textile industry. In *Sustainability in the textile industry*, ed. S. S. Muthu, 17–55. Singapore: Springer Singapore.
15. Farhana, K., et al. 2022. Energy consumption, environmental impact, and implementation of renewable energy resources in global textile industries: An overview towards circularity and sustainability. *Materials Circular Economy* 4 (1): 15.
16. Ramírez-Escamilla, H. G., et al. 2024. Advancing toward sustainability: A systematic review of circular economy strategies in the textile industry. *Recycling* 9 (5): 95.

17. Luo, Y., et al. 2021. Environmental sustainability of textiles and apparel: A review of evaluation methods. *Environmental Impact Assessment Review* 86: 106497.
18. Grace Annapoorani, S. 2017. Social sustainability in textile industry. In *Sustainability in the textile industry*, ed. S. S. Muthu, 57–78. Singapore: Springer Singapore.
19. Thorisdottir, T. S., and L. Johannsdottir. 2020. Corporate social responsibility influencing sustainability within the fashion industry a systematic review. *Sustainability* 12 (21): 9167.
20. Luque, A., and N. Herrero-García. 2019. How corporate social (ir) responsibility in the textile sector is defined, and its impact on ethical sustainability: An analysis of 133 concepts. *Corporate Social Responsibility and Environmental Management* 26 (6): 1285–1306.
21. Dainelli, F., T. Daddi, and L. Marrucci. 2024. Financial sustainability of circular innovations in SMEs. A case study from the fashion industry in Italy. *Journal of Cleaner Production* 451: 142042.
22. Rahaman, M. T., et al. 2024. Green production and consumption of textiles and apparel: Importance, fabrication, challenges and future prospects. *Journal of Open Innovation: Technology, Market, and Complexity* 100280.
23. Leal Filho, W., et al. 2019. A review of the socio-economic advantages of textile recycling. *Journal of Cleaner Production* 218: 10–20.
24. Shah, A., et al. 2024. Realized social economic benefits of industrial ecology: A study on textile industry. In *Industrial ecology: A fusion of material and energy in green supply chain context*, 213–230. Singapore: Springer Nature Singapore.
25. Oxford Economics. 2023. *The UK fashion and textile industry contributes £62bn to the UK economy*. 2023 November 14 [cited 2025 May 20]. https://ukft.org/industry-footprint-report/.
26. Seidu, R. K., et al. 2024. A systematic review of work-related health problems of factory workers in the textile and fashion industry. *Journal of Occupational Health* 66(1).
27. Akhtar, W. H., et al. 2022. A new perspective on the textile and apparel industry in the digital transformation era. *Textiles* 2 (4): 633–656.
28. Paudyal, P., et al. 2013. Low back pain among textile workers: A cross-sectional study. *Occupational Medicine* 63 (2): 129–134.
29. Biadgo, G. H., et al. 2021. Burden of neck pain and associated factors among sewing machine operators of garment factories in Mekelle City, Northern Part of Ethiopia, 2018, a cross-sectional study. *Safety and Health at Work* 12 (1): 51–56.
30. Parimalam, P., N. Kamalamma, and A. K. Ganguli. 2007. Knowledge, attitude and practices related to occupational health problems among garment workers in Tamil Nadu India. *Journal of Occupational Health* 49 (6): 528–534.
31. Memon, Q. U. A., et al. 2019. Health problems from pesticide exposure and personal protective measures among women cotton workers in southern Pakistan. *Science of The Total Environment* 685: 659–666.
32. Singh, M. B., R. Fotedar, and J. Lakshminarayana. 2005. Occupational morbidities and their association with nutrition and environmental factors among textile workers of desert areas of Rajasthan India. *Journal of Occupational Health* 47 (5): 371–377.
33. Singh, N. 2016. Safety and health issues in workers in clothing and textile industries. *Int J Home Sci* 2 (3): 38–40.
34. Lillypet, S., T. Jain, and B. Joseph. 2017. Health problems among garment factory workers: A narrative literature review. *Journal of Occupational Health and Epidemiology* 6 (2): 114–121.
35. Kabir, H., et al. 2019. Health vulnerabilities of readymade garment (RMG) workers: A systematic review. *BMC Public Health* 19 (1): 70.
36. Natasha, K. 2025. *Labor conditions in the Asian textile industry: A closer look at exploitation and global efforts for change*. 2025 February 25 [cited 2025 May 20]. https://www.globalsociety.earth/post/labor-conditions-in-the-asian-textile-industry-a-closer-look-at-exploitation-and-global-efforts-for.
37. United Nations. 2011. *Guiding principles on business and human rights*. [cited 2025 May 20]. https://www.ohchr.org/sites/default/files/documents/publications/guidingprinciplesbusinesshr_en.pdf.

38. Pauline, O. and T. Martje. 2014. *Fact Sheet- Child labour in the textile & garment industry: Focus on the role of buying companies*. Amsterdam: The Netherlands, 10.
39. POMP MMXIX. 2024. *5 challenges that sustainable fashion faces*. 2024 January 10 [cited 2025 May 19]. https://pomp.store/blogs/journal/sustainable-fashion-challenges#:~:text=The%20additional%20sustainable%20production%20methods,be%20paid%20a%20living%20wage!
40. Rahaman, M. T., et al. 2024. Green production and consumption of textiles and apparel: Importance, fabrication, challenges and future prospects. *Journal of Open Innovation: Technology, Market, and Complexity* 10 (2): 100280.
41. Chandanshive, V., et al. 2020. In situ textile wastewater treatment in high rate transpiration system furrows planted with aquatic macrophytes and floating phytobeds. *Chemosphere* 252: 126513.
42. Alegbe, E. O., and T. O. Uthman. 2024. A review of history, properties, classification, applications and challenges of natural and synthetic dyes. *Heliyon* 10 (13): e33646.
43. Zafar, S., D. A. Bukhari, and A. Rehman. 2022. Azo dyes degradation by microorganisms—An efficient and sustainable approach. *Saudi Journal of Biological Sciences* 29 (12): 103437.
44. Singha, K., et al. 2024. Chapter 11—Harmful environmental effects for textile chemical dyeing practice. In *Green chemistry for sustainable textiles*, eds. N. Ibrahim and C.M. Hussain, Woodhead Publishing, 153–164.
45. Dutta, S., et al. 2024. Contamination of textile dyes in aquatic environment: Adverse impacts on aquatic ecosystem and human health, and its management using bioremediation. *Journal of Environmental Management* 353: 120103.
46. Dwivedi, S. 2013. Effect of textile dyes on Spirulina platensis. *Journal of Chemical and Pharmaceutical Research* 5 (4): 66–80.
47. Hu, T.-L., and S. C. Wu. 2001. Assessment of the effect of azo dye RP2B on the growth of a nitrogen fixing cyanobacterium—Anabaena sp. *Bioresource Technology* 77 (1): 93–95.
48. Khan, S., and A. Malik. 2018. Toxicity evaluation of textile effluents and role of native soil bacterium in biodegradation of a textile dye. *Environmental Science and Pollution Research* 25 (5): 4446–4458.
49. Yalcin-Enis, I., M. Kucukali-Ozturk, and H. Sezgin. 2019. Risks and management of textile waste. In *Nanoscience and biotechnology for environmental applications*, ed. K. M. Gothandam, et al., 29–53. Cham: Springer International Publishing.
50. Nur Farzanah Binti, N., et al. 2022. A review on textile and clothing industry impacts on the environment. *International Journal of Academic Research in Business and Social Sciences* 12(10): 2137–2146.
51. Strähle, J., and K. Hauk. 2017. Impact on sustainability: Production versus consumption. In *Green fashion retail*, ed. J. Strähle, 49–75. Singapore: Springer Singapore.
52. Irenka, P. 2025. *5 environmental impacts of textile industry, and 3 ways to be more sustainable*. 2024 April 2 [cited 2025 May 20]. https://watertreatmentmagazine.com/en/impact-textile-industry/#:~:text=Emissions:%20textile%20industry%20carbon%20footprint,more%20than%2050%20per%20cent.
53. Wang, H., and M. Xu. 2011. Review on theoretical model of carbon emissions in textile industry chain. *Shanghai Text. Sci. Technol* 7: 1–4.
54. Wang, Z., et al. 2015. The agglomeration and transferring of Pan-Yangtze river delta's textile industry. *World Regional Studies* 24 (2): 123–130.
55. Zabaniotou, A., and K. Andreou. 2010. Development of alternative energy sources for GHG emissions reduction in the textile industry by energy recovery from cotton ginning waste. *Journal of Cleaner Production* 18 (8): 784–790.
56. Priambodo, A., and S. Kumar. 2001. Energy use and carbon dioxide emission of Indonesian small and medium scale industries. *Energy Conversion and Management* 42 (11): 1335–1348.
57. Ganesan, P., et al. 2015. Specific energy consumption and Co2 emission reduction analysis in a textile industry. *International Journal of Green Energy* 12 (7): 685–693.
58. Li, R., D. Wu, and H. Jiang. 2011. Carbon footprint and its measurement in dyeing and finishing. *Dyeing* 37 (18): 40–43.

59. Zhang, Y., C. Chen, and X. Zhang. 2024. The impact of environmental, social, and governance performance on the total factor productivity of textile firms: A meditating-moderating model. *Sustainability* 16 (16): 6783.
60. Colourful Socks. 2025. *Water consumption in textile industry statistics* [cited 2025 May 20]. https://bestcolorfulsocks.com/blogs/news/water-consumption-in-textile-industry-statistics.
61. Ozturk, E., and N. C. Cinperi. 2018. Water efficiency and wastewater reduction in an integrated woolen textile mill. *Journal of Cleaner Production* 201: 686–696.
62. Samanta, K. K., et al. 2019. 3—water consumption in textile processing and sustainable approaches for its conservation. In Water in textiles and fashion, ed. S. S. Muthu, Woodhead Publishing, 41–59.
63. Vajnhandl, S., and J. V. Valh. 2014. The status of water reuse in European textile sector. *Journal of environmental management* 141: 29–35.
64. Ethos. 2025. *The textile industry's climate reckoning: A path to circularity and low-carbon energy*. 2025 March 11 [cited 2025 May 19]. https://the-ethos.co/textile-circularity-and-low-carbon-energy-aii-unifi/.
65. Urska, T., et al. 2023. Fossil fashion: The hidden reliance of fast fashion on fossil fuels. *Changing Markets* 38.
66. Tang, K. H. D. 2023. State of the art in textile waste management: A review. *Textiles* 3 (4): 454–467.
67. Tomovska, E., et al. 2017. Attitudes towards managing post-industrial apparel cuttings waste. *The Journal of The Textile Institute* 108 (2): 172–177.
68. Sawant, J., et al. 2024. Sustainability in textiles: a critical review of eco—friendly practices and materials. *ShodhKosh: Journal of Visual and Performing Arts* 5(2): 85–97.
69. Tian, W., et al. 2022. Recent progress in biobased synthetic textile fibers. *Frontiers in Materials* 9:2022.
70. Jason, C. 2023. *Advances in bio-based fibers*. 2023 September 14 [cited 2025 May 19]. https://www.fiberjournal.com/advancesin-bio-based-fibers/.
71. Maduna, L. and A. Patnaik. 2023. 2—Introduction to sustainable fibres for fashion and textiles, in Sustainable Fibres for Fashion and Textile Manufacturing, ed. R. Nayak, Woodhead Publishing. p. 27–48
72. Eyupoglu, S. 2020. Sustainable plant-based natural fibers. In *Sustainability in the textile and apparel industries: sourcing natural raw materials*, ed. S. S. Muthu and M. A. Gardetti, 27–48. Cham: Springer International Publishing.
73. Okur, N., and C. Saricam. 2019. The impact of knowledge on consumer behaviour towards sustainable apparel consumption. In *Consumer behaviour and sustainable fashion consumption*, ed. S. S. Muthu, 69–96. Singapore: Springer Singapore.
74. Plakantonaki, S., et al. 2023. A review of sustainability standards and Ecolabeling in the textile industry. *Sustainability* 15 (15): 11589.
75. Zhang, L., et al. 2023. Advancing life cycle sustainability of textiles through technological innovations. *Nature Sustainability* 6 (3): 243–253.
76. DelRe, C., et al. 2021. Near-complete depolymerization of polyesters with nano-dispersed enzymes. *Nature* 592 (7855): 558–563.
77. Bataineh, K. M. 2020. Life-cycle assessment of recycling postconsumer high-density polyethylene and polyethylene terephthalate. *Advances in Civil Engineering* 2020 (1): 8905431.
78. Münkel, A., Y. S. Gloy, and T. Gries. 2017. Development and testing of a relay nozzle concept for air-jet weaving. *IOP Conference Series: Materials Science and Engineering* 254 (13): 132003.
79. Rohit Dev, S. 2025. *Overcoming challenges in sustainable textile logistics*. 2025 January 21 [cited 2025 May 19]. https://www.textileworld.com/textile-world/2025/01/overcoming-challenges-in-sustainable-textile-logistics/.
80. Mölsä, K. M., et al. 2022. Linear, reuse or recycling? An environmental comparison of different life cycle options for cotton roller towels. *Journal of Cleaner Production* 374: 133976.

81. Palma De Oliveira, D., et al. 2013. *Textile Dyes: Dyeing Process and Environmental Impact*, in *Eco-Friendly Textile Dyeing and Finishing*, ed. M. Gunay, IntechOpen: Rijeka.
82. Niinimäki, K., et al. 2020. The environmental price of fast fashion. *Nature Reviews Earth and Environment* 1 (4): 189–200.
83. Geissdoerfer, M., et al. 2017. The circular economy—A new sustainability paradigm? *Journal of Cleaner Production* 143: 757–768.
84. Torkayesh, A. E., et al. 2023. A systematic literature review of MABAC method and applications: An outlook for sustainability and circularity. *Informatica* 34 (2): 415–448.
85. Das, A. K., et al. 2025. Circular economy: A sustainable model for waste reduction and wealth creation in the textile supply chain. *SPE Polymers* 6 (1): e10171.
86. Akash, M. H., et al. 2025. Investigating the attributes for implementing circular economy in the textile manufacturing supply chain: Implications for the triple bottom line of sustainability. *Sustainable Horizons* 14: 100129.
87. Stahel, W. R. 2013. Policy for material efficiency—sustainable taxation as a departure from the throwaway society. *Philosophical Transactions of the Royal Society A: Mathematical, Physical and Engineering Sciences* 371(1986): 20110567.
88. Colucci, M., and A. Vecchi. 2021. Close the loop: Evidence on the implementation of the circular economy from the Italian fashion industry. *Business Strategy and the Environment* 30 (2): 856–873.
89. Debnath, B., et al. 2023. An integrated best-worst method and interpretive structural modeling approach for assessing the barriers to circular economy implementation. *Decision Analytics Journal* 7: 100250.
90. Goworek, H. 2011. Social and environmental sustainability in the clothing industry: A case study of a fair trade retailer. *Social Responsibility Journal* 7 (1): 74–86.
91. Dziubaniuk, O., L. Aarikka-Stenroos, and E.-L. Pohls. 2023. Catalysing the textile industry toward a circular economy: An ecosystem approach. In *The Routledge handbook of catalysts for a sustainable circular economy*, 67–87. Routledge.
92. Ranta, V., J. Keränen, and L. Aarikka-Stenroos. 2020. How B2B suppliers articulate customer value propositions in the circular economy: Four innovation-driven value creation logics. *Industrial Marketing Management* 87: 291–305.
93. Kumar, V., et al. 2019. Circular economy in the manufacturing sector: Benefits, opportunities and barriers. *Management Decision* 57 (4): 1067–1086.
94. Mishra, S., et al. 2019. Evaluating indicators for international manufacturing network under circular economy. *Management Decision* 57 (4): 811–839.
95. Ki, C.-W., S. M. Chong, and J. E. Ha-Brookshire. 2020. How fashion can achieve sustainable development through a circular economy and stakeholder engagement: A systematic literature review. *Corporate Social Responsibility and Environmental Management* 27 (6): 2401–2424.
96. Dissanayake, D. G. K., and D. Weerasinghe. 2022. Towards circular economy in fashion: review of strategies, barriers and enablers. *Circular Economy and Sustainability* 2 (1): 25–45.
97. Huang, Y.-F., et al. 2021. Exploring the decisive barriers to achieve circular economy: Strategies for the textile innovation in Taiwan. *Sustainable Production and Consumption* 27: 1406–1423.
98. Singh, P., and E. Giacosa. 2019. Cognitive biases of consumers as barriers in transition towards circular economy. *Management Decision* 57 (4): 921–936.
99. Aloini, D., et al. 2020. Driving the transition to a circular economic model: A systematic review on drivers and critical success factors in circular economy. *Sustainability* 12 (24): 10672.
100. Paras, M. K., P. Rudrajeet, and D. Ekwall. 2018. Systematic literature review to develop a conceptual framework for a reuse-based clothing value chain. *The International Review of Retail, Distribution and Consumer Research* 28(3): 231–258.
101. Mohr, L. A., D. J. Webb, and K. E. Harris. 2001. Do consumers expect companies to be socially responsible? The impact of corporate social responsibility on buying behavior. *Journal of Consumer Affairs* 35 (1): 45–72.

102. Abdallah, A. B., et al. 2024. Assessment of demographics and motivations of post-consumer textile waste management using a bibliometric approach. *Journal of Material Cycles and Waste Management* 26 (4): 2534–2545.
103. Joung, H.-M., and H. Park-Poaps. 2013. Factors motivating and influencing clothing disposal behaviours. *International Journal of Consumer Studies* 37 (1): 105–111.
104. Rogers, H. A., P. Deutz, and T. B. Ramos. 2021. Repairing the circular economy: Public perception and participant profile of the repair economy in Hull, UK. *Resources, Conservation and Recycling* 168: 105447.
105. Noh, M. 2021. Understanding the effect of information sources on college students' recycling/reuse behavior towards clothing and textile products. *Sustainability* 13 (11): 6298.
106. Laitala, K., and I. G. Klepp. 2018. Motivations for and against second-hand clothing acquisition. *Clothing Cultures* 5 (2): 247–262.
107. Kemi, A. P., and G. Zilahy. 2025. Sustainable post-purchase behaviour of consumers of fashion textile products. *Discover Sustainability* 6 (1): 94.
108. Sustainability Directory. 2025. *Why are certifications important for sustainable textiles?* 2025 May 3 May 18 2025]. https://fashion.sustainability-directory.com/question/why-are-certifications-important-for-sustainable-textiles/.
109. Lee, E. J., B. Joonheui, and K. H. Kim. 2020. The effect of sustainable certification reputation on consumer behavior in the fashion industry: Focusing on the mechanism of congruence. *Journal of Global Fashion Marketing* 11(2): 137–153.
110. Kim, K., and J. Kim. 2011. Third-party privacy certification as an online advertising strategy: An investigation of the factors affecting the relationship between third-party certification and initial trust. *Journal of Interactive Marketing* 25 (3): 145–158.
111. Chun, E., et al. 2021. Business transparency and willingness to act environmentally conscious behavior: Applying the sustainable fashion evaluation system "Higg Index." *Journal of Global Scholars of Marketing Science* 31 (3): 437–452.
112. Victor, D. 2025. *Sustainable textile production: Policies and certifications transforming the global landscape.* [cited 2025 May 19]. https://www.openaccessgovernment.org/article/sustainable-textile-production-policies-and-certifications-transforming-the-global-landscape/190123/.
113. Sustainability Directory. 2025. *What role does policy play in sustainable fashion and textile recycling?* 2025 May 19 [cited 2025 May 21]. https://fashion.sustainability-directory.com/question/what-role-does-policy-play-in-sustainable-fashion-and-textile-recycling/#:~:text=Policy%20can%20implement%20waste%20reduction,choices%2C%20and%20regulating%20misleading%20marketing.
114. Habib, A., et al. 2025. Comparative analysis of the effects of european union environmental policies on the textile and clothing sector of Turkey and Bangladesh: Implications for sustainable development goals. *Journal of Lifestyle and SDGs Review* 5 (2): e03840.
115. Accord on Fire and Building Safety in Bangladesh. 2023. *Bangladesh agreement on health and safety in the textile and garment industry.* 2023 November 1 [cited 2025 May 18]. https://internationalaccord.org/wp-content/uploads/2023/11/CSSP-Bangladesh-Agreement_public-version.pdf.
116. Salminen, J. 2018. The accord on fire and building safety in Bangladesh: A new paradigm for limiting buyers' liability in global supply chains? *The American Journal of Comparative Law* 66 (2): 411–451.
117. Shemona, S. 2023. *Türkiye apparel industry's road to green transformation.* 2023 February 1 [cited 2025 May 19]. https://www.just-style.com/news/turkiye-apparel-industrys-road-to-green-transformation/?cf-view.
118. Silia, L., I. Dbolia, and E. Lapkovska. 2024. Sustainable textile industry—wishful thinking or the new norm: A review. *Journal of Engineered Fibers and Fabrics* 19: 15589250231220360.
119. Castano Garcia, A., et al. 2021. High consumption, an unsustainable habit that needs more attention. *Energy Research and Social Science* 80: 102241.

120. Goel, V., and M. Michaelides. 2022. What is the economic impact of fast fashion?—Economics research question. *International Journal of Scientific and Technology Research* 9: 571–581.
121. Ribeiro, P. R., et al. 2023. Consumers' engagement and perspectives on sustainable textile consumption. *Sustainability* 15 (22): 15812.
122. Voola, R., et al. 2022. How are consumer behavior and marketing strategy researchers incorporating the SDGs? A review and opportunities for future research. *Australasian Marketing Journal* 30 (2): 119–130.
123. Ajzen, I. 1991. The theory of planned behavior. *Organizational Behavior and Human Decision Processes* 50 (2): 179–211.
124. Balasubramanian, M., and P. Sheykhmaleki. 2024. Comprehending the consumer behavior toward sustainable apparel. *Sustainability* 16 (18): 8026.
125. Bosnjak, M., I. Ajzen, and P. Schmidt. 2020. The theory of planned behavior: Selected recent advances and applications. *Europe's Journal of Psychology* 16 (3): 352–356.
126. Islam, Q. and S. M. F. Ali Khan. 2024. Assessing consumer behavior in sustainable product markets: A structural equation modeling approach with partial least squares analysis. *Sustainability* 16(8): 3400
127. Fishbein, M., and I. Ajzen. 1977. Belief, attitude, intention, and behavior: An introduction to theory and research. *Philosophy and Rhetoric* 10 (2): 130–132.
128. Leclercq-Machado, L., et al. 2022. Consumer patterns of sustainable clothing based on theory of reasoned action: Evidence from Ecuador. *Sustainability* 14 (22): 14737.
129. Chen, C.-C., C.-W. Chen, and Y.-C. Tung. 2018. Exploring the consumer behavior of intention to purchase green products in belt and road countries: An empirical analysis. *Sustainability* 10 (3): 854.
130. Joshi, Y., and Z. Rahman. 2015. Factors affecting green purchase behaviour and future research directions. *International Strategic Management Review* 3 (1): 128–143.
131. Saricam, C., and N. Okur. 2019. Analysing the consumer behavior regarding sustainable fashion using theory of planned behavior. In *Consumer behaviour and sustainable fashion consumption*, ed. S. S. Muthu, 1–37. Singapore: Springer Singapore.
132. Stern, P. C., et al. 1999. A value-belief-norm theory of support for social movements: The case of environmentalism. *Human Ecology Review* 81–97.
133. White, K., R. Habib, and D. J. Hardisty. 2019. How to SHIFT consumer behaviors to be more sustainable: A literature review and guiding framework. *Journal of Marketing* 83 (3): 22–49.
134. Sustainability Directory. 2025. *Why is consumer behavior important for sustainable textiles?* [January 27 April 25, 2025]. https://fashion.sustainability-directory.com/question/why-is-consumer-behavior-important-for-sustainable-textiles/.
135. Desore, A., and S. A. Narula. 2018. An overview on corporate response towards sustainability issues in textile industry. *Environment, Development and Sustainability* 20 (4): 1439–1459.
136. Sustainability Directory. 2025. *To what extent can consumer behavior influence sustainable textile adoption?* [January 26 April 18, 2025]. https://fashion.sustainability-directory.com/question/to-what-extent-can-consumer-behavior-influence-sustainable-textile-adoption/.
137. Rakic, B. and M. Rakic. 2015. Holistic management of marketing sustainability in the process of sustainable development. *Environmental Engineering & Management Journal (EEMJ)* 14(4)
138. D'Souza, C., M. Taghian, and P. Lamb. 2006. An empirical study on the influence of environmental labels on consumers. *Corporate Communications: An International Journal* 11 (2): 162–173.
139. Lin, S. H. 2009. Exploratory evaluation of potential and current consumers of organic cotton in Hawaii. *Asia Pacific Journal of Marketing and Logistics* 21 (4): 489–506.
140. Kang, J., C. Liu, and S.-H. Kim. 2013. Environmentally sustainable textile and apparel consumption: The role of consumer knowledge, perceived consumer effectiveness and perceived personal relevance. *International Journal of Consumer Studies* 37 (4): 442–452.
141. Nyilasy, G., H. Gangadharbatla, and A. Paladino. 2014. Perceived greenwashing: The interactive effects of green advertising and corporate environmental performance on consumer reactions. *Journal of Business Ethics* 125 (4): 693–707.

142. TerraChoice. 2010. *The sins of greenwashing: Home and family edition 2010*, 1–31.
143. Lu, X., et al. 2022. How does young consumers' greenwashing perception impact their green purchase intention in the fast fashion industry? An analysis from the perspective of perceived risk theory. *Sustainability* 14 (20): 13473.
144. Adamkiewicz, J., et al. 2022. Greenwashing and sustainable fashion industry. *Current Opinion in Green and Sustainable Chemistry* 38: 100710.
145. Akturan, U. 2018. How does greenwashing affect green branding equity and purchase intention? An empirical research. *Marketing Intelligence & Planning* 36 (7): 809–824.
146. Guo, R., et al. 2017. A path analysis of greenwashing in a trust crisis among Chinese energy companies: The role of brand legitimacy and brand loyalty. *Journal of Business Ethics* 140 (3): 523–536.
147. Gil-Cordero, E., et al. 2021. Measuring behavioural intention through the use of greenwashing: a study of the mediating effects and variables involved. *Sustainability* 13 (12): 6720.
148. Wang, H., B. Ma, and R. Bai. 2020. The spillover effect of greenwashing behaviours: An experimental approach. *Marketing Intelligence & Planning* 38 (3): 283–295.
149. Sustainability Directory. 2025. *To what extent does transparency impact consumer behavior?* [January 27 May 12 2025]. https://fashion.sustainability-directory.com/question/to-what-extent-does-transparency-impact-consumer-behavior/.
150. Ronda, L. 2024. Overcoming barriers for sustainable fashion: Bridging attitude-behaviour gap in retail. *International Journal of Retail & Distribution Management* 52 (1): 44–61.
151. Rausch, T. M., and C. S. Kopplin. 2021. Bridge the gap: Consumers' purchase intention and behavior regarding sustainable clothing. *Journal of Cleaner Production* 278: 123882.
152. Riesgo, S. B., M. Lavanga, and M. Codina. 2020. The consumption side of sustainable fashion: Price sensitivity, value and transparency demand. In *Proceedings of the Global Fashion Conference.*
153. Munro, P., S. Kapitan, and B. Wooliscroft. 2023. The sustainable attitude-behavior gap dynamic when shopping at the supermarket: A systematic literature review and framework for future research. *Journal of Cleaner Production* 426: 138740.
154. Adisorn, T., L. Tholen, and T. Götz. 2021. Towards a digital product passport fit for contributing to a circular economy. *Energies* 14 (8): 2289.
155. Tunn, V. S. C., et al. 2021. Consumer adoption of access-based product-service systems: The influence of duration of use and type of product. *Business Strategy and the Environment* 30 (6): 2796–2813.
156. Cruz, E. F. and A. M. Rosado da Cruz. 2023. Digital solutions for engaging end-consumers in the circular economy of the textile and clothing value chain—A systematic review. *Cleaner and Responsible Consumption* 11: 100138.
157. Adomavicius, G., and A. Tuzhilin. 2005. Personalization technologies: A process-oriented perspective. *Communications of the ACM* 48 (10): 83–90.
158. Das, D., and S. Paul. 2024. Promoting consumer engagement in sustainable fashion and circular textile through eco-gamification in Bangladesh. In *7th IEOM Bangladesh international conference on industrial engineering and operations management*, 241–251. Dhaka, Bangladesh: IEOM Society International.
159. Alice, N. 2019. *Why we should be asking #WhoMadeMyClothes? before every purchase.* [April 26 May 19 2025]. https://www.vogue.co.uk/article/who-made-my-clothes.
160. Fashion Revolution. *Fashion Transparency Index 2023.* 2023 July 12 May 19, 2025]; Available from: https://issuu.com/fashionrevolution/docs/fashion_transparency_index_2023_pages.
161. Ofori, E. A., et al. 2025. A review on upcycling as a strategic marketing tool for sustainability in the fashion industry. *Journal of Responsible Production and Consumption* 2 (1): 1–24.
162. Tayyab, M., et al. 2020. A sustainable development framework for a cleaner multi-item multi-stage textile production system with a process improvement initiative. *Journal of Cleaner Production* 246: 119055.
163. Naqvi, S. L. H., et al. 2024. Social and environmental impacts in textile production. In *Dye pollution from textile industry: Challenges and opportunities for sustainable development*, ed. P. Singh, 423–453. Singapore: Springer Nature Singapore.

164. Abbate, S., et al. 2024. Sustainability trends and gaps in the textile, apparel and fashion industries. *Environment, Development and Sustainability* 26 (2): 2837–2864.

Raphael Kanyire Seidu (Ph.D.) is a Research Assistant at the School of Fashion and Textiles at The Hong Kong Polytechnic University, Hong Kong. He holds a Ph.D. in Fashion and Textiles Design from The Hong Kong Polytechnic University, Hong Kong SAR. He was also a Visiting Research Student at Nottingham Trent University (NTU) in the UK. He has several publications in leading journals in textile, design, technology, and material science. His recent paper was adjudged the "Second-Best Paper" presented at the 12th Sustainable Education and Development Research Conference in Somanya, Ghana. His research interest focuses on retro-reflective materials, smart textiles, cloth culture, weaving, and print technologies.

Dr. Benjamin Tawiah holds a B.A. in Textiles and a Master of Engineering from Kwame Nkrumah University of Science and Technology, Ghana, and Jiangnan University, China, respectively. He also has a Postgraduate Diploma in Management Information Systems from GIMPA and a Ph.D. in Textile Technology (Chemistry) from Hong Kong Polytechnic University. Previously, he was a postdoctoral fellow at Hong Kong Polytechnic University and City University of Hong Kong. Currently, he lectures at Kwame Nkrumah University of Science and Technology, focusing on polymer nanomaterial composites, flame retardant polymer composites, and second-generation batteries for advanced grid energy storage.

Shouxiang Jiang is currently a Professor of fashion and textiles at The Hong Kong Polytechnic University. He teaches a broad range of subjects such as innovative textile design approach, advanced textile and fashion, and international fashion and textiles design. Prof. Jiang's research deals with innovative textile effects using physical and chemical treatment technologies. He is internationally renowned for his development of metallised textiles. Leading a team that has developed a sputtering coating technology with research funding from the HKSAR government and sponsorships from the textile industries, Prof. Kinor Jiang has designed and launched an anhydrous system that uses metallising technologies to produce functional and decorative textiles for the production line. He holds 30 granted and filed patents based on the coating technology. Prof. Jiang has published more than 200 journal papers and more than 10 books. His designs and art projects have been exhibited worldwide and received more than 30 awards.

Chapter 2
Strategies for Achieving Textile Sustainability

Raji Rafiu King

Abstract The strategies for achieving textile sustainability involve the design of new materials or modifying existing ones to meet the specific requirements for various applications. The design of materials for sustainable textiles combines advancements in both digital technologies and material sciences with environmental responsibility. This chapter deals with some key strategies that could be adopted to achieve textile sustainability. These strategies include but not limited to adoption of sustainable fibers and materials, electronic textiles (e-textiles) with eco-friendly components, energy harvesting and storage technologies in textiles, 3D printing and additive manufacturing, circular economy for textiles, smart textiles for reducing environmental impact, biomimicry and nature-inspired design, end-of-life and recycling considerations and blockchain for supply chain transparency. Achieving textile sustainability in the design and production of materials for digital technologies requires a multi-faceted approach that addresses the environmental, social, technological and economic aspects of sustainability.

Keywords Textile sustainability · Compostable textiles · Smart textiles · Circular economy · Digital technologies

2.1 Introduction

Designing materials for digital technologies entails creating new materials or enhancing existing ones to fulfil the unique demands of electronic devices, computing systems, and other digital applications. This interdisciplinary field draws from materials science, chemistry, physics, engineering and economics to create materials with properties that support the functioning of various digital devices, such as semiconductors, sensors, and displays.

R. R. King (✉)
Guangdong Provincial Key Laboratory of Electronic Functional Materials and Devices, Huizhou University, Huizhou, China
e-mail: mrkingraji@outlook.com

S.-X. Jiang et al. (eds.), *Advances in Textile Materials and Processing Techniques for Sustainability*, SDGs and Textiles, https://doi.org/10.1007/978-981-95-0469-5_2

With growing concerns about e-waste and the environmental impact of digital technologies, materials scientists are focusing on creating biodegradable materials, recyclable components, and reducing the reliance on traditional materials that tend to pose adverse environmental concerns. The design of materials for digital technologies, especially in the context of sustainable textiles, combines advancements in both digital technologies and material sciences with environmental responsibility. This path is aimed at creating textiles that not only meet the performance demands of modern smart textiles and wearable devices but also align with sustainability goals, including minimising environmental impact, enhancing recyclability, and promoting circular economy principles.

When the textile and apparel industry seeks to advance sustainability, the primary drivers of change are often associated with the adoption of eco-friendly materials and processes, as well as addressing ethical concerns in production practices [1]. Green electronics initiatives emphasise the use of sustainable materials and processes, such as organic semiconductors and eco-friendly packaging. The design of these materials requires a deep understanding of their structural, electrical, mechanical, and thermal properties, ensuring that they can meet the stringent performance demands of modern digital devices. As technology evolves, new materials will be essential for enabling faster, smaller, and more energy-efficient electronics.

2.2 Adoptable Strategies to Achieve Sustainability

Achieving textile sustainability in the design and production of materials for digital technologies requires a multi-faceted approach that addresses the environmental, social, and economic aspects of sustainability. As digital textiles and e-textiles become more prominent, integrating electronics and smart systems into fabrics, it becomes essential to ensure that their design, manufacturing, use, and end-of-life processes are sustainable.

Subsequent discussions deal with some key strategies that could be adopted to achieve textile sustainability. Figure 2.1 presents a host of adoptable strategies that could aid in achieving sustainability in the design and production of digital and e-textiles.

2.3 Sustainable Textile Materials Selection—Fibres, Colourants and Surfactants

Sustainable textile materials are produced with minimal environmental impact and can be recycled or biodegraded. These materials are either natural (cultivated organically), bio-based and or biodegradable. Bio-based fibres are produced either wholly or partly from renewable biological sources (renewable biological sources such as

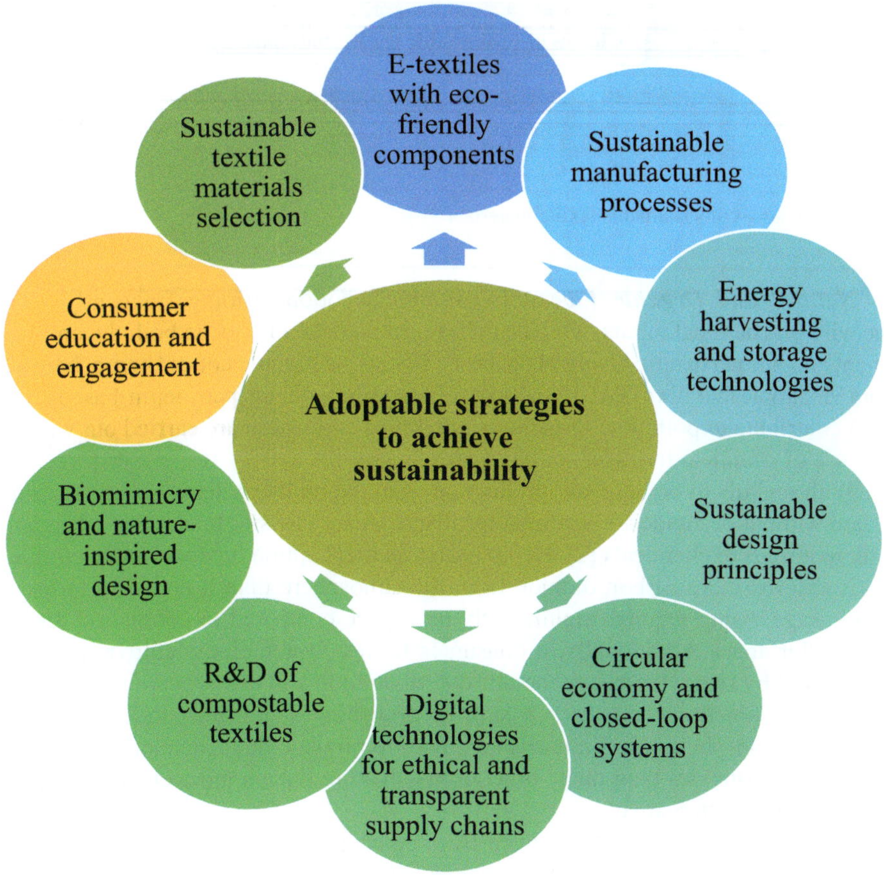

Fig. 2.1 Adoptable strategies to achieve sustainability

plants, algae, bacteria, and fungi). They may not necessarily be biodegradable or suitable for composting.

Natural Bio-Based Polymers are directly sourced from nature. Examples include cellulose-based fibers, Cotton, Linen (Flax), Hemp, Ramie, Jute, and Bamboo (if mechanically processed). Protein-based fibers including wool (Sheep), Silk (Silkworms, Spiders) hair (Alpaca, Cashmere, Mohair), Soy Protein Fibers and Milk Casein Fibers (e.g., Qmilk) Not all natural bio-based polymers are environmentally friendly. While they come from renewable sources, their impact depends on factors like resource consumption, processing methods, and end-of-life disposal. Also though most natural polymers degrade under the right conditions, some can persist in the environment due to their molecular structure or how they are processed. Figure 2.2 presents a schematic diagram of sustainable textile materials classes.

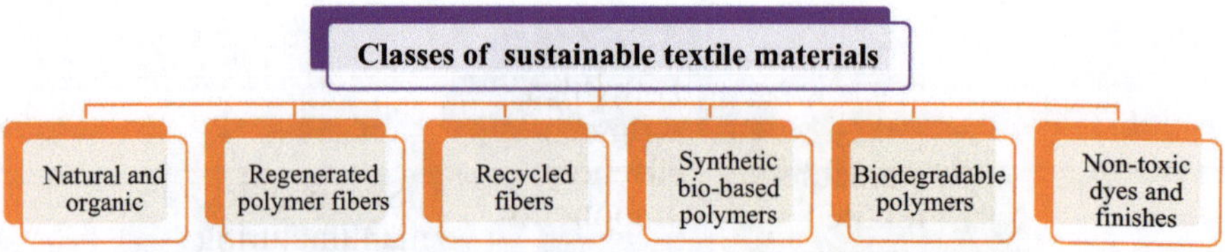

Fig. 2.2 Classes of sustainable textile materials

Natural and Organic Fibers: Conventional cotton farming frequently relies heavily on chemical inputs, which can have detrimental effects on both the environment and human health. Using plant-based fibres like organic cotton, hemp, bamboo and Pineapple leaf fibre (PALF) reduces reliance on synthetic, petroleum-based fibres that contribute to pollution. Organic cotton fibres cultivation are carried out without the use of chemical fertilisers or pesticides.

With regards to organic silk production also, the mulberry leaves used are cultivated without pesticides or artificial fertilisers. Also, raw silk treatments of organic silk require zero chemicals [2]. Hemp is also an ideal option for sustainable clothing as it absorbs more carbon dioxide from the atmosphere than it emits. In terms of land usage, hemp requires significantly less space compared to cotton and thrives in smaller areas. Additionally, it consumes far less water. Hemp cultivation helps purify soil by removing heavy metals, contributing to soil detoxification. Additionally, hemp functions as a carbon sink by absorbing significant amounts of CO_2 from the atmosphere [3]. Similarly, flax, used to produce linen, demands less water and fewer pesticides than many other crops, making linen a more environmentally friendly fabric alternative.

Natural fibres can also be derived from pineapple plant leaves (Pineapple Leaf Fiber or PALF), which are usually discarded after the fruit is collected. With millions of tons of pineapple leaves being discarded annually, thousands of tons of natural fibre could potentially be extracted, enabling the industrialisation and commercialisation of Pineapple Leaf Fiber (PALF) processing [4]. Consequently, PALF is recognised for its strong mechanical properties and can be extracted from pineapple leaf waste through various processing techniques [5].

Bamboo, on the other hand, is an incredibly sustainable plant due to its rapid growth, making it a highly renewable resource. Unlike traditional wood, bamboo can be harvested in just 3–5 years, while hardwood trees require several decades to reach maturity before they can be harvested [6]. Certain bamboo species can grow up to 1.2 m per day, with a yield that is up to 25 times greater than that of the fastest-growing timber trees [7]. These fibers are biodegradable and renewable, making them a better choice for eco-friendly digital textiles.

Regenerated Bio-based Polymers (Chemically Processed from Natural Sources): Regenerated fibers are derived from natural polymers (such as cellulose or protein) that are chemically processed into a new fiber form. Examples include Cellulose-Derived Fibers, Rayon (Viscose), Lyocell (TENCEL®), Modal, Acetate &

Triacetate, Bamboo Viscose (if chemically processed), Chitosan-Based Fibers, and Derived from chitin (found in shrimp shells, crab shells, and fungi).

Synthetic Bio-based Polymers: Synthetic bio-based polymer fibers are made from bio-based monomers, which are polymerized into new materials that do not exist naturally. These fibers mimic synthetic petroleum-based fibers but are produced from renewable resources. Examples include Polylactic Acid (PLA) Fibers—Made from corn starch or sugarcane (e.g., Ingeo®), Polyhydroxyalkanoates (PHA) Fibers—Biodegradable polyesters produced by bacteria. Bio-Based Polyamides (Nylon 11, Nylon 6, 10)—Derived from castor oil instead of petroleum. Bio-Based Polyethylene (Bio-PE)—Made from sugarcane ethanol. Bioengineered Spider Silk (Microsilk®)—Synthetic silk produced using genetically modified microbes.

Polylactic acid (PLA) or bio-nylon is one of the most popular among the examples and offers alternatives to conventional plastics and synthetics, providing renewable sources for materials used in digital textiles, particularly in flexible components. Polylactic acid (PLA) is a linear, non-aromatic polyester formed by linking repeating units of lactic acid, chemically known as 2-hydroxypropanoic acid [8]. PLA is a type of polyester made from fermented plant starch from corn, cassava, maize, sugarcane or sugar beet pulp.

Regenerated polymer fibres and synthetic bio-based polymer fibres are different. The main difference lies in their production process and the nature of the polymers used. Regenerated fibres are chemically processed from natural polymers (like cellulose or protein). Synthetic bio-based polymer fibres are fully synthetic polymers, but they come from renewable biological sources instead of fossil fuels.

Emerging and Innovative Bio-based Polymers: These polymers are derived from renewable resources and often designed with enhanced properties, such as biodegradability, biocompatibility, and reduced environmental impact. Algae-Based Fibers—Produced from seaweed and algae extracts (e.g., SeaCell®, Mushroom-Based Fibers—Derived from mycelium (fungal networks), for example is Mylo™ (Bolt Threads). Mylo™ is a material made from mycelium, the underground root structure of mushrooms. It is soft, supple and serves as an alternative to animal leather and has significantly lower environmental impacts than raising livestock. Bioengineered Spider Silk—Synthetic silk produced using biotechnology (e.g., Bolt Threads' Microsilk®) Orange Fiber—Made from citrus juice by-products. Banana Fiber (Abacá)—Sourced from banana plant stems.

Recycled Fibers (Fibers from Recyclable Materials): These fibers are created by reprocessing post-consumer or post-industrial waste into new materials. Examples include Recycled Polyester (rPET), recycled cotton, nylon, wool, acrylic and blends of various fibers. Incorporating recycled polyester, nylon, or fibers from post-consumer textiles (like recycled PET bottles) reduces the demand for virgin materials. Recycled cotton serves as a sustainable and environmentally friendly option compared to conventional cotton production. It is produced by gathering and reprocessing cotton waste from both industrial sources and consumer products, including fabric scraps and used clothing. This helps decrease the energy, water, and chemicals needed for raw material production.

Non-toxic Dyes and Finishes: Conventional dyeing techniques frequently rely on toxic chemicals and consume significant quantities of water, resulting in environmental pollution and excessive waste. In contrast, eco-friendly dyes (Fig. 2.3) are derived from natural or non-toxic substances and are crafted to reduce the environmental footprint of textile manufacturing. Natural sources offer sustainable and safe alternatives, such as plant-based options like indigo, turmeric dye, onion skin dye, and walnut dye.

Animal-based dyes, also referred to as animal source dyes, are derived from creatures such as insects and invertebrates. Historically, dyes like Tyrian purple, obtained from the mucus secretions of Murex sea snails, and cochineal dye, extracted from Dactylopius coccus insects [9], Lac dye (Kerria lacca) derived from secretions of lac insects, has been a prominent example and has been favoured as being non-toxic. However, obtaining these dyes typically involves drying and crushing insects, molluscs, and snails, raising significant ethical and ecological issues. A Canadian outfit *Maiwa Handprints Ltd.* markets a host of natural dyes from both plant and animal sources, from logwood to cochineal insect extracts [10]. Examples of non-toxic dye and finish sources are illustrated in Fig. 2.3.

As a result, sustainably sourced dyes should ideally be derived from waste materials, be non-toxic, and require minimal processing. These include rich black or brown dyes obtained from cephalopods, such as squid ink, [11] which is sourced from waste generated through sustainable fishing practices. Cephalopods, which include squid, octopus, cuttlefish, and nautilus [12], are a group of molluscs found in oceans worldwide at nearly all depths, with an evolutionary lineage extending over 500 million years [13]. Cephalopods release ink, composed of melanin and mucus, to frighten or confuse predators, allowing them to escape unnoticed or avoid being consumed [14].

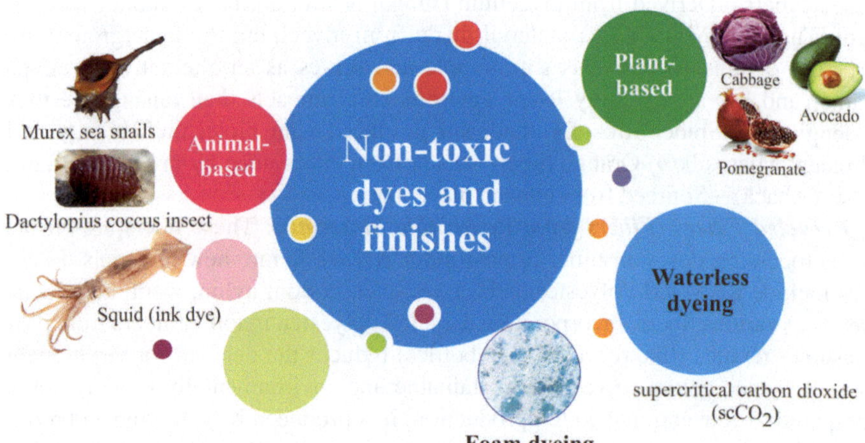

Fig. 2.3 Examples of non-toxic dyes and finishes sources

Waterless dyeing is a technique that eliminates the need for water in the dyeing process. Two widely recognised methods are air-dye and supercritical carbon dioxide (scCO2) assisted dyeing [15]. One of the major players in the industry DyeCoo™ emphasises that their patented technology uses CO_2 to dye fabrics with instead of water. The technology employs CO_2, a by-product from industrial processes, as the dyeing medium within a closed-loop system. Approximately 95% of the CO_2 is recycled within the machine for repeated use. This CO_2 dyeing process eliminates the need for additional chemicals to dissolve dyes. When pressurised, CO_2 reaches a supercritical state, existing between a liquid and a gas, which gives it a high solvent capacity, enabling the dyes to dissolve effortlessly and are then deeply infused into the fibres [16].

Another minimal water use method is foam dyeing, where foams are created using foaming agents, typically derived from an aqueous solution. This foam is then applied to the textile material for dyeing purposes [17]. Opting for natural dyes, low-water dyeing technologies, and eco-friendly finishes can reduce the environmental impact and minimise chemical runoff into water systems.

2.4 Electronic Textiles (E-textiles) with Eco-Friendly Components

The majority of sensors are made from petroleum-based materials derived from non-renewable resources, which are challenging to break down and often become non-biodegradable waste once they lose their functionality. Additionally, these discarded petroleum-based materials remain stable in the environment for extended periods, resisting degradation and contributing significantly to environmental pollution [18]. Eco-friendly components for electronic textiles (e-textiles) are materials and technologies that minimise environmental impact while maintaining functionality. Examples are as follows:

Bio-based Conductive Yarns: E-textiles require conductive elements to transmit signals or power. Graphene-coated fibres, silver nanowires, and carbon nanotube-based conductors can be integrated into fabrics to make them conductive without using traditional metals that are less sustainable and harder to recycle. Conductive threads coated with silver can be made from organic or recycled fibres, reducing the environmental impact of metal extraction. Polymers like PEDOT:PSS (poly(3,4-ethylenedioxythiophene) polystyrene sulfonate) [19] can be used as eco-friendly alternatives to metal-based conductors.

Organic Electronics: Organic semiconductors are being developed as alternatives to conventional silicon-based electronics. The environmental impact of silicon semiconductors is significant throughout their lifecycle, from raw material extraction to manufacturing, usage, and disposal. Silicon is derived from quartz (silicon dioxide, SiO_2), which requires high-energy processes to refine into pure silicon. Also, semiconductor fabrication plants (fabs) require massive amounts of ultrapure water

(UPW) for rinsing wafers [20], along with high electricity consumption. Organic semiconductors are a class of materials that combine the electrical properties of semiconductors with the mechanical flexibility and processing advantages of organic (carbon-based) materials.

Organic semiconductors encompass a variety of materials, from small molecules to polymers, capable of facilitating charge transport. In contrast to conductive materials, which allow electrons to move freely throughout the structure, organic semiconductors are built around a framework mainly consisting of carbon and hydrogen atoms. These atoms are structured to create overlapping π-orbitals, enabling electron delocalisation within individual molecules. This delocalisation supports the movement of electrons, or conductivity, along the molecule.

Moreover, the π-orbitals can overlap with those of adjacent molecules, establishing intermolecular electronic coupling. This allows delocalized electrons to "hop" between molecules within the organic semiconductor, facilitating charge transport. Often sourced from natural materials, these substances can be integrated into textiles to develop flexible, energy-efficient circuits, such as organic light-emitting diodes (OLEDs) and organic photovoltaics. Examples of such materials include Poly (3-hexylthiophene) (P3HT), polyaniline (PANI), and poly (3, 4-ethylenedioxythiophene) polystyrene sulfonate (PEDOT: PSS) [21].

Organic field-effect transistors (OFETs) serve as the essential building blocks for flexible displays and integrated circuits [22]. Instead of relying on expensive semiconductor materials like silicon or metal oxides, OFETs utilise more affordable organic materials, making them suitable for integration into large-area electronic applications.

Sustainable Material-Based Wearable Sensors: Textile-based sensors made from sustainable materials such as cellulose nanofibres, silk, or bio-based polymers can be used for monitoring health, environmental conditions, or performance, without adding to electronic waste at the end of their life. Fibres that generate electricity from mechanical stress (e.g., movement or pressure) using eco-friendly materials like polyvinylidene fluoride (PVDF).

Biodegradable, Transient or Recyclable Electronics: Materials like polylactic acid (PLA) or cellulose-based films can be used as biodegradable substrates for flexible circuits. Transient electronics, engineered to dissolve, degrade, or disintegrate either fully or partially, are unlocking novel opportunities in the emerging age of pervasive electronic technologies [23].

Numerous studies have utilised organic materials derived from natural sources, such as eumelanin, β-carotene, indanthrene yellow G, indanthrene brilliant orange RF, and perylene diimide, which have been employed as semiconductors. Additionally, sugars like glucose, lactose, and sucrose, as well as nucleobases such as adenine and guanine, and caffeine, have been used as dielectrics [24].

2.5 Sustainable Manufacturing Processes

Sustainable manufacturing involves using production methods that reduce harmful environmental effects while preserving natural resources for the benefit of future generations. The key elements of sustainable manufacturing are energy efficiency and emissions, waste management, sustainable supply chain management, circular product lifecycle, compliance & certifications and aligning profitability with sustainability.

Digital Fabrication of Textiles: Using 3D printing and other digital fabrication techniques, designers can create textiles with minimal waste. These methods can utilise recycled or biodegradable filaments, and 3D printing allows for customized textile production, reducing the need for mass production and excess material waste.

On-demand Manufacturing: By digitising the design and production processes, textiles can be made on-demand, reducing excess inventory and minimising waste associated with overproduction. The *Kornit Apollo DTG* system allows businesses to reduce production timelines, boost profitability, and expand into new markets through efficient short-to-medium volume manufacturing. Built on Kornit's proprietary Max technology and powered by an advanced automated system, the Apollo platform combines smart drying processes with the capability to manage multiple garment styles at once, achieving production speeds of up to 400 garments per hour [25].

Sustainable Packaging and Logistics: Reducing packaging waste by using biodegradable or recyclable materials. Optimising supply chains to reduce emissions from transportation. Using digital tracking to improve inventory management and reduce overproduction.

Worker Health and Safety: Ensuring ergonomic, safe, and ethical working conditions. Reducing exposure of workers to hazardous chemicals and pollutants, as well as encouraging employee engagement in sustainability initiatives. This approach not only promotes ethical labour practices but also enhances productivity, innovation, and long-term sustainability.

Compliance and Certifications: Adhering to environmental regulations and industry standards (e.g., ISO 14001, LEED). Conducting sustainability audits and reporting progress while maintaining clear and open communication with stakeholders about the environmental and social effects of operations.

2.6 Sustainable Design Principles

Growing environmental awareness among consumers, along with pressure from governmental and non-governmental organisations and the media, is pushing major players in the fashion industry to transform their production processes and offer garments and accessories that align with sustainable practices [26]. Sustainable design aims to provide enduring solutions that minimise adverse social and environmental effects. The succeeding discussion looks at some specific approaches.

Modular and Repairable Design: Modular product design offers a forward-thinking approach to innovation, serving as an effective strategy for extending product lifespans and enhancing material recyclability [27]. Designing smart textiles with modular components allows consumers to easily repair or upgrade certain parts of a product, extending its life. For example, detachable electronic components in e-textiles could be replaced when necessary, without discarding the entire textile.

Slow Design: A "slow design" approach aims to prolong a product's lifetime and to deepen product satisfaction. Slow fashion is designed to be used over a long time period, and it is made with high quality and high ethical values; it is durable and made of sustainable materials [1]. Ensuring that smart textiles are durable and long-lasting reduces the need for frequent replacements. High-performance materials that resist wear and tear, combined with durable electronics, can significantly extend the lifecycle of a product.

Multi-functional Design: Designing textiles that serve multiple purposes—such as being wearable, washable, and electronically functional—enhances their value and reduces the environmental impact of producing separate products for different uses.

Co-creation and Open Source Design: Co-creation and open-source design in the realm of smart textiles could leverage collaborative efforts among designers, engineers, researchers, and end-users to develop innovative, functional, and accessible textile-based technologies. A co-design process provides various stakeholders with the chance to collaboratively learn, address challenges, and create design solutions that lead to greater consumer satisfaction. In co-creation, the designer acts as an interpreter of user feedback, enabling the company and users to jointly generate value through shared innovation [28].

2.7 Energy Harvesting and Storage Technologies in Textiles

The creation of advanced energy harvesting and storage technologies represents a critical requirement for the next generation of low-power autonomous electronics. Sustainable energy solutions are essential for minimising environmental impact, and integrating energy harvesting and storage capabilities into textiles is pivotal to realising this objective. These innovations allow textiles to efficiently generate, store, and utilise energy, thereby decreasing dependence on non-renewable energy sources. Figure 2.4 shows some energy harvesting technology sources.

Energy Harvesting Mechanisms: Traditional clothing requires external power sources for functionalities like heating or health monitoring. Energy-harvesting textiles (such as piezoelectric, thermoelectric, and solar fabrics) convert mechanical, thermal, or solar energy into usable power, minimising external energy demand.

A myriad of energy generation/harvesting/storage mechanisms have been developed over the years and applied in fields other than smart textiles. These technologies can be employed to further the sustainability agenda. Examples of these technologies include Thermionic energy converters (TIC), triboelectricity, piezoelectricity,

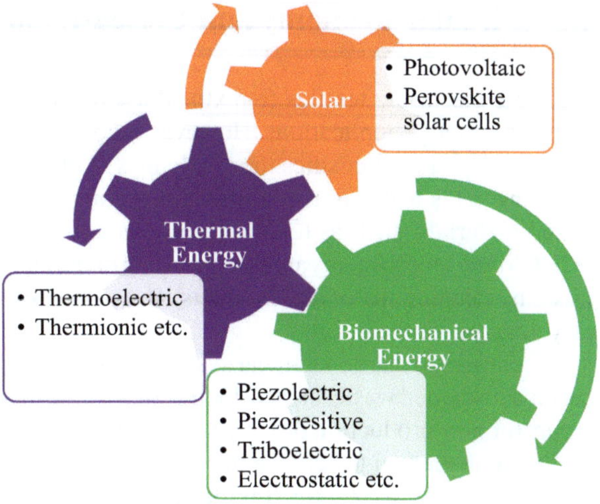

Fig. 2.4 Examples of energy harvesting technologies for digital textiles

Thermionic energy converters, also known as thermionic generators, are solid-state heat engines that, like thermoelectric devices, transform thermal energy directly into electrical energy [29]. The integration of piezoelectric or triboelectric nanogenerators into textiles allows them to harvest mechanical energy (from movement) to power small digital devices. These fibres can be designed from sustainable materials, such as natural rubber or biodegradable polymers. Eco-friendly outdoor gear (such as solar-powered backpacks and heated jackets) enhances sustainability in the fashion and outdoor industries.

Sustainable Batteries and Supercapacitors: As wearable devices require energy storage, solid-state batteries and bio-based supercapacitors such as biomass-derived activated carbon [30] are being explored as safer and more environmentally friendly options compared to traditional lithium-ion batteries. There are also micro energy storage devices (MESDs) [31]. These are small-scale energy storage systems designed to store and deliver power efficiently for miniature and portable electronics, sensors, and wearable technologies.

These devices are crucial for self-powered systems, enabling smart textiles, IoT devices, medical implants, and wireless sensors to function without bulky external power sources. Examples of MESDs are micro batteries, micro supercapacitors, micro fuel cells, and micro piezoelectric generators with storage and hybrid micro energy storage systems (combinations of batteries, supercapacitors, and energy harvesters for optimised performance).

Innovations in coating technologies have enabled textiles to store energy. These energy storage devices can be integrated directly into fabrics and made from eco-friendly materials. The integration of energy harvesting and storage technologies in textiles is a game-changer for sustainability, reducing energy waste, promoting a circular economy, and supporting self-powered systems in various industries. As research advances, these textiles will play a critical role in shaping a greener and more energy-efficient future.

2.8 Circular Economy and Closed-Loop Systems

Wearable electronic textiles (e-textiles) are revolutionising personalised healthcare with cutting-edge applications. However, incorporating electronics into textiles for e-textile production intensifies the challenges of electronic waste (e-waste) and textile recycling. This is due to the complex recycling and disposal processes required for mixed materials, such as textile fibres, electronic components, and other materials [32]. Closed-loop systems are a cornerstone of the circular economy, aiming to reduce waste by keeping materials, products, and resources in a continuous loop of reuse, recycling, or recovery.

In contrast to traditional linear systems, where products at the end of their life cycle become discarded waste, closed-loop systems repurpose these materials into valuable inputs for new production processes. Below, we explore some specific strategies that can be implemented to achieve this.

Design for Disassembly: Textiles designed for digital technologies can be engineered for easy disassembly, allowing electronic components to be separated from the fabric and recycled at the end of life. This approach minimises electronic waste and promotes the reuse of valuable materials.

Closed-loop Recycling: By designing textiles with mono-material components (using only one type of fibre throughout), the recycling process becomes more efficient, as different materials don't need to be separated. Closed-loop systems allow for fibres to be recovered and re-spun into new textiles. Research by Dulal et al. highlighted the development of conductive graphene-like material, such as recycled graphene powder, sourced from Tencel-graphene [32]. This innovation underscores the potential of reusing e-textile waste in manufacturing processes. The technique embodies a circular economy model for digital materials by repurposing thermal decomposition by-products into coating dispersions, thereby sustainably managing their lifecycle. Figure 2.5 presents a schematic diagram of closed-loop recycling and upcycling of digital wastes.

Upcycling Waste: Textile waste, as well as electronic waste, can be upcycled into new materials for digital textiles. For example, fibres from discarded e-waste or unused textiles can be transformed into conductive or high-performance yarns for new applications.

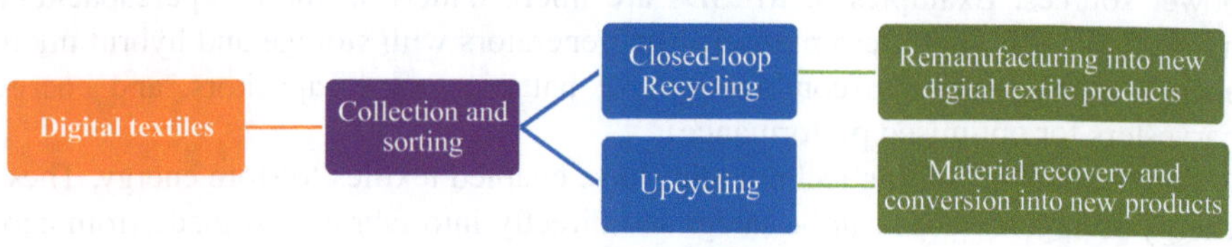

Fig. 2.5 Closed-loop recycling and upcycling of digital wastes

Closed-loop systems, in particular, ensure that digital textiles such as smart fabrics or wearable electronics are recovered, disassembled, and regenerated into new products, creating a continuous cycle of use and reuse. This not only addresses environmental concerns but also promotes innovation in sustainable design and production practices within the textile industry.

2.9 Digital Technologies for Ethical and Transparent Supply Chains

Over the past few decades, advancements in techniques for collecting, monitoring, exchanging, and disseminating information have facilitated the development of new strategies for managing sustainability in global commodity supply chains [33]. At the same time, sustainability challenges like deforestation, leading to carbon storage reduction and biodiversity loss, as well as human rights abuses and forced labour, are often most severe and damaging in the early stages of the supply chain, remaining largely out of view [34]. Ethical practices involve utilising materials and methods that minimise harm to the environment.

Traceability has become a vital element for ethical and responsible business practices within supply chains. At its essence, supply chain traceability refers to the capability to monitor the flow of goods, information, and materials across every stage of the supply chain. It creates a comprehensive record of a product's path from its source to its endpoint, ensuring transparency and accountability at every step.

Blockchain: Digital technologies such as blockchain can be incorporated into textile supply chains to monitor the source of fibers, the production process, and the environmental footprint of each phase. This level of transparency promotes sustainability and ethical standards in material sourcing.

Digital Tags: Smart textiles may incorporate RFID tags or QR codes woven into the fabric to hold information about the material's source, recycling possibilities, and care guidelines. Since assessing the lifespan and quality of textiles and garments at the time of purchase is challenging, producers could provide consumers with details about the expected durability and intended lifetime of the product [1]. This helps consumers make informed decisions about the sustainability of their products.

Satellite Monitoring and IoT: Satellite monitoring systems offer real-time insights into environmental factors like deforestation and water consumption, enabling businesses to verify that their suppliers follow ethical and sustainable standards [33]. This technology supports origin verification, compliance monitoring, and risk mitigation. Additionally, IoT devices can monitor the location, temperature, and humidity of goods in transit, ensuring ethical handling of products across the supply chain.

Product Impact Calculation Systems

United Legwear and Apparel Co. (ULAC), in partnership with *Worldly* (a public-benefit technology company), applies the Product Impact Calculator, a digital tool that tracks supply chain emissions in real time. ULAC is able to identify high-impact areas, such as synthetic fibre use, and shifts to recycled polyester and organic cotton. This tool supports their goal to reduce Scope 1–2 emissions by 42% by 2030 [35].

2.10 Research and Development of Biodegradable and Compostable Textiles

Synthetic fibres, mainly derived from petrochemicals, make up approximately 64% of the global textile fibre market by volume, with an annual production of 111 million metric tons [36]. Due to their hydrophobic nature and high crystallinity, most synthetic fibres, produced from industrially synthesised polymers, are resistant to environmental breakdown. This resistance contributes to the growing accumulation of plastic waste. As the global population continues to rise, waste disposal has become a significant issue, particularly over the past century.

According to the U.S. Environmental Protection Agency (EPA), 11 million tons of textile waste was sent to landfills in the United States in 2018. This figure is expected to rise significantly, reaching 4.5 billion tons by 2040 [37]. The growing reliance on petroleum-based fibers has detrimental impacts on the environment, water, and soil, raising significant global concerns.

In principle, biodegradation is straightforward, it describes how large or complex materials are broken down biologically into smaller, simpler, and ideally non-toxic components that can be reintegrated into the biosphere, preferably without the need for human involvement [38]. Since pure natural fibres, such as cotton, not only perform well in consumer textiles but also typically meet specific standardised biodegradability requirements, insights from the mechanisms behind natural biodegradation are driving new innovations in biologically enhanced textile waste remediation for both natural and synthetic fibres [38].

Three strategies for enhancing textile sustainability through fibre biodegradation are currently emerging. The first approach involves creating textiles capable of breaking down effectively in controlled microbial settings, such as composting or anaerobic digestion, offering a sustainable alternative to open dumping or landfilling. This strategy benefits from existing infrastructure and testing methods that can be modified to handle new types of waste streams.

The second approach focuses on creating biologically driven circular manufacturing systems that break down textile fibres into their basic monomers for recycling. This process utilises microorganisms, enzymes, or a combination of both, alongside process enhancements such as heating and grinding. These industrial methods depend on controlled environments to ensure precise and effective results.

The third approach focuses on reducing the environmental impact of textile waste released into natural ecosystems. Given that the natural environment is largely uncontrollable, any materials that enter it, whether intentionally or accidentally, should be designed to break down naturally without causing immediate or long-term harm to the ecosystem [39].

Utilising biodegradable fibres such as corn-based PLA or silk for the textile component, combined with bio-based conductive inks or materials, creates fully compostable products that align with circular economy principles.

2.11 Biomimicry and Nature-Inspired Design

The word biomimicry, also known as biomimetics, consists of bio, which means nature, and mimesis/mimicry, which means imitation. These terms refer to a way of learning from nature and incorporating its structures or solutions into human designs [40]. Biomimicry can be adopted to create innovative and smart products that are eco-friendly, efficient, and high-performing. Drawing inspiration from nature, textiles can be designed to mimic natural systems, such as self-cleaning surfaces inspired by lotus leaves or moisture-wicking capabilities inspired by spider silk. These designs reduce the need for chemical treatments or intensive washing, making them more sustainable over time.

Companies like *AMSilk* and *Bolt Threads* have developed bioengineered silk fibres that mimic spider silk, lightweight, strong, biodegradable, and produced without toxic chemicals. In their study, Tetsuya et al. demonstrated the mass production of fluorescent silk in three colours; green, red, and orange, using transgenic silkworms. This innovation leveraged a genetic vector derived from the fibroin H chain gene alongside classical breeding methods to achieve scalable, multi-coloured fluorescent silk [41]. Firefly-inspired bio-luminescent fabric is one example where researchers have explored how fireflies produce light with minimal energy loss. This innovation serves to deliver smart textiles for low-energy lighting solutions in safety gear and fashion.

The future projection is that textiles may even be designed to regenerate themselves after wear, using bio-based materials that can "heal" small tears or damage, extending the life of the product and reducing the need for frequent replacements. Cuttlefish are often referred to as the "chameleons of the sea" due to their remarkable ability to quickly change colours and seamlessly blend into their environment. They achieve this through chromatophores, specialised cells spread across their soft skin, which allow them to alter their skin colour and patterns. Inspired by this mechanism, Han et al. engineered a synthetic chromatophore that dynamically adjusts its colour and patterns in reaction to light.

The team accomplished this by 3D-printing a light-responsive smart gel, which morphs in shape upon illumination, mimicking an "artificial muscle" [42]. Chromatophores in nature, such as those in cephalopods, operate with remarkable energy efficiency. By replicating these mechanisms, artificial muscles can be designed to consume less energy compared to traditional synthetic actuators, reducing overall energy demand. Old Navy's new nano-cure tech fabric has unique resealing and repairing capabilities [43]. Self-healing properties of Squid ring teeth proteins have also been exploited to create self-sealing fabric effects. These proteins have components that reorient themselves and form strong molecular bonds when they come in contact of water. These bonds hold up when dried, thus working like an adhesive and eliminating any risk of cracking.

2.12 Consumer Education and Engagement

Consumer education and engagement are critical components in promoting the adoption and sustainable consumption of digital textiles. Achieving the required shift in consumption patterns depends heavily on forward-thinking consumer education that prioritises ethical consumption practices [44]. According to Sheth et al. present sustainability strategies suffer from three key shortcomings: they fail to prioritise the customer directly, overlook the growing risks posed by increasing global over-consumption, and lack a comprehensive, holistic perspective [45]. It is essential to educate consumers on the care and maintenance of digital textiles while also providing guidance on proper disposal or recycling. The following approaches can further help to deal with consumer education strategies.

Digital Care Labelling: One way of distinguishing sustainable fashion is by utilising eco-labels [26]. To promote product longevity and minimise waste, companies can provide digital care instructions accessible via QR codes (Digital Care Labels) embedded in the textiles.

Mindful Design Marketing and Consumption: Reducing over-consumption through mindful design marketing and consumption strategies, such as encouraging consumers to buy fewer but higher-quality products that last longer, aligns with sustainable consumption goals. The mindful design concept emphasises integrating human behaviour into product design in a way that complements and enhances daily life, rather than disrupting it. It focuses on making thoughtful decisions that prioritise the well-being and needs of the product users [46]. Mindful marketing is a comprehensive strategy that prioritises the well-being of both the marketer and the customer, fostering a balanced and thoughtful approach to marketing practices. On the other hand, mindful consumption is rooted in a consumer mindset that emphasises caring for oneself, the community, and the environment. This mindset leads to behaviours that curb the self-destructive tendencies often linked to excessive, repetitive, and status-driven consumption [45].

Awareness of Digital Textiles: Consumers need to be educated about what digital textiles are, including their benefits, customisation, and innovative designs. It is important to highlight how digital textiles differ from traditional textiles in terms of production and environmental impact.

Sustainability Benefits: The environmental advantages of sustainable digital textiles need to be emphasised, such as lower water and energy usage, reduced chemical waste, and minimised overproduction. How digital textile technologies align with sustainable fashion goals also needs elucidation.

Transparency in Production: There is a need for the provision of clear information about the production processes, materials used, and the lifecycle of digital textiles. Transparency builds trust and helps consumers make informed choices.

Digital Literacy: Education of consumers on how to interact with digital platforms for customising or purchasing digital textiles is encouraged. This includes understanding virtual fitting tools, 3D design software, and online marketplaces.

End-of-life Options: Information about proper disposal or recycling methods for digital textiles, especially if they incorporate electronic or synthetic components, needs encouragement. Encourage participation in take-back programs or circular fashion initiatives.

In terms of consumer engagement, the following strategies can be applied.

Interactive Experiences: Employment of augmented reality (AR) and virtual reality (VR) to create immersive experiences where consumers can visualise and customise digital textiles in real-time. This fosters a deeper connection to the product.

Gamification: Introduction of gamified elements, such as rewards for sustainable choices or challenges to design their own digital textiles, to make the process fun and engaging.

Community Building: Creation of online communities or platforms where consumers can share their experiences, designs, and ideas about digital textiles. This fosters a sense of belonging and encourages collaborative innovation.

Storytelling: Sharing of stories about the creators, designers, and technologies behind sustainable digital textiles as a means of humanising the process and building emotional connections with consumers.

Feedback Loops: Encouraging consumers to provide feedback on digital textile products and experiences. Deployment of this input to improve designs, functionality, and sustainability practices.

Educational Campaigns: Launching of campaigns that highlight the environmental and social impact of digital textiles, using social media, influencers, and workshops to reach a broader audience.

By combining education and engagement, consumers can become active participants in the shift toward digital textiles, driving demand for sustainable innovations and fostering a more responsible fashion industry.

2.13 Ethical and Social Considerations of Adopting Digital Technologies in Sustainable Textile Design and Production

The integration of digital technologies into sustainable textile production introduces critical ethical and social considerations that must be addressed to ensure equitable and responsible innovation.

2.13.1 Ethical Considerations

Ethically, the shift toward automation and advanced manufacturing processes, such as 3D printing or AI-driven design, risks displacing workers in traditional textile sectors, particularly in regions reliant on manual labour. Ensuring fair labour transitions through retraining programs and safeguarding worker health, especially in environments exposed to hazardous chemicals during dyeing or electronics integration, is critical to upholding human dignity and safety.

Additionally, the proliferation of e-textiles embedded with sensors or conductive components introduces challenges in managing electronic waste. Without closed-loop recycling systems or designs prioritising disassembly, discarded smart textiles could exacerbate global e-waste crises, disproportionately affecting marginalised communities near landfills or resource extraction sites.

Mining rare earth metals for digital components (e.g., batteries, semiconductors) can exploit marginalised communities. Ethical sourcing of materials like rare earth metals must also avoid ecological degradation and human rights abuses, while transparent claims about sustainability, supported by technologies like blockchain, are vital to combat greenwashing and maintain consumer trust.

2.13.2 Social Considerations

Socially, the high cost of advanced materials like bioengineered silk and organic semiconductors risks deepening global inequities by restricting access to sustainable textiles in low-income regions. Culturally, there is a need to balance innovation with the preservation of indigenous textile practices, ensuring digital advancements complement, rather than erase, traditional craftsmanship, such as integrating natural dye techniques with waterless CO_2 dyeing methods.

Consumer education plays a pivotal role; promoting awareness of proper care, recycling, and the environmental impact of digital textiles through tools like QR-enabled care labels can empower responsible usage. Simultaneously, fostering mindful consumption to counter fast fashion's tech-driven overproduction is essential.

At the community level, decentralised digital production models may disrupt local economies, yet hybrid approaches that blend digital tools with artisanal skills can sustain livelihoods and cultural heritage. Addressing these ethical and social dimensions holistically ensures that the transition to digital sustainability fosters equity, preserves cultural identity, and aligns technological progress with planetary and human well-being.

2.14 Challenges and Opportunities

The design and production of sustainable textiles through digital technologies present both significant challenges and transformative opportunities. Key challenges include difficulties in recycling mixed-material e-textiles that integrate electronics with fabrics. Energy-intensive processes, such as silicon-based electronics manufacturing and traditional dyeing methods, further strain resources, while high costs of sustainable alternatives like bioengineered silk or organic semiconductors and consumer resistance due to price and awareness gaps hinder adoption. The lack of standardised regulations and technologies for global traceability and circular systems poses another challenge. Additionally, managing non-biodegradable textile waste and ensuring ethical labour practices in global supply chains remain critical hurdles.

However, opportunities abound through innovations such as bio-based polymers (e.g., PLA, algae-based fibers), organic electronics (e.g., PEDOT: PSS), and biodegradable materials inspired by biomimicry, which reduce reliance on fossil fuels. Advanced production methods like 3D printing, waterless CO_2 dyeing, and closed-loop recycling systems minimise waste and energy use, while energy-harvesting textiles (e.g., piezoelectric fabrics) and bio-based supercapacitors enable self-powered wearables.

Digital tools like blockchain and IoT enhance supply chain transparency, and modular, repairable designs extend product lifespans. Collaborative efforts to standardise regulations, promote circular economies, and educate consumers through AR/VR platforms or digital care labels can drive systemic change.

By addressing these challenges through technological innovation, ethical practices, and cross-sector collaboration, the industry can advance toward eco-efficient, functional textiles that align with global sustainability goals. By merging digital technologies with eco-friendly approaches, businesses not only reduce their ecological footprint but also enhance operational efficiency and stimulate innovative growth [47].

2.15 Future Perspectives

Ongoing research into biodegradable electronics, recyclable conductive materials, and self-healing textiles can pave the way for more sustainable smart textiles. Materials that can repair themselves, for example, would significantly extend product lifespans.

Adoption of techniques and technologies such as nanotechnology can enhance the properties of textiles, making them stronger, more energy-efficient, or more conductive without the need for harmful chemicals. This can lead to lighter, more durable smart textiles that use fewer resources during production.

Many countries and regions have stringent regulations regarding sustainable practices. However, the lack of standardised processes and technologies for traceability can create challenges in implementation. It is important to develop an international consensus on regulations for more coherent results. Companies must stay informed about these regulations and ensure that their traceability systems are designed to meet compliance requirements. There is also a need to harness the development of innovative and sustainable technologies.

2.16 Conclusion

The continued use of environmentally unfriendly practices continues to adversely affect the health of common people. So, keeping in mind all the ongoing problems, there is an urgent need for researchers, technologists and all relevant stakeholders to switch to more sustainable practices.

While bast fibres like jute, flax, hemp, ramie, and kenaf are more sustainable than many other cellulose-based fibres, particularly cotton, substantial efforts are still needed to maximise their contribution to sustainability. Achieving a significant impact would require major advancements in technology development and substantial investments in production capacity. The path toward sustainable textiles in digital technologies involves a combination of innovation in materials science, responsible production methods, and forward-thinking design principles.

By incorporating eco-friendly materials, focusing on the recyclability of smart textiles, and integrating energy-efficient technologies, the digital textiles of the future can contribute to a more sustainable and circular economy. These efforts will not only reduce environmental impact but also enhance the functionality and versatility of textiles in the digital age.

While natural bio-based polymers have the potential to be more environmentally friendly than traditional plastics, their sustainability depends on responsible sourcing, production, and disposal practices. Not all bio-based polymers are equally eco-friendly, and their benefits should be evaluated on a case-by-case basis.

References

1. Niinimäki, K., and L. Hassi. 2011. Emerging design strategies in sustainable production and consumption of textiles and clothing. *Journal of Cleaner Production* 19 (16): 1876–1883. https://doi.org/10.1016/j.jclepro.2011.04.020.
2. Karthik, T., R. Rathinamoorthy, and P. Ganesan. 2015. Sustainable luxury natural fibers—production, properties, and prospects. In *Handbook of sustainable luxury textiles and fashion*, vol. 1, ed. M. A. Gardetti and S. S. Muthu, 59–98. Singapore: Springer Singapore.
3. Rana, S., S. Pichandi, S. Parveen, and R. Fangueiro. 2014. Natural plant fibers: Production, processing, properties and their sustainability parameters. In *Roadmap to sustainable textiles and clothing: eco-friendly raw materials, technologies, and processing methods*, ed. S. S. Muthu, 1–35. Singapore: Springer Singapore.

4. Malou, J. R., Tita, W., Perret, J., Singh, A., Roque, R. M., and Nair, G. R. 2017. Effect of pretreatments in the processing of pineapple leaf fibers. *Advances in Food Science and Engineering* 1(4): 196–205. https://doi.org/10.22606/afse.2017.14006
5. Kengkhetkit, N., T. Wongpreedee, and T. Amornsakchai. 2018. Pineapple leaf fiber: From waste to high-performance green reinforcement for plastics and rubbers. In *Lignocellulosic composite materials*, ed. S. Kalia, 271–291. Cham: Springer International Publishing.
6. Atanda, J. 2015. Environmental impacts of bamboo as a substitute constructional material in Nigeria. *Case Studies in Construction Materials* 3: 33–39. https://doi.org/10.1016/j.cscm.2015.06.002.
7. Muchiri, M. N., and M. O. Muga. 2013. A preliminary yield model for natural Yushania Alpina bamboo in Kenya. *Journal of Natural Sciences Research* 3 (10): 77–85.
8. Khouri, N. G., J. O. Bahú, C. Blanco-Llamero, P. Severino, V. O. C. Concha, and E. B. Souto. 2024. Polylactic acid (PLA): Properties, synthesis, and biomedical applications—A review of the literature. *Journal of Molecular Structure* 1309: 138243. https://doi.org/10.1016/j.molstruc.2024.138243.
9. Deveoglu, O. 2020. A review on cochineal (Dactylopius Coccus Costa) dye. *Research Journal of Recent Sciences* 9 (3): 37–43.
10. LTD., M. H. 2016. Maiwa's natural dyes what they are and how to use them. In *Naturaldyes.ca.* Maiwa Handprints Ltd.
11. Derby, C. D. 2014. Cephalopod Ink: Production, chemistry, functions and applications. *Marine Drugs* 12: 2700–2730. https://doi.org/10.3390/md12052700.
12. Jain, S. 2017. Cephalopods. In *Fundamentals of invertebrate palaeontology: Macrofossils*, ed. S. Jain, 31–102. New Delhi: Springer India.
13. Cosgrove, J. A. 2021. Introduction to the Cephalopoda. In Klinkenberg, B. (ed.) *Electronic Atlas of the Fauna of British Columbia*, E-Fauna BC.
14. Derby, C. D. 2014. Cephalopod ink: Production, chemistry, functions and applications. *Marine Drugs* 12 (5): 2700–2730. https://doi.org/10.3390/md12052700.
15. Mahmud, I. and S. Kaiser. 2020. Recent progress in waterless textile dyeing. *Journal of Textile Science and Engineering* 10: 1–3. https://doi.org/10.37421/jtese.2020.10.421
16. B. V., D. T. S. 2019. DyeCoo's water-free and process-chemical free-dyeing technology for textiles. In *Business Europe*
17. Senthil Kumar, P. and P. R. Yaashikaa. 2018. *Sustainable Dyeing Techniques.* In Muthu, S.S. (ed.) *Sustainable innovations in textile chemical processes.* Springer Singapore: Singapore. p. 1–29.
18. Ma, H., Z. Cheng, X. Li, B. Li, Y. Fu, and J. Jiang. 2023. Advances and challenges of cellulose functional materials in sensors. *Journal of Bioresources and Bioproducts* 8 (1): 15–32. https://doi.org/10.1016/j.jobab.2022.11.001.
19. Huang, H., L. Yang, Q. Zhang, Y. Yao, Y. Yang, Y. Li, Y. Peng, S. Wu, L. Qin, and P. Li. 2024. Application of poly(3,4-ethylenedioxythiophene):Poly(styrene sulfonate) based on multiple treatments with hot alcohol solvents in organic solar cells. *Materials Science in Semiconductor Processing* 184: 108834. https://doi.org/10.1016/j.mssp.2024.108834.
20. Wang, Q., N. Huang, Z. Chen, X. Chen, H. Cai, and Y. Wu. 2023. Environmental data and facts in the semiconductor manufacturing industry: An unexpected high water and energy consumption situation. *Water Cycle* 4: 47–54. https://doi.org/10.1016/j.watcyc.2023.01.004.
21. Wood, A. 2025. What are organic semiconductors? in Ossila. www.ossila.com.
22. Hu, W., and X. Ren. 2023. Single-crystal organic semiconductors. In *Encyclopedia of Materials: Electronics*, ed. A. S. M. A. Haseeb, 41–50. Oxford: Academic Press.
23. Shim, J.-S., J. A. Rogers, and S.-K. Kang. 2021. Physically transient electronic materials and devices. *Materials Science and Engineering: R: Reports* 145: 100624. https://doi.org/10.1016/j.mser.2021.100624.
24. Irimia-Vladu, M., P. A. Troshin, M. Reisinger, L. Shmygleva, Y. Kanbur, G. Schwabegger, M. Bodea, R. Schwödiauer, A. Mumyatov, J. W. Fergus, V. F. Razumov, H. Sitter, N. S. Sariciftci, and S. Bauer. 2010. Biocompatible and biodegradable materials for organic field-effect transistors. *Advanced Functional Materials* 20 (23): 4069–4076. https://doi.org/10.1002/adfm.201001031.

25. Digital, K. 2025. Kornit Apollo printer, in Kornit. www.kornit.com.
26. Henninger, C. E. 2015. Traceability the new eco-label in the slow-fashion industry?—consumer perceptions and micro-organisations responses. *Sustainability* 7: 6011–6032. https://doi.org/10.3390/su7056011.
27. Amend, C., F. Revellio, I. Tenner, and S. Schaltegger. 2022. The potential of modular product design on repair behavior and user experience—Evidence from the smartphone industry. *Journal of Cleaner Production* 367: 132770. https://doi.org/10.1016/j.jclepro.2022.132770.
28. Fuad-Luke, A. 2009. Design activism: Beautiful strangeness for a sustainable world. 1st ed.: Routledge. 272.
29. Zabek, D., and F. Morini. 2019. Solid state generators and energy harvesters for waste heat recovery and thermal energy harvesting. *Thermal Science and Engineering Progress* 9: 235–247. https://doi.org/10.1016/j.tsep.2018.11.011.
30. Saini, S., P. Chand, and A. Joshi. 2021. Biomass derived carbon for supercapacitor applications: Review. *Journal of Energy Storage* 39: 102646. https://doi.org/10.1016/j.est.2021.102646.
31. Duan, Y., G. You, K. Sun, Z. Zhu, X. Liao, L. Lv, H. Tang, B. Xu, and L. He. 2021. The advances in wearable textile-based micro energy storage devices: structuring, application and perspective. *Nanoscale Advances* 3. https://doi.org/10.1039/D1NA00511A
32. Dulal, M., S. Afroj, M. R. Islam, M. Zhang, Y. Yang, H. Hu, K. Novoselov, and N. Karim. 2024. Closed-loop recycling of wearable electronic textiles. *Small* 20. https://doi.org/10.1002/smll.202407207
33. Gardner, T. A., M. Benzie, J. Börner, E. Dawkins, S. Fick, R. Garrett, J. Godar, A. Grimard, S. Lake, R. K. Larsen, N. Mardas, C. L. Mcdermott, P. Meyfroidt, M. Osbeck, M. Persson, T. Sembres, C. Suavet, B. Strassburg, A. Trevisan, C. West, and P. Wolvekamp. 2019. Transparency and sustainability in global commodity supply chains. *World Development* 121: 163–177. https://doi.org/10.1016/j.worlddev.2018.05.025.
34. Heldt, L. 2024. Space technology and supply chain sustainability: satellite-based due diligence and the EU deforestation regulation, in space law principles and sustainable measures. In: H. Jahankhani, et al. (eds.) Springer. p. 109–134.
35. Shim, K. 2022. Other worldly human sculptures [cited 2022]. https://art-sheep.com/wp-content/uploads/2015/11/fimm2-1024x622.jpg.
36. Exchange, T. 2020. Preferred fiber & materials market report 2020 [cited 2025] https://textileexchange.org/wp-content/uploads/2020/06/Textile-Exchange_Preferred-Fiber-Material-Market-Report_2020.pdf.
37. Razzaq, S., S. Shahid, and Y. Nawab. 2024. Applications and environmental impact of biodegradable polymers in textile industry: A review. *International Journal of Biological Macromolecules* 282: 136791. https://doi.org/10.1016/j.ijbiomac.2024.136791.
38. Egan, J., and S. Salmon. 2021. Strategies and progress in synthetic textile fiber biodegradability. *SN Applied Sciences* 4 (1): 22. https://doi.org/10.1007/s42452-021-04851-7.
39. Banwell, E., M. Schuknect, B. Rattner, N. Hulst, and D. B. 2020. The nature of fashion [cited 2025] https://biomimicry.org/thenatureoffashion/
40. Cui, T. Z., R. K. Raji, J. L. Han, and Y. Chen. 2024. Avant-garde concept in footwear research and design: review of application trends. *Leather and Footwear Journal* 24(2): 1–10https://doi.org/10.24264/lfj.24.2
41. Iizuka, T., H. Sezutsu, K.-I. Tatematsu, I. Kobayashi, N. Yonemura, K. Uchino, K. Nakajima, K. Kojima, C. Takabayashi, H. Machii, K. Yamada, H. Kurihara, T. Asakura, Y. Nakazawa, A. Miyawaki, S. Karasawa, H. Kobayashi, J. Yamaguchi, N. Kuwabara, T. Nakamura, K. Yoshii, and T. Tamura. 2013. Colored fluorescent silk made by transgenic silkworms. *Advanced Functional Materials* 23 (42): 5232–5239. https://doi.org/10.1002/adfm.201300365.
42. Han, D., Y. Wang, C. Yang, and H. Lee. 2021. Multimaterial printing for cephalopod-inspired light-responsive artificial chromatophores. *ACS Applied Materials and Interfaces* 13 (11): 12735–12745. https://doi.org/10.1021/acsami.0c17623.
43. Adams, C. 2018. New old navy and gap clothes do not rip; self-healing fabric.... In: *Different who*, Differentwho.com

44. Jarva, V. 2011. Consumer education and everyday futures work. *Futures* 43 (1): 99–111. https://doi.org/10.1016/j.futures.2010.10.016.
45. Sheth, J. N., N. K. Sethia, and S. Srinivas. 2011. Mindful consumption: A customer-centric approach to sustainability. *Journal of the Academy of Marketing Science* 39 (1): 21–39. https://doi.org/10.1007/s11747-010-0216-3.
46. Riley, S. 2024. *Mindful design: A survival guide for responsible product designers*, 2nd ed. CA: Apress Berkeley.
47. Glogar, M., S. Petrak, and M. Mahnić Naglić. 2025. Digital technologies in the sustainable design and development of textiles and clothing—A literature review. *Sustainability* 17. https://doi.org/10.3390/su17041371

Raji Rafiu King received his Bachelor's degree in Industrial Art with specialization in Textiles from the Kwame Nkrumah University of Science and Technology, Kumasi Ghana, in 2006. In 2014 he graduated from the Lanzhou Jiaotong University in Gansu province, Lanzhou China with a Master's degree in Information and communication Systems. He enrolled in Jiangnan University, China in 2015 where he obtained a Ph.D. degree in Textile Science and Engineering, 2019. He is currently a faculty member of Huizhou University's Xuri Guangdong Fashion College. His research interests are in design and development of soft strain sensors based on conductive textile materials and polymer optical fibers and their application in smart apparels and other wearable devices. He has also researched into the development of biometric authentication systems. He has published extensively in the fields mentioned with over 25 SCI indexed publications and a Chinese patent to his credit.

Chapter 3
Innovation Through Eco-Friendly Animal Based Fibres: A New Era in Textile Sustainability

Vandana Gupta

Abstract This chapter deals with the importance of wool and other animal fibres within the context of the Sustainable Development Goals of the United Nations. In response to global demand for eco-friendly, safe, and multifunctional materials, the complete supply chain of textile and allied industries is shifting towards innovative materials by incorporating traditional and new technologies such as enzymatic recycling. Along with other natural fibres, animal fibres have shown their potential in the development of composites, functional fabrics. This chapter also suggests the requirement of ethical & holistic production, keeping the three pillars of sustainability (economic, social, and environmental) into consideration. The readers will develop an understanding of the types, properties, and applications of wool, hair, and other animal fibres in the current industrial world.

Keywords Wool and hair fibres · Composites · Silk fibre · Sustainability development goals · Initiatives

3.1 Introduction

We are living in an era where the utilisation of materials to develop new products is more and the consumption period is less. With changing marketing trends, consumers are inclined towards new products, offering attractive opportunities to save and spend money. In this process, the repercussions of purchasing and discarding goods are not considered, which leads to increased landfills. Statistics indicate a decline in the health of Mother Earth and the organisms living in it, including soil, water, air, humans, animals, and aquatic life. An important example is the presence of microplastics in water, harming the water bodies extensively [1]. With the invention of synthetic raw materials and new technologies, societies are leading comfortable lives. But, in the last few decades, humans have realised that such materials and

V. Gupta (✉)
Department of Fashion & Textile Design, Parul University, Vadodara, India
e-mail: vandana.g178@gmail.com

S.-X. Jiang et al. (eds.), *Advances in Textile Materials and Processing Techniques for Sustainability*, SDGs and Textiles, https://doi.org/10.1007/978-981-95-0469-5_3

technologies are not conducive to sustainable growth. Despite the negative effects, the utilisation of such products is still prevalent in our society as these products have become an integral part of our daily lives. In the year 2015, the United Nations announced the 17 Sustainable Goals, out of which 16 goals emphasise the steps each individual should take to make this earth a better place to live for the existing and future generations. The global population has experienced and witnessed the natural cleaning process of Mother Earth during COVID-19. This eye-opening event has increased consumer awareness about the importance of natural products, raw materials, and processes.

A change in the business approach of the textile and apparel industries is required, as these industries are some of the most polluting industries and, contributing to air pollution, noise pollution, water pollution, and landfills. A recent report indicated 7% increase in the volume of textiles used for apparel, footwear, and interior textiles from 2022 to 2023. As per the report, polyester was the widely used fibre with 57% of the total fibre production in the synthetic category. It is important to note that, regardless of the development of new technologies for recycling textiles, such as enzymatic recycling for synthetics (PET), less than 1% of the global fibre comes from pre & post-consumer recycled textiles waste. The report also highlighted the decline in the volume of cotton fibre and 6% increase in manmade cellulosic fibres. On the other hand, wool fibres reported a positive trend. Wool produced under standards such as the Responsible Wool Standard (RWS), and the Sustainable Wool Standard (SCWS) increased their use from 4.2% in 2022 to 4.8% in 2023, maintaining 6% for recycled wool. Also, certified hair fibres such as mohair and cashmere market share reported 47% growth [2].

Keeping the global sustainability goals into consideration, scientists and designers are making an effort to understand the potential of fibers such as camel hair, wool, silk, spider silk, horsehair, etc., by incorporating them into products developed for different sectors like apparel, home textiles, automobiles, upholstery, aerospace and medical. Their inherent properties, such as nontoxicity, flame retardant, water absorbency, and resilience, are ideal to meet current market demand [3]. In India, the sustainable fashion market is still in its early stages. Positive upward growth has been seen over the past few years due to consumer awareness, innovation, utilisation of sustainable materials, and government initiatives. As reported by Researchand-Markets.com, a growth of 10% CAGR is expected during 2021–2026 [4]. This book chapter provides insight into the utilisation of animal fibres for the development of textile-based products and shares current research along with initiatives taken aiming to encourage curiosity-driven research and challenge-based innovation among readers.

3.2 History and Value of Animal Based Fibers in Textiles

3.2.1 History of Animal Based Fibers

To understand the history of animal fibres, two important fibres which fall in the category of eco-friendly protein fibres are discussed below. These fibres are considered the backbone of textiles and many allied industries. Understanding the types, chemical properties, and applications of these fibres will help the readers to delve into the use of other similar fibres and innovations related to them. The history of animal fibres dates back to the prehistoric age when nomadic life prevailed. The most important animal fibres which came into existence are silk & wool, were collected from the animal body and converted into textile material with processes like felting & weaving. Two fibres, that is, wool and silk, will be discussed in detail, which will work as a base for understanding other animal fibres.

3.2.1.1 The Wool Story

To protect themselves from adverse environmental conditions, the Stone Age hunters utilized natural fibrous material available on animal bodies. They collected the fleece of wild sheep and tested its applications, in the form of body covering, making warm tents, etc. With more settled life, humans developed techniques to convert fibers into fabrics. These techniques included spinning, weaving, knitting, and felting. The tools used were more in a rudimentary state and simple looms facilitated fabric weaving. Wool fibers were used to develop garments like cloaks, and tunics along with other textiles like blankets, carpet which protected from harsh environments. Historical evidence suggests civilizations like Babylonians and Mesopotamia used wool textiles as early as 4000–2500 B. C. Romans contributed to the development of Spanish Merino wool which is known as one of the finest wool till date. With advancements in transportation, technology, and similar innovations, wool and hair fibers were utilized more extensively. Sheep are now domesticated worldwide for their fleece; other animals like yak, horse, ox, rabbit, and camel provide valuable hair fibres utilised in the production of creative and diversified textile products [5].

3.2.1.2 The Silk Story

The origin of this flamboyant fibre lies in the 4000-year-old tales shared by legends. One of the most important fibres of the textile industry was discovered accidentally when a silkworm cocoon fell into the tea cup of Xi-Ling-shi (Chinese princes). The hot liquid unravelled the fibre from the cocoon. This discovery led to the development of the sericulture process at a large scale during the Shang Dynasty (sixteenth-eleventh century BCE), making silk an important commodity in Chinese society. This process involves the cultivation of silkworms. The filament fibres produced by the silkworm are collected by dipping the cocoons in hot water to remove the gum. This process is known as degumming. The collected fibres were converted into fine, luxurious textiles. Till date, the Bombyx mori species are widely cultivated. This species fed on mulberry leaves [6]. Earlier, silk textiles were mainly the monopoly of the Chinese dynasty and simple patterns were created. With time, elaborate and unique patterns in the form of brocades were introduced in silk weaving. These textiles were highly prized and utilised as garments, ceremonial offerings, etc. The Han Dynasty (206 BCE-220 CE) explored the Silk Road and facilitated the exchange of goods, cultural influences as well as ideas between China and other countries like India, Persia, and Europe. Later, countries like India developed their silk industry in different states. Notable among them are Patola of Gujarat, also known as double ikat. This textile is prepared by dying the yarns according to patterns/designs, followed by weaving. Similarly, Kanchivaram saris of South India are known for their architectural designs and contrasting borders. On the other hand, Termeh is unique to Persia, its production flourished during the Sassanian dynasty.

3.3 Value of Animal Based Fibers

3.3.1 Wool and Its Types

Wool fibre, as previously discussed, is obtained from animals like sheep, yaks, horses, camels, etc. These wool and hair fibres are known for their softness, warmth, durability, fire retardant, water absorbent, and many functional properties. All hair fibres have a similarity in chemical composition due to the presence of keratin. Differences can be seen in their physical characteristics such as fineness, length, shape, and internal structure. Hair fibres of camel, rabbit, and horse are developed according to their environmental conditions and provide warmth and comfort to the animals in diverse climates. Textiles made from these fibres provide comfort to humans when used as garments [7]. One such fibre is obtained from Yak a native to the Himalayan region. During the molting season (spring and summer), the undercoat fibre is collected through the shearing process. The hair fibre is available in different colours, including brown, gray, black, and white, and exhibits warmth, breathability, lightness, hypoallergenic, and odour-resistance properties. On the other hand, Bison is native to North America and the hair fiber obtained from this animal is soft, warm, and durable. It sheds its winter coats naturally during the spring and summer seasons. These fibres range from black to white.

Another important fiber that was valuable during the nineteenth century and was an important part of the fashion industry is horsehair fiber. In the realm of historical fashion, it is extensively used due to its strength, and durability. Also, its stiffening property was utilized to provide shape and support for fashion garments like crinoline. The term crinoline came from a French word "crin" (horsehair) and "lin" (linen) as the horsehair was woven with linen fibre to weave a stiff textile that could be used to hold the shapes of Victorian garments. With changing times, the raw material which was used earlier for crinoline was later replaced by flexible steel hooped crinolines. But still, the exclusive properties of horse hair made it a desirable raw material for other fashion/textile and upholstery products.

Even today, designers, artists, and craftsmen are inspired by this unique fibre and look forward to reviving it by designing and developing suitable products for the existing market [8, 9]. Similar to horsehair, camel hair is considered an important part of the textile industry worldwide. Camel wool is derived from the Bactrian Camel. Among the different breeds of camel, the Bactrian (two-humped) camel is well known in Turkey, China, and Siberia for its hair fibres. Camel hair insulates the animal's body from both heat and cold. This unique property is due to the adaptive nature of the animal in adverse conditions, where the day is extremely hot and the nights are cold. Another important property that these camel hair fibres exhibit is water repellency [7] and as compared to other camels, Bactrian camels have longer hair, which exhibits the primary property required to convert a fibre into a textile material. The two types of hair fibre are obtained from this animal, a Guard coat and an Undercoat. The guard coat consists of thick, inflexible fibres, not suitable for apparel. These fibres are blended with wool fibre to develop fabric appropriate for other applications or allied industries. On the other hand, undercoat fibres are thin, fluffy, and easy to convert into yarns. The undercoat is to insulate the camel and keep them warm. This fine hair fibre is valued for its softness, warmth, and durability and finds its application in garments, luxury products, and insulating fabrics [10]. Camel hair is categorised into different grades and is utilised to develop products accordingly. Grade one hair is close to animal skin. They feel soft and silky. They are stapled fibres of 1 1/4 to 3 inches in length. These are used for high-quality hair fabrics. Grade two fibres are coarser outer hair exhibiting a longer length of 15 inches. Used only for cordage and low-quality rugs [11, 12].

3.3.2 Silk and Its Types

Silk fibre, being natural, is one of the earliest discoveries of human beings and is made up of a protein known as Fibroin. At the larva stage, which is one of the life stages of the silkworm, it secretes the protein 'fibroin' from glands along with a gum known as sericin, which helps the larvae to form cocoons. The degumming process is performed to obtain the fibre from cocoons by immersing them in boiling water, and later the filament is reeled for further textile development. The Inherent properties that make this fibre unique and valuable are natural sheen, affinity towards colours, good tensile strength, poor heat conduction, as well as an excellent drape. They make their place in home furnishing, fashion, home textiles, hospitality, and the medical industry. Historically, among silkworms, Bombyx mori is an important silkworm, which contributes around 90% of world production, but there are other species which are used to develop silk filament and provide value to the textile, fashion, and allied industries [13, 14]. The silk can be categorised into three categories as given below:

i. Mulberry silk: Produced by Bombyx Mori, it thrives on mulberry leaves.
ii. Non-Mulberry Silks: These silks are produced by silkworms other than Bombyx Mori. The types include Eri Silk, which feeds on castor oil plant leaves. Silk is produced without killing the silkworm. This textile is preferred by monks.; Muga Silk feeds on the Som plant; Tussar Silk feeds on oak leaves.
iii. Other wild Silk: These wild silk include Anaphe silk, Fagara Silk, Coan silk, Mussel Silk, and Spider Silk.

Among the above-given types of silk, unravelling the molecular configuration of spider silk is being attempted by scientists. Spiders are unique in themselves as they have remarkable mechanical properties and have been considered superior to nylon, Kevlar, and steel. They convert liquid protein into strong silk by dissolving the salt linkages into a new configuration suitable for developing silk fibre. Spider silk fibres are natural polymeric fibres exhibiting high tensile strength and toughness with distinct thermal, optical as well as biocompatible properties. It is interesting to know that spider silk utilises 7 different types of silk for their daily life activities, which include shelter, defence, reproduction, prey capturing, etc.

As scientists unravel the secrets of spider silk, it has become evident that a broad range of diverse fiber types is secreted by spiders to serve different ecological purposes. These threads have a host of different functions: locomotion, wrapping prey, protection of developing eggs and web construction. Anatomical studies of typical orb or cob-weavers reveal the presence of seven distinct silk-producing glands: the major and minor ampullate glands (locomotion and web frame), tubuliform (egg case silk), flagelliform (spiral capture silk), aciniform (prey wrapping silk), aggregate (sticky adhesives on spiral capture or gumfooted lines, and the pyriform glands (attachment disc and joining fibers). Major ampullate glands have been the principal focus for researchers investigating the natural silk production pathway. This bulb-shaped gland is found in pairs in spiders and is responsible for producing dragline silk. Dragline silk is referred to as the 'safety line' of spiders, and it is often spun in pairs and is accompanied by two minor ampullate silk threads. Dragline silk is five times stronger by weight than steel and three times tougher than the man-made fibre Kevlar, a synthetic fibre used in body armour. Correlations between the reeling speed and the tensile strength of dragline silk fibres have been observed experimentally, supporting that spiders can tailor their thread properties by adjusting the draw rate from their spinnerets.

By controlling the reeling speeds during extrusion, spiders can tune the beta-sheet crystalline size and orientation of the protein polypeptide chain network. These parameters are important factors that module the tensile strength of silks. Spider Dragline silk fibres are thermally stable to about 230 °C on thermal gravimetric analysis. In addition, these fibres have shown the ability to retain their morphology, structure, and mechanical properties when autoclaved [15]. Spider silk fibre is reported to be used in the South Pacific as dip nets, fishing lures as well as weaving ceremonial dresses. Their application has been reported as sutures for centuries because of their biocompatibility and durability properties [16, 17].

3.4 Sustainable Approach for Textile Products

With the advent of synthetic fibres, the application and utilisation of natural fibres declined significantly, leading to less sustainable patterns of production and consumption. In the last few decades, despite the inherent and unique properties of fibres like wool, silk, and other natural alternatives, consumers and industries have largely shifted their focus towards petroleum-based synthetic fibres. These synthetic materials were easier to mass-produce, more affordable, highly versatile, and could be easily modified to meet consumer demands. Due to such intrusion, many natural raw fibre materials were either preserved in museums or used by only a few niche brands and companies. One report suggested that in the last 15 years, production volume has doubled and garment life span has decreased by 36%. Despite much noise about sustainability, only 1% of fibres are used for recyclable clothing manufacturing, which suggests the existing dependency on fresh raw materials [18].

However, increase in health issues, environmental pollution, and a decline in heritage textiles, a change in perspective is seen. A large amount of synthetic clothing at low cost is offered by companies like Zara in the name of fast fashion, whereas slow fashion brands are looking forward to developing sustainable products using sustainable raw materials and technologies. One such initiative was taken by a young designer Ishrat Sahgal, who converted silk saree waste into carpets under her brand "Mishcat Co" [19]. This venture not only works towards reducing waste but also provides livelihood to many skilled craftsmen. Many others, such as environment enthusiasts, have worked on reducing water pollution by dyeing textiles using natural waste materials such as waste flowers collected from the temples.

Designers, through design intervention, are converting these raw materials into designer products for the current market and consumers. Such initiatives work towards one or all pillars of sustainability, which are in line with the sustainable development goals given by the United Nations, such as the 6th SDG- clean water, 9th SDG- industry innovation and infrastructure, 8th SDG decent work and economic growth, and 12th Responsible consumption and production. Looking at the initiatives taken by the government to promote and implement sustainable goals, especially in the textile and fashion industry, many issues are addressed by introducing policies focused on promoting sustainable consumption and production. Another important initiative is "Made in India", motivating Indian textiles to flourish, which inherently works on sustainable principles. Conventional manufacturers are adopting approaches such as eco-conscious dyeing & water management, recycling & upcycling, sustainable fibers as well as ethical manufacturing. Such initiatives and innovations are creating a sustainable atmosphere through technologies such as 3D printing, Plasma technology, nanotechnology, smart textiles, and materials, to name a few [18]. In light of these initiatives and SDG goals, sustainable transformation is witnessed, a glimpse of which is given below.

3.5 Innovation in Eco-Friendly Animal Based Fibres

Synthetic and natural fibres like cotton have been researched extensively as pure, blended, and composite materials. However, fibres collected from animals have gained attention in recent years. Fibers like silk, cashmere, camel, yak hair, fur, feathers, and wool are emerging as viable alternatives for consumer and industrial products [16].

i. **Green Composites**: In the realm of green composites, the utilisation of animal fibres for the development of composites was limited. However, current technological advancements, research, innovations, and their incorporation as raw materials are facilitated. For instance, composite panels are fabricated by mixing chicken feathers with wood fiber using formaldehyde resins as an adhesive. While these composites exhibited reduced mechanical properties compared to wool panels, significant improvement in water absorption resistance and thickness stability was observed. These enhancements are attributed to the presence of hydrophobic keratin in the feather fibre [20]. On the other hand, wool integrated with epoxy resulted in high tensile [21]; palm fiber with sheep wool in a 50/50 ratio demonstrated notable properties like mechanical tensile [22]; furthermore, the incorporation of sheep wool with polycaprolactone for 3D printing exhibited higher biodegradability and yield strength [23, 24]. Talabi et al. [25], suggested the advantages of utilising animal based fibres and other sources in the development of biocomposites. In this domain, polymer matrix complex exhibits excellent specific strength, heat resistance, lightweight, corrosion resistance, durability, specific stiffness, and fatigue resistance [25]. These attributes make them suitable for various applications, including chemical vessels, building construction, aerospace, marine, sport, civil infrastructure, automobile, biomedical, and household appliances. The advantage of these materials is the possibility to tailor their performance for specific applications by undergoing appropriate design. The PMC (Polymer Matrix Composites) is made up of matrix and reinforcement phases. Among these two, the reinforcement phase is stronger than the polymer matrix. Fiber-reinforced and particle-reinforced PMCs show a spectrum of new possibilities for different industries. Due to more focus on sustainable development, industries are looking for sustainable and eco-friendly alternatives to develop PCMs. Animal-based fiber waste is used as fillers such as wool, silk, spider webs, feathers, hair, shells, teeth, etc. are being researched. These animal-based fibers and materials are made up of proteins with varying structures, and chemistry, possessing functional groups like –OH, -COOH, -CONH- exhibiting the possibility of cross-linking with major modifications, as well as properties such as fire-retardance, elasticity, resiliency, etc. [25]. Another important factor is that these materials are least harmful to the environment, easily degradable, have low density, and are safe for handling. In the pursuit of sustainable production, polymer composite is prepared by using chicken feathers, leather, and cow hair in combination with polyester resin. ZnO was also incorporated as a fillers to increase the mechanical properties of the polymer composite [26]. The potential of horsehair is discussed earlier for the development of crinoline and, in the last few decades research suggests its advantage to develop composites by combining it with other fibers. The research involved a hot compression molding technique to prepare reinforcing with 20% wt. horsehair and resulted in superior mechanical properties with an increase in moisture absorption [27].

ii. **Waste Reduction**: Apart from using natural materials, the studies suggest the possibility of reducing the waste from the mother earth. An interesting study conducted by Szczepanik et al. [24], suggested the positive effect of the amalgamation of natural fibers on the degradation of PLA composites. Composites were made by mixing PLA with flax fiber and separately with wool fiber. The observation revealed a significant acceleration in the degradation process of the wool-based composite compared to the flax-based composite [24].

iii. **Medical Application**: Studies reveal that the silk structures are synthesized using limited means, specifically low energy and relatively simple renewable constituents namely silk proteins. If successfully controlled and adequately incorporated in biomaterials, some properties of natural silk could lead to innovative green materials that may contribute to reducing the ecological footprint of societies. Notably, recent advancements suggest that spider silk-based materials may lead to advanced functional materials [28]. Technical advancements like electro spinning have been utilized to develop silk scaffolds for applications in the medical industry [29].

iv. **Fashion and other industries**: It has been reported that 2 billion pounds of feathers are produced in the United States by domesticated chickens annually, much of which is considered waste due to bacterial contamination posing a potential environmental hazard. In response to this issue, research is being performed by Yiqi Yang (a young researcher) and his colleagues who have contributed to improve properties (such as cross-linking) of fibers derived from feathers. Their study focuses on saccharide aldehyde, a class of cross-linking agents, and demonstrated that modifying the molecular structure of keratin in the presence of aldehydes significantly improves the strength of the resulting fiber. From the sustainability point of view, the process does not yield any toxic chemicals and fibers exhibit potential as a low-cost alternative to petroleum-based fibers [30]. Similarly, looking into the current use of horsehair fiber, the literature reveals that only a limited number of industries and dedicated artisans are actively harnessing its unique properties to bring this exceptional fiber into mainstream textile applications. One notable example is Manufacture du Crin located in Challas, France, where just a handful of skilled weavers are still working with the horsehair and preserving the tradition of horsehair weaving art. This weaving has been practiced by weavers since 1814 by using handloom with each inch of fabric requiring several hours to complete. The weavers have learned the craft from their parents and grandparents. They need to practice the craft for two years before being able to truly master the skill. The woven horsehair fabric is now prepared for the high-end offering of the textile company Métaphores. In most cases, the horsehair is used in the weft direction of the weave. The manufacturing process involves several intricate steps, which include a bunch of fibers; hand combed to remove knots; strands are separated by using tweezers; weft threads are aid in a trough and inserted in warp; a special hooked shuttle used to catch them [31]. Similarly camel fiber has inherent properties that make it unique, but like other natural fibers it exhibits few limitations. Ogulmerjen [32] shared the possibility of blending viscose and camel hair fiber to mitigate the limitations at both ends. As viscose is a regenerated cellulosic fiber, having properties of breathability and moisture absorbency along with strength and elasticity similar to synthetics; therefore, by combining the fibers (camel & viscose) a raw material with a wide range of applications can be prepared. The key advantage of blending viscose with camel hair is, balancing the high water-absorbing property of viscose with the water-repellent property of camel hair fiber. This provides greater comfort to the wearer in varied climates, wicking moisture away and providing warmth and insulation. The temperature-regulating property along with the soft and smooth texture of viscose can improve the overall feel of the product made up of the combination. Such amalgamation provides wider opportunities in terms of applications such as winter apparel, outer gear, luxury home textiles, and fashion accessories. Looking into the sustainable aspect of the blend, as compared to other synthetic fibres viscose is close to eco-friendly as it is made up of cellulose, and eco-conscious consumers can introduce such innovation in their wardrobes [32].

v. **Luxury and Interior textiles**: Few brands such as Jaeger, Lois Vuitton, Prada, and Ralph Lauren are utilizing the unique properties of camel fibers in their products. The hair fibers find their way into the shoe lining where camel fibers, cotton jersey, and polymer glue are combined to create a non-woven lining [33]. The fabrics prepared for Métaphores are blends of horsehair with silk and other rare materials, imparting unique properties, and remarkable sensory and visual creativity to the textile structure. This careful selection and utilization of raw materials, combined with traditional weaving techniques, reflect the company's commitment to sustainable development [34]. Another significant area of innovation is in the development of upholstery materials. Various companies and brands such as Magniflex, Dormiente, Olbrish.de, and John Rayn by Design are utilizing horsehair and horsetail fiber to develop a mattress that offers support and comfort. Different raw materials are combined such as wool and horsehair or wool, cotton, and horsehair. According to these companies, the properties of the horsehair may altered when blend while preparing the mattress. However, these combinations result in products with distinctive characteristics suited for mattress production. The proportion of each fiber type is carefully considered as horsehair contributes structural support, while wool enhances softness. One such product is Vi-Spring, which consists of 1200 gsm of horsehair and Shetland Isle fleece as a blend [35–38]. Hastens is another company that shares its success story of integrating horsehair into its product development. Their legacy was built over six generations of master craftsmen. Their ideology centers on exceptional craftsmanship and the finest materials sourced and produced ethically. They believe material should come from nature and returned to nature by the end of their lifecycle. They suggest horsehair as their most exclusive material due to its built-in ventilation system. The curled strand of horsehair works as an internal spring system providing support and pliability. The hollow structure of horsehair fibers allows for effective air circulation, creating a built-in ventilation system. Thus, resulting in a healthy, comfortable, and hygienic sleeping environment. Horsehair is naturally hypoallergenic- possessing antibacterial, antifungal, and anti-mite properties. The brand uses different types of horsehair such as A-Lyx horsehair- the longest and strongest strands creating a springier and firmer bed. On the contrary, J-Horsehair- a soft, springy, and flexible blend–creates a more cushioned and adaptable sleeping surface [39].

Similarly, John Boyd Textiles, established in Castel Cary in 1837 has conserved the tradition of weaving textiles using horsehair. The warps are of cotton/silk and horsehair takes its place in the weft direction. The company informs that the horsehair fabrics were once an important part of the creations of furniture designers, such as Chippendale, Hepplewhite, Lutyens, and Rennie Mackintosh, and have consistently featured in prestigious interior design projects worldwide. This company is one of the few last companies to weave horsehair textiles and they continue doing it by introducing new designs and colors for the existing market. The amalgamation of fibers like cotton warp and horsehair weft or colors is part of their creation. Furthermore, the company has recently launched an establishment service for horsehair textiles, enhancing their aesthetic and functional value [40].

3.6 Eco-Friendly Animal Based Fibers Initiatives

Initiatives are not only restricted to research and development but many companies, NGOs as well as charitable trusts look forward to promoting the sustainability and uniqueness of animal fibers by instituting awards and recognitions. These awards were established to bring wool and other animal fibers to the forefront, which have been part of many civilizations and have proven sustainable, biodegradable, and renewable in many ways. These fibers can meet the demands of the twenty-first century in their original form and can be adapted to respond to evolving environmental and social needs. These awards encourage commercial projects that place wool at the center of innovation, creativity, and environmentally responsible practices, offering real-world solutions and applications. Each award agency has different criteria to unfold the potential of the artist, creator, and innovators. But the underlying principle of selecting the winners is somewhat related to the points given below:

1. Relevance of the wool industry and sheep: The project should be related to the industry and tied to sheep farming.
2. Involvement with the wool industry of that particular state, or country.
3. Global application as well as commercialization potential.

3.6.1 Awards for the Innovation in Animal Fibres.

To gain a deeper understanding of recent advancements in animal fiber-based products, let's examine the work of the winners selected for the Wool Awards for 2023 and 2024. Winners were selected considering the novelty and versatility they adopted in utilizing wool fiber to create sustainable products.

i. Brightland Innovation in Kid's Apparel: An innovation related to the kid's footwear was showcased by Brightland. This footwear was developed by using wool. The minimalistic wool shoes are made up of 100 percent biodegradable wool, both functional and stylish, providing the perfect blend of durability, comfort as well as sustainability. This effort ensures healthy and comfortable growth among kids, leaving smaller carbon footprints.
ii. Eco-friendly insulation packaging: After understanding the issues related to plastic packaging, "*Puffin Packaging*" developed sustainable insulation packaging by using wool as an alternative to polystyrene packaging. This wool-based packaging material is not only biodegradable but leverages the natural insulating properties of wool to maintain the appropriate temperatures during product transportation. The innovators blended British and European wool for such sustainable packaging.
iii. Cloudwool: Transforming Wool into Novel Textiles for Fashion and Interiors: Traditionally different methods are employed to fabricate materials for clothing and home furnishing purposes which include weaving, knitting, macramé, nonwoven, etc. This innovation by the innovator focused on utilizing British wool along with local manufacturing techniques (nonwoven) to develop materials suitable for a wide range of products, particularly for the fashion and interior sectors. Their innovation is committed to local, low-impact manufacturing methods, aiming to support environmentally responsible entrepreneurship.
iv. Solidwool: Turning wool into a beautiful composite material. Andy Guard introduced a new perspective of wool in design and art. The company (Solid Wood) is revolutionizing the wool industry by re-engineering wool into products that are both functional and appealing to a modern, design-conscious, and art-loving world. With a strong emphasis on sustainability, the innovator combined wool with bio-resin and created a composite material that falls into the classification of sustainability and functionality. The material is developed using Herdwick and Welsh Mountain wool, which is often considered waste and unvalued. This innovative material has the potential to be used to create different products like chairs, tables, etc. [41].
v. Twool: Sustainable Nets from Dartmoor Wool: Sustainable garden and marine nets made up of wool. This might sound unreal but Twool, led by Kim Stead has made this a reality by utilizing the wool of a rare breed "Dartmoor sheep". Such products not only contribute to the sustainability pool as these products are natural and biodegradable, but they also provide value outlets for local wool farmers thereby indirectly supporting the rural economy. The innovators have

noted that the company is already hitting the market with their product beyond gardening.

vi. Natural Dales Wool Products: Ruth Lindsay's has developed a range of wool-based products like wool logs and needled wool felt intended to be used for environmental restoration and have their application in sectors like geotextiles. Such products are an important part of achieving sustainability and play an important role in climate change mitigation strategies. The utilization of wool in this context highlights its versatility and potential to support ecological balance and restoration processes [42].

3.6.2 Awards for the Innovation in Silk.

The silk industry, with its long history and deep cultural significance, has been a crucial part of many economies. This industry holds importance not just for its economic contribution, but also for the rich tradition of sericulture passed down through generations. To recognize the exceptional individuals and organizations that have contributed to the advancement of silk production, two prominent awards stand out: the Louis Pasteur Award and the Pt. Deendayal Upadhyaya Silk Ratna Samman. These awards honor excellence in sericulture and innovation within the silk industry, promoting the growth and development of this vital sector.

i. The Louis Pasteur Award: This award is regarded as the "Nobel Prize of Sericulture" and was created in honor of renowned French scientist Louis Pasteur. His groundbreaking work in 1870 led to the successful mitigation of "Pebrine", a deadly silkworm disease. This disease had a devastating effect on the silkworm population. His contribution is remembered as the modern–day sericulture practices motivating many to innovate and research in this area. This award is presented once every three years at the ISC Congress, a significant international event for those in the sericulture and silk industry. Each laureate receives a Citation, a Medal, and a Certificate, in recognition of their outstanding contributions to the upliftment of sericulture and the silk industry. The awardees are selected by a 5- member International Expert Committee, which is composed of esteemed scientists from five different countries. The committee's recommendations are subject to final approval by the Executive Committee of the ISC. The selection process is rigorous and transparent, with member countries empowered to nominate candidates for the award. The nominations are reviewed by an expert committee, ensuring that the award is given to individuals who have made remarkable contributions to the advancement of sericulture. Up to three individuals can be honored with the Louis Pasteur Award in any given year. The award recognizes excellence in sericulture development, such as scientific research, innovations in silk production, and contributions to the global silk industry [43].

ii. The Pt. Deendayal Upadhyaya Silk Ratna Samman: It is a prestigious award conferred by the Uttar Pradesh Sericulture Department to celebrate the Indian

Silk Industry. This award is bestowed on individuals who exhibit innovation in silk production and diversification. This award is named after a prominent political thinker and leader Pt. Deendayal Upadhyaya. The areas acknowledged under this award include cocoon production, silk products, cocoon and yarn sales, and best design. The award seeks to inspire researchers, entrepreneurs, and other stakeholders within the industry to explore novel approaches and push the boundaries of traditional sericulture practices. It is presented during a ceremonial event organized by the Uttar Pradesh Sericulture Department. At this event, participants from across the state and beyond come together to celebrate the achievements of the silk industry and the individuals who have made a lasting impact in the above-mentioned areas [44].

iii. Ray of Hope Prize®: Bio-Inspired Innovation in Silk Technology: In addition to national honors, the Ray of Hope Prize ® is an international recognition awarded to *Spintex* in 2021 for developing environmentally sustainable silk. This achievement is grounded in a detailed understanding of the natural spinning process of spider silk, which the company emulated without using any toxic chemicals or high-energy techniques [17].

3.7 Research Initiatives by Research Institutions

Another significant sector/organization involved in research and innovation of textile-based products comprised of research associations that operate directly or indirectly under Ministry of Textiles such as NITRA (Northern India Textile Research Association), SASMIRA (Synthetic & Art Silk Mills Research Association), SITRA (South India Textile Research Association), BTRA (Bombay Textile Research Association), WRA (Wool Research Association), CSWRI (Central Sheep and Wool Research Institute) etc. Among these, WRA & CSWRI are dedicated to research in wool fibers, while SAI (Silk Association of India) for silk. As part of WRA, the association takes the initiative to research different projects related to wool, thus enhancing the application potential of this unique and sustainable fiber. The projects undertaken during 2024–2025 include the sustainable Enhancement of Aesthetic Values of Woolens by Digital Printing; the Development, Application, and Benchmarking of Eco-Friendly moth-repellent and Moth Proofing Finish for Woolen Textiles; the Development of sustainable and biodegradable mulching for agro textiles application using Deccani wool fiber with added functionality using plant micronutrient; Utilization of Indian Coarse Wool as insulating Material for Building Construction as a Green Building Material and Comparison of such coarse wool based developed products against conventional materials like glass wool/rock wool; Design & Development of Wool (Special fibers & blends) based Novel Denim Fabric with Multi-functional properties such as moisture management, anti-microbial, etc. using eco-friendly finishing techniques [45]. The association is also actively working towards sustainable development by utilizing coarse wool, which is predominantly produced in India. This type of wool is regarded as waste and is generally undervalued due to its coarse diameter, high medullation, inferior product quality, and, processing difficulties. However, sheep wool is recognized as one of the high-performing insulating natural fibers due to its thermo-hygrometric and acoustic properties. In the building sector, sheep wool holds considerable potential as it meets the requirements of green building design– being an eco-friendly, renewable, and fully recyclable material. When used as an alternative to conventional insulation materials (e.g., fiberglass, rock wool, polyurethane foam, polystyrene), sheep wool offers significant benefits for sustainability notably, reduction in the production costs for new insulating materials and environmental pollution [46]. Contributing to the concept of slow fashion, WRA has developed footwear using waste wool which is not only comfortable but also demonstrates good moisture-wicking properties. They are biodegradable and capable of regulating the humidity within the footwear. Such innovative ideas of utilizing waste coarse wool which is otherwise considered as waste or of inferior quality—can open up more avenues towards sustainable and natural production.

Similar to initiatives undertaken by WRA, Central Sheep Wool Research Institute (CSWRI), located in Avikanagar, Rajasthan is also engaged in innovating in the area of consumption of coarse wool fiber for the development of diverse products. Recently, Dr. Seiko Jose, a scientist at the institute, has developed a composite material by mixing coarse wool and natural rubber in equal proportion, by using a compression molding machine. Although, the natural product showed a reduction in tensile strength but areal density, hardness, and tear strength were improved [47]. This innovative material has been further converted into bags and shoes, demonstrating promising potential for commercial application [48].

Silk Association of India (SAI) is a voluntary organization registered under the Karnataka Societies Act, 1960, in the year 2006. The main objective of the association is to work in close coordination with the State departments involved in sericulture development and the Central Silk Board for the development and innovation of the sericulture industry. The association comprises approximately 300 active members, including cocoon growers, registered Seed producers (RSP), proprietors of chawki rearing centers (CRC), reelers, and weavers. In addition, several senior retired officers from the Central Silk Board and various State Sericulture Departments are also members of the association. SAI is governed by an Executive Committee consisting of the Honorary President, President, Vice-President, Secretary, Joint Secretary, Treasurer, and twelve Executive Committee members representing different sectors of the sericulture industry. The objective of the association is to develop awareness among all the stakeholders about consultancy services arrange training programs, and extend technology to support sustainable sericulture practices along with self-sufficiency in much-needed quality of bivoltine silk. The focus is also given to establishing links with different organizations; managing industry-related crises; an initiative to establish a silk museum etc. [49].

3.8 Conclusion

With growing consciousness for environmental sustainability, transformative evolution is seen in the production and consumption approach of the textile sector, emphasizing ethical sourcing, responsible manufacturing, and mindful consumption of natural resources. As evident throughout this chapter, eco-friendly fibers are still overlooked despite holding immense potential as sustainable substitutes in a world dominated by synthetic materials and fast fashion. The case studies and literature discussed affirm the growing realization of ethical practices that need to be incorporated in line with UN sustainable goals. The efforts made by different research organizations in conserving and optimum utilization of eco-friendly animal fibers support circular models of finding values in the materials that were once considered discarded waste. It would be right to state that the future of textile sustainability clearly lies in embracing the potential of eco-friendly animal fibers. Shifting from waste to resource through innovation and inclusive development strategies, industries can meaningfully contribute towards a greener and more resilient world.

References

1. Allary, T. 2021. Future fashion. *Industrial Biotechnology* 17(4):190–203.
2. Material Market Report 2024. 2024. Textile Exchange. https://textileexchange.org/knowledge-center/reports/materials-market-report-2024/. Accessed on 1st November 2024.
3. Shagwira, H., and F. M. Mwema. 2021. Advances in animal/plant–plastic composites: Preparation, characterization and applications. In *Plant and Animal Based Composites; Walter de Gruyter GmbH: Berlin, Germany*, 25–38.
4. The Sustainable Fashion Market in India: An Overview. https://timesofindia.indiatimes.com/blogs/voices/the-sustainable-fashion-market-in-india-an-overview/. Accessed on 2nd September 2024.
5. The Story of Wool. https://ttu-ir.tdl.org/items/271eae44-3d9b-4863-baf8-f69b4551d6ed. Accessed on 4th Jan 2025.
6. Harriet G. Brown. 2025. The story of silk. F.A. OWEN publishing, New York. https://www2.cs.arizona.edu/patterns/weaving/monographs/bhg_silk.pdf. Accessed on 19th January 2025.
7. Clothing from Animal Hair. 2020. Textile value chain. https://textilevaluechain.in/in-depth-analysis/articles/clothing-from-camel-hair. Accessed on 14th December 2024.
8. Creative Fashion Services. 2024. The story of horsehair: The fabric behind Crinoline. https://www.cfscreativefashionservices.com/post/the-story-of-horsehair-the-fabric-behind-crinoline.
9. Kay, F., and N. R. Storey. 2022. Victorian fashions for women. Pen and Sword History.
10. Khomeiri, M., and B. A. Z. Yam. 2015. Introduction to Camel origin, history, raising, characteristics, and wool, hair, and skin: A review. *Research Journal of Agricultural and Environmental Management* 4: 496–508.
11. Mengüç, G. 2018. Physical and microscopic properties of Anatolian camel fibers, vol. 25, 109.
12. Ansari-Renani, H. R., M. Salehi, Z. Ebadi, and S. Moradi. 2010. Identification of hair follicle characteristics and activity of one and two humped camels. *Small Ruminant Research* 90 (1–3): 64–70.
13. Babu, K. M. 2020. Animal Fibers: Silk. *Handbook of Fibrous Materials*, 75–94.
14. https://sangisathi.wordpress.com/2021/04/26/silk-the-queen-of-fibres/.
15. Hsia, Y., E. Gnesa, F. Jeffery, S. Tang, and C. Vierra. 2011. Spider silk composites and applications. *Metal, Ceramic and Polymeric Composites for Various Uses* 2: 303–324.

16. Mann, G. S., N. Azum, A. Khan, M. A. Rub, M. I. Hassan, K. Fatima, and A. M. Asiri. 2023. Green composites based on animal fiber and their applications for a sustainable future. *Polymers* 15 (3): 601.
17. Asknature. 2021. Textile Fiber Inspired by Spider Silk. https://asknature.org/innovation/textile-fiber-inspired-by-spider-silk/. Accessed on 1st October 2024.
18. India's Textile Industry Embracing Sustainability and Innovation. 2024. India Brand Equity Foundation. https://www.ibef.org/blogs/india-s-textile-industry-embracing-sustainability-and-innovation. Accessed on 11 November 2024.
19. Fashion Value Chain. 2023. https://fashionvaluechain.com/wp-content/uploads/2023/12/FVC-Sep-Oct-2023_03-1.pdf. Accessed on 11th November 2024.
20. Winandy, J. E., J. H. Muehl, J. A. Micales, A. Raina, and W. Schmidt. 2003. Potential of chicken feather fibre in wood MDF composites. In *EcoComp 2003, Queen Mary, University of London, September 1–2, 2003: 6 pages, 20*
21. Kishor Sharma, Y., A. Meena, M. Sahu, and A. Dalai. 2023. Experimental investigation on mechanical and thermal characteristics of waste sheep wool fiber-filled epoxy composites. *Materials Today: Proceedings*
22. Abdellah, M. Y., M. G. Sadek, H. Alharthi, G. T. Abdel-Jaber, and A. H. Backar. 2023. Characteristic properties of date-palm fibre/sheep wool reinforced polyester composites. *J. Bioresour. Bioprod.* 8: 430–443.
23. Haque, A. N. M. A., M. Naebe, D. Mielewski, and A. Kiziltas. 2023. Waste wool/polycaprolactone filament towards sustainable use in 3D printing. *Journal of Cleaner Production* 386: 135781.
24. Szczepanik, E., P. Szatkowski, E. Molik, and K. Pielichowska. 2024. The effect of natural plant and animal fibres on PLA composites degradation process. *Applied Sciences* 14 (13): 5600.
25. Talabi, S. I., S. O. Ismail, E. I. Akpan, and A. A. Hassen. 2024. Quest for environmentally sustainable materials: A case for animal-based fillers and fibers in polymeric biocomposites. *Composites Part A: Applied Science and Manufacturing*, 108216.
26. Farhad Ali, Md., Md Abdus Shabur, and A. M. Sarwaruddin Chowdhury. 2024. Gamma radiation's influence on the tensile characteristics of waste animal fiber reinforced biocomposites. Discover Materials: Springer. https://link.springer.com/article/10.1007/s43939-024-00106-5
27. Kumar, N., A. Singh, K. Debnath, and R. Ranjan. 2020. Mechanical characterization of animal fibre-based composites. *Indian Journal of Fibre and Textile Research (IJFTR)* 45 (3): 293–297.
28. Lefèvre, T., and M. Auger. 2016. Spider silk as a blueprint for greener materials: A review. *International Materials Reviews* 61 (2): 127–153.
29. Belbéoch, C., J. Lejeune, P. Vroman, and F. Salaün. 2021. Silkworm and spider silk electrospinning: A review. *Environmental Chemistry Letters* 19: 1737–1763.
30. Process Improves Strength, Color of Feather–Based Fibers. 2022. Nebraska Today. https://news.unl.edu/article/process-improves-strength-color-of-feather-based-fibers. Accessed on 23rd December 2024.
31. Rare Craft: Inside the Studio of the Lat Surviving Horsehair Weavers. 2024. Design Leadership Network. https://www.designleadershipnetwork.org/journal/le-crin/. Accessed on 19th November 2024.
32. Ogulmerjen, M. 2024. Making a fiber by adding syntetic viscoze to a camel wool. 1 (25): 226–228.
33. Mirzayev, N. B., S. S. Alimkhonova, and A. S. Rafikov. 2022. Obtaining, structure and properties of non-woven lining material for camel wool shoes. In *AIP Conference Proceedings*, vol. 2430, No. 1. AIP Publishing.
34. Metaphores. https://www.metaphores.com/en. Accessed on 1st Jan 2025.
35. All About Horsehair Mattresses and Beds. 2020. https://johnryanbydesign.co.uk/understanding-beds/all-about-horsehair-mattresses-and-beds/. Accessed on 22nd October 2024.
36. Magniflex. https://magniflexindia.com/wp-content/uploads/2024/08/Magniflex_brochure.pdf. Accessed on 12th January 2025.
37. https://olbrish.de/en/Shop/Horsehair/. Accessed on 13th October 2024.
38. https://www.dormiente.com/en/materials-science/horsehair/. Accessed on 13th October 2024.

39. Hasten. https://www.hastens.com/en/natural-materials. Accessed on 22nd September 2024.
40. Horse Hair Fibers Woven BY JOHN BOYD TEXTILES LIMITED. https://johnboydtextiles.co.uk/. Accessed on 19th Dec 2024.
41. Innovation in Wool Awards 2024. https://www.woolmen.com/innovation-in-wool-award-2024/. Accessed on 14th November 2024.
42. Innovation in Wool Awards: Changing the Perception of Wool. 2024. https://innovationinwool.com/ Accessed on 11 December 2024.
43. International Sericultural Commission. https://inserco.org/en/louis-pasteur-award#:~:text=ISC%20instituted%20the%20top%20International,the%20renowned%20Scientist%20of%20France. Accessed on 15th September 2024.
44. UP to Present Pt. Deendayal Upadhyaya Resham Ratna Samman to Outstanding Contributors of Silk Industry. 2025. The Statesman. https://www.thestatesman.com/india/up-to-present-pt-deendayal-upadhyaya-resham-ratna-samman-to-outstanding-contributors-of-silk-industry-1503350001.html. Accessed on 14th October 2024.
45. https://www.wraindia.com/central-wool-development-board-ongoing.php. Accessed on 23rd November 2024.
46. Wool Research Association. https://www.wraindia.com/. Accessed on 23rd November 2024.
47. Jose, S., P. S. Shanumon, S. Adithyan, S. Thomas, S. Bera, and N. Francis. 2024. Preparation and characterization of coarse wool reinforced natural rubber green composite. *Industrial Crops and Products* 222: 119727.
48. Wool-Rubber Composite: A Sustainable Alternative. www.CHEMTAN.com. Accessed on 2nd Feb 2025.
49. Silk Association of India. https://www.saoi.in/activities.html. Accessed on 13th March 2025.

Dr. Vandana Gupta is a distinguished academic and researcher currently serving as a Professor at the Parul Institute of Design, Vadodara, Gujarat. With a strong background in fashion design and textile science, she has made significant contributions to sustainable fashion, natural fibers, and textile recycling. Her academic journey includes a Master's degree and extensive research in areas such as natural dyes, UV-protective textiles, and the conservation of textile heritage. Dr. Gupta has authored numerous scholarly articles and book chapters, and her work is widely cited in the fields of textile sustainability and museum textiles. She is also actively involved in promoting ethical design practices and has contributed to the development of educational resources, including textbooks on textile museums and conservation. Since 2007, she has been part of the reputed institutions and Universities where she mentors students and leads research initiatives in fashion and textile innovation.

Chapter 4
Advancement in Natural Dyes Extraction and Application for Textile Finishing

Edward Apau, Benjamin Tawiah, and Morrison Appiagyei Boadi

Abstract Traditional dyeing methods rely on synthetic dyes and chemicals, which often generate large amounts of toxic waste harmful to flora and fauna. In contrast, natural dyes are biodegradable, non-toxic, and ecologically friendly. The resurgence of natural dyes in the textile industry in recent years is due to growing environmental concerns and a shift towards sustainable practices consistent with the United Nations Sustainable Development Goals (SDGs), particularly SDG 12 (Responsible Consumption and Production) and SDG 6 (Clean Water and Sanitation). Adopting eco-friendly dyeing practices reduces industrial water pollution and promotes sustainable resource use, thereby supporting the global efforts towards a sustainable world. This chapter discusses the innovative advancements in natural dye extraction and application for textile finishing, highlighting their significance in promoting sustainable practices in the textile industry. Techniques such as ultrasound-assisted extraction, enzyme-assisted processes, and supercritical fluid extraction are examined for their efficiency and effectiveness in obtaining vibrant colours from various plant sources. Also, methods of extracting natural dyes from insects and mineral sources are critically examined, focusing on their commercial viability. It also explores the application processes of these dyes in textile finishing, including various mordanting techniques and dyeing methods that enhance colour fastness and aesthetic appeal. Additionally, discusses recent advancements in the extraction and application of natural dyes and how such innovations have overcome traditional limitations, making them viable alternatives to synthetic dyes. Furthermore, the study addresses the standardisation of natural dyeing processes, colour reproducibility, and quality control measures essential for industrial adoption. More so, the safety and regulatory issues with regard to the extraction application of natural dyes are

E. Apau (✉) · M. A. Boadi
Department of Visual and Industrial Art, Sunyani Technical University, Sunyani, Ghana
e-mail: edwardapass@gmail.com

M. A. Boadi
e-mail: yabstorb2000@gmail.com

B. Tawiah
Department of Industrial Art, Kwame Nkrumah University of Science and Technology, Sunyani, Ghana
e-mail: benjamin.tawiah@knust.edu.gh

S.-X. Jiang et al. (eds.), *Advances in Textile Materials and Processing Techniques for Sustainability*, SDGs and Textiles, https://doi.org/10.1007/978-981-95-0469-5_4

further discussed. This chapter concludes by providing insights into future research directions that promote the development of natural dyes for sustainable industrial processes.

Keywords Natural dyes · Sustainable textiles · Eco-friendly dyeing · Colour fastness · Innovative applications

4.1 Introduction

4.1.1 Overview of Natural Dyes: Definition and Historical Context

Natural dyes are vegetable colours derived mostly from plant components, such as roots, stems, seeds, bark, leaves, and wood [53]. Jabar [59] adds that water-soluble colours derived from natural dyes are safer, more biodegradable, renewable, sustainable, and clean for use in electronics, food, cosmetics, leather, textiles, and medicines. Additionally, Chowdhury et al. [30] explains that natural dyes are colourants derived from natural sources, such as plants, insects, and minerals, used for dyeing textile fibres. Repon et al. [110] buttress that plant parts, such as roots, leaves, and flowers, are used to make natural dyes. These dyes are recyclable, biodegradable, and suitable for the environment.

It is believed that ancient civilizations extracted vibrant hues from sources like indigo (Indigofera) and madder (*Rubia tinctorum*) for practical and aesthetic purposes. These dyes were used to colour textiles, clay, wood, and even body paint [62]. It is further told that these dyes have been integral to human culture for millennia, with evidence of their use dating back to 2600 BC in the Indus Valley [46, 54].

Significantly, the history of natural dyes goes beyond mere colour; they often played a role in trade, culture, and art. For instance, in the ancient Mediterranean world, indigo was sold widely along with a traditional palette of valuable colours, with Tyrian purple emerging as the first worldwide dye [84]. Comparably, the Karbis tribe in Northeast India paints their traditional clothing using a variety of natural dyes, such as indigo, with particular hues exhibiting cultural significance [127]. Throughout history, natural dyes, especially indigo, have been important for creative expression, trade, and cultural identity. Although indigo-dyed textiles have long been a staple of the Landian Yao people's cultural identity in Southwest China, modernisation is causing this tradition to decline [72].

However, synthetic dyes' dependability, effectiveness, and wider colour spectrum propelled the 19th-century switch from natural dyes [23]. Synthetic dyes dominated the textile industry due to their ability to meet increasing demand and provide consistent results [23, 117]. Interestingly, the global textile dye market is experiencing growth. It was valued at approximately USD 11.78 billion in 2023 and is expected to reach USD 17.21 billion by 2030, growing at a CAGR of 5.5% from 2024 to 2030

[25]. Additionally, the synthetic dyes market is also growing, projected to reach USD 9.1 billion by 2029 (CAGR of 5.0% from USD 7.1 billion in 2024).

This growth is attributed to the increasing demand for vibrant colours and high-performance dyes in textiles, food, cosmetics, and other applications. Fast fashion trends also contribute to the demand [20, 96]. However, there is a growing emphasis on sustainability in the synthetic dyes sector due to rising environmental concerns. Manufacturers are investing in eco-friendly dyeing methods and bio-based dyes. Innovations like waterless dyeing technologies are gaining popularity [96].

In recent years, consumer behaviour has been increasingly driven by sustainability, fuelled by environmental awareness, climate change concerns, and demand for ethical business practices. For instance, younger generations are leading the charge in sustainable consumption. It is reported that 75% of Gen Z consumers prefer sustainable products over conventional ones, even if they cost more [55]. Additionally, millennials also show strong support for eco-friendly brands, with 73% expressing willingness to pay extra for sustainable offerings [35]. Holloway [52] adds that consumer preferences are shifting towards transparency and sustainability, giving brands that communicate their sourcing and production processes a competitive advantage.

The use of natural dyes has grown in popularity due to environmental and health concerns associated with synthetic dyes [2, 9]. Natural dyes are perceived as eco-friendly, non-toxic, and sustainable alternatives [9, 117]. Agustarini et al. [6] concludes that natural dyes offer non-toxic, biodegradable alternatives to synthetic dyes, providing a sustainable option for colouring textiles like cotton. In recent times, these dyes are gaining renewed interest due to their eco-friendly properties and reduced health risks compared to synthetic alternatives.

It is revealed that the trend toward sustainable consumption is undeniable, supported by extensive research and shifting market dynamics. In order to satisfy the growing demand from environmentally concerned consumers, companies, including those in the textile dyeing sector, must implement actual sustainability policies. The future success of most industries globally will depend on transparency, affordability, and continued consumer education. It is suggested that the textile industry should strive to balance the strengths and weaknesses of natural and synthetic dyes whilst focusing on sustainable practices to meet the dynamics of consumer preferences.

4.1.2 Importance of Natural Dyes in the Textile Industry

Natural dyes have been a cornerstone of human civilisation in colouring textiles for centuries. These dyes not only provide aesthetic value but also contribute significantly to sustainable and eco-friendly practices in the textile industry. Their resurgence in modern times is driven by growing environmental concerns and consumer demand for sustainable fashion.

Since natural dyes are more environmentally friendly and have less environmental impact than synthetic dyes, the textile industry has shown renewed interest in using them [40]. These dyes offer a more sustainable alternative. They are biodegradable, often non-toxic, and can be sourced from renewable materials, making them an appealing choice for environmentally conscious manufacturers and consumers.

Recent studies have highlighted the benefits of natural dyes in terms of environmental benefits and sustainability. For instance, Zerin et al. [136] reveals that compared with synthetic dyes, which can release toxic chemicals during production and use, natural dyes are less likely to cause skin allergies and other health issues. Moreover, due to their renewable nature, biodegradability, and reduction in water pollution, natural dyes are considered more eco-friendly than synthetic dyes [97]. Additionally, Karadag [63] explains that the use of natural dyes aligns with the growing consumer demand for environmentally conscious products and sustainable practices in the textile industry.

The health and safety benefits of natural dyes cannot be undermined. Natural dyes offer several health and safety benefits compared to synthetic dyes. Affat [3] explains that natural dyes are generally safer for human health and the environment, as these dyes may contain less or no heavy metals and carcinogenic compounds often found in synthetic dyes. Moreover, many natural dyes possess antioxidant and anti-inflammatory properties, contributing to their health benefits [80]. Natural dyes exhibit antimicrobial properties, reduced toxicity, and UV protection capabilities [1]. Some natural dyes, such as carotenoids, anthocyanins, curcumin, and betalains, have demonstrated various pharmacological activities [71].

Furthermore, it is explained that natural dyes hold deep cultural and historical value, often linked to traditional artisanal practices. The revival use of natural dyes could support rural economies by creating relatively high demand for dye-yielding plants and promoting traditional dyeing techniques. This shift is believed to have the potential to empower local communities, particularly in regions with a rich history of natural dye production, such as India and Peru [44].

Natural dyes offer significant environmental and health benefits, making them an attractive alternative to synthetic dyes in the textile industry. However, challenges such as poor fastness properties, reproducibility issues, and the need for eco-friendly mordants hinder their widespread adoption. However, continued research and technological advancements, along with the establishment of industry standards, are essential to overcoming these challenges and promoting the use of natural dyes on a larger scale. Nonetheless, natural dyes are expected to play a significant role in sustainable textile production [108].

4.1.3 Current Trends and the Shift Towards Sustainable Practices

In response to the growing awareness of environmental issues, there has been a significant shift in consumer behaviour towards sustainable practices in fashion and textiles. This trend is characterized by an increasing preference for products made from organic materials and dyed with natural colourants. Recent advancements in extraction techniques, substrate preparation, mordanting, and dyeing processes have improved the performance and sustainability of natural dyes [97].

The rise of sustainable fashion has prompted brands to re-evaluate their production methods, leading to a resurgence in the interest and application of natural dyes. It is explained that consumers exhibit stronger environmental concerns and place importance on security and relationships [70]. Some studies have identified different consumer groups based on clothing benefits sought, with eco-friendly oriented consumers showing the highest interest in natural-dyed apparel and traditional dyeing methods [88]. Generally, people are willing to pay more for sustainable clothing, such as naturally dyed clothes. This desire is influenced by aesthetics, demography, value orientation, cognitive understanding, attitudes toward green products, and green buying intentions [102].

This shift has encouraged companies to innovate in their dyeing processes, incorporating natural dyes into their collections and marketing them as part of a broader commitment to sustainability. Brands that successfully integrate natural dyes often highlight their environmental benefits, appealing to a growing demographic of eco-conscious consumers.

Moreover, advancements in dyeing technology have made it possible to utilise natural dyes more effectively. Aggarwal [5] explains that modern extraction methods, including microwave and ultrasonic extraction, have increased the yield of natural dyes, making them more viable for industrial use. This innovation, coupled with a renewed interest in traditional practices, positions natural dyes as a viable alternative to synthetic dyes in the contemporary textile industry.

As an aim of this chapter, the subsequent sub-sections highlight the types of natural dyes available, the methods of extraction, the application techniques in textile finishing, and the challenges faced by this sector. The content will also examine the innovations that are shaping the future of natural dyes, highlighting their potential to transform the textile industry into a more sustainable and responsible practice. This chapter not only emphasises the historical significance of natural dyes but also illustrates their critical role in shaping a sustainable future for textiles.

4.2 Types of Natural Dyes

This part will discuss types of natural dyes, which can be categorized into three major types: plant-based, insect-based, and mineral-based dyes. As each category draws on nature's palette of colours, its qualities, sources, and uses differ from one another.

4.2.1 Classification of Natural Dyes

This classification of natural dyes (Fig. 4.1) is based on the source of origin. Tarikul Islam and Hossain [123] affirms that, based on their source of origin, natural dyes are divided into three categories: plant/vegetable dyes, animal dyes, and mineral dyes.

However, Li et al. [71] is of the view that natural dyes can be extracted from plants, animals, microorganisms, minerals, and some other materials (Fig. 4.2). This classification of natural dyes looks elaborate than that of Tarikul Islam and Hossain [123].

(a) *Plant-Based Dyes*: These natural dyes are the most often used kind and are made from the leaves, roots, flowers, and fruits of plants. For instance, the leaves of the Indigofera plant are used to make indigo, while the roots of Rubia tinctorum are used to make madder dye. Jabar [59] explains that plant-based natural dyes represent the most abundant and sustainable category of colourants, comprising diverse compounds such as indigoids, terpenoids, quinonoids, and phenolics.

(b) *Animal-Based Dyes*: In contrast with plant-based dyes, animal-based dyes are derived from animal parts, such as shells, bones, scales, and secretions. The use of this substance has been documented for thousands of years across cultures. In addition to textiles, pottery, and other materials, these dyes offer a wide

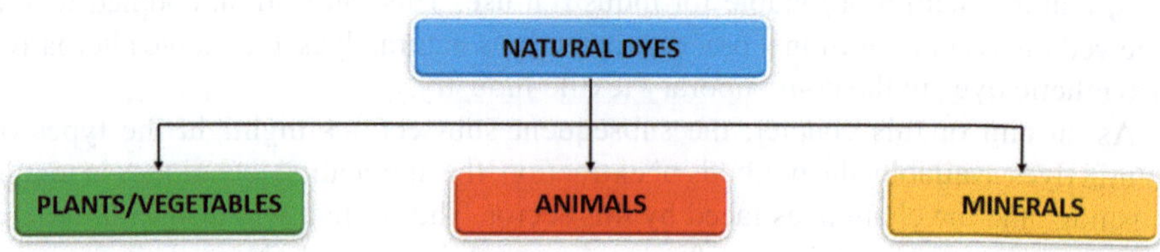

Fig. 4.1 Classification of natural dyes

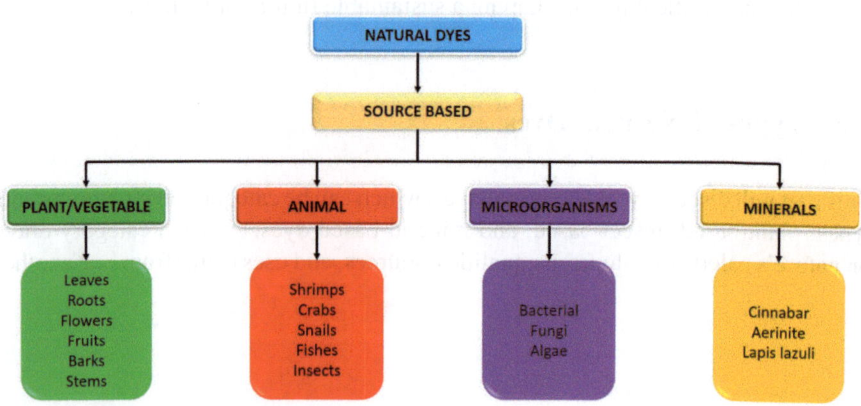

Fig. 4.2 Sources of natural dyes

spectrum of colours, from earthy tones to vibrant shades. Historically, they have played an important role in both trade and cultural practices. Cochineal, tyrian purple, sepia, oyster purple, guanine, bone black, and squid ink are a few examples of natural dyes. However, ecological issues have been highlighted by the overharvesting of sources such as Murex snails [9].

(c) *Mineral-Based Dyes*: Mineral-based natural dyes are derived from naturally occurring inorganic substances such as oxides, salts, and minerals, which serve as pigments or colourants in textiles. Common sources include iron oxides (red and yellow ochres), copper compounds like malachite and azurite (green and blue tones), and carbon (charcoal or soot) for blacks and greys. Minerals like lapis lazuli produce ultramarine blue, while clays and calcium carbonate offer earthy tones and whites. Alum (potassium aluminium sulphate) is widely used as a mordant and occasionally as a pale pigment base.

4.2.2 Common Sources of Natural Dyes

A variety of natural dyes are produced, including those derived from plants, animals, and minerals. These dyes have been used for centuries in textile production, art, and crafts due to their vibrant colours, cultural significance, and sustainability.

4.2.2.1 Plant-Based Natural Dyes

Plants are the most abundant source of natural dyes, providing a wide range of colours. They include different parts of plants such as roots, leaves, bark, flowers, fruits, and seeds. Hamdy and Hassabo [48] identifies the following common sources of plant natural dyes.

(a) *Indigo (Indigofera tinctoria)*: As one of the most prominent plant-based dyes, indigo produces a deep blue colour. This kind of dye is obtained from the indigo plant leaves. Indigo has been widely used in traditional dyeing practices, including the creation of denim.
(b) *Madder (Rubia tinctorum)*: The roots of the madder plant are a popular source of red dyes. Madder has been used since ancient times to create shades of red, ranging from pink to deep crimson.
(c) *Turmeric (Curcuma longa)*: Turmeric rhizomes are a source of vibrant yellow dye. While it offers bright colours, turmeric-dyed fabrics often have low colourfastness.
(d) *Henna (Lawsonia inermis)*: Henna leaves produce orange to reddish-brown dyes. Apart from textiles, henna is widely used for body art and hair dyeing.
(e) *Saffron (Crocus sativus)*: Saffron stigmas are used to create a brilliant golden-yellow dye. However, due to its high cost, saffron is more commonly used for culinary and ceremonial purposes.

(f) *Pomegranate (Punica granatum)*: The rind of pomegranate fruits produces yellow, brown, and greenish hues. It is also used as a mordant to enhance the fixation of other dyes.
(g) *Onion Skins (Allium cepa)*: Onion skins, especially the outer layers, provide yellow, orange, and brown dyes. They are a common household source for natural dyeing.
(h) *Cutch (Acacia catechu)*: The heartwood of cutch trees produces brown and tan dyes. It is widely used in traditional dyeing in India and Southeast Asia.

4.2.2.2 Animal-Based Natural Dyes

Animal-derived dyes are less common than plant-based dyes, but they have been historically significant. Karmakar et al. [64] explains that these sources of animal-based dyes are known for producing intense colours, such as the deep red from cochineal.

(a) *Cochineal (Dactylopius coccus)*: Cochineal is a red dye derived from the dried bodies of the cochineal insect, which lives on cactus plants. It produces carmine, a vibrant red colour, and has been widely used in textiles, cosmetics, and food colouring.
(b) *Tyrian Purple (Murex snails)*: Tyrian purple is a rare and highly prized dye obtained from the glands of sea snails (e.g., *Murex brandaris*). It was historically associated with royalty due to its vibrant purple hue and high cost.
(c) *Lac (Kerria lacca)*: Lac is a red dye derived from the resinous secretions of the lac insect. It has been used in textile dyeing and as a pigment in traditional art.

4.2.2.3 Mineral-Based Natural Dyes

Mineral-based dyes are inorganic substances extracted from natural minerals and ores. They are often used as pigments in artwork and traditional dyeing. Karmakar et al. [64] further explains that these mineral-based dyes are valued for their stability and unique hues. For instance, minerals like ochre and Prussian blue are used as natural mineral dyes.

(a) *Iron Oxides*: Minerals like hematite (red iron oxide) and limonite (yellow iron oxide) are used to create earthy red, yellow, and brown tones.
(b) *Malachite and Azurite*: Copper-based minerals such as malachite (green) and azurite (blue) provide vibrant colours. These minerals were historically ground into powders for dyeing and painting.
(c) *Lapis Lazuli*: This semi-precious stone was historically used to produce ultramarine blue, a rich and highly valued pigment.

4.2.2.4 Microorganism-Based Natural Dyes

Saeed et al. [115] explain that these dyes consist of fungi, algae and bacteria. They offer a sustainable alternative to synthetic dyes. However, fungi and algae are less common sources of natural dyes but offer unique possibilities for eco-friendly dyeing.

(a) *Lichens*: Lichens, which are symbiotic organisms consisting of fungi and algae, produce dyes in shades of orange, red, yellow, and purple. Lichen-based dyes are often used in traditional dyeing practices.
(b) *Algae*: Some species of algae produce pigments such as chlorophyll (green) and phycobilins (red and blue), which are being explored for use in textile dyeing.

However, Heer and Sharma [50] highlighted that microorganism-based dyes are emerging as valuable sources of natural dyes, offering advantages over synthetic alternatives. Bacteria, fungi, and algae can produce various pigments, including carotenoids, pyrroles, phenazines, and quinones.

4.2.3 Comparative Analysis of Natural and Synthetic Dyes in Terms of Environmental Impact and Textile Applications

Natural dyes and synthetic dyes differ significantly in their environmental impact, with natural dyes generally being more eco-friendly but less commercially viable. Natural dyes from plants, animals, and minerals offer a sustainable, non-toxic, and biodegradable alternative for the textile industry. The environmental impact of natural dyes is generally considered to be less harmful than that of synthetic dyes. Natural dyes are biodegradable and free from harmful chemicals, making them environmentally friendly. They do not release toxic effluents into water bodies, thus preserving aquatic ecosystems [3]. Contrarily, Muslimah et al. [87] explains that synthetic dyes are often derived from petrochemicals and can contain heavy metals and carcinogens as toxic substances. The production and disposal of synthetic dyes contribute significantly to water pollution, with the textile industry being a major source of contaminated wastewater.

Natural dyes are biodegradable and typically non-toxic, making them safer for both the environment and human health. They do not produce harmful by-products during their application, and many natural dyeing processes can utilize waste materials from other industries, promoting a circular economy. However, it is important to note that some natural dyes require mordant substances used to fix the dye to the fabric, which can sometimes be toxic. For example, certain metal mordants like chromium and tin can pose health risks if not handled properly. Comparison of natural dyes with synthetic dyes in terms of environmental impact is summarised in Table 4.1.

Table 4.1 A table showing a comparison of natural dyes with synthetic dyes in terms of environmental impact

Aspect	Natural dyes	Synthetic dyes
Raw material source	Derived from renewable sources (plants, animals, minerals)	Derived from non-renewable sources (petrochemicals)
Biodegradability	Biodegradable and decomposes naturally	Often non-biodegradable, leading to long-term pollution
Toxicity	Non-toxic and safe for the environment when untreated	Can release toxic chemicals and heavy metals into ecosystems
Water pollution	Minimal water pollution; dye runoff is generally non-harmful	Significant water pollution from dye effluents containing hazardous chemicals
Energy consumption	Low energy consumption during extraction and dyeing	High energy consumption during production and application
Carbon footprint	Low carbon footprint due to natural sourcing and minimal processing	High carbon footprint due to reliance on fossil fuels and energy-intensive processes
Waste generation	Produces organic, biodegradable waste	Produces hazardous waste that requires treatment
Sustainability	Sustainable if raw materials are harvested responsibly	Unsustainable due to dependency on finite resources
Health impact	Safe for human use; no harmful residues	It can cause allergic reactions and health issues due to chemical exposure
Ecosystem impact	Minimal disruption to ecosystems when properly managed	Can lead to severe ecological harm, such as aquatic toxicity

Additionally, the choice between natural and synthetic dyes in textile applications is driven by various factors, including colour vibrancy, fastness properties, sustainability, and economic considerations. While synthetic dyes dominate the market due to their cost-effectiveness and performance, natural dyes are increasingly being reconsidered for their eco-friendly properties.

Even though the textile industry's growing sustainability focus has reignited interest in natural dyes, synthetic dyes continue to challenge natural dyes in textile applications. For instance, a study in dyes and pigments found synthetic disperse dyes maintained colour grade 6–7 after accelerated fading tests [73], while even the best natural dyes (like madder root) scored only 4–5 [45]. The exception is natural indigo, which shows comparable light resistance to synthetic indigo. About wash fastness demonstration, synthetic reactive dyes on cotton retained 90% colour after 20 washes, whereas natural dyes averaged 60–70% retention even with mordants. Olua et al. [93] explains that reactive dyes are known for their excellent wash fastness, maintaining vibrant colours even after multiple washes. Table 4.2 compares

Table 4.2 A table showing a comparison of natural dyes with synthetic dyes in textile applications

Aspect	Natural dyes	Synthetic dyes
Light fastness	Moderate to poor; fades under UV exposure; prone to fading upon prolonged exposure to sunlight	Excellent; designed to resist fading from light exposure
Wash fastness	Variable; depends on mordants and fibre type, but still degrades faster	High, synthetic dyes bond chemically with fibres (e.g., vat/reactive dyes on cotton), ensuring high wash-fastness
Rubbing fastness	Generally low; colour may transfer when rubbed, especially on wet fabric	Highly designed to resist colour transfer under dry and wet rubbing conditions
Colour range and brilliance	Limited, derived from plants, insects, and minerals, mainly offering earthy and muted tones	Extensive; available in nearly unlimited shades, including bright and neon colours
Scalability	Low; limited availability of raw materials and labour-intensive processes restrict large-scale production	Highly synthetic dyes can be mass-produced efficiently with consistent quality
Cost	High due to extraction and processing requirements	Lower production cost due to large-scale manufacturing
Dye quality	Variable; depends on source material, extraction process, and mordanting; often lacks uniformity	Consistently high; chemically engineered for uniformity, brightness, and reproducibility
Dyeing effects	Produces unique and often uneven tones, creating artisanal aesthetics; influenced by natural variations	Provides uniform and predictable results with controlled intensity and shade consistency
Application methods	Mainly traditional techniques such as dye baths, hand-dyeing, and mordanting, requiring careful fibre preparation	Advanced and automated techniques like direct printing, continuous dyeing, and digital printing allow precise and efficient application
Application efficiency	Complex and labour-intensive; requires multiple steps	Simple, efficient, and scalable for mass production

the performance of natural and synthetic dyes, focusing on key parameters such as colour fastness, durability, and application efficiency.

It can be deduced that the classification of natural dyes based on their origin is plant, insect, and mineral. However, microorganisms are other viable sources of natural dyes, which are gaining attention in current research. These natural-based dyes have sources for obtaining the respective dyes. Plant-based dyes can be obtained from indigo, madder, turmeric, henna, etc. Cochineal, tyrian purple, lac can be sourced from animal-based dyes, whilst iron oxides, malachite and azurite, lapis lazuli are obtained from minerals. Moreover, fungi, algae and bacteria were identified as sources of natural-based microorganism dyes.

Apparently, it is observed that when compared to synthetic dyes, natural dyes generally present a more environmentally friendly alternative, although careful consideration of mordants and dyeing practices is essential to minimize any potential negative impacts. On the other hand, synthetic dyes outperformed natural dyes in colour fastness, durability, colour variety, scalability, dye quality, application efficiency, and cost-effectiveness. However, their limitations, such as a restricted colour range, poor fastness, labour-intensive processing, variable dye quality, and limited scalability, restrict widespread industrial use. Therefore, future research must focus on improving the performance of natural dyes through innovative mordanting techniques and sustainable extraction processes to bridge the gap between sustainability and performance. Interestingly, this subtopic of the chapter provides the steps to find out the various extraction methods of natural dyes coupled with their respective advantages and disadvantages in the textile industry.

4.3 Extraction Methods

The extraction of natural dyes involves separating the dye molecules from plant, animal, or mineral sources to impart colour to fibres and textiles. The choice of method depends on the dye source, its chemical nature, and the intended application. The efficiency and quality of natural dyeing heavily depend on the methods employed for extracting the dye from the dye source. The following points discuss various traditional and modern extraction techniques of natural dyes.

4.3.1 Traditional Extraction Techniques

Traditional natural dye extraction methods have been practiced for centuries and rely on simple, resource-efficient techniques to extract pigments from plants, animals, and minerals. These methods often vary based on regional practices, dye sources, and cultural traditions.

4.3.1.1 Aqueous/Boiling Extraction

Aqueous extraction is a foundational method for dye extraction, wherein dye materials are often first ground into small pieces or powder before being immersed in water. This method begins with boiling the dye source to loosen the cellular structure, promoting the release of dye components into the water. It is a sustainable approach, although it may be time-consuming due to the long extraction periods required and potentially low dye yields, as only water-soluble components are typically extracted.

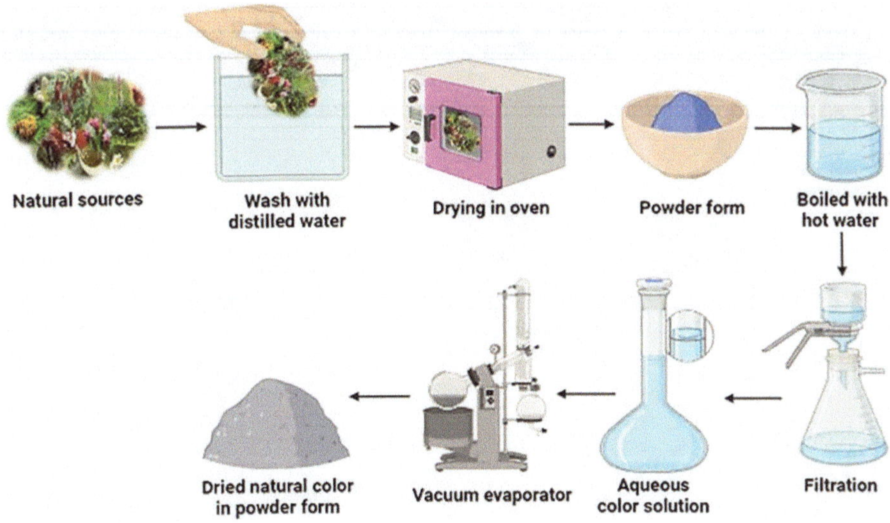

Fig. 4.3 Aqueous/boiling extraction method. Reproduced with permission from [110] Copyright (2024) from Springer Nature

Using this method, Saeed et al. [115] extracted dyes from beetroot, red cabbage, and turmeric. The study highlights that plant sources are viable natural alternatives to synthetic colourants in a variety of industries. This method is particularly suitable for extracting water-soluble dyes such as anthocyanins, betalains, and flavonoids, which are widely used in textile, food, and cosmetic industries. Figure 4.3 shows the standard procedure for obtaining an aqueous extract from dried leaves or fruits.

4.3.1.2 Solvent/Soaking Extraction

This extraction method of natural dyes involves using various organic solvents like ethanol, methanol, acetone, or water to extract colourants from natural sources such as plants, flowers, and leaves. Islam et al. [57] is of the view that ethanol has been shown to produce better extraction yields for organic dyes compared to water, as demonstrated in studies involving flower pigments. Additionally, this method is favoured for its ability to efficiently extract dyes while maintaining the integrity of the colour compounds. Some studies have shown the effectiveness of solvent extraction in obtaining high-purity dyes, such as anthocyanins from berries, curcumin from turmeric, and carotenoids from marigold flowers [37]. The choice of solvent and extraction conditions significantly impacts the efficiency and quality of the extracted dyes. However, a notable drawback is the potential presence of toxic residues from the solvents used, which necessitates careful handling and thorough rinsing to eliminate any harmful residues before application. Figure 4.4 shows the Solvent extraction method.

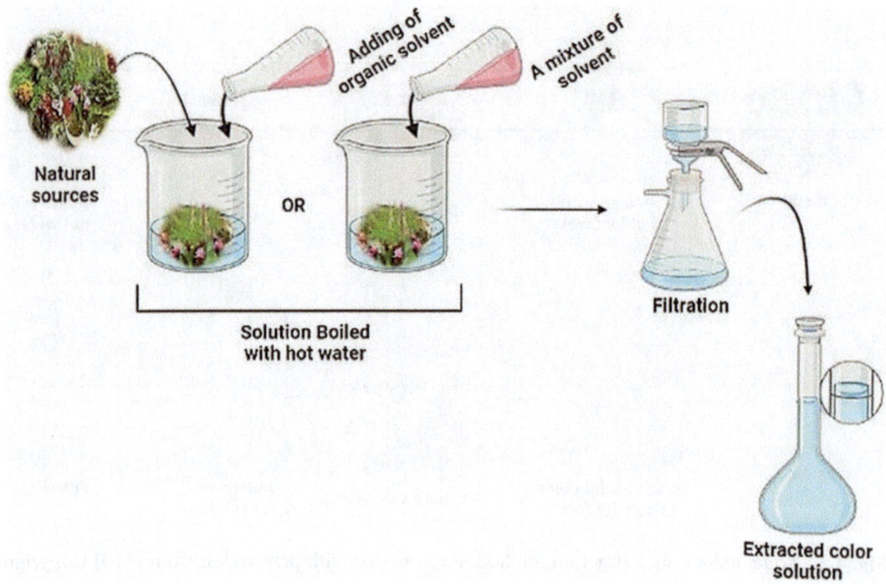

Fig. 4.4 Solvent extraction method. Reproduced with permission from [110] Copyright (2024) from Springer Nature

4.3.1.3 Fermentation Extraction

Fermentation extraction of natural dyes is gaining attention as a sustainable and eco-friendly method for obtaining colourants from plant and microbial sources. This process utilizes microorganisms such as bacteria, fungi, or yeast to break down complex substrates, releasing dye compounds like anthraquinones, flavonoids, and indigoids. For example, microbial fermentation has been effectively used to extract indigo from Indigofera species, where enzymatic action converts indican into indigotin, the blue dye molecule [26]. Recent studies highlight the advantages of fermentation extraction, including higher dye yields, improved colour fastness, and reduced environmental impact compared to conventional chemical methods [118]. Additionally, fermentation can enhance the bioavailability of dye precursors, making it a promising technique for industrial applications. Research by Teixeira-Guedes et al. [125] demonstrated that fermented natural dyes exhibit excellent dyeing properties on textiles, with superior adherence and vibrancy. Figure 4.5 shows the fermentation extraction process of Indigo naturalis.

Fig. 4.5 The fermentation extraction process of Indigo naturalis [103] (This is an open-access paper licensed under CC BY-SA 4.0)

4.3.2 *Modern Extraction Technologies*

Modern extraction technologies for natural dyes aim to enhance efficiency, yield, and sustainability while preserving the natural integrity of the dyes. These advanced methods address the limitations of traditional techniques, such as low dye yield, long processing times, and environmental concerns.

4.3.2.1 Ultrasound–Assisted Extraction

The extraction of natural dyes from plant materials using Ultrasound-assisted extraction (UAE) has become a popular and environmentally friendly method, providing significant advantages over traditional techniques, including reduced extraction time, lower solvent consumption, and improved yield. UAE utilises ultrasonic waves (20–100 kHz) to induce cavitation in the solvent, disrupting plant cell walls and releasing bioactive compounds, such as natural dyes. Several studies have demonstrated the effectiveness of UAE in extracting dyes from various plant sources. For instance, research by Ghareaghajlou et al. [43] highlighted the potential of ultrasound in enhancing the extraction efficiency of anthocyanins from red cabbage, achieving higher yields compared to traditional solvent extraction. Similarly, Yahya et al. [133] reported that UAE significantly improved the extraction of chlorophyll from spinach leaves, attributing the success to the mechanical effects of ultrasound that promote better solvent penetration and mass transfer. Furthermore, the UAE offers a sustainable approach due to its milder temperatures and reduced chemical usage,

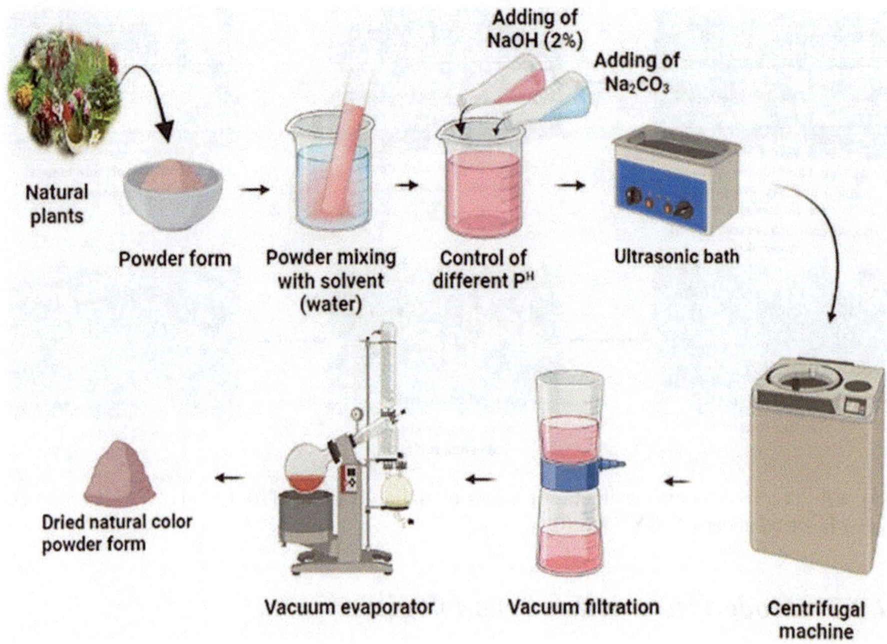

Fig. 4.6 Ultra-sound assisted extraction. Reproduced with permission from [110] Copyright (2024) from Springer Nature

supporting green chemistry principles. However, optimizing parameters like ultrasonic frequency, power, extraction time, and solvent type is crucial for maximizing dye yield and quality (Fig. 4.6).

4.3.2.2 Microwave-Assisted Extraction

Microwave-assisted extraction (MAE) (Fig. 4.7) has gained significant attention as a rapid and efficient method for extracting natural dyes from plant materials. MAE utilizes microwave energy to heat the solvent and plant matrix, causing localized heating and pressure build-up within the cells, which disrupts cell walls and facilitates the release of bioactive compounds, including natural dyes. This method is particularly advantageous for its ability to selectively target polar compounds, making it suitable for extracting a wide range of natural dyes such as anthocyanins, carotenoids, and chlorophylls. Zerin et al. [136] reported that MAE significantly improved the extraction efficiency of anthocyanins from purple corn, achieving higher yields in a fraction of the time required by traditional methods. Later on, Ngamwonglumlert et al. [90] highlighted the use of MAE for extracting carotenoids from marigold flowers, noting that the method preserved the stability and quality of the pigments. Additionally, because of its low energy consumption and little use of organic solvents, MAE is regarded as ecologically harmless and adheres to the

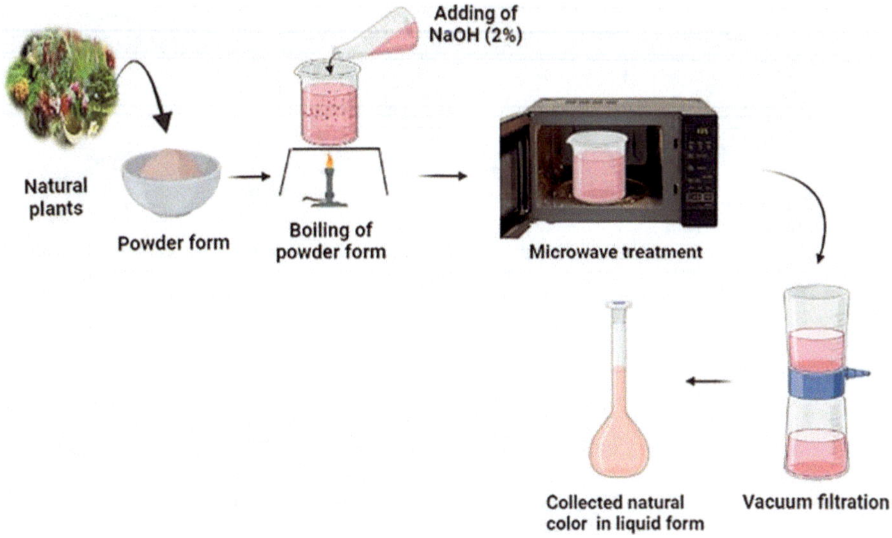

Fig. 4.7 Microwave-assisted extraction. Reproduced with permission from [110] Copyright (2024) from Springer Nature

principles of green chemistry. However, optimizing parameters such as microwave power, extraction time, solvent type, and temperature is critical to ensure maximum dye yield and quality. Recent advancements in MAE technology, such as the integration of response surface methodology (RSM) for process optimization, have further enhanced its applicability for industrial-scale extraction of natural dyes.

4.3.2.3 Enzymatic Extraction

Enzymatic extraction (Fig. 4.8) is an eco-friendly method that utilizes specific enzymes to break down plant cell walls and release natural dyes efficiently. Enzymes such as cellulases, pectinases, and proteases hydrolyse structural components like cellulose, pectin, and proteins, facilitating the diffusion of dye molecules into the solvent. Unlike conventional methods that require high temperatures and harsh chemicals, enzymatic extraction operates under mild conditions, preserving heat-sensitive pigments and reducing environmental impact. This method enhances dye yield, shortens extraction time, and minimizes solvent consumption, making it a sustainable alternative for natural dye production [126]. However, challenges such as enzyme cost, reaction time, and the need for precise pH and temperature control must be addressed for large-scale applications. Despite these limitations, enzymatic extraction offers a green, high-yield alternative for obtaining natural dyes with minimal environmental impact.

For sustainable textile production, natural dye extraction plays a key role in improving dye output, process efficiency, and environmental sustainability. It is

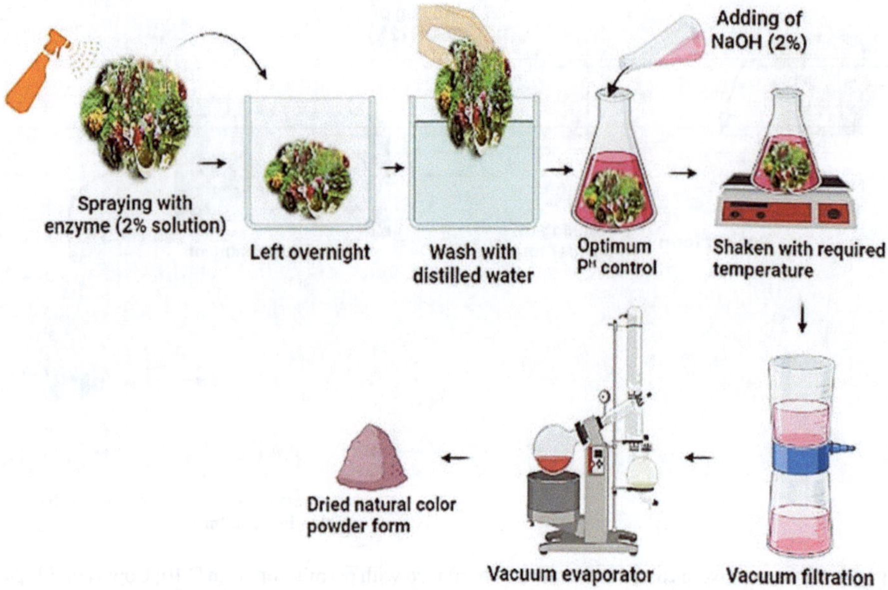

Fig. 4.8 Enzymatic extraction. Reproduced with permission from [110] Copyright (2024) from Springer Nature

important to maintain the ecological balance during the extraction process of natural dyes through various methods that optimize yield. Several techniques are used to extract natural dyes, including solvent extraction, enzymatic extraction, maceration, ultrasound-assisted, and microwave-assisted techniques. However, these methods vary in efficiency and sustainability, with newer techniques like ultrasound-assisted and microwave-assisted extractions offering faster and more efficient results compared to traditional methods.

Conventional natural dye extraction yields are significantly affected by solvent type, extraction time, temperature, and the solid-to-solvent ratio, even though they are more eco-friendly. These parameters critically impact both dye yield and quality. For instance, Gala et al. [42] explains that longer extraction times increase yield, but excessive duration can degrade dye quality. Microwave extraction of mahogany, for instance, is optimal at 30 min, while reflux requires 120 min. Aggarwal [4] explains that although conventional methods have advantages, modern techniques like microwave or ultrasonic extraction offer faster results with potentially better yields.

Interestingly, Taura et al. [124] supports that ultrasonic, microwave, and enzymatic extraction methods have been investigated for efficient natural dye extraction, offering faster processing at lower temperatures, conserving energy, and improving yield. Subsequently, Sooch et al. [119] adds that the extraction of dyes from Hibiscus sabdariffa has exhibited notable yield gains by the application of response surface methodology (RSM) in optimising extraction parameters, such as temperature and time. Table 4.3 shows the efficiency comparison of treated extraction methods under this sub-heading.

The above data is represented in a bar chart for further analysis (Fig. 4.9).

It can be deduced that microwave-assisted extraction (MAE) produces the highest dye yield (75%) with the shortest extraction time (30 min), making it highly efficient. However, ultrasound-assisted extraction (UAE) and enzymatic extraction (EE) also achieve high yields (65–70%), though EE requires significantly more time (240+ min). On the other hand, fermentation extraction (FE) has a moderate yield (55%) but is the slowest method, taking several days. Meanwhile, aqueous extraction (AE)

Table 4.3 A table showing dye yield, extraction time and energy consumption of some extraction methods

Extraction method	Dye yield (%)	Extraction time (min)	Energy consumption
Aqueous/boiling extraction	40	120	High (heating required)
Solvent/soaking extraction	50	360+	Moderate (solvent required)
Fermentation extraction	55	1440+ (days)	Low
Ultrasound-assisted extraction (UAE)	65	45	Low
Microwave-assisted extraction (MAE)	75	30	Moderate
Enzymatic extraction	70	240+	Moderate

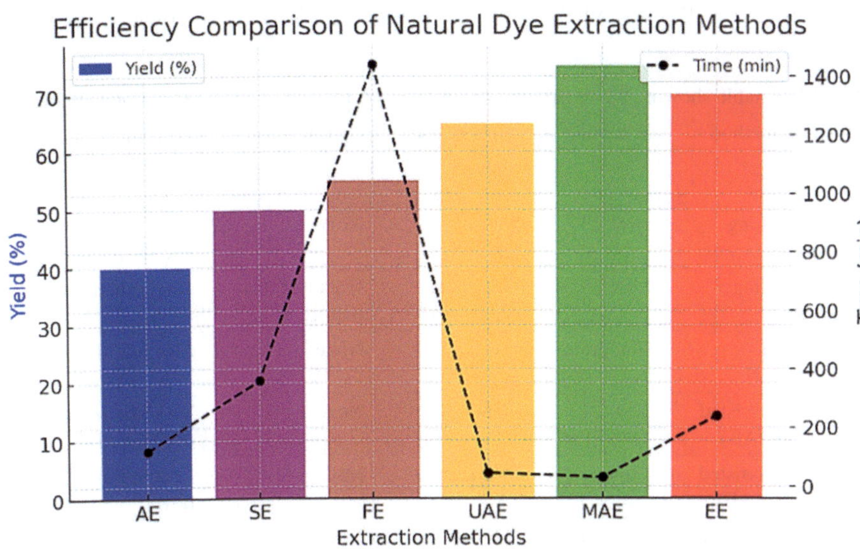

Fig. 4.9 A bar chart showing the efficiency comparison of some natural dye extraction methods

and solvent extraction (SE) yield the lowest results (40–50%) and demand higher energy or solvent use.

The environmental impact of natural dye extraction methods can be assessed by evaluating water and energy consumption, and chemical use. While natural dyes are generally more sustainable than synthetics due to better biodegradability and lower toxicity, extraction methods can vary greatly in their environmental footprint depending on the techniques and resources employed. Natural dye extraction methods, such as aqueous extraction and solvent extraction, can vary in water usage. Sk et al. [118] explains that the amount of water needed for aqueous extraction is usually significant, but typically less than for synthetic dye extraction, which has a high water consumption and pollution potential. It is revealed that the microwave-assisted extraction process (MAE) lowers water consumption by optimizing material-to-solvent ratios and minimizing extraction times, thereby reducing the overall water footprint [78]. Natural dye extraction often uses fewer chemicals than synthetic dye processes, which frequently employ dangerous chemicals. However, if not handled appropriately, many natural dye extractions may call for mordants or solvents that might harm the environment [20]. Table 4.4 shows the environmental impact comparison of these extraction methods.

From Table 4.4, it can be deduced that aqueous extraction (AE) consumes the most water and energy, making it the least eco-friendly. While solvent extraction (SE) has moderate energy consumption, it generates significant chemical waste due to solvents. In contrast, fermentation extraction (FE) is the most natural method, with minimal energy use but an extended processing time. Notably, ultrasound-assisted extraction (UAE) has the lowest environmental impact, requiring minimal water and energy. Although microwave-assisted extraction (MAE) is efficient and sustainable, its moderate energy consumption necessitates optimization. Finally, enzymatic

Table 4.4 A table showing the environmental impact comparison of some extraction methods

Extraction method	Water use	Energy use	chemical use	Overall environmental impact
Aqueous/boiling extraction	High	High	None	High (water and energy-intensive)
Solvent/soaking extraction	Low	Moderate	High (solvent required)	High (chemical waste)
Fermentation extraction	Moderate	Low	None	Low (eco-friendly)
Ultrasound-assisted extraction (UAE)	Low	Low	None	Very low (most eco-friendly)
Microwave-assisted extraction (MAE)	Low	Moderate	None	Low (efficient and fast)
Enzymatic extraction	Moderate	Moderate	None	Moderate (requires enzymes)

extraction (EE) is relatively sustainable but depends on enzyme production, which may introduce additional environmental costs.

4.3.3 Advantages and Disadvantages of Each Method

Extracting natural dyes can be achieved through various methods, each with its own set of advantages and disadvantages. Sousa [120] explains that there are several extraction methods available, each with unique benefits and disadvantages, including solvent extraction, aqueous extraction, enzymatic extraction, and supercritical fluid extraction. From the discussions under Sect. 4.3.2, these methods direct researchers on how to advance research in the textile dyeing industry. Table 4.5 shows the advantages and disadvantages of these natural dye extraction methods.

It can be explained that natural dye extraction from plant materials involves various methods with their advantages and limitations. Aqueous extraction is eco-friendly and efficient, but less effective for non-polar dyes. Solvent extraction is environmentally and health-conscious due to toxic organic solvents. Fermentation offers a sustainable method, but requires precise control over conditions. Although they need specialised equipment and parameter optimisation, advanced approaches such as ultrasound-assisted extraction and microwave-assisted extraction shorten extraction periods and use less solvent. Enzymatic extraction preserves natural dye stability and bioactivity, but its high cost and scalability limit its widespread adoption.

The extraction method depends on the application, dye type, and production scale. Advanced techniques like ultrasound-assisted extraction, microwave-assisted extraction, and enzymatic extraction are being adopted due to sustainability and efficiency. Future research should optimize these methods, reduce costs, and scale up processes while adhering to green chemistry principles. This balance can lead to broader applications in textiles, food, cosmetics, and beyond. The next sub-topic will focus on factors influencing natural dye extraction.

4.4 Factors Influencing Dye Extraction

The efficiency and quality of natural dye extraction depend on various factors that affect the release, solubility, and stability of dye pigments from their natural sources. These factors must be carefully considered to optimize the extraction process and achieve maximum dye yield while maintaining the desired colour quality.

Table 4.5 A table showing advantages and disadvantages of some extraction methods of natural dyes

Extraction method	Advantages	Disadvantages
Aqueous extraction	– Eco-friendly and non-toxic (uses water as a solvent) – Simple and cost-effective – Safe for food and cosmetic applications	– Low extraction efficiency for non-polar compounds – Longer extraction time – Limited to water-soluble dyes
Solvent extraction	– High extraction efficiency for both polar and non-polar compounds – Versatile and widely used	– Use of toxic organic solvents (e.g., methanol, acetone) – Environmental and health hazards – Requires solvent recovery
Fermentation extraction	– Eco-friendly and sustainable – Can enhance dye yield and quality – Utilizes microbial activity	– Time-consuming process – Requires controlled conditions (pH, temperature) – Risk of contamination
Ultrasound-assisted extraction (UAE)	– Reduced extraction time – Lower solvent consumption – Enhanced yield and quality of dyes – Eco-friendly	– High initial equipment cost – Optimization of parameters (frequency, power) required – Limited scalability
Microwave-assisted extraction (MAE)	– Rapid extraction process – High efficiency and yield – Reduced solvent usage – Energy-efficient	– Risk of thermal degradation of heat-sensitive dyes – Requires specialized equipment – Limited to small-scale applications
Enzymatic extraction	– Mild and eco-friendly conditions – High selectivity and specificity – Preserves dye stability and bioactivity	– High cost of enzymes – Requires optimization of enzyme concentration and conditions – Limited scalability

4.4.1 *Impact of Plant Part Used (Leaves, Roots, and Flowers)*

The plant part used for dye extraction significantly influences the dye yield, composition, colour intensity, and extraction efficiency. Different parts of a plant—leaves, roots, and flowers contain varying concentrations of pigments, tannins, and other bioactive compounds, affecting their suitability for dyeing applications.

4.4.1.1 Leaves

Leaves are a common source of natural dyes, containing chlorophyll (green pigment) and may also contain carotenoids (yellow-orange pigments), anthocyanins and flavonoids [121]. They generally produce green, yellow, and brown shades, depending on the plant species. Since leaves have a high surface area and thin cell walls, they allow for easier solvent penetration, leading to efficient dye extraction. For instance, Varjani et al. [130] explains that Malus domestica leaves contain diverse pigments, including dihydrochalcones, flavan-3-ols, and flavanols, which can be used for wool dyeing.

4.4.1.2 Roots

Roots often contain highly concentrated pigments, such as betalains, anthraquinones, and tannins, making them excellent sources of red, purple, and yellow dyes. Chowdhury et al. [29] explains that hydroxyanthraquinone dyes derived from plants like madder (*Rubia tinctorium* L.) are known for producing red, orange, and yellow shades. Alizarin and purpurin are notable examples used in textile dyeing. Due to their tougher structure (rich in lignin and starch), root extraction requires longer processing time and often benefits from techniques like fermentation, ultrasound, or enzymatic extraction. Roots generally provide more stable and long-lasting colours compared to leaves and flowers.

4.4.1.3 Flowers

Flowers are rich in anthocyanins, carotenoids, and flavonoids, producing vibrant shades of red, purple, yellow, and blue. The extraction process is relatively easier since flower tissues are soft and contain high amounts of water-soluble pigments. However, flower-based dyes are often pH-sensitive and prone to degradation, requiring careful handling and storage. Amir-Al Zumahi et al. [12] explains that these natural dyes demonstrate stability at room temperature but may degrade at higher temperatures.

4.4.2 Effect of Solvent Type and Concentration

The kind and concentration of solvent used a critical factors in the effectiveness, yield, and quality of natural dye extraction. Different solvents interact uniquely with plant pigments, affecting their solubility, stability, and extraction rate. The selection depends on the chemical nature of the dye compounds and the desired application.

Natural dye extraction is influenced by solvent type and conditions, with various solvents like water, ethanol, and acetone being studied for extracting pigments from plant sources [134]. Solvents are essential carriers in plant tissue extraction, facilitating the release of pigments. The choice of solvent depends on the polarity of the dye, with polar dyes like anthocyanins and betalains best extracted with water or ethanol, while non-polar dyes require non-polar or moderately polar solvents like acetone, hexane, or methanol. Water is the most eco-friendly and cost-effective solvent, but its efficiency is limited to water-soluble dyes. Ethanol is widely used due to its low toxicity.

The concentration of the solvent also plays a crucial role; higher solvent concentration generally improves dye extraction by increasing solubility, but excessive amounts may lead to pigment degradation or unnecessary dilution. Diluted ethanol (50–70%) is often preferred over absolute ethanol, as it enhances the extraction of both water- and alcohol-soluble compounds. Highly concentrated acids or bases can break down cell walls but may degrade sensitive pigments, reducing dye quality.

4.4.3 Temperature and Time Parameters

In natural dye extraction, temperature and time are two crucial factors that impact the effectiveness, yield, and colour qualities of extracted dyes. Optimal temperature is essential because it affects the solubility of pigments, the rate of diffusion, and the stability of heat-sensitive compounds. By improving the solvent's ability to penetrate plant tissues, higher temperatures often make dyes more soluble and speed up the extraction process. However, excessive heat can degrade thermolabile pigments such as anthocyanins and carotenoids, leading to colour fading or loss of bioactivity.

Extraction time is equally important, as it determines the duration of contact between the solvent and plant material. Longer extraction times can improve yield by allowing more complete dissolution of pigments, but they may also lead to the degradation of sensitive compounds or the extraction of undesirable impurities. Conversely, traditional methods like maceration or Soxhlet extraction often require several hours to achieve comparable yields, making them less efficient. Balancing extraction time with temperature is crucial; shorter extraction times at moderate temperatures are often preferred to preserve dye quality and reduce energy consumption. In a study, Lambrecht et al. [69] explains that temperatures above 70 °C led to polyphenol degradation. Moreover, it was found that extending the extraction time beyond two hours did not increase yield.

4.4.4 pH Levels and Their Influence on Colour Extraction

The colour output of most natural dyes can vary due to their sensitivity to pH changes. Natural dyes, particularly those derived from plant sources, often contain pH-sensitive compounds such as anthocyanins, betalains, and flavonoids, which can undergo structural changes and colour shifts depending on the acidity or alkalinity of the extraction medium. pH plays a vital role in the extraction of pigments from plant materials, as it can significantly alter the chemical structure of the dye molecules. For example, anthocyanins from purple sweet potato exhibit colour changes from red to yellow as pH increases, suggesting potential use as pH indicators [91]. In acidic conditions (pH 1–3), anthocyanins appear red, while in neutral to slightly alkaline conditions (pH 7–8), they turn blue or purple. At highly alkaline pH levels above 8, anthocyanins degrade, leading to colour loss.

The pH of the extraction medium also affects the solubility and release of pigments from plant tissues. Acidic conditions can enhance the extraction of certain dyes by breaking down cell walls and increasing the solubility of pigments. Conversely, alkaline conditions can improve the extraction of other pigments, such as flavonoids and tannins, which are more soluble at higher pH levels. However, excessive alkalinity can lead to the degradation of sensitive pigments and the co-extraction of unwanted compounds, reducing dye purity. Berlanga-Del Pozo et al. [22] highlights that extreme pH conditions, such as highly acidic or highly alkaline, can trigger chlorophyll degradation.

4.5 Application Techniques in Textile Finishing

The application of natural dyes in textile finishing involves several techniques to ensure proper dye penetration, colourfastness, and uniformity. These methods vary based on fabric type, dye solubility, and desired colour intensity.

4.5.1 Pre-treatment of Textile for Better Dye Uptake

Pre-treatment of textiles in the dyeing process is a very important step, as it significantly enhances the uptake of natural dyes, improves colour fastness, and ensures uniform dye distribution. Pre-treatment methods prepare the textile fibres by removing impurities, increasing surface roughness, and creating active sites for better dye adsorption. Raw fabrics often contain natural impurities like wax, pectin, and oils, which hinder dye penetration. Proper pre-treatment enhances fibre reactivity, allowing for better bonding between the dye and the textile structure. The choice of pre-treatment method depends on the fibre type and the dye being used. Various pre-treatments can enhance dye uptake in fabrics, particularly cotton. Alkaline scouring and peroxide bleaching are conventional methods, but enzymatic alternatives show promise [15]. Enzymatic scouring and bleaching offer comparable whiteness and dye exhaustion while reducing environmental impact through lower BOD, COD, and TOC levels [15]. Combined pre-treatment and causticization can improve dye uptake in cotton fabrics [122]. Cationic pre-treatment is another effective approach, enhancing the attraction between negatively charged dye molecules and cotton fibres [49]. This method not only increases dye uptake but can also impart antibacterial properties to the fabric. Enzymatic and cationic pre-treatments emerge as promising alternatives to conventional methods, offering improved dye uptake, reduced environmental impact, and potential for additional functional properties [49].

4.5.2 Mordants

Mordants are chemical agents used in natural dyeing to enhance dye uptake, improve colourfastness, and modify the final colour of fabrics [21]. They are substances, typically metal salts or organic compounds, that form complexes with natural dyes, enhancing their affinity for the fabric and improving colour fastness, brightness, and durability. Mordants help bind the dye molecules to the fabric to prevent fading and enhance resilience to light and washing because many natural dyes don't have a strong affinity for fibres. The use of mordants is particularly important for achieving vibrant and long-lasting shades, especially on protein-based fibres like wool and silk, as well as cellulose-based fibres like cotton. Mordants can be of different types, including metal salts, tannins, and organic substances, each affecting the final hue and durability of the dye.

4.5.2.1 Types of Mordants

Mordants can be classified into three main categories: metal mordants, tannins, and oil mordants. However, in recent times, bio-mordants derived from natural sources have been explored as eco-friendly alternatives to metal salts [21]. Metal mordants, such as alum (potassium aluminium sulphate), iron (ferrous sulphate), tin (stannous chloride), and copper (copper sulphate), are the most commonly used. Alum is particularly popular due to its ability to produce bright and clear colours without significantly altering the natural hue of the dye. Iron, on the other hand, tends to darken colours, making it suitable for achieving deeper shades like blacks and grays. Tannins, which are naturally occurring polyphenolic compounds found in plants, are often used as pre-mordants to improve dye uptake. Oil mordants, such as tannic acid, are primarily used for dyeing protein fibres. Table 4.6 shows types of mordants, examples and effects on dyeing.

Mordants work by forming coordination complexes with the dye molecules and the functional groups present in the textile fibres [79]. For example, alum dissociates in water to release aluminium ions, which bind to both the dye and the fibre, creating a stable bond. In addition to improving colour retention, this procedure makes the dyed cloth more washable and light-fast.

Mordants can be applied through three main methods: pre-mordanting, meta-mordanting, and post-mordanting [107]. Pre-mordanting is the most popular and efficient technique, which entails preparing the cloth with the mordant before dyeing. Wanyama et al. [132] explain that this process often results in better dye uptake and more intense colours. For example, pre-mordanting with aluminium sulphate

Table 4.6 A table showing types of mordants, examples and effect on dyeing

Types of mordant	Examples	Effects on dyeing
Metal salt mordants	Alum (potassium aluminium sulphate), Iron (ferrous sulphate), copper sulphate, tin chloride, chrome (potassium dichromate—less used due to toxicity)	Improves dye absorption and colourfastness. Different metals shift the final colour (e.g., iron darkens shades, alum brightens colours)
Tannin-based mordants	Gall nuts, myrobalan, oak bark, sumac	Used mainly for plant fibres (cotton, linen) as they lack natural protein-binding sites. Helps fix dyes and provides a base for metal mordants
Natural acid mordants	Vinegar (acetic acid), citric acid, tartaric acid	Adjusts pH and enhances dye solubility. Used mainly for plant-based dyes like cochineal and turmeric
Alkaline mordants	Soda ash (sodium carbonate), wood ash, ammonia	Alters pH, affecting colour variation and intensity. Used for certain dyes like indigo
Oil-based mordants	Soy milk, linseed oil	Helps natural dyes adhere to cellulose fibres like cotton. Commonly used in traditional Asian dyeing techniques

achieved the highest dye uptake in cotton dyed with turmeric extract. Other studies have shown that pre-mordanting with citric acid enhances colour durability in cotton fabrics dyed with natural dyes like Cudrania javanensis [34]. On the other hand, in meta-mordanting or simultaneous mordanting, the mordant is added straight to the dye solution, enabling instantaneous interaction between the dye and the cloth. Mai et al. [76] explains that this method can achieve a broad range of colourations and is particularly effective when using wood extracts as dyes. Lozano et al. [75] adds that this method can lead to different colour outcomes depending on the dye and fabric used. Lastly, to fix the colour, post-mordanting refers to treating the dyed cloth with the mordant. Moreover, Önal et al. [94] explains that this method can improve certain fastness properties, such as wash fastness, but may result in lower colour depth compared to pre-mordanting and meta-mordanting. Again, Lozano et al. [75] explains that applying the mordant after dyeing can also affect the final colour. In the case of onion skin dye, post-mordanting with iron sulphate provided the best colour fastness. Interestingly, Dulo et al. [37] explains that the choice of mordant and method can affect the final colour and fastness properties, making it crucial to select the appropriate combination based on the desired outcome. In the next sub-topic, the impact of different mordants on colour shades and fastness on dyed fabrics will be considered.

4.5.2.2 Impact of Different Mordants on Colour Outcomes and Fastness Properties

The type of mordant used significantly impacts the colour and fastness of natural dyes. Mordants bind dyes to fabric, improving dye absorption and enhancing fastness properties like wash, light, and rub resistance. However, different mordants can produce a variety of shades from the same dye, and they also impact the durability of the colour against washing and light exposure. In natural dyeing processes, this is essential for achieving reliable and durable results.

Mordants are typically categorized into three main types: metallic salts, tannins, and natural mordants. Each type of mordant interacts differently with both the fibre and the dye, leading to variations in the final colour and fastness. Lozano et al. [75] explains that metallic mordants can significantly alter the colour and intensity of the dye. For instance, iron sulphate tends to produce darker shades, such as dull yellow-green on alpaca fibres and darker colours on cotton fabrics dyed with onion skin extract. Furthermore, their study explains that natural mordants like aloe vera and lemon generally result in lighter, softer colours. They are less effective in producing intense shades compared to metallic mordants.

Zubairu and Mshelia [137] explains that mordants not only influence the colour shade of dyes but also enhance the colourfastness properties of dyes. Fabrics treated with mordants generally show improved wash and light fastness compared to untreated fabrics. In a study, the use of ferrous sulphate as a mordant was found to produce darker colours and improve fastness to rubbing, washing, and sweating, as well as enhance the physical properties of the dyed material [135]. Natural mordants like myrobalan, lodhra, and pomegranate have been shown to improve colour fastness while also offering environmental benefits. For instance, lodhra allows dyes to display their true colours, whereas pomegranate provides excellent antibacterial activity [60]. Additionally, it is noted that mordants like copper sulphate, potassium dichromate, and aluminium potassium sulphate are frequently employed to enhance dye absorption and fastness qualities. These mordants enhance the chemical bonding between the dye and the fabric, resulting in more vibrant and durable colours [56, 82]. Table 4.7 summarises how different mordants influence colour outcomes and fastness properties in textile dyeing.

The choice of mordant greatly influences the colour outcomes of dyed fabrics as well as their fastness properties. Even though metallic mordants like aluminium, iron, copper, and tin can improve aspects like colour intensity, wash fastness, and light fastness, they also have health and environmental risks. It is important to select mordants carefully, based on dye source and desired properties, in order to achieve both high-quality colour results as well as satisfactory fastness properties.

Table 4.7 A table showing the effects of different mordants on colour outcomes and fastness properties

Mordant	Colour outcome	Wash fastness	Rubbing fastness	Light fastness	Example with common dyes
Alum (potassium alum)	Light, pastel tones, clear and bright colours	Good wash fastness	Moderate rubbing fastness	Good light fastness	Madder → Reddish-orange; Turmeric → Golden yellow
Iron (iron sulphate)	Dark, muted shades (e.g., grays, blacks)	Moderate wash fastness	Poor rubbing fastness (especially with moisture)	Poor light fastness (fades under sunlight)	Logwood → Charcoal grey; Oak galls → Dark brown
Copper (copper sulphate)	Rich, vibrant shades (especially in blue-green tones)	Moderate wash fastness	Moderate rubbing fastness	Good light fastness (resistant to fading)	Weld → Olive green; Onion skins → Mossy yellow
Tin (stannous chloride)	Bright, vivid colours (yellows, oranges, reds)	Good wash fastness	Good rubbing fastness	Poor light fastness (fades quickly under sunlight)	Cochineal → Vibrant red; Brazilwood → Deep pink
Tannins (natural bio-mordant)	Earthy tones, muted colours (dependent on source)	Moderate wash fastness	Moderate rubbing fastness	Moderate light fastness	Pomegranate rind → Mustard yellow; Myrobalan → Warm beige

4.5.3 Dyeing Methods (Batch Dyeing, Continuous Dyeing)

Dyeing methods are critical in determining the efficiency, uniformity, and sustainability of natural dye application on textiles. It can be performed using batch (discontinuous) or continuous methods, with various techniques such as exhaust, pad, and printing [119]. The dyeing process typically consists of preparation, dyeing, and finishing stages, with temperature and time being critical factors. These primary dyeing methods have their respective advantages and limitations. The kind of fibre, the dye's composition, the size of manufacturing, and the intended results all influence the process selection. Additionally, the choice of method influences dye uptake, colour uniformity, and overall efficiency of the process.

4.5.3.1 Batch Dyeing

Batch dyeing, also known as exhaust dyeing, is the most common method for applying natural dyes to textiles. In this process, the fabric is immersed in a dye bath for a specific period, allowing the dye to gradually penetrate the fibres. The dye bath typically contains water, natural dyes, and mordants (if required), pH level. Additionally, controlled temperatures are used during the process [74]. Temperature, pH, and mordants play crucial roles in achieving the desired shade and colourfastness. Batch dyeing is highly versatile and suitable for small to medium-scale production, making it ideal for artisanal and craft-based dyeing. The batch dyeing process can be further categorized into different methods, such as cold pad-batch dyeing, which is widely used for its energy efficiency and environmental benefits.

4.5.3.2 Continuous Dyeing

Continuous dyeing is a large-scale industrial process where the fabric passes through a series of dye application and fixation stages continuously. This method is typically used for synthetic dyes, but can be adapted for natural dyes with modifications. This method is ideal for large-scale production, improving efficiency and reducing water and dye consumption compared to batch dyeing [66]. The process typically involves padding, where fabric is passed through a dye solution, squeezed between rollers to remove excess dye, and then subjected to heat or chemical fixation. While less common for natural dyes, continuous dyeing offers potential for high-volume production.

4.5.4 Post-dyeing Processes (Fixation, Washing, and Finishing)

Post-dyeing processes are very considerable in natural dyeing of textiles, as they ensure the stability, durability, and quality of the dyed fabric. These processes include fixation, washing, and finishing, each of which plays a critical role in enhancing colour fastness, removing excess dye, and improving the overall performance of the dyed material. It is explained that natural dyes often have poor affinity for textile substrates and require mordants or fixers for proper fixation [47]. Therefore, proper post-dyeing treatment is particularly important for natural dyes, which can be less stable than synthetic dyes and require additional care to achieve satisfactory results.

4.5.4.1 Fixation

Fixation is the process of ensuring that dye molecules are firmly bonded to the textile fibres, enhancing colour fastness and preventing colour fades during subsequent use and washing. Natural dyes often exhibit poor bonding with textile fibres, necessitating effective fixation methods. For natural dyes, fixation often involves the use of mordants or fixing agents that form stable complexes with the dye and the fibre. However, the use of certain mordants has raised environmental and health concerns due to their toxicity [10]. In order to mitigate this problem, researchers are exploring sustainable fixation methods. For instance, Hossain et al. [53] in a study examined a sustainable, green way to enhance the fixation of a natural functional dye onto cotton fabric, minimising the environmental effect of conventional mordanting methods. Additionally, heat treatment, such as steaming or curing, is also commonly used to enhance fixation. Ali et al. [10] highlight that heat treatments, such as steaming or curing, are used to enhance the fixation of natural dyes on textiles. These processes help in the penetration of the dye into the fibre and improve the overall colour performance.

4.5.4.2 Washing

Post-dyeing washing is intended to remove unfixed dye molecules, mordants, and any residual chemicals from the dyed fabric. Proper washing ensures that excess dye does not bleed during subsequent laundering and helps in achieving the desired colourfastness. In a study, it was revealed that naturally dyed wool and silk fabrics with olive peel extract demonstrated satisfactory colourfastness when appropriate mordants and washing conditions were applied [16]. The washing process typically involves rinsing the dyed fabric in cold water to remove surface dye, followed by washing in warm water with a mild detergent or soaps to eliminate any remaining unfixed dye. However, it is important to avoid harsh detergents and high temperatures, as these can cause the natural dyes to fade.

4.5.4.3 Finishing

Finishing processes are applied to enhance the fabric's appearance, handle, and performance characteristics. In the context of natural dyeing, finishing can involve several steps:

(a) *Softening*: Natural dyed fabrics may feel stiff after dyeing and washing. Applying natural softeners, such as aloe vera or natural oils, can improve the fabric's hand feel [109].
(b) *Calendering*: Through the use of heated rollers, this mechanical technique gives the cloth a smooth, glossy surface that increases its lustre. Ali et al. [11] adds that

natural dyes produce duller finishes compared to synthetic dyes. Calendaring helps in overcoming this limitation by imparting a glossy finish.

(c) *Sanforizing*: A pre-shrinking process that ensures the fabric maintains its dimensions during use and laundering.

4.5.5 Advantages of Using Natural Dyes in Textile Finishing

Natural dyes have been integral to textile colouration for centuries. In recent times, there has been a resurgence in their use due to growing environmental and health concerns associated with synthetic dyes. The advantages of using natural dyes in textile finishing are multifaceted, encompassing environmental sustainability, health benefits, unique aesthetic qualities, and cultural significance.

4.5.5.1 Environmental Sustainability

One of the primary benefits of natural dyes is their eco-friendliness. They are biodegradable and derived from renewable resources, which significantly reduces the ecological footprint of the textile industry. Unlike synthetic dyes, which often rely on petrochemicals and can contribute to water pollution, natural dyes minimize environmental impact [33]. Additionally, many natural dye sources can be cultivated without synthetic fertilizers or pesticides, promoting sustainable agricultural practices.

4.5.5.2 Health Benefits

Natural dyes are safer for both textile workers and customers since they are often non-toxic and non-allergenic. The absence of harmful chemicals in the dyeing process reduces the risk of skin irritations and other health issues. For instance, natural dyes like henna have been shown to impart antibacterial properties, and certain natural dyes contain antioxidants that can neutralise free radicals, highlighting the health advantages of natural dyes [28].

4.5.5.3 Unique Aesthetic Qualities

Textiles finished with natural dyes often exhibit unique colours and textures that are difficult to replicate with synthetic dyes. The natural variations in hue and tone contribute to the distinctiveness and appeal of naturally dyed fabrics. Natural dyes, sourced from plants, insects, and microorganisms, offer a diverse colour palette, like dragon fruit extract used for dyeing wool, ensuring good colour fastness [114].

This uniqueness is particularly valued in traditional crafts and high-quality textile products.

4.5.5.4 Functional Properties

Beyond aesthetics, certain natural dyes impart additional functional properties to textiles. For example, some plant-based dyes have been found to provide ultraviolet (UV) protection and antibacterial properties. Pizzicato et al. [97] explains that the functional value of cotton fabrics dyed with plant extracts was increased by their exceptional UV protection and excellent insect repellency.

4.5.5.5 Cultural and Historical Significance

The use of natural dyes is deeply rooted in various cultural traditions worldwide. They play a crucial role in preserving cultural heritage and traditional craftsmanship. For instance, the recent inauguration of Pakistan's first dedicated textile museum in Karachi showcases centuries-old textile traditions, many of which utilize natural dyes, highlighting their cultural importance [39].

The integration of natural dyes in textile finishing presents significant benefits, including environmental sustainability, health advantages, unique aesthetics, and the preservation of cultural heritage. As the textile industry continues to seek sustainable and responsible practices, natural dyes offer a compelling alternative to synthetic counterparts. While natural dyes offer numerous advantages, there are challenges to their widespread adoption. The next subtopic seeks to discuss some major challenges in natural dye application.

4.6 Challenges in Natural Dye Application

In textile applications, natural dyes provide a more environmentally friendly option than synthetic dyes. Additionally, natural dyes offer numerous advantages, such as environmental sustainability, cultural significance, and unique aesthetic qualities, etc., however, their adoption is accompanied by several challenges that impact their practicality and scalability in modern textile production. The following pointers explain some major challenges in natural dye application.

4.6.1 Colour Consistency and Reproducibility

Achieving uniform and predictable colours with natural dyes is challenging due to inherent variations in raw materials. Factors such as plant species, soil conditions, climate, and harvest time can lead to significant differences in dye composition, resulting in inconsistent hues across dye batches. This variability complicates quality control in textile manufacturing. Ahmed and Anwar [7] affirm that consistency in colour reproduction is another significant challenge, as natural dyes can vary greatly between batches due to differences in raw material sources and extraction methods.

4.6.2 Colour and Lightfastness Issues

Natural dyes often exhibit inferior colour fastness compared to synthetic dyes. Exposure to light, washing, and rubbing can lead to significant fading over time. For instance, many natural dyes have a lightfastness rating below BS grade 5, indicating a propensity to fade when exposed to sunlight. It is further explained that most natural dyes, except for a few like indigo, exhibit poor light fastness, which limits their use in applications when exposed to sunlight [24]. This limitation affects the longevity and appearance of naturally dyed textiles.

4.6.3 Resource Intensity and Environmental Impact

Producing natural dyes can be resource-intensive. Large quantities of plant material are often required to extract sufficient dye, leading to concerns about land use and the sustainability of sourcing practices. Doty et al. [36] explain that the limited availability of raw materials and the need for large quantities to achieve desired colour intensity further exacerbate cost issues. Additionally, many natural dye sources are seasonal or region-specific, making it difficult to ensure a consistent and reliable supply. Overharvesting of certain plants for dye extraction can also lead to ecological imbalances and threaten biodiversity.

4.6.4 Scalability and Economic Viability

Scaling natural dye production to meet industrial demands presents significant challenges. The labour-intensive nature of cultivating, harvesting, and processing natural dye sources, coupled with variability in dye yield and quality, makes it difficult to achieve economies of scale compared to synthetic dyes [71]. Consequently, naturally dyed textiles often come at a higher cost, limiting their competitiveness in the mass market.

4.6.5 Limited Colour Range and Market Acceptance

Natural dyes offer a narrower colour palette than synthetic dyes, which boast a wide spectrum of vibrant colours [97]. This limitation may not meet the aesthetic preferences of consumers accustomed to the bright and consistent colours achievable with synthetic dyes [81]. Additionally, the perception of natural dyes as less durable or less fashionable can hinder their acceptance in the mainstream market.

Although natural dyes are a more eco-friendly option than synthetic ones, addressing these challenges is essential to their wider use in the textile sector. Despite these hurdles, the increasing demand for environmentally friendly and sustainable fabrics offers natural dyes a chance to establish a larger presence in the market. This situation leads to the next topic, where these challenges are being addressed to promote natural dyes application in the textile industry.

4.7 Innovations and Advancements

In recent years, innovations and advancements in natural dye production have addressed many of the challenges associated with their application, making them more viable for the textile industry. These advancements focus on improving extraction efficiency, enhancing colour fastness, reducing environmental impact, and scaling up production.

4.7.1 Recent Research and Developments in Natural Dye Technology

Recent research and developments in natural dye technology have focused on overcoming the limitations of traditional natural dyes, such as limited colour range, low colour fastness, and scalability issues. These advancements aim to make natural dyes more competitive with synthetic dyes while maintaining their eco-friendly and sustainable characteristics.

4.7.1.1 Advanced Extraction Techniques

Recent studies have explored advanced extraction techniques to address environmental concerns and improve the efficiency and yield of natural dyes. Enzyme-assisted extraction, ultrasound-assisted extraction (UAE), and microwave-assisted extraction (MAE) are gaining popularity due to their ability to reduce extraction time, increase pigment yield, and decrease solvent consumption. Ultrasonic and microwave-assisted extractions have shown promise in accelerating extraction rates at lower temperatures compared to conventional methods. Additionally, enzymatic extraction using cellulase, amylasc, and pectinase has also been explored to enhance dye release from plant tissues [51]. Other techniques include solvent extraction, supercritical fluid extraction, and acid/alkali extraction [106]. These methods aim to maximize colour yield while reducing solvent use, time, and energy consumption [5].

4.7.1.2 Development of Eco-Friendly Mordants

Traditional mordants, such as alum and iron, can pose environmental and health risks. Recent research has focused on developing bio-mordants derived from plant sources, such as tannins, pomegranate rind, and myrobalan, as well as enzyme-based mordants that are safer and more sustainable. Benli [21] explains that bio-mordants can be classified into three main groups: natural or bio-mordants, oil mordants, and new-generation non-vegetable-based mordants. The use of these eco-friendly mordants in combination with natural dyes offers a sustainable approach to textile coloration.

4.7.1.3 Genetic Engineering and Microbial Production

Advances in biotechnology have enabled the production of natural dyes through genetic engineering and microbial fermentation. Genetically modified microorganisms, such as bacteria and yeast, can produce high yields of pigments like indigo, carotenoids, and anthocyanins, offering a scalable and sustainable alternative to plant-based extraction. Additionally, genetically engineered microorganisms (GEMs) are essential for attaining total dye degradation and resolving environmental issues related to synthetic dyes [131]. While challenges remain, such as understanding complex metabolic pathways, genetic engineering continues to advance the field of natural dye production and degradation. Genetic engineering and microbial production offer promising avenues for natural dye production. Metabolic engineering of microorganisms has enabled large-scale, sustainable manufacturing of natural colourants [101].

4.7.2 Use of Nanotechnology in Enhancing Dye Performance

Nanotechnology improves natural dye performance, notably colour fastness and durability. Nano-encapsulation of natural dyes and the use of nanoparticles as mordants have shown promise in improving dye fixation, colour vibrancy, and durability to fabrics. Moreover, nanostructured carriers facilitate a more uniform dispersion of dyes on fibres, leading to consistent coloration and improved binding between the dye and textile substrate [116]. These nanotechnology applications not only improve dyeing performance but also offer additional functional properties such as UV protection, deodorizing, and antibacterial qualities. For instance, Cellulose nanocrystals (CNCs) used as mordants improved colour fastness to light, washing, and rubbing for cotton fabrics dyed with natural extracts. Furthermore, advancements in nanotechnology also allow for the controlled release of dye molecules, which can tailor the intensity and longevity of colours on fabrics [97].

4.7.3 Case Studies of Successful Applications in Sustainable Fashion

The integration of innovations and advancements in natural dye technology into sustainable fashion has led to several successful case studies. These examples demonstrate how cutting-edge techniques and eco-friendly practices can be applied to create sustainable, high-quality textiles that meet consumer demand for environmentally responsible products.

4.7.3.1 Patagonia: Waste Valorisation and Natural Dyes

Patagonia, a leader in sustainable fashion, has incorporated natural dyes and waste valorization into its production processes. Pomegranate rinds and onion peels are examples of agricultural by-products that the firm utilises to make natural colours for its apparel collections. Pelargonium graveolens hydrodistillation solid waste is utilized as a natural wool fabric dye, achieving desirable colour strength and polyphenol content, resulting in shades of brown and beige on wool fabrics [86]. Additionally, by adjusting the waste-to-product ratio and using different mordants, coffee waste achieved high colour fastness and reduced environmental waste, offering a sustainable alternative for the denim industry [17].

4.7.3.2 Stella McCartney: Biotechnology and Microbial Dyes

Stella McCartney, a pioneer in sustainable luxury fashion, has partnered with biotechnology companies to develop microbial-based dyes. These dyes are produced through fermentation using genetically engineered microorganisms, offering a scalable and sustainable alternative to traditional dyeing methods. The company has successfully created vibrant, durable colours without the environmental impact associated with synthetic dyes. In a study, Quijano et al. [104] highlight that biotechnology provides the fashion business and society with several methods to establish connections with nature, proposing “taming,” “rewilding,” and “speculating” as some of its functions in fashion practice.

These case studies demonstrate the successful application of innovations and advancements in natural dye technology within the sustainable fashion industry to create high-quality, sustainable textiles. In addition to lessening the harmful effects of fashion manufacturing on the environment, these initiatives also satisfy the rising demand from consumers for eco-friendly goods. By continuing to innovate and adopt sustainable practices, the fashion industry can play a significant role in promoting environmental responsibility and sustainability.

4.8 Environmental and Socio-economic Impacts

The textile industry's usage of natural dyes has an effect on the environment and the economy. Natural colours provide an alternative to synthetic dyes that can reduce negative socioeconomic and environmental effects. These impacts are discussed in the following points.

4.8.1 Benefits of Natural Dyes in Reducing Pollution

In comparison to synthetic dyes, which are frequently associated with health risks and environmental degradation, natural dyes offer notable advantages in reducing pollution. Natural dyes offer a sustainable alternative to synthetic dyes in textile production, contributing to reduced pollution. They are biodegradable and produce fewer pollutants compared to synthetic dyes. The use of natural dyes can contribute to a cleaner and more sustainable textile industry by minimizing the release of harmful chemicals, reducing water pollution, and promoting eco-friendly practices [40]. The use of safer mordants, such as aluminium sulphate, further minimizes pollution and health hazards compared to more toxic alternatives [61]. Additionally, Rosyida and Masykuri [111] explain that the use of natural dyes can minimize water pollution, as they produce lower levels of biological oxygen demand (BOD), chemical oxygen demand (COD), and total suspended solids (TSS) compared to synthetic dyes.

4.8.2 Economic Implications for Local Communities and Artisans

Natural dyeing is deeply rooted in cultural heritage and traditional craftsmanship. Reviving these practices not only preserves indigenous knowledge but also offers income opportunities for rural and marginalized communities. Significantly, natural dyes are gaining popularity in niche markets [8], and this has given small-scale artisans and weavers the economic benefits from niche markets that value handcrafted, eco-friendly textiles. For instance, in Indonesia, developing Indigofera spp. cultivation for blue dye production has the potential to increase community incomes and employment, especially in areas with suitable land [6].

4.8.3 Role of Natural Dyes in Promoting Biodiversity

When produced and harvested sustainably, natural dyes can promote biodiversity. Dye plants such as indigo, madder, and turmeric can be grown in organic agricultural systems that improve soil health and support local ecosystems. Ali et al. [11] explains that growing plants that produce dyes can support biodiversity by preserving ecological balance and providing habitat for a variety of species.. Furthermore, these dye plants often encourage regenerative agricultural practices, which can restore degraded land and reduce reliance on chemical fertilizers and pesticides. Andriamanantena et al. [13] explains that the exploration of underutilized plants, such as the munj sweet cane and mangrove species, for dye production helps in reducing the pressure on commonly used dye plants and supports biodiversity by diversifying the plant species used in the industry. Moreover, natural dyes often require less water and result in fewer pollutants compared to synthetic dyes, leading to reduced environmental impact. This eco-friendly aspect of natural dyes contributes to healthier ecosystems and supports overall biodiversity [112].

Natural dyes present a promising route toward reducing the environmental burden of textile dyeing and empowering local economies. They lower toxic chemical inputs, support biodiversity, and revive traditional crafts that offer socio-economic benefits to rural communities. Although challenges in scalability, resource management, and process efficiency remain, ongoing research and sustainable practices are gradually mitigating these issues. In this way, natural dyes not only offer a more environmentally friendly alternative to synthetic dyes but also contribute to cultural preservation and economic upliftment in developing regions. However, the safety and regulatory issues relating to natural dyes are a major concern that will be discussed in the next subtopic.

4.9 Safety and Regulatory Issues

The use of natural dyes, while considered more sustainable and environmentally friendly than synthetic dyes, still involves safety and regulatory challenges. The potential risks associated with natural dyes primarily stem from their chemical composition, application methods, and the presence of allergens. It is emphasised that before market entry, natural dyes must undergo toxicity tests, quality assessments, and regulatory approvals [71].

4.9.1 Health Hazards Associated with Natural Dyes

Natural dyes are not completely devoid of health risks, even though they are typically thought to be safer and more ecologically friendly than synthetic dyes. The potential health risks associated with natural dyes arise from the handling of dye materials, the use of mordants, and the presence of naturally occurring toxic compounds in some dye sources. Some natural dyes can be contaminated with harmful substances during extraction or processing. For instance, the use of metallic mordants, which are often required to fix the dye to fabrics, can introduce toxicity. For instance, mordants like potassium dichromate are known to be toxic [27, 92]. Additionally, respiratory issues may arise from breathing in mordant fumes or powders when dyeing. Furthermore, certain mordants, including tin and chromium, are extremely poisonous and can have serious negative health effects if ingested or breathed.

4.9.2 Environmental Impact of Extraction Processes

In general, natural colours are thought to be more eco-friendly than synthetic dyes, however, the extraction processes involved can still pose environmental challenges if not managed properly. The dyeing process, whether using natural or synthetic dyes, is water-intensive. Large quantities of water are required for dye extraction, fabric preparation, and rinsing. This extensive water use can lead to significant environmental impacts, especially in regions where water is scarce [100]. Although natural dyes are biodegradable, the dyeing process can still produce coloured wastewater. Wastewater is coloured by even small amounts of residual dyes, which, if improperly handled, can have an impact on aquatic ecosystems [105]. Furthermore, cultivating dye-producing plants requires land, water, and other agricultural inputs. If not managed sustainably, this cultivation can lead to deforestation, habitat loss, and soil degradation, negatively impacting biodiversity.

4.9.3 Risk of Allergens and Irritants in Natural Dyes

The safety of some natural dyes and mordants varies according to the extraction technique and source, and some of them might trigger allergic responses. Natural dyes can cause allergic reactions in some individuals. For example, cochineal red, a natural dye, has been associated with IgE-mediated hypersensitivity, leading to allergic reactions such as anaphylaxis in sensitive individuals [14]. Additionally, some natural dyes, like henna, derived from the Lawsonia inermis plant, have been associated with allergic reactions. While pure henna is generally considered safe, adulterated forms, especially "black henna" containing para-phenylenediamine (PPD), can cause severe allergic responses [129]. In other studies, it is believed that some natural dyes and mordants can release volatile compounds when heated, which may cause respiratory irritation. For instance, direct contact with metal-based mordants can cause skin irritation and allergic reactions.

The textile industry's shift towards natural dyes is fueled by environmental concerns and consumer preference for eco-friendly options. Varying regional regulations on natural dye use in industries like textiles and food significantly impact industry practices and consumer choices. For example, the Registration, Evaluation, Authorisation, and Restriction of Chemicals (REACH) law in the European Union requires businesses to identify and control the hazards related to the chemicals they produce and sell. Thiyagarajan et al. [127] highlights that in regions like Europe, strict eco-standards have been imposed to mitigate the adverse effects of synthetic dyes, encouraging the use of natural dyes. While REACH primarily addresses synthetic chemicals, its stringent standards have prompted textile manufacturers to explore natural dyes as safer alternatives. This shift aligns with the EU's emphasis on environmental sustainability and consumer safety.

In the United States, the Food and Drug Administration (FDA) oversees the use of colour additives in food products. Recent scrutiny over artificial dyes, such as Red Dye No. 3, has led to bans and increased consumer demand for natural alternatives. For instance, Aldi has proactively removed synthetic colours from its products, opting for plant-derived colourings like annatto extract to meet consumer preferences and address health concerns. Additionally, in Asia, regulatory approaches to dyes vary by country. Thailand, for example, has implemented standards for synthetic dyes, reflecting a commitment to regulating dye use in the textile industry. These standards, effective since August 2023, aim to ensure product safety and environmental protection. Such regulations encourage manufacturers to consider natural dyes as viable alternatives to comply with environmental and safety standards. Ferreira et al. [41] explain that China, Japan, and Korea are actively working to bridge the regulatory gap with the EU's REACH system. Japan has a well-established chemical control system, while Korea is incorporating REACH-like aspects into its Toxic Chemicals Control Act. China is also making strides with its measures on the Environmental Management of New Chemical Substances.

The evolving regulatory landscape has led industries to adopt more sustainable dyeing practices. Manufacturers are investing in research and development to improve the stability and vibrancy of natural dyes, addressing challenges such as colour consistency and production cost. Innovations include exploring dyes derived from bacteria, algae, and food waste, as well as adopting technologies like digital and waterless dyeing methods to reduce environmental impact.

While regional regulations and consumer preferences are driving industries toward the adoption of natural dyes, the textile industry's use of natural dyes raises a number of safety and regulatory concerns that must be resolved to safeguard customers, employees, and the environment. Health risks from mordants and dye materials, environmental safety concerns, and regulatory challenges are key issues that require attention. When best practices are adopted, safety measures implemented, with developed clear and consistent regulations are developed, the natural dye industry can overcome these challenges and promote the safe and sustainable use of natural dyes.

Moreover, continued research and innovation in natural dye technology in the future will further enhance safety and compliance, making natural dyes a viable and responsible choice for the textile industry. The next subtopic guides readers to the future directions of natural dye production and application.

4.10 Future Directions

The resurgence of natural dye manufacturing is driven by increased customer demand for sustainable textiles, offering a viable alternative to synthetic dyes in the textile industry's efforts to reduce its environmental footprint. The shift towards natural dyes is not only motivated by their environmental benefits but also by their potential to impart additional functionalities to textiles, such as antimicrobial and UV protection properties. This renewed interest is fostering innovation and shaping future directions in the industry.

4.10.1 Emerging Trends in Natural Dye Research

Recent research in natural dyes is focusing on enhancing sustainability, improving application techniques, and expanding the range of sources. The following points give the emerging developments in natural dye research in the textile industry.

(a) *Eco-friendly Extraction and Mordant Techniques*: It is explained that the use of natural mordants is being investigated to replace harmful metallic mordants, enhancing the eco-friendliness of the dyeing process [28]. Researchers are developing sustainable extraction techniques that minimize environmental impact, such as ultrasonic and gamma irradiation, to reduce the use of harmful solvents and lower energy consumption during the dye extraction process [77, 97]. Furthermore, advances in green chemistry and extraction methods are improving the efficiency and sustainability of natural dye production [98].

(b) *Exploration of Novel Natural Sources*: There is a growing trend of using agricultural waste and by-products as sources of dyes. This broadens the range of natural colours accessible and gives waste materials a sustainable use [10, 28]. Moreover, research innovations include extracting pigments from food waste, contributing to waste reduction and offering new avenues for natural dye production [18].

(c) *Multifunctional Applications*: Natural dyes are being developed for applications beyond textiles, including cosmetics and food colouring, due to their non-toxic and biodegradable nature [65]. Additionally, natural dyes are increasingly being used for electrical applications, such as photosensitizers in dye-sensitized solar cells (DSSCs) [32].

4.10.2 *Potential for Integration with Modern Textile Technologies*

There is a lot of potential for producing sustainable, superior, and inventive textile goods through the combination of natural dyes with contemporary textile technology. The integration of natural dyes with advanced technology can improve market appeal, performance, and efficiency as the textile industry strives to minimise its environmental impact.

(a) *Biotechnology and Microbial Dyeing*: The use of microorganisms to produce dyes is gaining attention. This biotechnological approach offers a sustainable alternative to traditional dye sources, potentially reducing reliance on agricultural resources and decreasing environmental impact [28]. Genetically modified microorganisms can be engineered to produce and fix dyes directly onto textiles, providing a scalable and sustainable alternative to traditional dyeing methods.
(b) *Digital Printing and Plasma Technology*: Emerging technologies like digital textile printing with biodegradable inks and plasma technology are being explored to enhance the application of natural dyes in modern textile processes [29]. Digital printing technology allows for the precise application of dyes, reducing waste and enabling intricate designs. Integrating natural dyes with digital printing can create eco-friendly, high-resolution patterns on textiles.
(c) *Nanotechnology Applications*: The incorporation of nanotechnology, such as the use of zeolites, is being researched to improve the colour stability and fastness of natural dyes, making them more viable for commercial applications. Nanotechnology has emerged as a promising field for dye remediation in the textile and leather industries, addressing environmental and health concerns associated with dye effluents [68]. Nano-encapsulation and the use of nanoparticles as mordants can enhance dye stability and adhesion to fabrics. Nanoparticles can enhance the binding of natural dyes to textile fibres, improving dye uptake and durability.
(d) *Renewable Energy*: In dye-sensitized solar cells (DSSCs), a promising renewable energy technology, natural dyes are used as photosensitisers [32]. Dyes derived from sources such as anthocyanins, chlorophyll, and carotenoids exhibit high light absorption and eco-friendliness, making them suitable for sustainable energy generation. Meneghetti et al. [85] explains that natural dyes in DSSCs offer conversion efficiencies comparable to synthetic dyes, with anthocyanin-sensitized cells reaching up to 0.64% in some studies. Additionally, Mejica et al. [83] explains that chlorophyll and carotenoid dyes extracted from plants, such as ivy gourd leaves and turmeric, effectively absorb light, with chlorophyll demonstrating the strongest absorbance. However, research is needed to improve the stability and energy conversion efficiency of natural dyes in DSSCs.
(e) *Smart Textiles and Functional Materials*: Natural dyes are being integrated into smart textiles to provide additional functionalities, such as UV protection, antimicrobial properties, and thermochromic or photochromic effects. Pargai [95] explains that natural dyes can improve the UV protection of textiles, making

them ideal for outdoor clothing. For example, silk fabrics dyed with natural dyes exhibit significantly enhanced UPF values.

The integration of natural dyes with advanced technologies is opening new frontiers in textile innovation. Modern research is exploring their application in smart textiles, nanotechnology, biotechnology, and sustainable energy solutions. Natural dyes are increasingly being incorporated into wearable technology, where they function as eco-friendly sensors for temperature and pH detection. For example, natural dyes are used as light-absorbing materials in dye-sensitized solar cells (DSSCs), providing a sustainable substitute for their synthetic equivalents. This innovation aligns with SDG 7 (Affordable and Clean Energy) by supporting renewable energy development. DSSCs offer low-cost production, flexibility, and reduced environmental impact due to the potential use of non-toxic materials, making them a promising alternative to conventional photovoltaics. Bartolome et al. [19] explains that in contrast to silicon-based solar cells, which need high temperatures and high vacuums, DSSCs can be produced easily and with little energy. Furthermore, reducing the environmental impact of dyeing processes and supporting renewable energy applications like DSSCs supports SGD 13 (Climate action).

Another significant development is nano-enhanced natural dyeing, where nanomaterials improve the colour fastness, antimicrobial properties, and UV protection of dyed fabrics. This has practical implications for medical textiles (such as antimicrobial wound dressings) and functional clothing, contributing to SDG 3 (Good Health and Well-being). Additionally, enzymatic and microbial dyeing techniques are enhancing the sustainability of natural dye applications, reducing water and chemical consumption, and supporting SDG 12 (Responsible Consumption and Production).

These findings suggest that industries adopting natural dyes can gain competitive advantages by aligning with environmental regulations and consumer preferences for sustainable products. Sabrina [113] suggests that due to customer preferences for eco-friendly and non-toxic products, there is a developing niche market for fabrics dyed with natural dyes. Unlike synthetic dyes, which can be carcinogenic and non-biodegradable, natural dyes are safe for both humans and the environment. Additionally, their use in textiles, cosmetics, and food products contributes to healthier lifestyles and reduces exposure to harmful chemicals, advancing SDGs 3 and 12.

Furthermore, the growing restrictions on hazardous synthetic dyes push industries toward safer alternatives, impacting regulatory frameworks and business strategies, thereby promoting SDG 8 (Decent Work and Economic Growth). Chungkrang et al. [31] revealed that this shift encourages innovation in dye production and wastewater treatment technologies, creating new business opportunities and potentially leading to economic growth. It also boosts local economies by increasing demand for dye-producing plants and the use of agricultural waste, encouraging SDG 15 (Life on land).

The future of natural dyes in high-tech applications is promising, with advancements in smart textiles, sustainable energy, and nanotechnology enhancing their viability. However, further research is needed to optimize scalability, durability, and industrial adoption. Natural dyes have the potential to transform the textile sector and support global environmental and economic goals by fusing scientific advancement with sustainability principles.

4.10.3 Advocacy for Policy Changes Supporting Natural Dye Use

Advocacy for policy changes supporting the use of natural dyes is gaining momentum due to their environmental and health benefits compared to synthetic dyes. This advocacy is supported by research highlighting the therapeutic properties and antimicrobial activities of natural dyes, which further enhance their appeal [128]. The shift towards natural dyes is driven by increasing consumer awareness and environmental regulations, which have already led to bans on certain synthetic dyes in countries like Germany and India [31].

(a) *Promoting Eco-Friendly Practices*: Policies promoting natural dyes in textiles are being pursued to lessen pollution from synthetic dyes. A proposal emphasizes the need for commercializing natural eco-dyes to mitigate the adverse environmental impacts of traditional dyeing processes [58]. Convincingly, the European Union is expected to soon restrict some toxic dyes, and India has already prohibited the use of specific azo dyes and provided enough legal teeth for a panel to take action against individuals who use them [31].
(b) *Addressing Health Concerns*: Unlike synthetic colours, which are frequently non-biodegradable and carcinogenic, natural dyes are biodegradable and do not provide toxic byproducts [67]. Additionally, there is a growing movement to eliminate artificial dyes from consumer products due to potential health risks. For instance, protesters have demanded the removal of artificial colours from popular cereals, citing concerns over behavioural issues in children. This advocacy underscores the need for policies that favour natural alternatives in food and textiles [38].
(c) *Supporting Local Economies*: Natural dyes offer economic opportunities, especially in developing countries, through the sustainable harvest and sale of dye-producing plants [128]. For instance, projects like the Baltimore Natural Dye Initiative explore how growing and utilising natural dyes affects culture and the economy. Such initiatives advocate for policies that support local agriculture and small businesses involved in natural dye production, promoting economic development and cultural heritage [99].
(d) *Encouraging Sustainable Practices*: Educational institutions are proposing projects like natural dye gardens to address global challenges and demonstrate a commitment to sustainability. These initiatives advocate for policy support in integrating natural dye education into curricula and funding research on sustainable dyeing practices [89].

The future of natural dye production is bright, with numerous advancements on the horizon. It can be explained that the future of natural dye production is being shaped by technological innovations, sustainable practices, evolving market dynamics, and increased community engagement. These developments aim to make natural dyes a more viable and attractive option for both producers and consumers, contributing to a more sustainable textile industry. However, by addressing present issues and embracing these future paths, the natural dye sector can significantly contribute to the advancement of sustainable textile manufacturing and satisfy the rising demand for environmentally friendly goods. Additionally, further research and cooperation will be necessary to fully utilise natural dyes in the textile industry.

4.11 Conclusion

Natural dyes in textile finishing, while historically significant, environmentally beneficial, and promising for innovation, still face challenges in widespread adoption. Natural dyes are a sustainable substitute for synthetic dyes and may be obtained from a range of sources, including plants, animals, minerals, and microbes. This shift aligns with the growing global emphasis on eco-friendly textile production and the desire for more sustainable practices in various industries.

Natural dyes offer several key benefits that make them attractive for both consumers and manufacturers. Firstly, they are biodegradable, breaking down safely in the environment and reducing pollution and waste. Additionally, natural dyes are generally non-toxic, making them safer for both users and the environment, as they are less likely to contain harmful chemicals that pose health risks. Furthermore, the use of natural dyes supports traditional practices and promotes cultural heritage, fostering a deeper connection to local artisans and communities. Recent advancements in extraction methods, such as ultrasound-assisted extraction, microwave-assisted extraction, and enzymatic techniques, have further enhanced the efficiency and yield of natural dyes, making them more competitive with synthetic alternatives. These innovations not only increase the availability of vibrant colours but also improve the overall viability of natural dyes in the modern textile industry.

Despite their advantages, the adoption of natural dyes is hindered by several challenges. Colour inconsistency can arise from the natural variations in dye sources, leading to unpredictable results, which is problematic for manufacturers seeking consistency. Furthermore, natural dyes often struggle with limited fastness, particularly regarding wash and light resistance, reducing the longevity of dyed textiles. The processes for extraction and application can also be resource-intensive and time-consuming, raising production costs. Additionally, scaling natural dye production to meet the demands of the global textile market poses significant logistical and economic hurdles. However, innovative solutions are emerging, including the use of bio-mordants to enhance dye bonding, nanotechnology to improve dye properties, and microbial dyeing techniques that leverage bacteria and fungi for sustainable dye production to address these challenges. Furthermore, the potential integration of natural dyes with smart textiles and renewable energy applications opens new avenues for their use beyond traditional methods.

While natural dyes present advantages in sustainability, regulatory and safety concerns must also be addressed. Issues such as allergen risks associated with certain natural dyes and the environmental impact of sourcing and extraction processes require careful management. As the industry progresses, advocacy for supportive policies and increased consumer education will be crucial in positioning natural dyes as viable alternatives within the textile landscape.

Significantly, natural dyes hold considerable promise for promoting sustainable practices in textile production. However, their large-scale industrial adoption will depend on overcoming the technical and economic barriers that currently exist. There is a need for continued research and innovation that is essential to achieve a balance between ecological responsibility and performance. This will allow the textile industry to foster a future where sustainability and tradition thrive in unison. This holistic approach will ensure that natural dyes become a cornerstone of a more responsible and culturally enriched textile industry. It must be stressed that the future of natural dyes depends on the combined efforts of researchers, manufacturers, and consumers. Sustainable textiles require continuous innovation, responsible production, and conscious consumption. In conclusion, these stakeholders can ensure that natural dyes become a mainstream and viable solution for eco-friendly fashion.

References

1. ANM, S., F. Ahmed, P. Saha, and B. Talukder. 2021. A review on natural dyes: Raw materials, extraction process, and their properties.
2. Ado, A. D., H. K. Yahaya, A. A. Kwalli, and R. S. Abdulkadir. 2015. Dyeing of textiles with eco-friendly natural dyes: A review.
3. Affat, S. S. 2021. Classifications, advantages, disadvantages, toxicity effects of natural and synthetic dyes: A review. *University of Thi-Qar Journal of Science* 8 (1): 130–135.
4. Aggarwal, S. 2024. Advancement in extraction and characterization techniques of natural dyes from dye-yielding plants sources: A review. *Pigment & Resin Technology*. https://doi.org/10.1108/prt-06-2023-0056.
5. Aggarwal, S. 2025. Advancement in extraction and characterization techniques of natural dyes from dye yielding plants sources: A review. *Pigment & Resin Technology* 54 (2): 293–301.
6. Agustarini, R., Y. Heryati, Y. Adalina, W.C. Adinugroho, D. Yuniati, R. A. Fambayun, and A. Perdana. 2022. The development of Indigofera spp. as a source of natural dyes to increase community incomes on Timor Island, Indonesia. *Economies* 10 (2): 49.
7. Ahmed, U., and A. Anwar. 2022. Application of natural dyes in dye-sensitized solar cells. In *Dye-sensitized solar cells*, 45–73. Academic Press.
8. Ahsan, R., A. Masood, R. Sherwani, and H. Khushbakhat. 2020. Extraction and application of natural dyes on natural fibers: An eco-friendly perspective. *Review of Education, Administration & LAW* 3 (1): 63–75.
9. Alegbe, E. O., and T. O. Uthman. 2024. A review of history, properties, classification, applications and challenges of natural and synthetic dyes. *Heliyon* 10.
10. Ali, N., E. H. El-Khatib, and F. A. Bassyouni. 2022. Utilization and characterization of natural products pretreatment and dyeing wool fabric by natural dyes with economical methods. 8(6): 178–183. https://doi.org/10.15406/jteft.2022.08.00319.
11. Ali, H. M., K. Abdelzaher, U. Rashad, Azza El-Halwagy, and H. El-hennawi. 2020. New nature dye printing paste functional on Varies fabric kinds enhanced by plasma irradiation.

The Egyptian Journal of Chemistry 63 (2): 5–6. https://doi.org/10.21608/EJCHEM.2020.18694.2153.

12. Amir-Al Zumahi, S. M., N. Arobi, H. Taha, M. K. Hossain, H. Kabir, R. Matin, and M. M. Rahman. 2020. Extraction, optical properties, and aging studies of natural pigments of various flower plants. *Heliyon, 6* (9).
13. Andriamanantena, M., S. Pithon, M. Dijoux, M. Hoareau, C. Fontaine, J. Ferrard, and Y. Caro. 2023. A survey on the potential contribution of Reunion Island dye plant species diversity to the market demand for bioactive plant-based dyes and pigments. *Journal of Ethnobiology and Ethnomedicine* 19 (1): 8.
14. Antonini, L. M. 2023. *Natural and artificial dyes: Its health risks*. Seven Editora.
15. Asaduzzaman, M. M., F. Hossain, X. Li, and H. Quan. 2016. A study on the effects of pre-treatment in dyeing properties of cotton fabric and impact on the environment. *Journal of Textile Science & Engineering* 6 (2).
16. Asghar, A., A. Tajammal, S. Ashiq, S. Hussain, M. R. Ahmad, K. Sian, N. Al-Zaqri, A. Boshaala, B. Iqbal, and M. A. Bodlah. 2023. Eco-friendly dye of olive fruit peel and its color fastness applications on wool/silk fabrics. *Polish Journal of Environmental Studies*. https://doi.org/10.15244/pjoes/158789.
17. Aslan, F., C. Alperen, N. Cinperi, and U. K. ahin. 2023. The effect of blend, mordant and waste ratio on dyeing of denim products with coffee waste. *Avrupa Bilim ve Teknoloji Dergisi* 47: 61–63.
18. Baillie, C. 2024. Hottest design trends for 2024—from orange peel lamps to wheatgrass fabrics. Available at: https://www.thetimes.com/life-style/property-home/article/hottest-design-trends-for-2024-from-orange-peel-lamps-to-wheatgrass-fabrics. Accessed 4 Feb 2025.
19. Bartolome, G. J. C., J. P. S. de Mesa, J. A. C. Adoña, and A. E. L. Torres. 2020. Performance of dye-sensitized solar cells with natural dye from local tropical plants. 18 (1). https://mjst.ustp.edu.ph/index.php/mjst/article/download/395/144.
20. Ben Slama, H., A. C. Bouket, Z. Pourhassan, F. N. Alenezi, A. Silini, H. Cherif-Silini, T. Oszako, L. Luptakova, P. Goliska, and L. Belbahri. 2021. Diversity of synthetic dyes from textile industries, discharge impacts and treatment methods. *Applied Sciences* 11 (14): 6255. https://doi.org/10.3390/APP11146255.
21. Benli, H. 2024. Bio-mordants: A review. *Environmental Science and Pollution Research* 31 (14): 20714–20771.
22. Berlanga-Del Pozo, M., L. Gallardo-Guerrero, and B. Gandul-Rojas. 2020. Influence of alkaline treatment on structural modifications of chlorophyll pigments in NaOH—treated table olives preserved without fermentation. *Foods* 9 (6): 701.
23. Bide, M. J. 2014. *Sustainable dyeing with synthetic dyes*.
24. Buliga, D.-I., A. Diacon, I. Calinescu, I. Popa, E. Rusen, A. Ghebaur, O. Tutunaru, and C. Boscornea. 2022. Enhancing the light fastness of natural dyes by encapsulation in silica matrix. 432: 114085. https://doi.org/10.1016/j.jphotochem.2022.114085.
25. Chakravarty, S. 2024. Exploring textile dyes industry: SWOT analysis and market trends for 2024 and beyond. Available at: https://www.marketresearchreports.com/blog/2024/09/19/exploring-textile-dyes-industry-swot-analysis-and-market-trends-2024-and-beyond. Accessed: 4 April 2025.
26. Chandel, N., B. B. Singh, C. Dureja, Y. H. Yang, and S. K. Bhatia. 2024. Indigo production goes green: A review on opportunities and challenges of fermentative production. *World Journal of Microbiology and Biotechnology* 40 (2): 62.
27. Chaurasia, P. K., and S. L. Bharati. 2022. Applicability of plants in detoxification of dyes. In *Bioremediation*, 249–262. CRC Press.
28. Che, J. N., and X. Yang. 2022. A recent (2009–2021) perspective on sustainable color and textile coloration using natural plant resources. *Heliyon* 8 (10): e10979. https://doi.org/10.1016/j.heliyon.2022.e10979.
29. Chowdhury, J., J. Ferdous, J. Lihavainen, B. R. Albrectsen, and J. Lundberg-Felten. 2023. Fluorogenic properties of 4-dimethylaminocinnamaldehyde (DMACA) enable high resolution imaging of cell-wall-bound proanthocyanidins in plant root tissues. *Frontiers in Plant Science* 13: 1060804.

30. Chowdhury, M. R., A. N. Khan, and S. Sultana. 2024. Study on the effect of natural dyes and synthetic dyes on textile fabric. *Journal of Materials Science Research* 13 (2): 25. https://doi.org/10.5539/jmsr.v13n2p25.
31. Chungkrang, L., S. Bhuyan, and A. R. Phukan. 2021. Natural Dyes: Extraction and Applications. *International Journal of Current Microbiology and Applied Sciences* 10 (01): 1669–1677. https://doi.org/10.20546/IJCMAS.2021.1001.195.
32. Cui, H., W. Xie, Z. Hua, L. M. Cao, Z. Xiong, Y. Tang, and Z. Yuan. 2022. Recent advancements in natural plant colorants used for hair dye applications: a review. *Molecules* 27 (22): 8062. https://doi.org/10.3390/molecules27228062.
33. Cyril, M. 2024. Natural dyes in textiles: Beauty, benefits, and challenges. Available at: https://www.qforquinn.com/blogs/news/natural-plant-based-dyes. Accessed 1 Feb 2025.
34. Darmawan, A., A. Riyadi, H. Muhtar, and S. Adhy. 2024. Enhancing cotton fabric dyeing: Optimizing mordanting with natural dyes and citric acid. *International Journal of Biological Macromolecules* 276: 134017.
35. Doshi, V., and S. Noble. 2023. Consumers are in fact buying sustainable goods: Highlights from new research. Available at: https://www.mckinsey.com/industries/consumer-packaged-goods/our-insights/consumers-are-in-fact-buying-sustainable-goods-highlights-from-new-research. Accessed: 3 April 2025.
36. Doty, K., D. N. Green, and S. J. Haar. 2022. Natural dyes in the United States: An exploration of natural dye use through the lens of the circuit of style-fashion-dress. *Breaking Boundaries*. https://doi.org/10.31274/itaa.13506.
37. Dulo, B., K. Phan, J. Githaiga, K. Raes, and S. De Meester. 2021. Natural quinone dyes: A review on structure, extraction techniques, analysis and application potential. *Waste and Biomass Valorization* 12 (12): 6339–6374.
38. Durbin, D. 2024. Protesters demand Kellogg remove artificial colours from Froot Loops and other cereals. Available at: https://apnews.com/article/kellogg-artificial-colors-dyes-cereal-c167f3c51f03d8f43612fc6afe9b2fdd. Accessed: 5 Feb 2025.
39. Ellis-Petersen, H. 2025. Stories woven in cloth in Pakistan's first textile museum. Available at: https://www.theguardian.com/world/2025/jan/27/stories-woven-in-cloth-in-pakistans-first-textile-museum?utm_source=chatgpt.com. Accessed: 1 Feb 2025.
40. Elsahida, K., A. M. Fauzi, I. Sailah, and I. Z. Siregar. 2019. Sustainability of the use of natural dyes in the textile industry. In *IOP Conference Series: Earth and Environmental Science*, vol. 399, No. 1, 012065. IOP Publishing.
41. Ferreira, M. T., A. S. Matos, A. Couras, J. Marto, and H. Ribeiro. 2022. Overview of cosmetic regulatory frameworks around the world. *Cosmetics* 9 (4): 72. https://doi.org/10.3390/cosmetics9040072.
42. Gala, S., S. Sumarno, and M. Mahfud. 2020. Comparison of microwave and conventional extraction methods for natural dyes in wood waste of mahogany (*Swietenia mahagoni*). 18 (4): 618–623. https://doi.org/10.5937/JAES0-23695.
43. Ghareaghajlou, N., S. Hallaj-Nezhadi, and Z. Ghasempour. 2021. Red cabbage anthocyanins: Stability, extraction, biological activities and applications in food systems. *Food Chemistry* 365: 130482.
44. Gupta, D., and A. Laha. 2007. Antimicrobial activity of cotton fabric treated with *Quercus infectoria* extract.
45. Gökta, O., E. Özen, E. Baysal, R. Mammadov, M. H. Alma, and A. Sönmez. 2009. Colour stability of wood treated with madder root (*Rubia tinctorium* L.) extract after lightfastness test. *Wood Research* 54 (1): 37–44. https://www.cabdirect.org/cabdirect/abstract/20093121312.
46. Gürses, A., M. Açkyldz, K. Güne, and M. Gürses. 2016. Historical development of colourants.
47. Haji, A., and M. Naebe. 2020. Cleaner dyeing of textiles using plasma treatment and natural dyes: A review. *Journal of cleaner production* 265: 121866.
48. Hamdy, D. M., and A. G. Hassabo. 2021. Various natural dyes from different sources. *Journal of Textiles, Colouration and Polymer Science* 18 (2): 171–190.

49. Hanbing, W., H. Haase, and B. Mahltig. 2020. Cationic pretreatment for reactive dyeing of cotton and its simultaneous antibacterial functionalisation. *Tekstilec* 63 (1).
50. Heer, K., and S. Sharma. 2017. Microbial pigments as a natural colour: A review. *International Journal of Pharmaceutical Sciences and Research* 8 (5): 1913–1922.
51. Helmy, H. M. 2020. Extraction approaches of natural dyes for textile coloration. *Journal of Textiles, Coloration and Polymer Science* 17 (2): 65–76.
52. Holloway, S. 2024. Unveiling customer perceptions: A qualitative study on the role of supply chain transparency in brand trust. https://doi.org/10.20944/preprints202406.1738.v1.
53. Hossain, M. Y., T. Jiang, W. Zhu, S. Sarker, M. N. Pervez, M. I. U. Hoque, and V. Naddeo. 2022. Green and sustainable method to improve fixation of a natural functional dye onto cotton fabric using cationic dye-fixing agent/D5 microemulsion. *Journal of Natural Fibers* 19 (15): 11283–11298. https://doi.org/10.1080/20511787.2021.1956782.
54. Hossain, I. 2021. Sources and applications of natural dyes. Available at: https://textilefocus.com/sources-applications-natural-dyes/. Accessed: 5 Nov 2024.
55. Insight, F., and B. R. Center. 2021. The state of consumer spending: Gen Z influencing all generations to make sustainability-first purchasing decisions.
56. Islam, T., K. M. Islam, S. Hossain, M. A. Jalil, and M. M. Bashar. 2024. Understanding the fastness issues of natural dyes. *IntechOpen*. https://doi.org/10.5772/intechopen.1005363.
57. Islam, M.T., M. S. Farhan, and M. H. Mahmud. 2024. Green extraction of dyes and pigments from natural resources. In *Renewable dyes and pigments*, 19–36. Elsevier.
58. Iyer, V. V. 2018. Embrace a gift of nature: A proposal for commercializing natural eco-dyes for the textile industry. *Geo. Envtl. L. Rev.* 31: 725.
59. Jabar, J. M. 2024. Classification of natural dyes for sustainable exploitation. In *Natural dyes and sustainability*, 153–191. Springer Nature Switzerland.
60. Jain, H., and M. Vasantha. 2016. Eco-friendly dyeing with natural dye—Areca nut; enhancing colour fastness with natural mordants (Myrobalan, Lodhra and Pomegranate) and increasing the antibacterial activity. *Archives of Applied Science Research* 8 (8): 1–7. https://www.scholarsresearchlibrary.com/articles/eco-friendly-dyeing-with-natural-dye--areca-nut-enhancing-colour-fastness-with-natural-mordants-myrobalan-lodhra-and-pom.pdf.
61. Jordeva, S., M. Kertakova, S. Zhezhova, L. S. Golomeova, and K. Mojsov. 2020. Dyeing of textiles with natural dyes. *Tekstilna industrija* 68 (4): 12–21.
62. Kadolph, S. J. 2008. Natural dyes: A traditional craft experiencing new attention. *The Delta Kappa Gamma bulletin* 75: 14.
63. Karadag, R. 2023. Establishing a new international standard for natural dyed textile goods [Natural Organic Dye Standard (NODS)]. *Journal of Natural Fibres*. https://doi.org/10.1080/15440478.2022.2162187.
64. Karmakar, S., Q. A. U. Zaman, M. S. Hossain, A. Siddika, S. Tabassum, A. S. M. Ibrahim, and S. Huque. 2022. Locally available natural dyes for dye-sensitized solar cells. *Dhaka University Journal of Applied Science and Engineering* 7 (1): 38–44.
65. Kartini, I., and A. D. Hatmanto. 2021. Natural dyes: From cotton fabrics to solar cells. In *Dyes and pigments-novel applications and waste treatment*. IntechOpen.
66. Katheresan, V., J. Kansedo, and S. Y. Lau. 2018. Efficiency of various recent wastewater dye removal methods: A review. *Journal of environmental chemical engineering* 6 (4): 4676–4697.
67. Khan, B., R. Sindhyan, A. Divan, and S. Rathod. 2018. Extraction, characterization & applications of natural dyes. *Annals of Plant Sciences* 7 (11): 2463–2467. https://doi.org/10.21746/APS.2018.7.11.4.
68. Krishnan, S. K., K. Subbiah, S. Kandasamy, and K. Subramaniam. 2020. Application of metal nanoparticles for textile dye remediation. In *Sustainable development in energy and environment: Select proceedings of ICSDEE 2019*, 217–223. Springer Singapore.
69. Lambrecht, L., J. Gisbert-Payá, E. Bou-Belda, and M. Á. Bonet. 2020. Optimization of tea extracts composition to dye cotton. Time and temperature influence. *Journal of Applied Research in Technology & Engineering* 1 (1): 3–7.
70. Laroche, M., J. Bergeron, and G. Barbaro-Forleo. 2001. Targeting consumers who are willing to pay more for environmentally friendly products. *Journal of Consumer Marketing* 18: 503–520.

71. Li, N., Q. Wang, J. Zhou, S. Li, J. Liu, and H. Chen. 2022. Insight into the progress on natural dyes: Sources, structural features, health effects, challenges, and potential. *Molecules (Basel, Switzerland)* 27 (10): 3291. https://doi.org/10.3390/molecules27103291.
72. Li, S., A. B. Cunningham, R. Fan, and Y. Wang. 2019. Identity blues: The ethnobotany of the indigo dyeing by Landian Yao (Iu Mien) in Yunnan, Southwest China. *Journal of Ethnobiology and Ethnomedicine* 15.
73. Liu, Q. 2015. Synthesis and dyeing properties of a reactive disperse dye. *Journal of Dalian Polytechnic University.*
74. Loum, J., R. Byamukama, and P. Wanyama. 2021. Application of natural dyes from selected indigenous plants on cotton and silk fabrics. *Journal of Textile Engineering and Fashion Technology* 7: 71–77.
75. Lozano, F., S. Tapía-Pacci, R. Ccopa, A. Quispe-Quispe, and J. C. Salazar-López. 2024. Effect of mordants on colour variation in alpaca fibre dyed with dye extracted from the tankar stem (*Berberis boliviana* L.). *Coloration Technology*. https://doi.org/10.1111/cote.12760.
76. Mai, T. H., T. Grethe, and B. Mahltig. 2024. Wood extracts for dyeing of cotton fabrics—special view on mordanting procedures. *Textiles* 4 (2): 138–164.
77. Mamun, A. A., A. Haji, Md. H. Mahmud, Md. R. Repon, and M. T. Islam. 2023. Bibliometric evidence on the trend and future direction of the research on textile coloration with natural sources. *Coatings* 13 (2): 413. https://doi.org/10.3390/coatings13020413.
78. Manhita, A., T. Ferreira, A. Candeias, and C. Dias. 2011. Extracting natural dyes from wool—an evaluation of extraction methods. *Analytical and Bioanalytical Chemistry* 400 (5): 1501–1514. https://doi.org/10.1007/S00216-011-4858-X.
79. Manian, A. P. 2023. The role of mordants in fixation of natural dyes. *Handbook of natural colorants* 507–515.
80. Mansour, R. 2018. Natural dyes and pigments: Extraction and applications. *Handbook of renewable materials for colouration and finishing* 9: 75–102.
81. Martins, N., C. L. Roriz, P. Morales, L. Barros, and I. C. Ferreira. 2016. Food colorants: Challenges, opportunities and current desires of agro-industries to ensure consumer expectations and regulatory practices. *Trends in food science & technology* 52: 1–15.
82. Medina-Borges, R.-M. 2023. The role of mordants in fixation of natural dyes. 507–515. https://doi.org/10.1002/9781119811749.ch23
83. Mejica, G. F. C., Y. Unpaprom, P. Khonkaen, and R. Ramaraj. 2020. Extraction of anthocyanin pigments from malabar spinach fruits as a potential photosensitizer for dye-sensitized solar cell. 5–9. https://doi.org/10.37516/GLOBAL.J.SCI.ENG.2020.006.
84. Melo, M. J. 2009. History of natural dyes in the ancient mediterranean world.
85. Meneghetti, M., A. Talon, E. Cattaruzza, E. Celotti, E. Bellantuono, E. Rodríguez-Castellón, S. Meneghetti, and E. Moretti. 2020. Sustainable organic dyes from winemaking lees for photoelectrochemical dye-sensitized solar cells. *Applied Sciences* 10 (6): 2149. https://doi.org/10.3390/APP10062149.
86. Moussa, I., I. Ghezal, and F. Sakli. 2023. Valorization of Pelargonium g raveolens L'Hér. Hydrodistillation solid waste as natural dye for wool fabrics. *Journal of Natural Fibers* 20 (1): 2156966.
87. Muslimah, E., M. Reynara, and R. Fitriadi. 2022. environmental impact analysis on dyes in handwritten Batik. In *International conference on emerging smart cities*, 421–428. Springer Nature Singapore.
88. Nam, M., and J. Chung. 2007. The purchasing behaviour of natural dyed apparel products according to clothing benefits sought.
89. Natural Dye Garden Project. n.d. Available at: https://www.fitnyc.edu/about/initiatives/sustainability/campus-groups/student-initiatives/dye-garden.php?utm_source=chatgpt.com. Accessed 5 Feb 2025.
90. Ngamwonglumlert, L., S. Devahastin, and N. Chiewchan. 2017. Natural colorants: Pigment stability and extraction yield enhancement via utilization of appropriate pretreatment and extraction methods. *Critical reviews in food science and nutrition* 57 (15): 3243–3259.

91. Nhut Pham, T., T. Quoc Toan, T. Duc Lam, H. Vu-Quang, D. V. N. Vo, T. Anh Vy, and L. M. Bui. 2019. Anthocyanins extraction from Purple Sweet Potato (*Ipomoea batatas* (L.) Lam): The effect of pH values on natural color. In *IOP conference series: Materials science and engineering*, vol. 542, No. 1, 012031. IOP Publishing.
92. de Oliveira, M. E. P., W. M. B. da Silva, D. R. Alves, J. B. de Andrade Neto, S. M. de Morais, and S. de Oliveira Pinheiro. 2021. Dyes and pigments used in foods: An integrative literature review. *Research, Society and Development* 10 (10): e316101018925–e316101018925.
93. Olua, E. B., A. O. Oforghor, and C. N. Shailong. 2023. Dying properties of natural dyes extracted from *Parkia biglobosa* tree bark on cotton fabric.
94. Önal, A., S. Nached, and M. Ylmaz. 2023. Determination of colour properties of dyed textile products using different mordanting methods. 27–34. https://doi.org/10.58692/jotcsb.1251980.
95. Pargai, D. 2021. Application of smart and functional dyes in textiles. *IntechOpen*. https://doi.org/10.5772/INTECHOPEN.96045.
96. Paul, A. 2024. The synthetic dyes market: Trends, analysis, and future outlook. Available at: https://vocal.media/journal/the-synthetic-dyes-market-trends-analysis-and-future-outlook. Accessed: 2 April 2025.
97. Pizzicato, B., S. Pacifico, D. Cayuela, G. Mijas, and M. Riba-Moliner. 2023. Advancements in sustainable natural dyes for textile applications: A review. *Molecules*. https://doi.org/10.3390/molecules28165954.
98. Popescu, V. 2024. *New trends in the application of natural dyes in textile dyeing*, 111–137. Elsevier BV. https://doi.org/10.1016/b978-0-443-15213-9.00004-1.
99. Populoh, V., and mlará Williams McCallister. 2021. Reflections on the baltimore natural dye initiative and cultivating communities of care. *Journal of Textile Design Research and Practice* 9 (3): 359–380.
100. Poratelli, F. 2021. Fabric dyeing processes and their impact on the environment. Available at: https://www.cikis.studio/en/article/fabric-dyeing-processes-and-their-impact-on-the-environment?. Accessed 4 Feb 2025.
101. Prabowo, C. P. S., H. Eun, D. Yang, D. Huccetogullari, R. Jegadeesh, S. J. Kim, and S. Y. Lee. 2022. Production of natural colorants by metabolically engineered microorganisms. *Trends in Chemistry* 4 (7): 608–626.
102. Pratama, A. Z., E. Liquiddanu, and E. Pujiyanto. 2023. Factors affecting the consumers' willingness to pay for sustainable fashion products (study case: Natural dyed batik). In *Proceedings of the international conference on industrial engineering and operations management*.
103. Qi-Yue, Y., Z. Ting, H. Ya-Nan, H. Sheng-Jie, D. Xuan, H. Li, and X. Chun-Guang. 2020. From natural dye to herbal medicine: A systematic review of chemical constituents, pharmacological effects and clinical applications of indigo naturalis. *Chinese Medicine* 15: 1–13.
104. Quijano, L., R. Speight, and A. Payne. 2021. Future fashion, biotechnology and the living world: Microbial cell factories and forming new 'oddkins.' *Continuum* 35 (6): 897–913.
105. Radhakrishnan, S., R. Rajangam, and P. Peruran. 2024. *Social and environmental impact of natural dyeing. in natural dyes and sustainability*, 127–152. Springer Nature Switzerland.
106. Ragab, M. M., A. G. Hassabo, and H. Othman. 2022. An overview of natural dyes extraction techniques for valuable utilization on textile fabrics. *Journal of Textiles, Coloration and Polymer Science* 19 (2): 137–153.
107. Ramli, Q. H., R. M. Hassan, and N. M. Nor. 2021. Dyeing of textile using different mordants, mordanting techniques and their effects on fastness properties. *ASEAN Journal of Life Sciences* 1 (2): 64–67.
108. Ratnapandian, S. 2020. Natural colourants and its recent developments. In *Sustainable technologies for fashion and textiles*, 189–208. Woodhead Publishing.
109. Rehman, A., M. ur Rehman, M. Naveed, K. Javed, and A. Khan. 2024. Eco-friendly dyeing of cotton with natural colorants using natural mordants obtained from aloe vera. *Journal of the Indian Chemical Society*. https://doi.org/10.1016/j.jics.2024.101182.

110. Repon, M. R., T. Islam, T. K. Paul, S. Jurkonien, A. Haji, S. Shukhratov, and G. F. I. Toki. 2024. Natural dyes in textile printing: Parameters, methods, and performance. *Environmental Science and Pollution Research* 31 (35): 47552–47583.
111. Rosyida, A., and M. Masykuri. 2022. Minimisation of pollution in the cotton fabric dyeing process with natural dyes by the selection of mordant type. *Research Journal of Textile and Apparel* 26 (1): 41–56.
112. Russell, S. 2024. Why natural dyes are the future of fashion. Available at: https://environment.co/natural-dyes/?utm_source=chatgpt.com. Accessed 03 Feb 2025.
113. Sabrina, N. 2024. The influence of internal environmental capabilities on sustainable competitive advantage through green innovation with corporate reputation moderation: A study on natural dye batik SMEs in Yogyakarta. *Enrichment* 2 (9). https://doi.org/10.55324/enrichment.v2i9.238.
114. Sadannavar, M. K., A. P. Periyasamy, S. R. Islam, F. Shafiq, D. Xue, and T. Zhao. 2024. Natural dyeing and antimicrobial functionalization of wool fabrics dyed with Chinese dragon fruit extract to enhance sustainable textiles. *Sustainability* 16 (16): 6832. https://doi.org/10.3390/su16166832.
115. Saeed, A. A., M. S. Al-Salimi, M. T. Qashqoosh, G. A. Al-Dahbali, W. S. A. Yusef, T. H. A. Mohammed, and W. A. S. Nagi. 2024. Exploiting eco-friendly natural dyes from plant sources: Extraction and diverse applications. *Electronic Journal of University of Aden for Basic and Applied Sciences* 5 (3): 320–330.
116. Samanta, P. 2020. A review on application of natural dyes on textile fabrics and its revival strategy. In *Chemistry and technology of natural and synthetic dyes and pigments*, 1–25.
117. Saxena, S., and A. S. Raja. 2014. Natural dyes: Sources, chemistry, application and sustainability issues.
118. Sk, S., R. Mia, M. A. Haque, and A. M. Shamim. 2021. Review on extraction and application of natural dyes. *Textile & Leather Review* 4 (4): 218–233. https://doi.org/10.31881/TLR.2021.09.
119. Sooch, B., N. Sandhu, M. Mann, and R. C. Ray. 2024. *Valorization of beetroot waste for extraction of natural dye for textile and food applications*, 237–260. https://doi.org/10.1007/978-981-99-8266-0_11.
120. Sousa, C. 2022. Anthocyanins, carotenoids and chlorophylls in edible plant leaves unveiled by tandem mass spectrometry. *Foods* 11 (13): 1924.
121. Tadesse Abate, M., and V. Nierstrasz. 2017. Combined pre-treatment and causticization of cotton fabric for improved dye uptake. *Advance Research in Textile Engineering* 2 (1).
122. Tamilarasi, A., and M. Banuchitra. 2021. Classification and types of natural dyes: A brief review. *International Journal of Creative Research Thoughts (IJCRT)* 9 (11): 527–532.
123. Tarikul Islam, K. M. R. I., and S. Hossain. 2024. Understanding the fastness issues of natural dyes. *Dye chemistry-exploring colour from nature to lab: Exploring colour from nature to lab*, 21.
124. Taura, U. H., M. A. Abubakar, A. M. Abubakar, and M. U. Kurgiya. 2024. Extraction and characterisation of natural dye from orange peel for textile applications. *Journal of Applied Life Sciences and Environment*. https://doi.org/10.46909/alse-571130.
125. Teixeira-Guedes, C., J. S. Gomes-Dias, S. A. Cunha, M. E. Pintado, R. N. Pereira, J. A. Teixeira, and C. M. Rocha. 2023. Enzymatic approach for the extraction of bioactive fractions from red, green and brown seaweeds. *Food and Bioproducts Processing* 138: 25–39.
126. Teron, R., and S. K. Borthakur. 2012. Traditional knowledge on herbal dyes and cultural significance of colours among the Karbis, an Ethnic Tribe in Northeast India. *Ethnobotany Research and Applications* 10: 593–603.
127. Thiyagarajan, S., K. Balakrishnan, and R. Venkadachalam. 2016. Extraction and application of eco-friendly natural dye obtained from flower of *Acacia eburnea* (L.f.) willd on cotton fabric. 4 (3): 1–4. https://doi.org/10.15226/SOJMSE.2016.00137.
128. Toscano, C. 2024. Henna saved my strands after i fried them with bleach. Available at: https://www.allure.com/story/henna-hair-color?. Accessed 4 Feb 2025.

129. Valkovska, V., and L. Orola. 2021. Characterization of pigments from *Malus domestica* leaves for wool dyeing. *Key Engineering Materials* 903: 63–68.
130. Varjani, S., P. Rakholiya, H. Y. Ng, S. You, and J. A. Teixeira. 2020. Microbial degradation of dyes: An overview. *Bioresource Technology* 314: 123728.
131. Vieira, E. F., S. Souza, M. M. Moreira, R. Cruz, A. B. D. Silva, S. Casal, and C. Delerue-Matos. 2022. Valorization of phenolic and carotenoid compounds of *Sechium edule* (Jacq. Swartz) leaves: Comparison between conventional, ultrasound-and microwave-assisted extraction approaches. *Molecules* 27 (21): 7193.
132. Wanyama, P. A. G., B. T. Kiremire, P. Ogwok, and J. S. Murumu. 2015. The effect of different mordants on strength and stability of colour produced from selected dye-yielding plants in Uganda.
133. Yahya, A., S. N. A. Sulaiman, and M. Maryudi. 2024. Effect of operating condition of natural dye production from red spinach (*Amaranthus dubius*) using solvent extraction. *ASEAN Journal of Chemical Engineering* 24 (3): 342–352.
134. Yang, Z., and W. Zhai. 2010. Optimization of microwave-assisted extraction of anthocyanins from purple corn (*Zea mays* L.) cob and identification with HPLC–MS. *Innovative food science & emerging technologies* 11 (3): 470–476.
135. Zarkogianni, M., E. Mikropoulou, E. A. Varella, and E. Tsatsaroni. 2011. Colour and fastness of natural dyes: Revival of traditional dyeing techniques. *Coloration Technology* 127 (1): 18–27. https://doi.org/10.1111/J.1478-4408.2010.00273.X.
136. Zerin, I., N. Farzana, A. Sayem, D. Anang, and J. Haider. 2019. Potentials of natural dyes for textile applications. https://doi.org/10.1016/b978-0-12-803581-8.11668-6.
137. Zubairu, A., and Y.M. Mshelia. 2015. Effects of selected mordants on the application of natural dye from onion skin (*Allium cepa*). https://doi.org/10.5923/j.scit.20150502.02.

Edward Apau A lecturer at the Department of Visual and Industrial Art, Sunyani Technical University, Sunyani, Ghana and a Ph.D. Candidate in Textile Design Technology at Kwame Nkrumah University of Science and Technology (KNUST). I am a graduate with B.A. Industrial Art (Textiles) and Master of Fine Arts (MFA) in Textile Design from KNUST, respectively. My research interest focuses on weaving technology, sustainable dyeing, design thinking and innovation, textile surface designs, and printing technology.

Benjamin Tawiah (Ph.D.) is a lecturer in the Department of Industrial Art (Fashion and Textiles), Kwame Nkrumah University of Science and Technology (KNUST). He holds a B.A. Industrial Art (Textiles), KNUST, Ghana, Master of Engineering in Textile Engineering from Jiangnan University, China and a Ph.D. in Textile and Clothing Technology from the Hong Kong Polytechnic University. His research interest focuses on sustainable polymer nanomaterial composites, sustainable textile colouration and sustainable fashion.

Morrison Appiagyei Boadi Morrison Appiagyei Boadi is an Assistant Lecturer at the Department of Visual and Industrial Art, Sunyani Technical University, Sunyani, Ghana. He holds a Bachelor of Arts (B.A) in Industrial Art in Textiles, and a Master of Fine Arts (MFA) in Textile Design from Kwame Nkrumah University of Science and Technology (KNUST), Kumasi. His research attention is on indigenous textile production methods, textile design innovation and textile art.

Chapter 5
Waterless Dyeing: Redefining Colour in Sustainable Textiles

Bernice Akweley Djangmah, George Kwame Fobiri, Sylvia Asheley Gyampoh, Edem Kwami Buami, and Benjamin Eghan

Abstract The colouration of textile materials through dyeing requires large amounts of water. This wet dyeing process in the textile industry has contributed to the largest consumption of fresh water in the environment, hence making the industry the second largest global polluter of water. The use of heavy colouring materials such as synthetic dyes and toxic or non-biodegradable dye auxiliaries in the dyeing process poses significant health and safety threats to the ecosystem. Due to the high concentration of heavy metals, hydrocarbons, chlorine, formaldehyde, and other chromium-based chemicals in the dye materials, the wastewater from dyeing, when discharged indiscriminately, affects the quality of the water and the environment. To help reduce the level of pollution from dye wastewater to the environment, significant approaches or technologies have been adopted, most especially waterless dyeing for textile colouration. This technology has several benefits, such as reducing the use of water, hence the non-use of water in the dyeing process, no generation of wastewater, reduced time for dyeing, and less environmental impact. Drawing on these advantages, this chapter discusses key waterless dyeing technologies such as supercritical CO_2 dyeing, air dyeing, plasma dyeing, and electrochemical dyeing. Key case studies of real-world applications of these waterless dyeing technologies are also discussed. We further discuss the challenges and regulatory issues concerning the application of waterless dyeing technologies in the textile industry. Lastly, this chapter concludes with insights

B. A. Djangmah
Department of Industrial Art, Kwame Nkrumah University of Science and Technology, Kumasi, Ghana

G. K. Fobiri
Department of Fashion Design and Textiles Studies, Kumasi Technical University, Kumasi, Ghana

S. A. Gyampoh
Department of Fashion Design, Radford University College, Accra, Ghana

E. K. Buami
Department of Industrial Art, Ho Technical University, Ho, Ghana

B. Eghan (✉)
Faculty of Art, Design and Humanities, De Montfort University, Leicester, UK
e-mail: bennelson2224@gmail.com

S.-X. Jiang et al. (eds.), *Advances in Textile Materials and Processing Techniques for Sustainability*, SDGs and Textiles, https://doi.org/10.1007/978-981-95-0469-5_5

into potential areas for future studies and advancement in the field of sustainable dyeing practices in the textile industry.

Keywords Waterless dyeing · Supercritical CO_2 dyeing · Air dyeing · Plasma treatment · Electrochemical dyeing · Sustainable dyeing practices

5.1 Introduction

The textile industry has played a significant role in countries and manufacturers to advance the creation of jobs and economic growth to curtail poverty and unemployment [1]. In fact, the textile industry is estimated to employ about 35 million workers globally, with a $1 trillion generation [2]. Aside from these benefits, the industry is, however, regarded as a major contributor to the pollution of the environment. These pollutions have come from various production processes such as dyeing, bleaching and washing [3], that utilises heavy chemical and machines which consumes fuel. Most especially, dyeing, which is a colouration process, is a major contributor to polluting the environment through its waste, which has accounted for about 20% of the pollution of water bodies [4, 5]. Heavy metals, dyes, salt and other chemicals are contained in the wastewater [6]. Other negative impacts of wastewater from conventional dyeing processes include dissolved levels of oxygen, a reduction in photosynthesis due to the non-penetration of light in water [7] and health hazards posed by the toxic nature of some dyes [8]. The issue of high energy and water consumption also remains a critical challenge in using the conventional dyeing process. This is because the colouration process relies heavily on using large volumes of water to prepare and dye textile materials. For example, it is estimated that annually, about 90–93 m^3 of water are used in the dyeing process [9]. To address these problems, sustainable and cost-effective approaches have been advanced to ensure reduced negative effects from the dyeing process in the textile industry.

Most importantly, waterless dyeing practices are advanced for a responsible dyeing process for textile materials. This dyeing process is an innovative technique that limits the utilisation of water to colour different textile materials. Some of these techniques are dope dyeing, air dyeing, solvent-based systems, foam dyeing and supercritical carbon dioxide, amongst others [10]. Recent studies have reported on advanced techniques such as supercritical CO_2 [11, 12], dope dyeing [13–15] and electrochemical dyeing [16, 17] amongst others for colouring textile materials. These approaches are applied to colour fabrics, which is different from the conventional dyeing method [9]. Furthermore, the key benefits of waterless dyeing include a reduction in water usage, hence water conservation, less energy consumption and reduced operational costs. Others include reduced carbon emissions and waste [18], shorter dyeing periods and less concentration of chemicals [19]. However, these benefits have several challenges, which affect the full utilisation in the textile industry. With further advancement in various waterless dyeing methods, the effectiveness and full benefits of this dyeing method will contribute significantly to redefining the colouring of sustainable textiles. Herein, this chapter discusses essential waterless dyeing technologies such as supercritical CO_2 dyeing, air dyeing, plasma dyeing, and electrochemical dyeing. Further discussion of real-world applications of these waterless dyeing technologies is presented. We further discuss the challenges and regulations concerning the use of waterless dyeing technologies in the textile industry. This chapter concludes with insights into potential areas for future studies and advancement in the field of sustainable dyeing practices in the textile industry.

5.2 Environmental Impacts of Conventional Dyeing Methods

Conventional dyeing processes in the textile industry rely heavily on freshwater resources, non-renewable energy, and non-sustainable chemicals such as colourants and processing aids. This leads to environmental and health problems linked to the dangers posed by unfixed dyes and chemicals that pollute ecosystems and endanger human health. Ibrahim and Eid [20] revealed that unfixed dyes and other chemicals discharged as effluents damage the environment and pose major dangers to human health, including sensitivity, toxicity, and carcinogenicity [21]. These dyestuffs, especially synthetic dyes, are used for the colouration and finishing of textile materials, which pose significant health risks for humans and the environment. The evaluation of the risks or negative effects (to human health and environment) of these dyestuffs to understand the ecological, toxic, chemical and physical properties, is imperative [22]. Due to these effects, stringent safety protocols should guide the handling of synthetic dyes. Alternatively, a common health concern, a respiratory problem, results from the inhalation or handling of dye particles, which leads to sore eyes, an irritated nose and skin amongst others [23]. Fibre and yarn processing produces some lint and dust particles which when inhaled causes a variety of respiratory illnesses [24].

Conventional dyeing methods are associated with environmental impacts due to the excessive discharge of wastewater, consumption of water and energy. The discharge of wastewater from dyeing processes contains a mix of large quantities of chemicals, heavy metals, salts and dyes, which are toxic and non-biodegradable. These pollutants can significantly impact key quality parameters of water, such as chemical oxygen demand (COD), biological oxygen demand (BOD) and pH, amongst others, which influence the aquatic ecosystems [25]. Furthermore, there is a disruption of oxygen and photosynthesis, and the penetration of sunlight that affects aquatic life [22]. Another area of concern is soil contamination. When wastewater from dyestuff is disposed of into the environment, dye residues reduce soil fertility because the pores are clogged. This leads to soil nutrient loss and the contamination of underground water, which finds its way into the food chain, with significant threats to human and animal existence [26].

Subsequently, there are global impacts due to water and air pollution, where toxic pollutants are found in the air and water. Air pollution refers to chemicals, particles, or biological air contaminants that harm humans or other species [27]. These pollutants that cause air pollution include exhaust from boilers, solvent mists, acid vapour, oil fumes, lint, and dust, amongst others. Formaldehyde, often known as methane, and acetic acid are also major causes of emissions [23]. During the carbonisation of wool, corrosive acid vapours are produced, as well as acid volatilisation. Vapours of hazardous substances, such as kerosene or mineral turpentine oil, formaldehyde, mono- and dichlorobenzene, ethyl acetate, hexane, styrene, and others are produced after dyeing and printing. The odour is connected to oil mists or solvent fumes. Odour is generated by carriers used in polyester dyeing, bleaching with sodium hypochlorite, and colour reduction in cotton sulphur dyeing with hydrosulphite [28]. Several potential ideas and innovative solutions have been developed and implemented to address the difficulties of sustainable colouration and after-finishing processes, novel greener textile goods, and efficient use of natural resources.

5.3 Waterless Dyeing Methods

A textile dye is a chemical compound used to colour clothes. It could be natural or synthetic. Despite efforts to reduce water input, such as changing conventional equipment, recycling water, and reusing wastewater, water consumption remains high in the textile business. Non-aqueous dyeing processes can significantly minimise or eliminate the need for water. Water conservation improves the environment while also saving money. Some revolutionary techniques to advance textile wet processing include the use of ultrasonic waves, microwave dyeing, plasma technology, supercritical carbon dioxide, and electrochemical dyeing [29]. Table 5.1 details some waterless dyeing methods that improve the environmental challenges and economic benefits of recent times.

5.3.1 Benefits of Waterless Dyeing Over Traditional Dyeing

As the textile industry strives to implement greener and more sustainable practices, advances in unit operations, optimised dyeing conditions, and the incorporation of recycling technology have emerged as critical advancements. This section will look at major breakthroughs in waterless dyeing techniques, such as the air-dye method and supercritical carbon dioxide ($scCO_2$) dyeing, among others, that considerably cut water usage and waste while lowering energy consumption and greenhouse gas emissions. With traditional wet chemical procedures still in use, the need to migrate to these novel technologies is highlighted by their ability to solve environmental concerns related to textile processing. This section also looks at individual case

Table 5.1 Different types of waterless dyeing methods

Dyeing method	Description	References
Ultrasonic waves	Utilises high-frequency sound waves to enhance dye penetration and reduce dyeing time, minimising water usage	[29]
Plasma technology	Employs plasma to modify textile surfaces, allowing for better dye adhesion and less water consumption	
Supercritical carbon dioxide	Uses supercritical CO_2 as a solvent for dyeing, enabling vibrant colours without the need for water	[24]
Air dyeing	Involves applying dye in a gas phase, allowing for colour application without liquid solvents	[30]
Solvent dyeing	Uses organic solvents to dissolve dyes, allowing for application without the use of water	[31]

studies and technological breakthroughs that demonstrate the efficacy and sustainability of waterless approaches in current textile manufacture. Recent innovations in textile unit operations, well-organised processing with optimised parameters for dyeing and bioprocessing, and the incorporation of reusing and recycling technologies are all welcome advances in making textile processing greener and more sustainable [32]. As a result, there is an urgent need to keep water clean and pollution-free, particularly during textile dyeing, and to develop innovative techniques. Most dyeing enterprises presently employ the traditional wet chemical technique to colour fabrics; however, a few companies have lately adopted waterless dyeing. The air-dye technique and the supercritical carbon dioxide ($scCO_2$) aided method are two examples of waterless dyeing that have recently attracted a lot of interest. Waterless dyeing processes reduce water use, waste, energy consumption, and greenhouse gas emissions compared to conventional dyeing methods.

A large number of research publications and scientific reports have recently been published, addressing and discussing numerous elements and issues of waterless dyeing technology [33]. Colorep, a California-based sustainable technology startup, has presented a novel airflow dyeing technique [34]. In an airflow dyeing machine, colouring liquor is first atomised, then blended with high-pressure airflow, and then sprayed onto the fabric to be dyed. Mahmud et al. [18] stated that while water serves as a solvent for dyeing spirits and the dye chemicals come into direct contact with the fabric, just a small amount of water is wasted. The air-dye technique does not harm water because it transports dye by air rather than water, emits no hazardous waste, and wastes no water. Instead of dipping textiles in a typical bath of water and dye, Air Dye method uses printing equipment. This technology transfers precisely prepared colours from paper to polyester fabric [34]. Air-Dye utilises 95% less water, depending on the fabric, 86% less energy, and 84% less greenhouse gases than traditional dyeing. Air-dye has considerable advantages, such as reducing hazardous waste emissions and water usage, and many elements are covered in numerous scientific studies and journals. However, this technology only works on synthetic materials such as polyester; natural fibres such as cotton are not suitable for the air-dyeing process. Several textile converters in Europe, Asia, and North America now use Air-Dye [18]. Carbon dioxide (CO_2) is commonly used as a fluid since it is non-flammable, harmless, and readily available. Furthermore, the critical temperature ($Tc = 31.1$ °C) and pressure ($Pc = 74$ bar) are lower than those of many other fluids, making CO_2 more appealing [35]. At temperatures exceeding 31.1 °C and pressures more than 74 bar, supercritical CO_2 functions as both a liquid and a gas. The liquid-like densities are useful for dissolving hydrophobic dyes, whereas gas-like low viscosities and diffusion capabilities can result in quicker dyeing durations than water. Using $scCO_2$ as a solvent for synthetic and natural dyes offers time and cost savings of up to 50% over traditional water dyeing methods. This approach may also result in high solubility, diffusivity, and zero surface tension. Furthermore, CO_2 and dyes can be reused, making the process both economically viable and environmentally friendly.

5.3.2 *Fibre Types for Waterless Dyeing*

In the context of developing waterless dyeing processes, fibre type selection is critical in determining the success and usability of these techniques. This section focuses on essential synthetic fibres, specifically polyamide (PA) types like nylon, and their benefits in waterless dyeing applications. Studies have revealed that synthetic fibres have better characteristics, making them perfect candidates for novel dyeing procedures aimed at conserving water and reducing environmental impact. Below is a table illustrating significant studies conducted on different fibre types and their corresponding dyeing techniques. Polyamide (PA), a polymer bound together by amide bonds (–NH–CO–), popularly known as nylon, is a key form of synthetic fibre. Although several forms of PA are created, PA6 (polyamide 6, $(–HN–(CH_2)5–CO)n$) and PA6.6 (polyamide 6.6, $(–HN–(CH_2)6–NHCO–(CH_2)4–CO–)n$) are the most important varieties of PA fibres that can be used in waterless dyeing [1]. Aramids are another major kind of PA fibre, consisting entirely (or at least 85%) of link-adjacent aromatic monomers [30]. Synthetic fibres are significant in the textile industry because of their versatility and usability in a wide range of applications, whereas natural fibres such as cotton fall short due to inherent limitations. Synthetic textiles offer a variety of benefits, including greater strength, flexibility, less wrinkling, lightweight qualities, and shorter drying times. These characteristics make synthetic materials suitable for sportswear and other performance-related applications [36]. Table 5.2 discusses the importance of synthetic and natural fibres in waterless dyeing processes, highlighting their advantages and promising outcomes from recent research.

Polyester, the most commonly used synthetic fibre, has unique qualities due to its aliphatic and aromatic molecular components and regular molecular structure. These fibres offer low moisture absorbency, high resilience, dimensional stability, great wear, weather, and light resistance, good abrasion resistance, and may be blended effectively with cotton [30]. Poly(ethylene terephthalate) (PET) is the major polyester

Table 5.2 Synthetic and natural fibres in waterless dyeing processes

Fibre type	Dyeing techniques	Benefits	Results	References
Synthetic fibres				
Polyamide (PA)	Supercritical CO_2 Dyeing	Excellent dye affinity, high strength and durability, resistance to abrasion	High dye uptake with reduced processing time and water usage	[18]
Nylon	Air dye method	Lightweight and strong, elasticity and resilience, Quick drying properties	95% less water and 86% less energy consumption compared to traditional dyeing	[34]
Aramid	Ultrasonic wave dyeing	Exceptional durability and heat resistance, lightweight with high tensile strength, Resistance to chemicals	Enhanced dye penetration and faster dyeing times	[30]
Polyester	Microwave dyeing	Good colour retention, quick-drying and moisture-wicking properties, easy to care for	Significant reduction in water use and improved colour vibrancy	[37]
Natural fibres				
Cotton	Supercritical CO_2 Dyeing	Biodegradable, soft feel, and good dye affinity	Successful dyeing with minimal water and environmental impact	[18]
Hemp	Air dye method	Strong, durable, and environmentally friendly	Reduced water usage and improved dye uptake compared to traditional methods	[34]
Linen	Plasma technology	Breathable, lightweight, and moisture-wicking properties	Effective dyeing with enhanced colour vibrancy and reduced processing time	[37]
Tencel (Lyocell)	Ultrasonic wave dyeing	Sustainable, soft, and biodegradable	High dye retention with significantly lower water need	[30]

used in fibre manufacture, and it is preferred not only for its end-use qualities and cost-effectiveness, but also for its ease of physical and chemical modification, which can improve good attributes while suppressing negative ones. Furthermore, polyester finds use in composite materials due to its inexpensive cost, dimensional stability, and low curing temperature. These properties allow for its usage in simplified moulds, which contributes to its involvement in composite material development alongside other thermoset polymer matrices [38]. Given these environmental concerns, waterless dyeing of fibres is on the rise. The use of supercritical carbon dioxide ($scCO_2$) is a commercially viable method for waterless dyeing. In textile dyeing, replacing water with supercritical carbon dioxide ($scCO_2$) can save energy, reduce water consumption, and prevent pollution [39]. Various fibres have been dyed in supercritical carbon dioxide ($scCO_2$), including polyester [40], wool [41], cotton [42], cellulose acetate [43], and cellulose acetate [44]. However, only polyester is currently commercially available. Abou et al. [45] and Penthala et al. [35] have also reported on dyeing PA in supercritical carbon dioxide ($scCO_2$). Furthermore, Abou et al. [45] dyed the PA6 cloth with a series of dispersion azo dyes that have potential antibacterial effects, utilising supercritical carbon dioxide ($scCO_2$) as the medium.

5.4 Supercritical CO_2 Dyeing

Large volumes of water are required for the dyeing process in the textile industry, a situation which leads to the generation of wastewater. The latter heavily contains various harmful chemicals and dyestuffs, which are not eco-friendly. This recurring practice has influenced the innovation of new alternative methods, such as supercritical CO_2 for the dyeing of fabrics. This is a waterless process that requires the use of carbon dioxide in its supercritical state for effective dyeing. Herein, beyond a critical value of temperature and pressure, CO_2 is used under a supercritical conditions [46] as shown in Fig. 5.1a. At a critical point under high pressure and temperature, water is replaced with CO_2 to effectively dye fabrics [44, 47]. A super-critical compressed liquid or fluid is developed when the CO_2 is subjected to a pressure above 74 bar and 31 °C of temperature [48]. Herein, the CO_2 behaves as a solvent for the dyes to effectively carry the colouring matter evenly into the pores of the fibres. This is made possible due to the solvation characteristics and high diffusion properties of a liquid and gas, respectively [44].

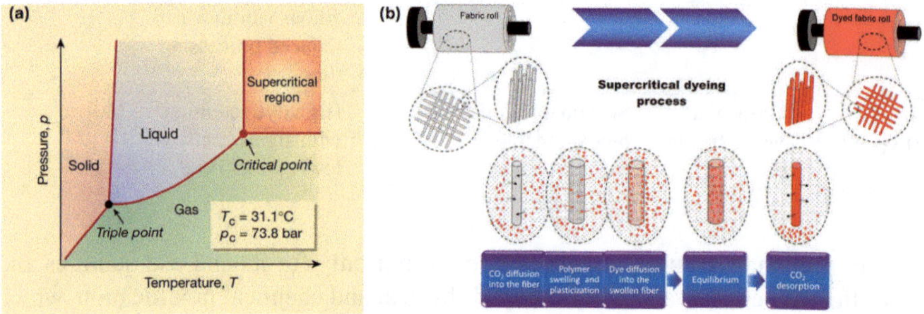

Fig. 5.1 **a** Supercritical CO_2 phase diagram [46] Copyright 2000 Macmillan Magazines Ltd., **b** Supercritical CO_2 dyeing processes. Reproduced with permission from [51] Copyright 2021 Elsevier

Table 5.3 Summary of important features of the supercritical CO_2 process [18]

Key components	Description
Main power source	Supercritical CO_2 as a fluid
Working condition	High pressure and temperature are necessary
Synthetic and natural dye	The technology developed and optimised for the dyeing of polyester, acetate and nylon, while some experimental studies on natural fibres are available
Colourfastness	Good colourfastness
Design potential	The dye is distributed evenly over the fabric
Speed and efficiency	Due to favourable diffusion properties, the times needed for the dissolution of the solid dyestuff will be cut to a negligible minimum
Uniformity of dye distribution	The non-uniformity of the fluid may result in staining of the textile and affect the evenness of the dye
Co-solvent	The addition of small amounts of a co-solvent can considerably increase the solubility of solutes in $scCO_2$
Water and energy use	CO_2 dyeing is a dry process, eliminating the need to evaporate water. Technology is very energy efficient
Reuse	CO_2 can be vacuumed out after use, allowing for 95% recovery and reuse
Environmental sustainability	Ecologically harmless, non-toxic and non-explosive. A green method for the sustainable and eco-friendly textile industry

The key principle of supercritical CO_2 is solubility and transportation of dyes. This method directly dissolves and ensures a uniform absorption of dyes without additional chemicals. As such, $scCO_2$ promotes deep dye penetration even in polyester fibres for improved dye fixation. In principle, $scCO_2$ dissolves and diffuses dye into the fibre matrix for penetration and absorption. The CO_2 returns to a gas state upon depressurisation, a situation which remove the CO_2 from the fibre matrix for easy recovery and reuse [49]. In this process, there is elimination of wastewater treatment due to the recoverability, purification and reuse of $scCO_2$. Figure 5.1b illustrates the physiochemical phenomena with the interaction of fibres and CO_2 for the effective dyeing process. Due to this, fluids in their supercritical state are widely applied in chemical reactions, impregnation, sterilisation, purification, concentration, separation and extraction stages of the industrial processes [50]. This ensures energy efficiency where operational costs and energy consumption are reduced.

The application of the supercritical CO_2 dyeing process (Table 5.3) leads to great savings on energy and water consumption as compared to conventional wet dyeing [51]. For example, to dye textile materials of 1 kg weight, large volumes of clean water (approximately 100–150 L) is required [52]. The recovery of residual dyes from the process is also possible to help limit any environmental impacts. With less time for dyeing and no pollution to the air, this process does not use any harmful chemicals to facilitate the penetration of the dyes into the fibres [49]. Aside from this, no required cost is incurred for drying the fabrics since the release of the CO_2

already dries the fabrics. With the use of a small dye concentration, the eventual colour shade is achieved [47]. Based on the benefits of supercritical CO_2 dyeing, it has found various applications to dye natural and synthetic fibres to understand the related effectiveness of the process. There limitations when applying supercritical CO_2 to dye cellulose-based fabrics. For example, due to their hydrophilic nature, there is reduced active compounds, diffusion of dyes and swelling of fibres to ensure the penetration of colouring matter. This further affects fibre hydrophobization and molecular adaptation [53]. However, a study by [53] showed the effectiveness of using supercritical CO_2 to dye cotton with reactive disperse dyes containing mono- and bi-acyl fluoride groups. This dyeing process required no co-solvents and pretreatments. To improve dyeing ability and solubility, butyl grafted with curcumin in the presence of dimethyl sulfoxide (DMSO) was fabricated for the supercritical dyeing process in silk [54]. This process solves the challenges of the low solubility of curcumin, which yields low colour depths in natural fabrics [55].

5.5 Air Dyeing

The wet processes involved in textile embellishment contribute to significant water consumption and contamination, with the industry estimating about 17–20% of today's pollution due to colouring treatment, using about 72 toxic chemicals in water supplies. To reduce these water contaminations, a new technology, air-dyeing, has emerged. This involves the application of colour to textiles/fibres in the absence of water [56]. According to Mahmud et al. [18], Colorep, a sustainable technology organisation, proposes an air-dyeing process that requires atomisation of the dyeing liquor, which is then mixed with high-pressure airflow. This is then conveyed onto the fibre substrate to impart colour embellishment. This demonstrates a sustainable approach with benefits including 95% less water usage, producing less or no wastewater, emitting 84% less greenhouse gases and 87% less energy [57], unlike the conventional method [58]. This innovative dyeing technique reduces fabric damage from 10% of the conventional method to 1%. Table 5.4 illustrates the key components of the air-dyeing process. Nevertheless, it is advantageous to wash fastness at any temperature and allows for new designs where both sides of a single piece of fabric can be dyed differently [58].

5.6 Plasma Dyeing

Plasma dyeing is an innovative and eco-friendly method generally applied to enhance the uptake of dyes from the textile materials. For example, in a recent study by Serrano-Martinez [59], an oxygen plasma of low pressure was used to activate the surface of a cotton material to improve dye uptake (Fig. 5.2). Because it modifies the materials' surface and alters their chemical composition, plasma application

Table 5.4 Summary of important features of the air-dyeing process [18]

Key components	Description
Main power source	Uses air instead of water
Working condition	Dyeing liquor is first atomised, then mixed with high-pressure airflow, and finally sprayed on fabric to be dyed
Synthetic and natural dye	The technique works on synthetic, dyeing of fabrics with specialised treatment of raw cotton is reported
Colourfastness	Good colourfastness
Design potential	Print or dye both sides of the fabric simultaneously
Speed and efficiency	Colouring and printing can be done in a short time with full accuracy and efficiency. A reduction in the overall process time of approximately 25%
Uniformity of dye distribution	Unlimited flexibility concerning all-fibre except for pure wool
Co-solvent	Dyes penetrate deeply into the filaments of the yarn to create rich, deep, and brilliant colours
Water and energy use	Use 90–95% less water and save 86% of the energy
Reuse	Air-dye recycles paper used in the process, and the dyes are inert, which can go back to their original state and be reused
Environmental sustainability	Causes minimum harm to the environment and reduces the industry's share of global warming by 84%

to textile materials is a pre-treatment and finishing technique that has generated a lot of interest. A novel and environmentally friendly substitute for the traditional wet-chemical method used in the textile industry is plasma treatment [60]. It offers a more efficient and sustainable approach, inducing physicochemical alterations on textile surfaces when subjected to a low-temperature, ionised gas (like oxygen, argon, nitrogen, etc.,) to improve dyeability [61], while conserving energy due to its rapid treatment times [62–64].

5.6.1 *Plasma Types*

In and of itself, plasma is a gas whose atoms become ionised when heat is applied because one or more of their electrons have disengaged. However, the detached electrons are still present in the gas volume, which is electrically neutral overall. Therefore, plasma is any ionised gas made up of almost equal amounts of positive and negative ions. Plasma is basically of two types; thus, low and high temperature plasma or classified into cold/non-thermal and hot/thermal plasma depending on the temperature of the plasma zone, in Jelil's [65] assertion. It can be classified on different basis, including pressure—low (0.01 kpa) and atmospheric (100 kpa), temperature of electrons and ions—hot (10,000+0 °C) and cold plasma (100–0 °C), frequency of power supply—low (40 kHz), radio (13.56 MHz) and microwave

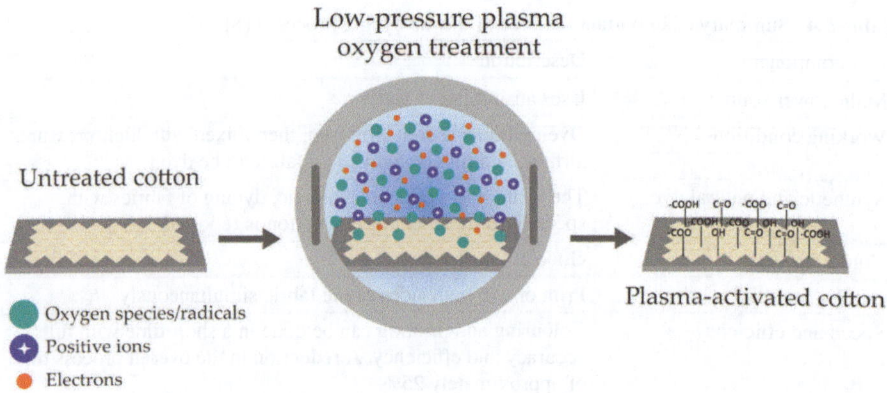

Fig. 5.2 Plasma activation treatment for cotton [59] (This article is an open access article distributed under the terms and conditions of the Creative Commons Attribution (CC BY) license)

frequency (2.56 GHz). Low temperature plasmas (LTPs), which can be used to modify the surface of heat-sensitive polymeric and textile materials [65], are ionised gases or occasionally liquids that are produced at gas pressures ranging from 0.1 to 2 Torr [66]. LTP can be produced by creating a radiofrequency resonant current in a coil or by putting an electrical field across two electrodes with a gas between them. The reactive species found in LTP, including fast-moving electrons, ions, free radicals, metastables, and photons, can start physicochemical reactions on the textile surface and give it the desired functionalities without changing the bulk characteristics of the fibres that make it up [67]. In contrast, a high voltage, high temperature arc can be used to create high temperature plasma artificially. This is the basis for both the corona discharge process and the plasma torch that is used to vaporise and redeposit metals.

5.6.2 Plasma Application to Increase Dyeability

Syibyan et al. [68] reported on the glow discharge plasma surface modification of cassava starch-based bioplastic sheets for enhanced hydrophilicity and adhesion properties, to improve their dye absorption. Various treatments involving oxygen (O_2), air, and nitrogen (N_2) were analysed in this work, along with different times and powers of treatment. Absorption tests performed indicate that the plasma-treated sheets retain natural dyes, indigofera at 657 nm and tegeran at 414 nm-better. Durability tests confirm that plasma treatment improves dye adherence and colour intensity; this is attributed to increased surface roughness, with the formation of polar functional groups enhancing wettability and chemical interactions [69]. Among these, the oxygen plasma treatment proves most effective, and the O_2/30 W/3 min sample showed the highest absorbance values. Another example by Mostafa et al.

[70] used low-temperature N_2 plasma to create N_2-containing groups on the cotton fabric surface to enhance cotton dyeability with Acid Blue 284, which does not normally have a direct affinity for cotton. In the FTIR results, N_2-containing groups significantly improved the treated cotton textiles' wettability, dyeability, and overall fastness qualities. However, when compared to untreated fabric, the mechanical and optical qualities of the treated cotton fabrics declined.

Vankar and Gangwar [71] further investigated the effectiveness of plasma in the natural dyeing of polyester fabric using four natural dyes in the presence of benign mordant alum. The fibre showed wettability increase due to surface modification, leading to better dye uptake and adherence. They concluded that Plasma-mediated coupled with alum-mordanted dyeing method exhibited optimal natural dyeing of polyester, attaining remarkable fastness properties against untreated alternatives, showing desirable results in 2 min of dyeing action. Meng et al. [72] similarly enhanced cotton-dye uptake synergising diallyldimethylammonium chloride (DADMAC)—with its tiny molecular weight that can easily enter the fibre interior to give a uniform and level dyeing experience; and an oxygen plasma for the graft polymerisation of DADMAC onto cotton fabrics to change the surface of the cotton, making it more receptive to dye and increase in colour increase (see Fig. 5.3).

Although all research agrees that plasma treatment improves dyeing on various textile and bioplastic substrates, the mechanisms and results differ. Syibyan et al. [68] and Meng et al. [72] highlight the importance of surface roughness and functional group creation in the better adhesion of the dye. Mostafa et al. [70] pointed out, however, the trade-offs in nitrogen plasma application, where gains in dyeability are made at the cost of mechanical and optical fabric properties. Vankar and Gangwar

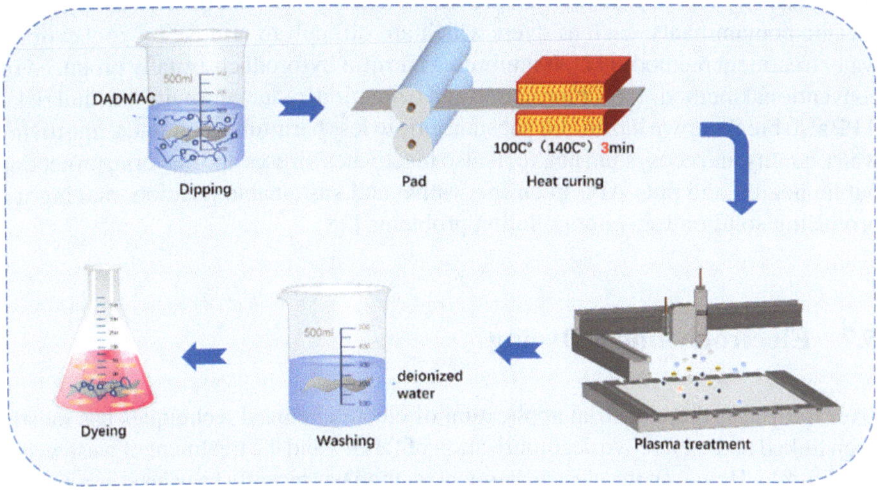

Fig. 5.3 Cationic modification of cotton for dyeing. Reprinted with permission from [72] Copyright 2021 The Korean Fiber Society for Fibers and Polymers and Springer

[71] illustrate plasma treatment combined with mordants for natural dyeing optimisation. Together, these investigations illustrate the flexibility of plasma technology for altering surface characteristics to improve dyeing uptake, yet material degradation and process optimisation issues need to be more thoroughly investigated.

5.6.3 *Environmental Impacts*

Plasma technology represents an interesting possible alternative to traditional textile dyeing processes [73]. This would otherwise allow partially treated and untreated effluent to be discharged into waterways and land, which influences the soil, surface water, groundwater, biodiversity and the natural ecosystem at large. However, the application of plasma technology exhibits sustainable impacts like saving a large quantity of water, electrical energy and chemicals by avoiding the generation of effluent or hazardous by-products [74]. Additionally, plasma technology converts biomass waste to hydrogen and methane, reducing the negative impact of petroleum-based fuel [75]. Oost [76] concludes that thermal plasma treatment generally impacts positively. It converts organic waste to synthesis gas (syngas) and inorganic materials to inert, non-leaching vitrified slag, removing the risk of toxic leachates. It is also a closed-loop process, minimising toxic emissions and optimising energy efficiency. In comparison with traditional waste treatment processes, plasma treatment releases fewer exhaust gases, i.e., less air pollution. Its compatibility with a large variety of wastes and chemical compositions further adds to its environmental advantages, rendering it a green and environmentally friendly waste treatment technology. Nevertheless, atmospheric pressure plasma (APP) is effective in degrading organic contaminants, such as dyes, which are difficult to remove by conventional water treatment methods [77]. It minimises harmful by-products usually produced by conventional methods like chlorination and ozonation, reducing environmental risks. APP also breaks down hazardous substances into less harmful compounds, improving water quality and ecosystem health. It also inactivates viruses and bacteria, protecting public health, and puts APP in an innovative and sustainable position, making it a promising solution for water pollution problems [78].

5.7 Electrochemical Dyeing

Over the years, the industrial application of electrochemical techniques has mostly been linked to the effective decolourisation of fabrics and the treatment of wastewater [17, 79–81]. However, the electrochemical method has recently been used as a significant dyeing approach in the textile industry [82]. In indigo dyeing, a reduction process is essential to render the dye water-soluble, which is necessary to have a good affinity for cellulose fibres, as the indigo dye is water-insoluble in its original state [83, 84]. The use of traditional approaches such as sodium dithionite in the reduction process is reported to have ecological challenges, as it releases harmful sulfur-related chemicals such as sulfites, sulfates, and sulphides, hindering sustainable dyeing [85–87]. In this instance, the electrochemical approach to indigo reduction is deemed an appropriate method for sustainable dyeing [87, 88]. In the electrochemical dyeing process, the dye reduction is done by transferring electrons from electrodes with the help of an electric current toward eco-friendly dyeing [89]. This technology requires no water, uses lower chemical concentration and a shorter dyeing period being the underlying advantages [19]. The electrochemical dyeing process was used to dye a denim fabric as a piece of evidence to apply this method in a sustainable textile dyeing approach [82]. Figure 5.4 shows the electrochemical method of dyeing textile materials. In the experiment, as described in the study [82], the exhaustion method was used in the dyeing process of a 5 g denim fabric, using a liquor ratio of 20:1. Using a cathode electrolyte, the fabric was dyed for a minute, and subsequently oxidised for 5 min in the air, and washed in distilled water for effective neutralisation.

The principle of the electrochemical method of textile dyeing is explained in a previous study as a simple replacement of a reducing agent with electrons to

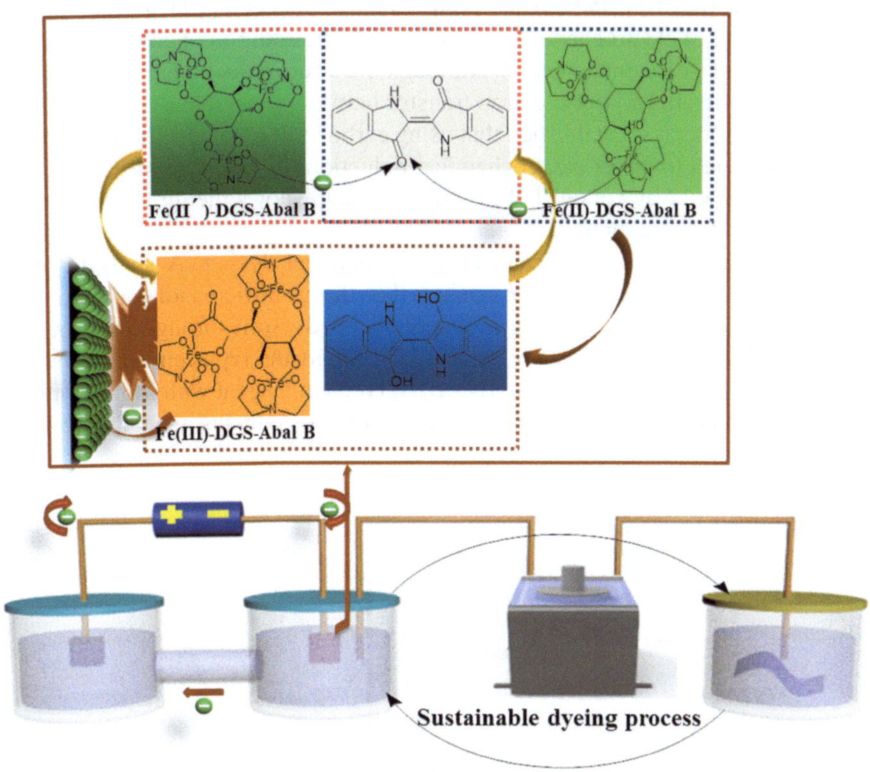

Fig. 5.4 The electrochemical method of textile dyeing. Reprinted with permission from [82], Copyright 2024 Elsevier B.V.

facilitate a process that does away with wastewater for environmentally safe dyeing [90]. This indicates the need to adopt this eco-friendly approach to dyeing by textile industries to mitigate the harmful ecological effects of conventional dyeing. The two main approaches to electrochemical dyeing, especially for vat dyes, are the direct and indirect reduction methods. The direct method of electrochemical dyeing involves the direct reduction of organic dyestuffs through contact between the electrode surface and the dye [91]. This method was applied to dye cotton fabric as an alternative way to sustain the environment in the textile dyeing process, with two major aims: applying various electrical potentials in the cotton fabric dyeing and embedding the fabric onto an electrode to improve its dyeability [92]. The study concludes by establishing that the experiment therein is an environmentally safe approach and a sustainable energy method worth adopting for cotton fabric dyeing. On the other hand, the indirect method of electrochemical dyeing does not involve dye reduction at the electrode surface. Instead, a reducing agent is added, which is oxidised after dye reduction. It is then reduced again at the cathode surface and reused in a cyclic operation [91]. This method is said to be a cost-effective and excellent approach to recycling effluents toward sustainable dyeing [82, 93]. In treating wastewater containing reactive dyes and reusing such effluents, Sala et al. [94] posited that the indirect method of electrochemical oxidation with the addition of chlorine provides the best effluent treatment approach. Although the direct reduction method has been reported to suffer from poor electron transfer between insoluble dye particles and the solid electrode, it is established that using carbon felt as an electrode material is the best approach to direct electrochemical reduction of vat dyes in textile dyeing, especially with CI Vat Yellow 1 [95]. Figure 5.5a–b shows the direct and indirect electrochemical dye reduction.

Reducing agents are significant in electrochemical dyeing. A study to investigate appropriate reducing agents posited that the iron-triethanolamine complex could be used as a reducing agent in dispersed vat dyestuff through indirect electrolysis [96]. Also, Bechtold and Turcanu [97] report the replacement of environmentally harmful sodium dithionite ($Na_2S_2O_4$) reducing agent with iron-complexes with triethanolamine and Na-*d*-gluconate as a cathodic reducing agent that could be

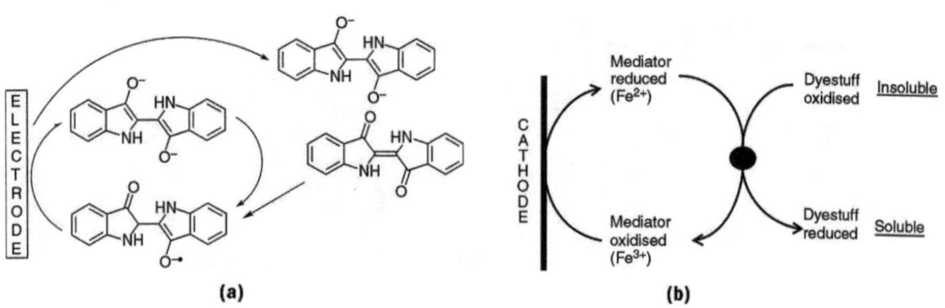

Fig. 5.5 Electrochemical dyeing. **a** Direct dye reduction. **b** Indirect dye reduction. Reprinted with permission from [96] Copyright 2007 Wiley VCH

regenerated through electrochemical means for eco-friendly dyeing process. Findings indicate that a significant amount of chemical consumption and wastewater volume was decreased through the ultrafiltration method of dyebath regeneration. It is also presented in a study that the Ca^{2+}–Fe^{3+}-triethanolamine redox system is an effective reduction approach for multiple applications in indirect electrochemical dyeing, including the reduction of indigo, vat dyes, pollution emissions, and electrolyte recycling [98]. As a step to substituting $Na_2S_2O_4$ reducing agents with sustainable ones in indigo dyeing, Saikhao et al. [99] investigated the use of monosaccharides such as galactose, glucose, and fructose coupled with reducing disaccharides including lactose and maltose, as sustainable reducing agents for cotton fabric dyeing. Findings revealed the suitability of reducing sugar in indigo reduction, as less than −700 mV reduction–oxidation of the dye liquor was observed (remained stable for an hour) when reducing sugars were added within 10 min. It could be said that employing the right reducing agent facilitates smooth and effective eco-friendly dyeing in the electrochemical textile dyeing method.

5.8 Case Studies of Real-World Industry Applications

The electrochemical reduction approach has been adopted in several areas, especially the textile industry for sustainable manufacturing, since it is an ideal method of reducing dyes and minimising pollution caused by harmful chemicals [95]. The textile industry observes that the electrochemical dyeing method is the safest technique for ensuring eco-friendly dyeing where the volume of effluents and chemical consumption are minimised [97]. As an alternative to conventional dyeing, the electrochemical approach promises eco-friendly dyeing and a cost-effective method that saves about 63.5% of manufacturing costs [82]. This has led to the continuous adoption of the technique in several industries for wastewater treatment, including the environmental sector, carbon dioxide (CO_2) reduction for fuel production, metal refining and recovery in the metallurgy/materials industry, printed circuit boards manufacturing in the electronics industry, etc. Pollution of the environment with unsafe chemicals through the actions of industries like the textile industry could be addressed through electrochemical technology as it is an effective solution to chemically polluted environments such as water bodies [100, 101]. The real-world industry applications of the electrochemical reduction technique, including its effective use in textile dyeing as reported in previous studies, are summarised in Table 5.5.

Past studies also report using an electrochemical approach in metal refining and recovery. A study exploring the extraction of metals and alloys with electrochemical techniques concludes that the technique is used to produce metals that are of high purity and require no further purification processes [113]. The electrochemical technique is also employed in a recent study to extract metals from primary and secondary sources for an eco-friendly metal recovery due to its high selectivity property for complex mixtures [114]. As presented in Fig. 5.6, the ability of specific compounds to be selected out of mixed materials calls for the use of the electrochemical process

Table 5.5 Summary of case studies of industry application of electrochemical reduction technique

Application	Description	References
Textile industry	Electrochemical dyeing with direct and indirect and indirect approaches for sustainable textile colouration in the textile industry	[96]
	The use of electrochemical techniques in textile dyeing (with vat and sulfur dyes), fibre manufacturing, fabric decolourisation, and wastewater treatment through colour removal from effluents	[17]
	Electrochemical dyeing of cotton fabric with reactive blue X-BR using nickel electrode, sodium carbonate (fixing alkali), and anodic regions. This is regarded as a good alternative to traditional reactive dyeing of cotton fabric	[102]
	Using the electrochemical technique to treat wastewater containing reactive dyes in the textile industry through colour removal and generating hydrogen	[103]
	Degradation of reactive dyes, dye mixture, and treatment of textile effluents through the electrochemical oxidation approach for decoloured water that could be used for agricultural purposes	[104]
Environmental sector	Treatment of soil-washing effluent with the electrochemical technique through the use of boron-doped diamond electrodes for the removal of pollutants, including hydrocarbons, dyes, pesticides, etc.	[105]
	Using electrochemical advanced oxidation processes to treat environmentally harmful soil-washing solutions	[106]
	Soil remediation through electrochemically-assisted thermal-based technologies	[107]
	Degradation of reactive dyes in wastewater from the textile industry in Ethiopia with an electrochemical approach for a safe environment. As reported in the study, the electrochemical method effectively removed 94.89% of dyes from the wastewater with a sodium chloride electrolyte concentration of 5.1 g/L, 17.0 min reaction duration, and a pH of 6.9	[108]
	An exploration of various substrates for preparing an electrode (dimensionally stable anode) for electrochemical treatment of wastewater containing reactive dyes	[100]
	For removal of nutrients such as nitrogen and phosphorus in wastewater with advanced electrode materials including carbon-based high-performance metal-oxides	[109]
Fuel production	Using an electrochemical approach to produce fuels and chemicals from a biomass-based feedstock for sustainable production in the fuel industry	[110]
	Storage of renewable energy in chemical form through carbon dioxide reduction with an electrochemical method	[111]
	Using reactive oxygen from water electrolysis species to remove sulfur compounds from fuel under the electro-oxidation desulfurization method	[112]

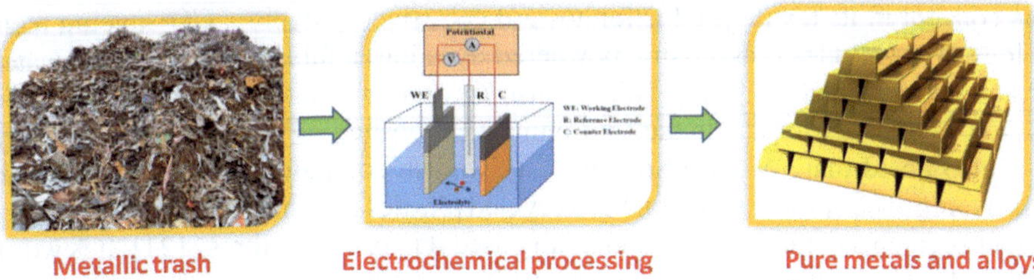

Fig. 5.6 Electrochemical processing of pure metals and alloys from primary and secondary sources. Reprinted with permission from [114] Copyright 2024 Elsevier B.V.

to obtain pure and high-quality metals from primary and secondary resources. In this instance, metallic trash is carefully treated with three main electrodes, including working electrodes, reference electrodes, and counter electrodes, in dissolution of the trash into electrolyte solution and reducing metal ions on the electrode. The resulting metals and alloys are regarded as pure, as confirmed by Shamsuddin and Sohn [113] who buttress their assertion that the electrochemical process is the best metal recovery technique, requiring no further purification.

From the application of the electrochemical method, the textile industry stands as a potential beneficiary due to the eco-friendly nature of the approach in industrial applications. The textile industry is known for its massive generation of pollutants [115–117]. Adopting the electrochemical dyeing approach would mitigate wastewater generation and reduce chemical application in the industry.

5.8.1 *Key Brands and Companies Adopting Waterless Dyeing Technologies*

It is observed that the unsustainable dyeing approach, where $Na_2S_2O_4$ is used as a reducing agent, has been acted upon by some brands who have thought of going green in their manufacturing process in the textile industry. With the conventional textile dyeing method, effluents from textile industries contain harmful chemicals that are unsafe for the environment due to the use of water in the process. Recently, many companies have focused on going waterless in their textile dyeing operations to ensure sustainable production. Brands adopting or developing sustainable dyeing techniques, specifically waterless technologies, recently include Adidas, Nike, IKEA, ColorZen, We aRe SpinDye, etc. Adidas is known for its partnership with DyeCoo, a Dutch company that, together with Adidas, is championing the waterless dyeing technology to sustain the environment [118]. In waterless dyeing technology, especially with CO_2, water is substituted for CO_2 in textile dyeing by supercritical CO_2 formation, where it is heated above its critical point to be transformed into a state that renders it both gas and liquid-like characteristic [119]. IKEA has employed this concept in its textile production toward eco-friendly dyeing. CO_2 dyeing technology also includes pressurising powdered dyes into a fabric without using water, as adopted by DyeCoo [120]. Nike is also a company that has adopted waterless dyeing technology to sustain the environment, as posited by Hapburn's [120] study. It is also noted that 'We aRe SpinDye', a Swedish company, is engaged in waterless dyeing. The company first melted recycled plastics and then adding the preferred colour pigments to the material, and finally used it to produce fibres [121]. It is worth noting that the technology of waterless dyeing is gradually being adopted by many textile brands for sustainable development, although challenges lie with production costs.

5.9 Challenges and Regulatory Issues

The issue of challenges and regulatory concerns affects the widespread adoption of the various waterless dyeing technologies in the textile industry to help limit environmental impacts. The waterless dyeing process has potential benefits of reduced chemical usage, savings on energy and reduction in the release of harmful substances or effluents to the environment. Aside from these benefits, several challenges have influenced the adoption of this innovative process to reduce environmental impacts. These include technical challenges to handle the technology, limited compatibility of fabrics and the related cost and investment required for the waterless technologies in the industry. The waterless dyeing process requires the use of specialised equipment under optimal conditions to ensure the desired outcomes are achieved. As a result, this equipment will require trained technical personnel to operate and maintain it under the appropriate conditions. This is because any changes in the optimal conditions could influence the consistency and quality of the fabric colours. The price and cost of the appropriate waterless dyeing technologies for the dyeing process also raise serious concerns. Due to the technical nature of the machines, the cost of purchasing and installation for the dyeing process is higher as compared to conventional dyeing machinery [119]. This would require heavy investment from both small and medium companies in the industry [122], a situation that would limit these companies from fully using these innovative technologies in their dyeing processes. It is thus imperative for a reduction in the cost of purchase and installation of these machines to ensure that companies and the environment benefit. The compatibility of fabrics for the dyeing process is another challenge limiting the full utilisation of waterless dyeing technologies. This dyeing process cannot be applied to all fabric types, hence limiting the operations of the dyeing process in the industry. The cost implications would mean that companies, after investing heavily in the appropriate machinery, can only dye certain types of fabrics. For example, supercritical CO_2 is widely applied for dyeing polyester fabrics than natural fabrics [49]. Such difficulty can limit the adoption and application of some waterless dyeing technologies in certain areas of the textile industry.

Aside from these challenges, several regulatory issues limit the full realisation of the benefits of various waterless dyeing techniques. Key regulatory issues include chemical handling and safety, emissions and use of energy, market certifications, international standard compliance and safety of workers. For example, the effects of chemical handling for dyeing without proper protective measures could endenger workers safety upon exposed. Hence, critical measures are required. There should be regulatory measures which mandate strict safety protocols to protect workers. Certifications of these waterless dyeing methods also remain another major regulatory issue. This is to demonstrate their compliance with safety and environmental standards.

5.10 Conclusions and Future Research Directions

The advancement of waterless dyeing techniques has the potential to contribute significantly to sustainable practices in the textile industry. These techniques seek to replace the conventional dyeing process that utilises large volumes of water and dyes, consumes high energy and produces large quantities of wastewater that have serious environmental impacts. These waterless dyeing methods aim to reduce the negative impacts of conventional dyeing methods to ensure energy efficiency, reduced water usage, and reduced environmental threats. Thus, future research directions must advance the development, utilisation and application of these waterless dyeing methods for a more sustainable dyeing practice. Key research directions could be advanced under innovations, standardisation and industry collaborations, scalability and awareness. Innovations in waterless dyeing technologies will revolutionise the colouration process in the textile industry. These dyeing technologies offer a sustainable approach to ensure the carbon footprint is reduced. This will further lead to significant contributions to sustainability in the environment, a reduction in the release of toxic waste. For example, future research studies focused on areas to ensure material compatibility and optimisation of processing for supercritical CO_2 will lead to enhanced parameters to dye various fibre types. Other areas of innovation, such as solubility behaviour, synthesis of new dyes and modification of chemicals for supercritical CO_2, are important research focuses [123]. Additionally, adopting these green and sustainable approaches will lead to reduced use of harmful chemicals and materials, which pose environmental and health threats [123]. Developing standards and regulations for the various waterless dyeing technologies is critical to enhance their usage and safety. It is thus imperative for collaboration between regulatory bodies, industry and researchers to develop key standards governing the utilisation of these technologies for improved sustainability, safety and quality. For effective usage and scalability of these technologies, there should be energy efficiency and lower production costs. This will ensure that manufacturers have a more cost-effective dyeing process line for large-scale use.

References

1. Hasan, K., et al. 2016. Role of textile and clothing industries in the growth and development of trade & business strategies of Bangladesh in the global economy. *International Journal of Textile Science* 5 (3): 39–48.
2. Desore, A., and S. A. Narula. 2018. An overview on corporate response towards sustainability issues in textile industry. *Environment, Development and Sustainability* 20 (4): 1439–1459.
3. Hossain, M. S., et al. 2018. Reuse of textile mill ETP sludge in environmental friendly bricks—effect of gamma radiation. *Radiation Physics and Chemistry* 151: 77–83.
4. Ammayappan, L., S. Jose, and A. Arputha Raj. 2015. Sustainable production processes in textile dyeing. In *Green Fashion*, vol. 1, 185–216. Springer.
5. Nayak, R., T. Panwar, and L. V. T. Nguyen. 2020. 1—Sustainability in fashion and textiles: A survey from developing country. In *Sustainable technologies for fashion and textiles*, ed. R. Nayak, 3–30. Woodhead Publishing.
6. Bilińska, L., and M. Gmurek. 2021. Novel trends in AOPs for textile wastewater treatment. Enhanced dye by-products removal by catalytic and synergistic actions. *Water Resources and Industry* 26: 100160.
7. Hassan, M. M., and C. M. Carr. 2018. A critical review on recent advancements of the removal of reactive dyes from dyehouse effluent by ion-exchange adsorbents. *Chemosphere* 209: 201–219.
8. Khatri, J., et al. 2018. Advanced oxidation processes based on zero-valent aluminium for treating textile wastewater. *Chemical Engineering Journal* 348: 67–73.
9. Pathak, B. 2025. What is waterless dyeing? Types of waterless dyeing techniques. March 30, 2025; Available from: https://www.textilesresources.com/articles/what-is-waterless-dyeing-types-of-waterless-dyeing-techniques/
10. Catarino, M. L., F. Sampaio, and A. L. Gonçalves. 2025. Sustainable wet processing technologies for the textile industry: A comprehensive review. *Sustainability* 17 (7): 3041.
11. Allay, A., et al. 2025. Optimizing ethanol-modified supercritical CO_2 extraction for enhanced bioactive compound recovery in hemp seed oil. *Scientific Reports* 15 (1): 8551.
12. Lai, J., et al. 2025. Fluid simulation and dyeing effect of a multilayer dye dissolver for supercritical waterless dyeing. *ACS Sustainable Chemistry & Engineering* 13 (5): 1984–1992.
13. Nitu, N. A., et al. 2024. Wearable colorful nanofiber of thermoplastic polyurethane (TPU) mechanical and colorfastness properties by dope dyeing. *Fibers and Polymers* 25 (7): 2485–2502.
14. Gurudatt, K., et al. 2005. Dope-dyed polyester fibers from recycled PET wastes for use in molded automotive carpets. *Journal of Industrial Textiles* 34 (3): 167–179.
15. Zhang, L., et al. 2017. Dope dyeing of lyocell fiber with NMMO-based carbon black dispersion. *Carbohydrate Polymers* 174: 32–38.
16. Xiong, W., et al. 2021. Electrochemical dyeing of cotton fiber for environmental protection. In *Hydraulic and civil engineering technology VI*, 602–607. IOS Press.
17. Sala, M., and M. C. Gutiérrez-Bouzán. 2012. Electrochemical techniques in textile processes and wastewater treatment. *International Journal of Photoenergy* 2012 (1): 629103.
18. Mahmud, I., and S. Kaiser. 2020. Recent progress in waterless textile dyeing. *Journal of Textile Science and Engineering* 10 (6): 1–3.
19. Babu, K. M. 2022. Overview on significant approaches to waterless dyeing technology. In *Emerging technologies for textile coloration*, 171–188. CRC Press.
20. Ibrahim, N. A., and B. M. Eid. 2018. *Emerging technologies for source reduction and end-of-pipe treatments of the cotton-based textile industry*, 185–202. Handbook of textile effluent remediation.
21. Gürses, A. 2019. 2—Sustainable colorants. In *The impact and prospects of green chemistry for textile technology*, ed. I. Shahid ul and B. S. Butola, 21–55. Woodhead Publishing.
22. Khattab, T. A., M. S. Abdelrahman, and M. Rehan. 2020. Textile dyeing industry: Environmental impacts and remediation. *Environmental Science and Pollution Research* 27 (4): 3803–3818.

23. Islam, M. T., et al. 2022. Synthetic dyes for textile colouration: Process, factors and environmental impact. *Textile and Leather Review* 5: 327–373.
24. Kim, G., et al. 2022. Application of best available techniques to remove air and water pollutants from textile dyeing and finishing in South Korea. *Sustainability* 14 (4): 2209.
25. Pokhriya, P., et al. 2020. Impact assessment of textile effluent on health and microbiota of agricultural soil in Bhagwanpur (Uttarakhand), India. *SN Applied Sciences* 2 (9): 1539.
26. Rudnicka-Kępa, P., and A. Zaborska. 2021. Sources, fate and distribution of inorganic contaminants in the Svalbard area, representative of a typical Arctic critical environment–a review. *Environmental Monitoring and Assessment* 193 (11): 724.
27. Özkan Buzğan, E. B., et al. 2022. Properties of textile air conditioner waste dust-filled polyurethane rigid foam composites. *Plastics, Rubber and Composites* 51 (7): 352–362.
28. Gupta, B. G., J. K. Biswas, and K. M. Agrawal. 2017. Air pollution from bleaching and dyeing industries creating severe health hazards in Maheshtala textile cluster, West Bengal, India. *Air, Soil and Water Research* 10: 1178622117720787.
29. Chena, Y., et al. 2011. Atmospheric pressure plasma vapor treatment of thermo-sensitive poly (N-isopropylacrylamide) and its application to textile materials. *Journal of Fiber Bioengineering & Informatics* 4 (3): 285–290.
30. Eren, S., and İ. Özyurt. 2024. Waterless dyeing of polyamide 6.6. *Polymers* 16 (11): 1472.
31. Saleem, M. A., et al. 2020. Sustainable dyeing of nylon with disperse dyes in Decamethylcyclopentasiloxane waterless dyeing system. *Journal of Cleaner Production* 276: 123258.
32. Saxena, S., A. S. M. Raja, and A. Arputharaj. 2017. Challenges in sustainable wet processing of textiles. In *Textiles and clothing sustainability: Sustainable textile chemical processes*, ed. S. S. Muthu, 43–79. Singapore: Springer Singapore.
33. Banchero, M. 2020. Recent advances in supercritical fluid dyeing. *Coloration Technology* 136 (4): 317–335.
34. Bomgardner, M. M. 2018. These new textile dyeing methods could make fashion more sustainable. *Chemical & Engineering News* 96 (29): 28–33.
35. Penthala, R., et al. 2022. An ecofriendly dyeing of nylon and cotton fabrics in supercritical CO_2 with novel tricyanopyrrolidone reactive disperse dye. *Journal of CO_2 Utilization* 60: 102004.
36. Lee, S. H., and W. S. Song. 2010. Surface modification of polyester fabrics by enzyme treatment. *Fibers and Polymers* 11 (1): 54–59.
37. Kanelli, M., et al. 2017. Surface modification of polyamide 6.6 fibers by enzymatic hydrolysis. *Process Biochemistry* 59: 97–103.
38. Maradini, G. S., et al. 2020. Characterization of polyester nanocomposites reinforced with conifer fiber cellulose nanocrystals. *Polymers* 12 (12): 2838.
39. Eren, H. A., et al. 2020. Sustainable textile processing with zero water utilization using super critical carbon dioxide technology. In *Sustainability in the textile and apparel industries: Production process sustainability*, ed. S. S. Muthu and M. A. Gardetti, 179–196. Cham: Springer International Publishing.
40. Yiğit, İ, et al. 2021. An investigation of process parameters on colour during the dyeing of polyester in supercritical carbon dioxide media. *Coloration Technology* 137 (6): 625–644.
41. Zheng, H., et al. 2017. An ecofriendly dyeing of wool with supercritical carbon dioxide fluid. *Journal of Cleaner Production* 143: 269–277.
42. Beltrame, P. L., et al. 1998. Dyeing of cotton in supercritical carbon dioxide. *Dyes and Pigments* 39 (4): 335–340.
43. Gao, D., et al. 2015. Supercritical carbon dioxide dyeing for PET and cotton fabric with synthesized dyes by a modified apparatus. *ACS Sustainable Chemistry & Engineering* 3 (4): 668–674.
44. Broadbent, P. J., et al. 2023. Supercritical carbon dioxide (SC-CO) dyeing of cellulose acetate: An opportunity for a "greener" circular textile economy. *Coloration Technology* 139 (4): 475–488.
45. Elmaaty, T. A., et al. 2015. Eco-friendly disperse dyeing and functional finishing of nylon 6 using supercritical carbon dioxide. *Fibers* 3 (3): 309–322.

46. Leitner, W. 2000. Designed to dissolve. *Nature* 405 (6783): 129–130.
47. Abou Elmaaty, T., et al. 2020. Development of textile dyeing using the green supercritical fluid technology: A review. *Materials International* 2: 373–390.
48. SciMed. 2024. What is a supercritical fluid? 2 Feb 2025; Available from: https://www.scimed.co.uk/education/what-is-a-supercritical-fluid/
49. Agrawal, B. J. 2015. Supercritical carbon-dioxide assisted dyeing of textiles: An environmental benign waterless dyeing process. *International Journal of Innovative Research and Creative Technology* 1 (2): 201–205.
50. Nikolai, P., et al. 2019. Supercritical CO_2: Properties and technological applications—a review. *Journal of Thermal Science* 28 (3): 394–430.
51. Goñi, M. L., N. A. Gañán, and R. E. Martini. 2021. Supercritical CO_2-assisted dyeing and functionalization of polymeric materials: A review of recent advances (2015–2020). *Journal of CO_2 Utilization* 54: 101760.
52. Eren, H. A., O. Avinc, and S. Eren. 2017. Supercritical carbon dioxide for textile applications and recent developments. *IOP Conference Series: Materials Science and Engineering* 254 (8): 082011.
53. Yang, D.-F., et al. 2017. Dyeing of cotton fabric with reactive disperse dye contain acyl fluoride group in supercritical carbon dioxide. *Dyes and Pigments* 139: 566–574.
54. Liu, M., et al. 2018. Eco-friendly curcumin-based dyes for supercritical carbon dioxide natural fabric dyeing. *Journal of Cleaner Production* 197: 1262–1267.
55. Zhan, S., et al. 2017. Measurement and correlation of curcumin solubility in supercritical carbon dioxide. *Journal of Chemical & Engineering Data* 62 (4): 1257–1263.
56. Rani, L., and N. Arya. 2019. Dyeing and finishing of fabrics with low or without water. *The Pharma Innovation Journal* 8 (6): 1159–1165.
57. Kant, R. 2012. Textile dyeing industry an environmental hazard. *Natural Science* 4 (1): 5.
58. Kumar, D., et al. 2010. Dyeing without water. *Journal of Textile Association* 71: 171–174.
59. Serrano-Martínez, V. M., et al. 2024. Improving cotton fabric dyeability by oxygen plasma surface activation. *Surfaces* 7 (4): 1079–1095.
60. Eyupoglu, C., et al. 2024. Natural dyeing of air plasma-treated wool fabric with *Rubia tinctorum* L. and prediction of dyeing properties using an artificial neural network. *Coloration Technology* 140 (1): 91–102.
61. Nourin, S., et al. 2024. Dyeing jute-cotton blended fabric with natural dyes: A chemical-free approach of LPGD plasma technology using O_2 and N_2 gases. *Colloids and Surfaces A: Physicochemical and Engineering Aspects* 691: 133773.
62. Nageswaran, G., L. Jothi, and S. Jagannathan. 2019. Chapter 4—plasma assisted polymer modifications. In Non-*thermal plasma technology for polymeric materials*, ed. S. Thomas, et al., 95–127. Elsevier.
63. Naveed, M. 2018. Plasma treatment as green technology for dyeing of textile fabrics. *Trends Textile Engineering Fashion Technology* 2: 168–173.
64. Shakeri, M., and A. Bashari. 2022. Plasma technology for textile coloration: solutions for green dyeing of textiles. In *Emerging technologies for textile coloration*, 203–216. CRC Press.
65. Jelil, R. A. 2015. A review of low-temperature plasma treatment of textile materials. *Journal of Materials Science* 50 (18): 5913–5943.
66. Adamovich, I., et al. 2017. The 2017 plasma roadmap: Low temperature plasma science and technology. *Journal of Physics D: Applied Physics* 50 (32): 323001.
67. Shishoo, R. 2007. *Plasma technologies for textiles.* Woodhead Publishing.
68. Syibyan, F. L., et al. 2024. Plasma functionalization for enhanced natural dye affinity on bioplastics: A novel approach. *Plasma Processes and Polymers* 21 (12): 2400184.
69. Arolkar, G. A., et al. 2015. The study of air-plasma treatment on corn starch/poly(ε-caprolactone) films. *Polymer Degradation and Stability* 120: 262–272.
70. Mostafa, K., H. Ameen, and A. Medhat. 2024. Harnessing of low-temperature nitrogen plasma technique as an eco-friendly approach for dye-ability of cotton fabric with acid dye. *Pigment & Resin Technology* 53 (4): 425–433.

71. Vankar, P. S., and A. Gangwar. 2024. Natural dyeing mediated by atmospheric air pressure plasma treatment of polyester. *Pigment & Resin Technology* 53 (5): 569–575.
72. Meng, X., et al. 2021. Enhanced dyeability and wash fastness through a salt-free plasma-induced grafting of cationic monomers on cotton fabrics. *Fibers and Polymers* 22 (12): 3378–3384.
73. Zille, A. 2020. 6—Plasma technology in fashion and textiles. In *Sustainable technologies for fashion and textiles*, ed. R. Nayak, 117–142. Woodhead Publishing.
74. Yacoub, A. A., and B. H. Patel. 2024. Atmospheric plasma technology for sustainable wet processing of textiles. *Materials Today: Proceedings.*
75. Tsai, C.-H., and K.-T. Chen. 2009. Production of hydrogen and nano carbon powders from direct plasmalysis of methane. *International Journal of Hydrogen Energy* 34 (2): 833–838.
76. Oost, G. V. 2022. Applications of thermal plasmas for the environment. *Applied Sciences* 12 (14): 7185.
77. Sanito, R. C., S.-J. You, and Y.-F. Wang. 2022. Degradation of contaminants in plasma technology: An overview. *Journal of Hazardous Materials* 424: 127390.
78. Yin, Y., et al. 2023. Recent progress in applications of atmospheric pressure plasma for water organic contaminants' degradation. *Applied Sciences* 13 (23): 12631.
79. Doğan, D., and H. Türkdemir. 2005. Electrochemical oxidation of textile dye indigo. *Journal of Chemical Technology & Biotechnology* 80 (8): 916–923.
80. Dube, A., et al. 2025. Electrochemical water treatment: Review of different approaches. *Journal of Environmental Management* 373: 123911.
81. Muddemann, T., et al. 2019. Elektrochemische Reaktoren für die Wasserbehandlung. *Chemie Ingenieur Technik (CIT)* 91 (6).
82. Li, X., et al. 2020. Sustainable electrochemical dyeing of indigo with Fe(II)-based complexes. *Journal of Cleaner Production* 276: 123251.
83. Kabish, A. K., et al. 2023. The importance of natural indigo dye and its revitalization and Ethiopian potential for indigo growing. *Advances in Materials Science and Engineering* 2023 (1): 2135014.
84. Fobiri, G. K. 2022. Synthetic dye application in textiles: A review on the efficacies and toxicities involved. *Textile & Leather Review* 5: 180–198.
85. Blackburn, R. S., T. Bechtold, and P. John. 2009. The development of indigo reduction methods and pre-reduced indigo products. *Coloration Technology* 125 (4): 193–207.
86. Silva, L. M. C., et al. 2024. Insights into the dyeing using natural indigo (*Indigofera tinctoria*): Toward an environmentally friendly garment. *Coloration Technology* 140 (1): 114–124.
87. Yi, C., et al. 2020. Practical and environment-friendly indirect electrochemical reduction of indigo and dyeing. *Scientific Reports* 10 (1): 4927.
88. Bechtold, T., and T. Pham. 2023. Production scale test for indirect cathodic reduction of indigo opens a route for greener denim. *Sustainable Materials and Technologies* 37: e00699.
89. Madhu, A. 2022. Insights into electrochemical technology for dyeing of textiles with future prospects. In *Emerging technologies for textile coloration*, 57–88. CRC Press.
90. Schrott, W. 2004. Electrochemical dyeing. *Textile Asia* 35 (2): 45–47.
91. Kulandainathan, M. A., et al. 2007. Potentiostatic studies on indirect electrochemical reduction of vat dyes. *Dyes and Pigments* 73 (1): 47–54.
92. Kalapriya, K., and H. G. Prabu. 2022. Electrochemical method of dyeing cotton fabric using direct dyes. In *American institute of physics conference series*. AIP.
93. Lara, L., I. Cabral, and J. Cunha. 2022. Ecological approaches to textile dyeing: A review. *Sustainability* 14 (14): 8353.
94. Sala, M., V. López-Grimau, and C. Gutiérrez-Bouzán. 2014. Photo-electrochemical treatment of reactive dyes in wastewater and reuse of the effluent: Method optimization. *Materials* 7 (11): 7349–7365.
95. Yang, Z., et al. 2021. Direct electrochemical reduction and dyeing properties of CI Vat Yellow 1 using carbon felt electrode. *Dyes and Pigments* 184: 108835.
96. Kulandainathan, M. A., et al. 2007. Review of the process development aspects of electrochemical dyeing: Its impact and commercial applications. *Coloration Technology* 123 (3): 143–151.

97. Bechtold, T., and A. Turcanu. 2009. Electrochemical reduction in vat dyeing: Greener chemistry replaces traditional processes. *Journal of Cleaner Production* 17 (18): 1669–1679.
98. Tan, X. D., et al. 2018. Indirect electrochemical reduction of indigo and dyeing. *Key Engineering Materials* 773: 379–383.
99. Saikhao, L., et al. 2018. Green reducing agents for indigo dyeing on cotton fabrics. *Journal of Cleaner Production* 197: 106–113.
100. Patel, D., et al. 2024. Electrochemical degradation of reactive dyes mixture over DSA electrodes: A sustainable approach. *Materials Today: Proceedings* 111: 111–120.
101. Yasri, N. G., and S. Gunasekaran. 2017. Electrochemical technologies for environmental remediation. In *Enhancing cleanup of environmental pollutants: Volume 2: non-biological approaches*, ed. N. A. Anjum, S. S. Gill, and N. Tuteja, 5–73. Cham: Springer International Publishing.
102. Hou, Y., et al. 2022. The influence of electrode for the electrochemical dyeing of cotton fabric. *Advances in Transdisciplinary Engineering* 24. IOS Press.
103. Cuesta-Mota, D., et al. 2024. Hydrogen production from electrochemical treatment of textile dyeing wastewaters. *Journal of Water Process Engineering* 64: 105628.
104. Santos, D. H. S., et al. 2020. Electrochemical degradation and toxicity evaluation of reactive dyes mixture and real textile effluent over DSA® electrodes. *Chemical Engineering and Processing - Process Intensification* 153: 107940.
105. de Oliveira Silva, K. N., M. A. Rodrigo, and E. V. dos Santos. 2021. Electrochemical treatment of soil-washing effluent with boron-doped diamond electrodes: A review. *Current Opinion in Solid State and Materials Science* 25 (6): 100962.
106. Mousset, E., C. Trellu, and M. A. Oturan. 2021. Treatment of soil washing solutions by electrochemical advanced oxidation. In *Electrochemically assisted remediation of contaminated soils: Fundamentals, technologies, combined processes and pre-pilot and scale-up applications*, ed. M. A. Rodrigo and E. V. Dos Santos, 89–109. Cham: Springer International Publishing.
107. Lanzalaco, S., and I. Sirés. 2021. Electrochemically assisted thermal-based technologies for soil remediation. In *Electrochemically assisted remediation of contaminated soils: Fundamentals, technologies, combined processes and pre-pilot and scale-up applications*, ed. M. A. Rodrigo and E. V. Dos Santos, 369–400. Cham: Springer International Publishing.
108. Gezahegn, G. Y., and W. B. Eremed. 2024. Application of electrodes for degradation of reactive dyes in the case of textile industry wastewater. *Chemistry Africa* 7 (7): 3949–3958.
109. Lee, J., et al. 2024. Electrochemical methods for nutrient removal in wastewater: A review of advanced electrode materials, processes, and applications. *Sustainability* 16 (22): 9764.
110. Mathison, R., et al. 2024. Accelerated analysis of the electrochemical production route for biomass-derived adiponitrile. *Chem Catalysis* 4 (5).
111. Resasco, J., and A. T. Bell. 2020. Electrocatalytic CO_2 reduction to fuels: Progress and opportunities. *Trends in Chemistry* 2 (9): 825–836.
112. Zhang, T., et al. 2021. Review of electrochemical oxidation desulfurization for fuels and minerals. *Fuel* 305: 121562.
113. Shamsuddin, M., and H. Y. Sohn. 2023. Role of electrochemical processes in the extraction of metals and alloys—a review. *Mineral Processing and Extractive Metallurgy* 132 (3–4): 193–209.
114. Sanad, M. M. S., and A. Toghan. 2024. Industrial perspectives for electrochemical extraction of metals from primary and secondary resources. *Inorganic Chemistry Communications* 167: 112710.
115. Wojnowska-Baryła, I., et al. 2024. The growing problem of textile waste generation—the current state of textile waste management. *Energies* 17 (7): 1528.
116. Khan, W. U., et al. 2023. A critical review of hazardous waste generation from textile industries and associated ecological impacts. *Journal of the Indian Chemical Society* 100 (1): 100829.
117. Chand, S., S. Chand, and B. Raula. 2023. Textile and apparel industries waste and its sustainable management approaches. *Journal of Material Cycles and Waste Management* 25 (6): 3132–3143.

118. Heida, L. 2025. *Can waterless dyeing processes clean up the clothing industry?* 2 Feb 2025; Available from: https://e360.yale.edu/Features/Can_waterless_dyeing_processes_clean_up_clothing_industry_pollution?Utm_source=chatgpt.Com
119. Mahmud, I., and S. Kaiser. 2020. Recent progress in waterless textile dyeing. *Journal of Textile Science & Engineering* 10 (6): 1–3.
120. Hepburn, S. 2015. Nike and Adidas show cautious support for eco-friendly dye technology. 2 Feb 2025; Available from: https://www.theguardian.com/sustainable-business/sustainable-fashion-blog/2015/apr/24/nike-and-adidas-show-cautious-support-for-eco-friendly-dye-technology
121. Klingelhöfer, C. 2016. "We are SpinDye" revolutionizes textile dyeing and stops high water consumption. 2 Feb 2025; Available from: https://www.ispo.com/en/companies/we-are-spindye-textile-dyeing-becomes-sustainable
122. Expo., S. M. 2025. Environmental impacts of traditional dyeing processes in the textile industry. 2 Feb 2025; Available from: https://www.sustainablemanufacturingexpo.com/en/articles/environmental-impacts-dyeing.html
123. de Oliveira, C. R. S., et al. 2024. Supercritical fluid technology as a sustainable alternative method for textile dyeing: An approach on waste, energy, and CO_2 emission reduction. *Journal of Environmental Sciences* 140: 123–145.

Bernice Djangmah Akweley is a valedictorian awardee for the College of Art and Built Environment and a Fashion Designer from Ghana, who completed her undergraduate studies at the Kwame Nkrumah University of Science and Technology in 2023. She is the second author for an Emerald publication on 'A review of upcycling as a strategic marketing tool for sustainability in the fashion industry'. Bernice is a focus-driven young woman who has research interests in interactive fashion, smart textiles, sustainability, upcycling, educational and organisational studies, sewing, and just to mention a few and plans on furthering her education in these fields.

George Kwame Fobiri is an Assistant Lecturer at the Department of Fashion Design and Textiles Studies at Kumasi Technical University. He holds a bachelor's degree in Industrial Art (Textiles) and a master of Fine Art (MFA) in Textile Design from the Department of Industrial Art, Kwame Nkrumah University of Science and Technology (KNUST). He is currently pursuing a Ph.D. in Textile Design Technology at KNUST. His research interests cover textiles, fashion, computer-aided design (CAD), e-textiles, smart textiles, textile printing, textile dyeing, etc.

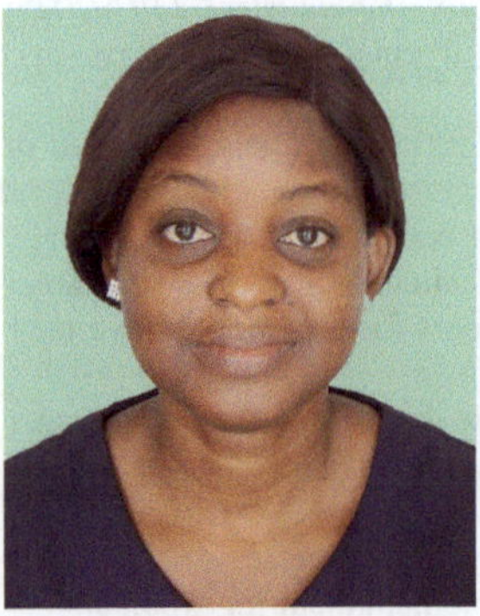

Sylvia Asheley Gyampoh is a lecturer and Acting Head of Fashion Design at Radford University College, Ghana, with over 12 years in the education sector of the fashion and textiles industry. She's pursuing a Ph.D. at KNUST and holds multiple degrees in fashion and education. An award-winning educator, she received recognition for her online teaching and sustainability efforts. Sylvia's mentor's youth supports fashion start-ups and advocates for sustainable fashion. She is also a guest lecturer and leadership coach in the industry. Her research spans design innovation and technology, fashion marketing, entrepreneurship, consumer behaviour, and the sociological and sustainable aspects of fashion.

Mr. Edem Kwami Buami is a lecturer in the Department of Industrial Arts at Ho Technical University, in the Volta Region of Ghana. He holds a Master of Fine Arts (MFA) degree in Textile Design and is currently pursuing a Ph.D. in Textile Design Technology at Kwame Nkrumah University of Science and Technology (KNUST) in Kumasi, Ghana. His research focuses on areas such as design innovation, textile printing and finishing technologies, sustainable design practices, including efficient production patterns, resource use, pollution and waste reduction particularly as it aligns with Sustainable Development Goal 12.

Benjamin Eghan is a Ph.D. candidate at DMU, UK, where he is based in the Faculty of Arts, Design and Humanities (ADH). His research focuses on sustainability in the textiles and fashion industry. Specifically, his doctoral work explores innovative techniques and technologies for transforming textile and fashion waste into new products, to generate social and environmental impact. Benjamin currently holds a position as a Research Assistant in Art and Design within the ADH Faculty at DMU. He also serves as a peer reviewer for the Fashion and Textile Review journal, contributing his expertise to the academic community.

Chapter 6
Sustainable Finishing Processes: Enhancing Performance with Minimal Impact

Jyoti Rani, Sakthivel Santhanam, Ramratan Guru, and Raphael Kanyire Seidu

Abstract The textile industry is undergoing a transformative shift toward sustainability, driven by the urgent need to minimise environmental impact while enhancing fabric performance. Using dyes derived from plants, minerals, and insects is a more sustainable alternative to synthetic dyes, which often contain harsh chemicals. Natural dyes are also slower and more traditional, and they minimise water waste. This paper reviews innovative sustainable finishing processes that exemplify this change, including enzyme-based finishing, which reduces water and chemical usage, and plasma treatment and supercritical CO_2 dyeing, which eliminate the need for water and hazardous substances. Additionally, techniques such as foam finishing, digital methods, and ozone finishing minimise resource consumption and waste. In contrast, advancements in nano finishing, waterless dyeing, and mechanical processes enhance fabric properties with minimal chemical inputs. Notable industry examples, such as Levi's Water < Less™ technology and Adidas' partnership with Parley for the Oceans, demonstrate significant water savings and the use of recycled materials. Brands like Nudie Jeans promote circularity through repair initiatives, and KOTN has achieved a remarkable 70% reduction in water use. The long-term use of nanotechnology in the textile sector can minimise the use of hazardous and toxic chemicals that affect the environment. Water and beverage-repellent protective coating is applied to nano-processed garments. Their protective layer is not visible to the naked eye. Pangaia natural pigment dyes and Zara's closing the Loop initiative

J. Rani (✉)
Assistant Professor, Parul Insitute of Design, Parul University, Vadodara, Gujarat, India
e-mail: Sawantravi976@gmail.com

S. Santhanam
Assistant Professor, School of Design, Presidency University, Bangalore, India

R. Guru
Assistant Professor, School of Design, Mody University of Science and Technology, SIkar, Rajasthan, India

R. K. Seidu
School of Fashion and Textiles, The Hong Kong Polytechnic University, Hung Hom, Kowloon, Hong Kong, China

S.-X. Jiang et al. (eds.), *Advances in Textile Materials and Processing Techniques for Sustainability*, SDGs and Textiles, https://doi.org/10.1007/978-981-95-0469-5_6

further illustrate the commitment to non-toxic processes and responsible production. Collectively, these sustainable strategies address pressing environmental concerns and respond to consumer demand for ethical fashion, positioning the textile industry toward a more sustainable future and a circular economy. Sustainable fashion is based on three fundamental principles: environmental, social, and economic. The goal is to minimise the negative impact of fashion on the planet while still creating stylish and high-quality garments.

Keywords Sustainable · Finishing · Textile · Brands · Nano-processed

6.1 Introduction

The textile industry's sustainability has to deal with environmental, economic, and social aspects that can be seen as a complex but crucial dilemma. It is a well-known resource-intensive industry, especially in terms of water and chemicals used in dyeing, necessitating green practices to conserve water and energy, minimise waste, and reduce carbon footprints [1]. Although often confused with textiles, the clothing segment has a comparatively lesser impact on the environment [2] due to the cleaner manufacturing procedure and the effective recycling of waste by-products, this industry has, in fact, helped in the sustainability efforts. The circular economy (CE) paradigm is progressively being included in textile recycling while ring advantages of competition, as well as the implementation of sustainable quality administration treatments, have never entirely paid off, particularly in emerging economies [3]. The shift toward sustainable textile makers includes the choice of sustainable fibres, the upgrade of production machinery, and working on problems such as microplastics that are essential to achieve true sustainability [4]. Social sustainability, which is often ignored, is one important issue with others like gender pay gaps and collaboration with NGOs and other groups to improve social responsibility, starting to become new metrics for this purpose [5]. Innovative recycling solutions are showing promise to minimise environmental impacts, keeping product quality to drive the transition from a linear to a circular textile chain [6]. Studies on resource use reveal significant potential for reductions in water, chemicals and energy in the various textile sub-sectors with the urgent need for cleaner production methods [7]. In the face of obstacles such as fast fashion, consumer demand and sustainable production techniques, there are several methods, such as the use of sustainable supply chain management and sustainable material use in the production of food, textiles and many other products [8]. Recycling, especially mechanical recycling, is an important part of sustainability initiatives, with closed-loop processes having higher recycling rates than open loops, but are less industrially scaled than open loops [9]. Lastly, the consumer demand and regulatory pressures for sustainable textiles require the development of sustainability standards and ecolabeling, which demands a transition to greener manufacturing technologies. What we need to do for a greener and cleaner textile industry is to combine technological, managerial, and policy innovations throughout the entire supply chain.

Considering the textile industry's significant impact on pollution and resource consumption, we all face the challenge of minimising the environmental impact of the activity. The industry is a major water and energy consumer, especially in processes such as dyeing, which also lead to greenhouse gas emissions and water pollution. Nevertheless, some innovative methodologies, including digital twin technology, have demonstrated significant potential in enhancing dyeing procedures, leading to an energy consumption and emission reduction of around 12.1% [10]. Another consideration is to support green industry standards by reusing and recycling wastewater, as well as improving the energy efficiency of the refrigeration systems to ensure that the textile sector is more carbon–neutral and more sustainable in future [1]. As a research group, we can find a sustainable solution to early identify and sort post-consumer textile waste at the industrial level for conversion to thermal insulation materials, which reduces environmental impacts [11], and by doing so, recycling becomes an essential process. Studies of the life cycle of garments produced with recycled fibres further indicate that their environmental benefits relative to conventional cotton also derive from the establishment of a circular textile value chain [12]. Further approaches like improving the wear life of clothes can provide continuous environmental progress, minimising demand for new clothing, meeting demand and decreasing overall environmental impacts [13]. The ability to separate and recycle mixed fabric classes, such as cotton and polyesters, further contributes to sustainability by facilitating the recovery of valuable materials and encouraging a circular economy, which has been cited [14]. Sustainable practices play a crucial role in the lacing of new and old materials, which have been recycled into textiles, as well as recycling and using advanced technologies to meet these challenges. This chapter focuses on sustainable textile practices, with an emphasis on eco-friendly dyeing and finishing methods such as enzyme-based finishing, plasma treatment, and waterless dyeing. It showcases technological innovations like nanotechnology and AI-based solutions that minimise the environmental footprint. It encompasses research on industry best practices, circular economy models, and the changing consumer attitude towards sustainability. It also highlights future opportunities through policy support and innovation, while addressing challenges like high costs and complexity in supply chains. The chapter provides insights into sustainability, exploring how industry stakeholders can spark a new paradigm of a conscious textile and fashion sector.

6.2 Issues with Synthetic Dyes and Their Environmental Impact

The quest for different colours and the availability of dyes which are non-existent for natural dyes led to the advancement of synthetic dyes with improved colouring matter [15]. Historically, Perkin in the mid-nineteenth century, accidentally discovered mauveine (the first synthetic dye) from coal tar when synthesising quinine. This discovery provided a consistent range of colours that were unavailable in natural dyes for the textile industry [16]. Due to their ease of use, uniform colouring ability or colour stability, high intensity and good colour fastness properties, synthetic dyes are widely available on the market [17] for use in colouring textile materials. Synthetic dyes are classified as basic dyes, disperse dyes, azo dyes, sulphur dyes, vat dyes, reactive dyes, direct dyes and acid dyes [18]. These dyes are used appropriately with their auxiliaries to ensure the dispersion and diffusion of dye molecules into the fibres. Thus, large quantities of dyes and chemicals are used during the dyeing process hence the release of large volumes of wastewater or effluents, which are harmful to the environment, water, humans and air (Fig. 6.1). These effluents, which are non-treated, contain organic pollutants such as dyes, which pose some threats [19]. For example, these untreated wastewaters, when released into water bodies, are used by some farmers in developing countries for agricultural purposes, hence having a serious impact on crop and soil quality [19].

The environment is made up of soil that forms a key foundation for agricultural purposes. The indiscriminate discharge of dye effluents into the environment significantly harms plants and the soil. Accidental spillage and leakage from industry contain substances with heavy chemicals and metals (from additives and dyes or pigments) that cause serious soil pollution [21]. For example, the soil pores are clogged after the spread of wastewater of dye effluents which affects the soil [22]. Even though biological processes are used to treat textile effluents from industrials, this process is unable to remove harmful chemical compositions and dyes[23]. The release of these textile effluents causes three critical problems such as groundwater pollution, affected soil productivity and pollutant accumulation in the food chain. the quality of groundwater in the soil over time is affected by the large deposit of untreated textile effluents which contain various harmful chemical compositions [24]. Additionally, these unchanged chemical compositions in textile effluents affect groundwater quality, making drinking water unsafe for individuals. For example, reports have revealed the detection of mutagenic activity from the water in the Cristais River, Brazil which is caused by textile effluents [25]. Commercially available dyes like Dispersol Black Dye (CVS) composed of Disperse Violet 93 (DV93), C.I. Disperse Orange 37 (D037) and C.I. Disperse Blue 373 (DB373), all disperse azo dyes present high strengths of mutagenic activity that affects water quality underground [26]. Subsequently, untreated textile effluents contain serious heavy metals, bad odour, suspended solids, and high chemical oxygen and biological oxygen demand, affecting the water quality and making it unsuitable for irrigation and drinking. The presence

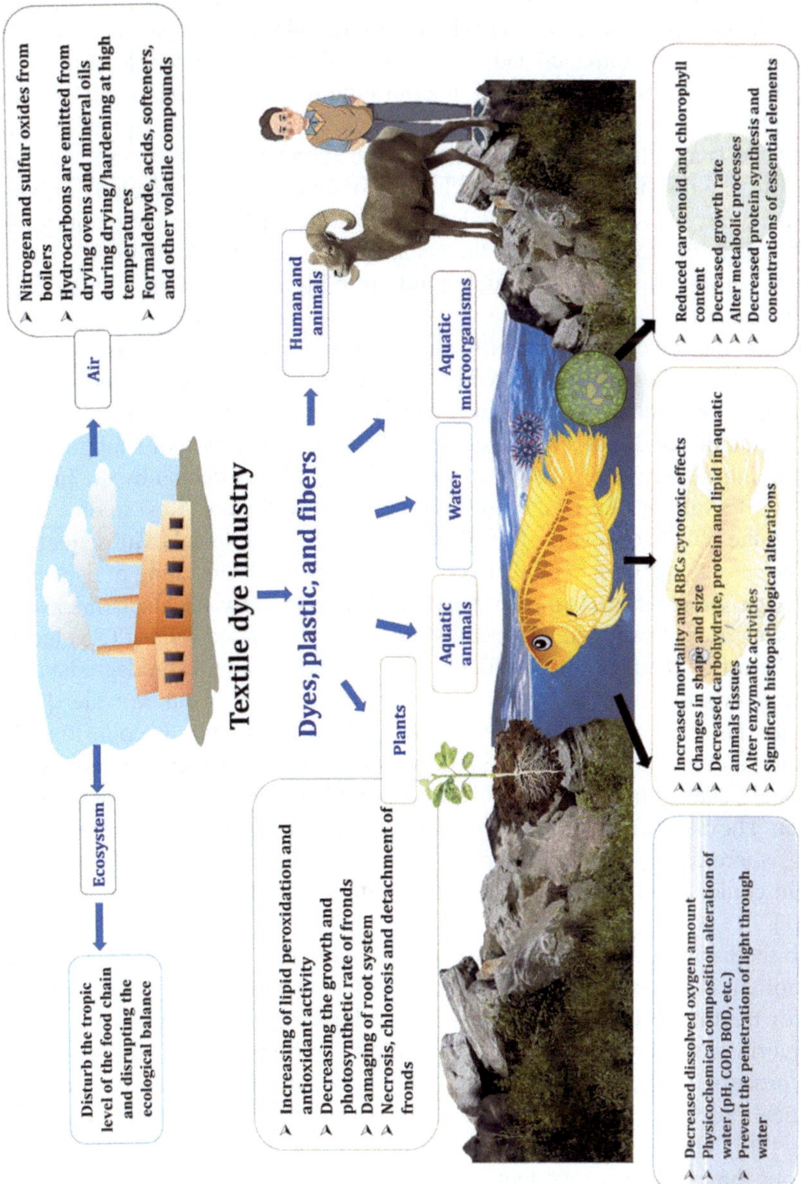

Fig. 6.1 The ecotoxicological threats from dye effluents [20] (This is an open-access paper available under the Creative Commons CC-BY-NC-ND license)

of these contaminants further affects the biological, chemical and physical properties of thc soil hence the soil quality is compromised. There is a reduction in thc dehydrogenase enzyme, microbial biomass carbon, bacterial population, and soil respiration in environments affected by textile effluents. These indicators are used to identify the health and quality of soil [27] for agricultural purposes. Such continuous practice of disposing of untreated and contaminated wastewater or textile effluents is however, discouraged to help limit soil deterioration that could affect plant growth. Harmful substances contained in dye effluents inhibit the growth of plants where photosynthesis is affected, protein content is reduced and oxidative stress is caused. Alternatively, the discharge of textile effluents contains pollutants such as organic compounds, dyes, salts and heavy metals when accumulated within the plants, pose serious threats to the food chain and the health of humans. These pollutants in the soil are taken up by plant tissues which build up in high concentration in the food products produced.

Aside from the effects of textile effluents on the environment, the aquatic life of the various water bodies is also affected by the harmful pollutants contained in these effluents. The latter contains high turbidity and dark colours that impact the pH levels, reduce the quantity of dissolved oxygen and affect the penetration of sunlight [28]. The photosynthesis activities of aquatic algae and plants are affected by the limited penetration of sunlight due to the presence of textile effluents. Key studies have reported on the effects of textile effluents on aquatic life (Table 6.1). Subsequently, changes in water temperature could affect aquatic activity like hatching of fish eggs, hence posing a serious threat to their population [29]. Ultimately, the biochemical and physical properties of water are affected by high total suspended solids (TSS) levels [30]. The harmful effects of textile effluents further lead to the low biodegradability of aerobic microorganisms and photosynthesis inhibition by plants in water [31]. This implies that these factors alter the biodiversity and ecosystems, become toxic to life in water, and lead to bioaccumulation, where harmful pollutants accumulate in the tissues of aquatic animals, which has significant health implications for humans when eaten. The direct contact with dyes and their chemicals leads to some respiratory problems, whereas the prolonged indirect contact of contaminated products from the food chain could lead to heart disease, gene mutations and haemorrhage amongst others [31].

Due to the harmful effects of these textile effluents, it is imperative to note that key technological processes like chemical, biological, hybrid and physical could be applied for treatment. For example, key studies have outlined various techniques such as hybrid processes [36], membrane filtration, ion exchange and adsorption for physical processes [37], electrochemical, photolysis and coagulation-flocculation for chemical processes [37], and fungal, bacteria and algae-assisted for biological processes [38]. These techniques help limit the challenges of textile effluents that pose serious threats to the ecosystem.

Table 6.1 Some studies on the effects of textile effluents on aquatic life

Description	References
Findings reported that physiochemical properties such as chloride, phosphate, calcium, potassium, total dissolved solid (TDS), sodium, total suspended solid, and salinity among others, were higher and above recommended levels	[32]
High pH levels in textile effluents affected aquatic life	[33]
Reactive Red 239 dye and textile effluents with high ecotoxicological effects impacted aquatic organisms	[34]
Textile effluents lead to high reactive oxygen species, which affect the peroxidation of algal cells of aquatic organisms	[35]

6.3 The Case for Sustainable Dyeing Practices

The textile industry has contributed negatively to the ecosystem through the continuous discharge of wastewater from the dyeing process. The latter is effectively used to transfer colouring materials (pigments or dyes) onto a textile material either by machine or by hand for uniform colouration effects [39]. Furthermore, this process involves the adsorption of the colouring materials onto the fibres and the subsequent diffusion into the fibre, for temporary and permanent bonds. Wastewater from the dyeing process contains harmful chemicals from dyestuffs and their auxiliaries, threatening the environment, aquatic life, and human survival. Conventional dyeing practices involve the consumption of high energy, the usage of large volumes of water and harmful chemicals, which, when disposed of into the environment, pose serious threats [40]. This has underscored the significant need for sustainable dyeing practices which seek to limit energy consumption, pollution of water bodies and the environment, and water usage for colouring textile materials. These benefits are influenced by contributory factors such as the type of dyes and additives used, dyeing methods and management approaches.

Effective management approaches play a critical role in ensuring the dyeing practices are sustainable to limit any threats. This places key responsibility on industry management to adopt and implement core protocols and strict standards that aim at reducing the chemicals used and energy consumption [41]. For example, efficient practices such as automated chemical dosing and dyes, reuse of dyestuff and effective washing plants are used to reduce energy and pollution in the textile dyeing industry [42]. These practices ensure that the appropriate quantities of water, dyes and chemicals are applied in the dyeing process. Subsequently, the use of standardised automated waste treatment plants aims at reducing the number of chemicals and dyes in wastewater to help limit pollution. Aside from this, internal educational programs by industry management are critical to train and educate employees in-house on sustainable practices. This approach fosters effective knowledge acquisition on key initiatives that drive sustainability within the dyeing industry.

Another key contributory factor to sustainable dyeing practices is the advancement of dyeing methods to ensure the removal of effluent or pollutants, reduced

Table 6.2 Some sustainable dyeing methods or technologies are applied in the dyeing industry

Dyeing methods	Description	References
Supercritical CO_2 dyeing	This process utilises less energy where CO_2 causes swelling, penetration and diffusion of dyes into the fibre structures. Herein, at a 250 bar pressure and 120 °C temperature, the CO_2 is activated and becomes supercritical for the dyeing of fibres. During dyeing, the supercritical CO_2 is used to dissolve the dye where the colouring matter is transferred to the fibres for colouring	[45]
Ultrasonic	The production of high-frequency waves is critical when applied in the dyeing process. Such ultrasonic wave disperses dye molecules in individual form, where the activated energy of dye molecules is increased for rapid diffusion into the fibres. This implies that there is dispersion, degassing and diffusion of dye molecules into the fibres for effective colouring	[46]
Plasma treatment	This is a surface modification process that increases the receptibility, uptake and dyeability of the fibres. Plasma treatment is an eco-friendly process that uses less chemicals, energy and water hence the non-production of harmful waste products	[47]
Electrochemical dyeing	This process adopts an electrochemical process for the reduction and fixing of dyes into the fibres. For example, sulphur and vat dyes applied in electrochemical dyeing undergo reduction and oxidation to bond well with cellulosic fibres. This process eliminates the use of sodium hydrosulphite or sodium dithionite as reducing agents during dyeing	[48]

energy consumption and effective maintenance systems, which are associated with the conventional dyeing process [43]. These dyeing methods significantly promote a healthy workspace, reduce exposure to harmful chemicals and discharge of wastewater that cause serious carbon footprints [44]. For example, there is effective optimisation of resources to ensure the right quantities of energy, water and dyestuffs are applied for reduced waste generation. Additionally, adopting key sustainable dyeing methods such as electrochemical dyeing, enzyme-based finishing, plasma treatment and supercritical CO_2 dyeing, amongst others, helps the textile industry to be compliant with regulations, and demonstrate their corporate responsibility to limit any threats to the environment. Table 6.2 outlines some dyeing methods that promote sustainable dyeing practices in the textile industry for ecosystem benefits. It is also imperative to add that the rise in awareness of consumers on environmental issues from the dyeing industries influenced the dyeing companies to adopt these dyeing methods to effectively meet their demands for sustainable products.

Lastly, the choice of dyes and additives also plays a critical role in adopting sustainable dyeing practices to help limit environmental threats. Natural dyes have gained attention as a colouring matter that drives sustainable dyeing with fewer threats to humans and the environment. This is due to their non-toxicity, affordability, availability and eco-friendly nature [15]. These dyes are applied with bio-mordants that ensure the quality and sustainability of the dyeing process with less harm [49]. The application of natural dyes, which are sourced from nature, is set to replace synthetic dyes that are used in conventional dyeing; however, they contribute negatively to the ecosystem.

The following sections provide a detailed description of the environmental impact of synthetic dyes, the use, benefits, challenges and scalability of natural dyes or dyeing process.

6.4 Natural Dyes: Sources and Benefits

Due to the harmful effects of synthetic dyes, natural dyes have gained attention for colouring textile materials and application in various areas because of its less health and environmental threats. Natural dyes are colouring matter derived from natural sources such as animals, minerals and plants [50]. For plant sources, natural dyes are extracted from the fruits, flowers, roots and leaves of the plant to form different colouring matter for textile materials [51]. Others are derived from the husks, hulls, rinds, wood shavings, bark, heartwood, stems, and twigs. Also, tannins, indigoids, carotenoids, anthocyanins, flavonoids and anthraquinones are the main colouring compounds for plant dyes [52]. Saxena et al. [53] further report on some colour extracts from natural dyes as shown in Table 6.3.

For animal sources, natural dyes are extracted from animals, most especially, cochineal insect, tyrian purple and kermes. A red dye is derived from the bodies of the female cochineal insect from the *Dactylopius coccus* species, where carminic acid is the principal colouring matter and the good light and wash fastness properties [53]. The colour spectrum of this dye can be varied from brown to red by mordanting the

Table 6.3 Examples of colour extracts from natural dyes

Natural source	Dye colour
Plant *indigofera tinctoria* (from the leaves)	Blue dye or Indigo
Water lilly (from the Rhizomes)	Blue colour
Rabia species (from the roots)	Madder or red colour
Wood of *Caesalpina sappan*	Red colour
Red sandalwood (from the wood)	Red
Morinda citrifolia	Shades of red
Turmeric	Yellow colour
Plant *Crocus sativus* (Saffron)	
Jack fruit tree (*Atrocarpus heterophyllus*)(*from the bark*)	
Plant barberry (*Berberis aristata*) (from the stems, bark and roots)	

cochineal dye with copper, iron, chromium and alum [54]. Furthermore, a deep violet coloured dye (Tyrian purple) is produced from the secretion of the sea mollusc *Murex*. The crimson red dye is obtained from the *Kermes licis* insect, which is widely used for colouring animal fibres. Alternatively, further animal natural dyes are lac, dyes, astaxanthin and carmine acid [55]. Lastly, for mineral dyes, these are extracted from mineral sources from the earth, are derived from hydrated oxides or oxides. These natural dyes are extracted through advanced processes such as aqueous extraction, solvent extraction, acid or alkaline extraction, microwave-assisted or ultrasound-assisted extraction, supercritical fluids, and enzymatic extraction.

Natural dyes have several benefits that are important for human health, aquatic life and the environment. These include key benefits such as its non-toxic nature and performance qualities for colouring textiles. Natural dyes are biodegradable, eco-friendly and non-toxic hence has limited harmful threats to the environment. These are derived from natural sources that are biodegradable with less environmental impact, non-allergic or non-carcinogenic to the human health as compared to synthetic dyes that are toxic. Natural dyes have significant performance qualities or properties such as antimicrobial, antifungal, and antibacterial (as shown in Table 6.4) hence when applied to textile materials can be used for different applications. Subsequently, natural dyes are good antioxidant agents and insect repellent ability, and ultraviolet protection ability for effective protection from ultraviolet lights. It is important to note that, bio-mordants or mordants are used to improve the fastness and poor substantivity of natural dyes. The dyeing process with natural dyes generates less harmful wastewater that are non-toxic and reduced risk for pollution in the waterbodies. Alternatively, the energy required to extract natural dyes are lower hence less energy is consumed for reduce carbon emission. Dyeing with natural dyes provides an opportunity to promote sustainable dyeing practices with less environmental impacts that threats animals, plants, aquatic life and human health.

6.5 Challenges and Scalability of Natural Dyeing

The harmful threats of the synthetic dyeing process to the ecosystem have advanced efforts to use natural dyes from different natural sources to colour textile materials. Aside from it benefits of being an eco-friendly dye that promotes sustainable dyeing in the textile industry, it has several challenges and scalability issues that limit the advancement of natural dyeing of textile materials. As far as challenges are concerned, the issue of colour fastness, cost of production, colour range or variety and stability influence the full utilisation of natural dyes.

The colour fastness property of a dye is the ability to resist and maintain its colour shade even after subjection to environmental conditions. The fastness performance of natural dyes to light, rubbing, perspiration and washing is critical and falls short of the standard requirements on the market, hence could limit their application. The overexposure of natural dyes to natural sunlight or artificial light could result in fading effects of the colour [54]. Alternatively, due to the frequent wear of clothing

Table 6.4 Some studies on the functional performance of natural dyes

Natural dye source	Textile material	Findings	References
Chinese gall	Wool	Good antibacterial properties when the textile is dyed with 98 °C temperature, 2.5% mordant concentration and a 200% dye liquor concentration	[56]
Plant sources such as *Nerium oleander, Trigonella foenum graecum* and *Curcuma longa*	Cotton	Good antimicrobial activity against pathogenic microbes	[57]
Dyes extracted from henna, saffron petals, turmeric, madder, green tea with aluminium sulfate as mordant	Wool	Good antibacterial properties when applied without mordants Increased antibacterial properties when mordanted with aluminium sulfate	[58]
Safflower yellow	Silk	Excellent UV resistance properties with a UPF of 73.79	[59]
Orange peel	Wool	Good UV-protection property	[60]

and use of other products coloured with natural dyes, the ability of the dyes to maintain their colour shade and limit any colour transfer when rubbed against any object is important. Even though the rubbing performance of natural dyes varies from different fibre types, natural dyes have a low rubbing fastness where colour is easily transferred to another via friction. However, these rubbing and light fastness properties of natural dyes could be improved by adopting different mordants on various fibre types. For example, 3-chloro-2-hydroxy propyl trimethylammonium chloride was used to cationise cotton fabric and coloured in dyes extracted from Henna to improve the colour fastness to light, rubbing, colour change and washing. Natural dyes from strawberry, China rose, beetroot and rose was used to form stable covalent bonds with cellulose fabrics in an alkaline medium, where colour fastness performance was stable and good [61]. Subsequently, the ability of the colour from natural dyes to be stable and not fad when in direct contact with perspiration from the human body is important. This implies that, dyes must be firmly bonded and fixed onto the fibres to prevent any fading effects when exposed to perspiration. These permanent bonds could be affected by the temperature levels, the type of fibres and the type of natural dyes. For example, there were low fastness properties for acid and alkaline perspiration when fabrics (cotton, polyester and cotton-polyester blends) were dyed with onion skin dye and henna dye. The colour change to alkaline and acid perspiration was however, good and fair respectively, when silk fabrics were dyed in a natural colourant (lac) with mordants such as stannous chloride, copper sulphate, ferrous sulphate, aluminium potassium sulphate [62].

Lastly, the ability of a dye to resist washing is an important property to determine its durability for various applications. Even though, natural dyes are eco-friendly, the ability of the colourant to wash fastness is, however of critical concern as compared to synthetic dyes. Such fastness to washing could be influenced by the dyeing methods and the mordants applied in the dyeing process. For example, silk fabrics dyed in a natural colourant (lac) with and without mordants such as stannous chloride, copper sulphate, ferrous sulphate, aluminium potassium sulphate, showed poor wash fastness properties when pad-dry and pad-batch techniques were used in the dyeing process [62]. Further studies by Ahmed et al. [63] revealed that, with or without mordants (potash alum and ferrous sulphate), the fabric samples showed poor wash fastness properties with dyed with natural dye extracts from *Pyllanthus emblica*, *Terminalia bellirica*, *Terminalia chebula*, and *Terminalia Arjuna*. This draws on the need for advancement of different mordants with various conditions on fibre types to yield good wash fast textile materials dyed with natural dyes.

Low yield of natural dyes is also another challenge that affects their full utilisation for colouring textile materials. In fact, the cost-effectiveness and efficiency are affected by the low yield of natural dyes. This is because the resistance performance and colour longevity are influenced by the high yield ability of natural dyes. Thus, the resistance to or fastness of washing, rubbing, light and perspiration are improved when a dye has a high colour yield. However, with changes on the pH values, the dye yield of natural dyes was good as observed when pH was 3.4 and 12 in acidic and alkaline media, respectively [64]. This implies that the stability of the colours are also important for consumer products since the brightness or shade of colour over time plays a role in consumers' purchasing and using a product. The problem of brightness and stability of colours is widely associated with natural dyes since they are easily affected by factors such as pH, temperature and light [58]. This makes it extremely difficult to maintain the colour stability on any textile material. Furthermore, the colour range for natural dyes is, however, limited as compared to synthetic dyes. For example, indigo is a great source for blue natural dyes which is widely applied in colouring textile materials. Also, the mixing compatibility of natural dyes are limited, hence making it difficult to achieve a wide range of colours.

Aside from the challenges associated with natural dyes, the issue of scalability becomes apparent, even though they are eco-friendly with less environmental impact as compared to synthetic dyes. The scalability of natural dyes is affected by key factors such as regulatory support, market demand and advancement in technology. For effective adoption and widespread scalability of natural dyes for industry use, it is important for regulatory support. Due to the numerous benefits of natural dyes most especially less harmful effects to the environment, policies are necessary to recognise and support the use of natural dyes. Based on this, Karadag proposed a Natural Organic Dye Standard that catalogues key analysis and test methods, mordants, natural colouring compounds and resources for natural dyes. Support and compliance with key regulations by regulatory bodies and the Governments will further foster long-term advancement in natural dyes for improved properties and trust from consumers on products coloured with natural dyes. With an increase in consumer trust coupled with increased demand on the market, there will be significant growth and scalability for natural dyes. This is due to the non-toxic nature and fewer environmental impacts of natural dyes. Lastly, the issue of technological advancement is critical to improve the scalability of natural dyes for various applications. Traditional natural dye extraction processes are time-consuming and labour-intensive, with numerous inconsistencies in colour yield. Based on this, new extraction methods are important to increase the scalability of natural dyes. For example, new approaches like microwave-assisted and ultrasonic extractions methods have been advanced for higher colour yields with limited threats to the environment [65]. Furthermore, the advancement in nano natural dyes with increased surface area and smaller particle sizes has enhanced the scalability and performance of natural dyes with improved fastness properties and colour strength. It is, however, recommended that further advancement in these areas is required for improved properties and scalability of natural dyes to remain synthetic dyes which have serious environmental threats.

6.6 Innovations in Sustainable Finishing Processes

6.6.1 Enzyme-Based Finishing

Enzyme-based finishing is helping people save water and chemicals, and to become greener in nature, and the textile industry is no exception. Enzymes such as amylase, cellulase, catalase, protease, pectinase, laccase, and lipase are increasingly used for the complete removal of this connecting material (which is a source of organic impurities) since they can substitute traditional harsh chemicals used in textile processing, lead to lower pollution, and better quality of the product [66]. Ultrasound treatment of cellulase and pectinase enzymes has already been successfully used in the utilisation of textile processing, while shown to be more productive at micro-scale and able to reduce expensive enzyme additions without losing fibre (matrix) integrity of some fabrics [67]. The significant role of enzymes in wool shrink-proofing—for example, through protease or L-cysteine treatments–can render obsolete outdated methods based on chlorine, meaningful for reducing environmental pollution while ensuring the preservation of wool quality [68, 69]. This can be performed through the use of amylase derived from *Bacillus cereus*, which serves as a better alternative to chemical desizing while not damaging the fabric. Another example is the application of cellulase enzymes from sugarcane bagasse for the biopolishing of cotton fabrics which further underscores the potential of an enzyme-based approach to replace excessive chemical consumption and consequent environmental burden [70]. A combination of enzyme and chemical processes is widely applied even in the leather industry, which uses certain kinds of enzymes for tanning, which reduces the intake of chemicals and avoids pollution, simultaneously decreasing the emission of chromium and other pollutants while assisting in maintaining the leather quality [71]. In addition, adding enzymes to laundry detergents can enhance the cleaning ability for the low-temperature wash cycles and reduce the energy input and the water consumption (Yau et al. 2024). Overall, the traditional mechanisms bridged with goals of global environmental conservation and in general, enzyme-based finishing mechanisms in

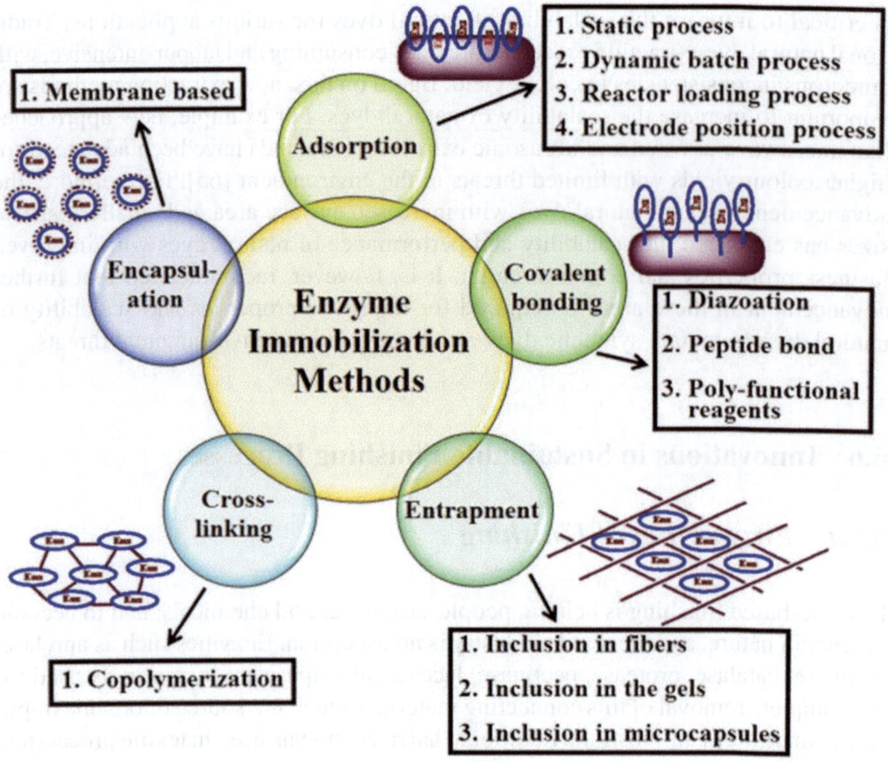

Fig. 6.2 Sustainable textile finishing processes and pollution control based on enzyme technology. Reprinted with permission from [72] Copyright 2016 Elsevier B.V.

textile industries reduce water and chemical use, which could have made these a sustainable and economical alternative (Fig. 6.2).

6.6.2 *Plasma Treatment and Supercritical CO_2 Dyeing*

One of them, plasma treatment and supercritical CO_2 ($scCO_2$) dyeing, are some recently developed and environmentally friendly processes in the textile industry, as traditional dyeing processes have caused significant ecological problems. Additionally, as plasma treatment increases the dye uptake, the process can reduce the consumption of water and chemicals as well. Plasma-treated cotton and bamboo rayon fabrics show increased dye uptake, and can also provide antibacterial properties to the absorbent material when dyed with natural extracts from *Fallopia japonica* [73]. A less detrimental alternative to high effluent loads is the combination of cationisation with plasma treatment, which has been demonstrated to improve cotton fabrics' dyeability with reactive and acid dyestuffs [74]. If this is compared, dyeing with

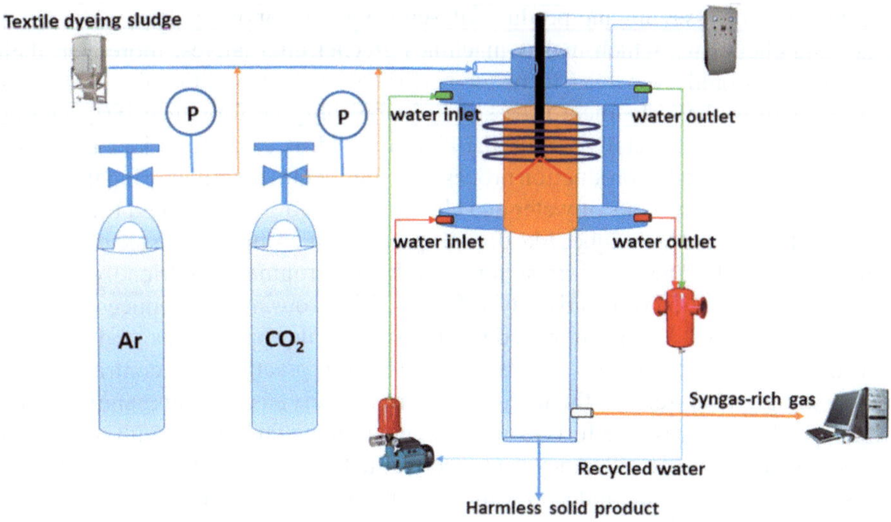

Fig. 6.3 Highly efficient treatment of textile dyeing sludge by CO_2 thermal plasma gasification. Reprinted with permission from [79] 2019 Elsevier Ltd.

scCO_2 is a water-free dyeing technology with high potential to decrease environmental impact due to the absence of water and effluent discharge, capture of CO_2 from the atmosphere and lower energy consumption [75]. Disperse reactive dyes were used to dye numerous textiles, including polyester, nylon, silk and wool, with excellent fixation and colour strength [76]. The new classes of dyes, e.g. tricyano pyrrolidone reactive disperse dyes that are scCO_2 based-dye tools to offer significantly stronger chemical interactions with fabric, thus resulting in improved dye fixation and superior colour strength. Moreover, assist-dissolving agents in scCO_2 dyeing have been studied to enhance solubility and promote uptake of slightly dissolved dyes to obtain better dyeing effects [77]. Although they exhibited promising efficacy, plasma treatment and scCO_2 dyeing have major challenges to overcome with respect to scalability and cost-effectiveness; therefore, additional studies must be conducted to tackle these challenges and facilitate widespread application in the textile manufacturing process [78]. Sustainable development through environmental impact assessment using life cycle assessment with examples for carbon footprints of diverse cellulose-based fibres (Fig. 6.3).

6.6.3 Foam Finishing, Digital Printing, and Ozone Finishing

Foam finishing, digital printing, and ozone finishing in the textile industry are becoming concerted efforts to help reduce environmental footprints and make the process gradually more sustainable. Unlike conventional padding, foam finishing is a sustainable, cost-saving and productivity-enhancing technology. It requires fewer water and chemicals, which makes machines greener alternatives; moreover, their optimisation would even lead to better performance than traditional methods by 84.61% both in their chemical use as well as finishing performance [80]. Digital printing, however, especially within the denim world, is part of a wider transition toward more sustainable production processes. Now, denim manufacturing that traditionally uses huge amounts of water and chemicals is being investigated for its environmental costs, and techniques like digital spray and microbially assisted dyeing are being studied [81]. The ozone finishing is another environmentally friendly solution for the decolourisation of reactive-dyed fabrics. Bio-scouring also reduces the need for harsh chemicals, can be optimised for industrial-scale applications, and provides a sustainable route for textile processing [82]. These practices are contributing to a continuing global trend in the textile sector re towards sustainable manufacturing (SM), which encompasses efforts to minimise environmental impacts with ecological protection and resource efficiency principles [83]. The advent of these technologies is also consistent with the industry's movement toward a circular economy and the triple bottom line. These textile processing innovations, therefore, provide important opportunities that not only address the key issue of environmental sustainability but also the broader objectives of the textile industry to continue reducing resource consumption and make textile products more sustainable throughout the full lifecycle [4, 6].

6.7 Advancements in Sustainable Technologies Nano Finishing

The art of sustainable nanotechnologies has indeed transformed the textile industry, providing new possibilities to improve the properties of many fabrics, reducing the environmental impact whilst uniting graceful and functional design. These technologies are based on applied nanoparticles on textiles with functional properties like antibacterial, UV shielding, water repellency, and flame retardancy that meet green chemistry. Nanotech in textiles is also part of a broader trend toward sustainability by reducing the need for harmful chemicals and energy-intensive processes. Here are some key innovations and examples of sustainable nanotechnologies in the textile industry.

6.7.1 Nanoparticle Applications in Textiles

- Antibacterial and Flame Retardant Properties: Nanoparticles such as metal oxides are used to impart antibacterial and flame-retardant properties to textiles. These particles can be grafted onto fabrics using green methods like plasma pre-treatment and sol–gel processes, which activate fibrous surfaces for better nanoparticle adherence [84].

- UV Protection and Water Repellency: Janus nanoparticles, which combine hydrophilic and hydrophobic properties, have been developed to enhance UV protection and water repellency in textiles. These nanoparticles are synthesised using waste-derived polyphenols, making the process both cost-effective and environmentally friendly [85].
- Self-Cleaning and Durability: The use of zinc-silica core–shell nanoparticles has been shown to improve the self-cleaning properties and wash durability of fabrics. This approach not only enhances fabric performance but also reduces the need for frequent washing, thereby conserving water [85].

6.7.2 *Sustainable Nanomaterial Synthesis*

- Green Synthesis Techniques: The synthesis of nanoparticles using green methods, such as the use of bio-based materials and waste-derived substances, is gaining traction. This approach reduces the environmental footprint of textile processing and aligns with the Sustainable Development Goals [86].
- Nanocellulose Applications: Nanocellulose, derived from renewable sources, is used to reinforce textiles, improve dyeing processes, and impart hydrophilicity. Its application reduces the need for synthetic materials and lowers effluent loads, contributing to a circular economy.

6.7.3 *Environmental and Economic Benefits*

- Effluent Treatment: Nanotechnology offers advanced solutions for treating textile effluents, addressing challenges such as colour removal and pollutant degradation. Nanomaterials like nanocomposites and nanocatalysts enhance wastewater treatment efficiency, promoting the recycling and reuse of water.
- Energy and Resource Efficiency: Techniques like sonoprocesses, which utilise low temperatures and pressures, are being explored to reduce energy consumption in textile processing. These methods support the sustainable application of nanomaterials, minimising the ecological footprint of the textile industry [87].

While the integration of sustainable nanotechnologies in textiles presents numerous benefits, challenges remain in terms of scalability and cost-effectiveness. The development of reliable and economically viable methods for nanoparticle synthesis and application is crucial for widespread adoption. Additionally, interdisciplinary collaborations are essential to optimise these technologies and ensure they meet both environmental and industrial standards. As research progresses, the potential for nanotechnology to revolutionise the textile industry continues to grow, promising enhanced fabric functionalities and reduced environmental impact.

6.8 Waterless Dyeing and Mechanical Processes

Innovations such as mechanical in the textile industry and waterless dyeing technologies are a significant turning point in environmental challenges and also improving its sustainability. Their practices include ways to decrease water as well as energy usage and prevent waste, and attach a smaller carbon footprint compared with conventional textile dyeing methods. This change is led by the incorporation of new technologies like supercritical carbon dioxide ($scCO_2$) dyeing, deep eutectic solvents, and nano bubble dyeing. Not only do these methods provide environmental advantages, but they also enhance the efficiency and quality of textile manufacturing. Here's a look at key developments and their significance in the textile sector.

6.8.1 Waterless Dyeing Technologies

- Supercritical Carbon Dioxide ($scCO_2$) Dyeing: This technique utilises $scCO_2$ as a dye solvent, thus avoiding the usage of water, and decreasing effluent output. It has been demonstrated to yield colour values similar to those obtained with conventional dyeing, and to provide darker shades at lower temperatures and pressures. It also permits the recycling of the dealing with non-adsorbed dyes and the saving in energy consumption and CO_2 emissions, being thus a low-impact alternative to the traditional dyeing [75, 88].
- Deep Eutectic Solvents (DES): The use of DES as a water-free dyeing medium presents a green alternative to water usage while also enhancing the aesthetic characteristics of the fabric. This method has exhibited a considerable amount of energy savings; ultrasonic dyeings showed a decrease of 58% and 25% of electrical and thermal energy, respectively, when compared to conventional methods [89].
- Nano Bubble Dyeing: An innovative technology, achieves a low liquor ratio of 1:1, tremendously saving water, dye and chemicals. It maintains uniform dyeing, acceptable fastness properties, with zero liquid discharge to minimise the environmental impact [90].

6.8.2 Mechanical Process Innovations

- Anhydrous Dyeing Systems: Supercritical fluids and solvents, including, but not limited to, ethanol and dimethyl sulfoxide, these systems enhance energy efficiency while minimising waste discharge. They facilitate colour application, optimise dye penetration and intensity, and harmonise with sustainability objectives [91].
- Cold Pad Dyeing Technology: This technology utilises a salt-free, cationic pigment hybrid latex for a near-zero-discharge dyeing process. It significantly reduces water and energy use, with 88.9 and 58.5% savings for water and energy,

respectively, versus currently common methods. The dyebath can be recycled, which further reduces wastewater emissions [92].

6.8.3 *Environmental and Economic Impacts*

- Zero Liquid Discharge Technologies: Technologies such as Adiabatic Sonic Evaporation and Crystallisation (ASEC) allow for all water to be recycled, eliminating the water footprint altogether. This strategy is aligned with both a circular economy and decarbonization goals and leads to a considerable reduction in water consumption and CO_2 emissions.
- Energy Efficiency and Cost Savings: Implementing AI and machine learning in process optimisation saved energy significantly. For example, the installation of solar power plants in textile factories shows that 60% energy saving is achieved in summer and 25% in winter, indicating the economic viability of these developments [1].

Although these new technologies provide revolutionary solutions for textiles to become sustainable, these technologies need to be scaled up into systems that are widely adopted. Resistance to digitalisation. Also, the initial investment costs and the technology/vanguard of the equipment required can be roadblocks for some manufacturers. Yet the potential long-term environmental and economic advantages and regulatory pressures to advance green initiatives, combined with consumer demand for sustainable products, will likely continue to spark new ideas and products in this industry.

6.9 Industry Examples and Best Practices

6.9.1 *Levi's Water < Less™ Technology*

The textile dyeing industry requires large volumes of water to colour textile materials for consumer usage. In response to this issue, Levi's developed an innovative approach to colour and produce denim using less water in its finishing process. This Water < Less technology uses "¼ cup of water" to produce their quality jeans, hence saving about 96% of water which would have been used in conventional dyeing industries. Through this open-sourced technology, Levi's aim to reduce the high consumption of water common in denim production in the industry hence addressing the issue of water scarcity in the world. Aside from its impact, Levi's further encourages consumers to engage with the application of air-drying and cold water when washing their jeans, a practice which reduces consumption of energy and water.

6.9.2 Adidas' Partnership with Parley for the Oceans

The effects of plastic pollution in the oceans remain a major problem that threats aquatic life in the world. Key harmful effects of plastic pollution include ingestion and entanglement that threats the life of marine or aquatic life. For example, ingestion of these plastic debris causes intestinal blockage, reduced stomach capacity which leads to injuries internally and hinders growth of aquatic animals. To help limit these threats, Adidas partnered with Parley to draw the needed awareness and fight against plastic pollution for the oceans. In this initiative, they presented a shoe concept produced from reclaimed plastics waste from the ocean. Through this initiative, Adidas and Parley have advanced this awareness to operate key movements such as Run for the oceans, 747 Warehouse St., and Parley Ocean School, to drive the concept of reclaiming plastic waste for alternative products. This partnership seeks to transform plastic waste from the ocean into innovative sportwear with high quality and performance to foster a sustainable agenda and strategy for reduce environmental impacts.

6.9.3 Nudie Jeans' Circularity and Repair Initiatives

With the continuous threats from fast fashion, circularity and circular strategies have gained attention in the fashion industry to help reduce the bulk of waste disposed of in the environment. The production of denim requires large quantities of resources and water that have environmental impacts and a later carbon footprint when disposed of. To help address the end life, Nudie jeans started a free repair initiative that encourages their consumers to fix or repair their old jeans clothing rather than discarding them into the landfill. Herein, by operating three key approaches, repair kits, mobile repair stations and repair shops, consumers can easily donate their jeans for recycling in-house, or repair for use. This approach by Nudie jeans fosters awareness for recycling and repair services for consumers, which contributes a more responsible and sustainable future.

6.9.4 KOTN's Water Use Reduction Achievements

The use of large volumes of water in the production processes in the textile industry has been of major concern hence resulting in water scarcity and pollution. This has led to innovative approaches being advanced to improve production processes where less water is used to produce equally high-quality products for consumers and the market. Additionally, with the high rise in micro-plastic pollution, it poses serious threats to the environment, the quality of water and aquatic life. To help address the threats of micro-plastics and water consumption in the production processes, KOTN advanced a sustainable approach which adopts organic cotton over conventional cotton as its raw material source due to the less water required, hence reducing water consumption in the manufacture of their products for consumers. KOTN further adopts non-toxic dyes to reduce the effects of harmful wastewater on the ecosystem. The manufacturing process utilises less water and energy to contribute to a reduced consumption of water and energy, hence reducing their carbon emissions and footprints. Through their sustainable approach, KOTN promotes environmental conservation and reduces water usage in the industry. Subsequently, their sustainability efforts further include the use of recyclable paper bags and tissues, where their products are distributed in eco-friendly packages for consumers.

6.9.5 Pangaia's Natural Pigment Dyes

The effects of the conventional dyeing process, which utilises synthetic dyes for colouring textile materials, cannot be overemphasised. These dyes, which use harmful chemicals, contribute heavily to environmental footprints. To address this, Pangaia's "natural dye capsule" creates dyes from botanical sources like thyme plants and chestnut trees, which are used, coupled with Tintex's Colorau technology in their dyeing process. Herein, the process adopts no toxic chemicals that are harmful to the ecosystem. This process extends the sustainability agenda in the industry, where there is a reduction in the use of excess water and harmful chemicals to colour textile materials for the markets.

6.9.6 Zara's Closing the Loop Initiative

The harmful effects of fast fashion to the ecosystem have influenced brands to advance innovative approaches to help reduce the recurring threats. As such, Zara introduced a closed-loop initiative that significantly promotes three key pillars of collect, reuse and recycle of textile products. This initiative extends the life span of textile products where recycling and reuse is encouraged for consumers to limit garments disposed of into landfills. It is important to note that about 12,228 tons of garments were collected in 2017 from different outlets around the World. Through Zara outlets, consumers are encouraged to donate used garments for reuse or recycling, hence contributing to the sustainable approach of the brand. These garments are recycled to produce garments, hence reducing heavy use of virgin resources.

6.10 The Principles of Sustainable Fashion

The definition of sustainable fashion is continually evolving as it is all about the integration of environmental, social and economic sustainability into the fashion industry. Its objective is to reduce the harm of fashion manufacturing and buying on the world and foster good practices. Sustainable Fashion Principle [93] emphasized that the principles of sustainable fashion aim to reduce waste, improve resource efficiency, and promote a circular economy. Safety, Circularity, and Fairness- These are the basis for working on sustainability in the fashion industry and combating the industry's quantitative contribution to pollution and natural resource depletion. This article covers some essentials of sustainable fashion based on the given research papers.

6.10.1 Waste Management and Resource Efficiency

- The fashion sector plays a significant role in global waste generation, which requires sound waste management policies. Waste Flow Mapping (WFM) and Life Cycle Assessment (LCA) are used to identify and quantify waste production and environmental impacts at each stage of the fashion supply chain. We can achieve 25–30% reductions in environmental impacts and 20–25% improvement in resource efficiency by implementing these tools [94].
- Recycled materials also play a critical role in sustainable fashion, as studies indicate that around 40–50% of materials can be obtained from recycled sources, leading to decreased demand for virgin materials [94].

6.10.2 Material Innovation and Education

- Material innovation underpins many usages of sustainable fashion. Bacterial cellulose (BC) for instance, has been investigated as an eco-friendly substitute for conventional materials used in fashion, showing compatibility with current production processes [95].
- If done properly, education can be a key provider for more sustainable behaviour. Incorporating sustainability into fashion design/technology curriculum can prepare the students with the knowledge and skill set to instigate the next innovation in sustainable textiles [95].

6.10.3 Sustainable Alternatives and Consumer Behaviour

- To minimise waste in the leather goods sector, eco-friendly alternatives like vegan leather are in demand. Nonetheless, synthetic alternatives may also come with further environmental costs than such as whether they show a higher environmental impact than leather produced the conventional way [96].
- There is a direct correlation between consumer choices and the acceptance of sustainable fashion. Consumer awareness, preferences, and willingness to pay for sustainable products are essential drivers for a shift to more sustainable practices in the recent market [97].

6.10.4 Business Models and Industry Practices

- Fashion business models that result in more sustainable systems combine social, environmental, and economic value. Nordic fashion industry, for instance, is known to lead sustainability with a focus on product durability, gender neutral fashion and timeless fashion [98].
- To align brand practices with sustainability goals, sustainable merchandising strategies are a crucial aspect. It essentially covers the incorporation of social responsibility in merchandising processes to make brands more competitive while laying emphasis on sustainability [99].

While sustainable fashion has some promising solutions to limit the environmental impact, it is not without its challenges. Hurdles exist in the form of a complex fashion supply chain, consumer bias towards fast fashion, and the absence of standardised sustainability metrics. It would also require collaborative effort in the industry and continued investment in sustainable practices. However, with consumer demand for more sustainable products increasing and so much more awareness around environmental issues, this is forcing fashion to innovate and find more sustainable pathways.

6.11 Social Sustainability

The concepts of sustainable fashion, in particular, the practice of social sustainability, extend beyond ecological and environmental considerations to include ethical methods that safeguard the rights of humans and communities along the manufacturing life cycle of fashion. Fair working conditions, equitable economic opportunities, and the engagement of community aspects that fall under social sustainability in fashion. It works to alleviate the negative social consequences that have long been associated with the fashion industry, such as labour exploitation and unfair economic practices. Key themes in social sustainability in fashion are based on the above research papers.

6.11.1 Fair Labour Practices

- It also allows a sustainable lifestyle to flourish. These include safe working conditions, fair wages, and respect for workers' rights.
- Prior research points to significant difficulties in ensuring that social standards are implemented effectively because of incompatible goals and asymmetric information across multi-tier supply chains, resulting in agentive behaviours that erode social standards [100].
- Such issues can be addressed through developing stronger corporate social responsibility (CSR) departments and incorporating social compliance in sourcing decisions [100].

6.11.2 Community Engagement and Economic Opportunities

- Community engagement and equitable economic opportunities are vital aspects of sustainable fashion practices as well. This means investing in local economies and ensuring that fashion production has positive impacts for the local community.
- The Nordic fashion industry is a pioneer of a holistic sustainability approach combining social, environmental, and economic values while producing long-lasting and gender-neutral designs that support local economies [98].
- Experiential learning workshops such as these stimulate sustainable garment procurement and act as a catalyst for alternative social behaviours, community engagement and skills development [101].

6.11.3 Transparency and Accountability

- Social sustainability is about transparency in the fashion supply chain. It includes transparent implementation of sustainability practices and the social effects of fashion products.
- The absence of standardised sustainability reporting and the widespread greenwashing practices constitute a serious challenge. Common key performance indicators (KPIs) and clear reporting can mitigate these issues [102].
- Adoption of sustainability scoring systems, as is done in other sectors, can increase transparency and provide consumers with important information to inspire trust [102].

6.11.4 Consumer Awareness and Behaviour

- Consumer consciousness and conduct in fashion a key factors in social sustainability. Teaching the consumers about the social impact of the industry they are buying into can change their consumption habits.
- Some strategies to accelerate consumer acceptance of sustainable fashion include micro-marketing sustainable products to larger audiences and advertising awareness for conscious fashion choices.
- Improved consumer communication and education could change consumers' perception of priorities, moving away from the focus on fast fashion towards an emphasis on quality and durability [96].

The concepts of social sustainability in fashion are beginning to gain traction, yet fully integrating these practices across the industry is still a challenge. This makes it challenging to implement consistent social standards due to the complexity of global supply chains and the myriad stakeholders involved. Final thoughts: Aspects such as consumer behaviour and market benefits of fast fashion can compromise social sustainability. But with industry collaboration, transparency and consumer education, the fashion industry can finally make meaningful progress towards social sustainability.

6.12 Economic Sustainability

Developing fashion is a concept continuously growing, where economic sustainability is placed alongside our environmental and social responsibility. Economic sustainability in the fashion industry is when business models are developed that allow businesses to continue to grow while lowering their impact on the environment, while pursuing social equity. It will take a new approach rather than continuing with how things have always been done, finding ways to evolve out of the practices with proven benefits to the economy and the environment. This section examines principles of economic sustainability in the fashion industry, informed by best practices from multiple research studies.

6.12.1 Economic Sustainability in Fashion

- Sustainable Business Models: The Nordic fashion scene knits together socio-environmental and economic values. These new business models prioritise the durability of their products, include genderless designs, and provide timeless fashion, which also promotes economic sustainability through the reduction of waste and the enhancement of long-lasting customer engagement [98].

- Circular Economy: Adopting circular economy principles is vital for economic sustainability. This includes creating products for durability, reusing materials, and minimising waste. Many brands are increasingly using recycled materials and new strategies such as the Reduce, Reuse, and Recycle (RRR) framework to advance sustainability as well as economic viability [103].
- Technological Innovations: Smart factories or automated warehouses powered by technologies such as IoT, AI, machine learning, etc. make use of systems to minimise resource consumption and support economic sustainability. These innovations reduce production costs as well as meet consumer demand for sustainable products.

6.12.2 *Case Studies and Examples*

- Kering Group: Kering, the group behind brands such as Gucci and Saint Laurent, is in the lead when it comes to economic sustainability as it embeds a strict set of sustainability standards in the way its entities operate. This not only protects brand reputation but also safeguards economic performance over time.
- Nordic Fashion Industry: The Nordic fashion industry is a powerful example of sustainable practices and shows how the generation of value, a key perspective from the 3D perspective, can lead to economic sustainability to a greater degree [98].

The fashion industry continues to show progress towards economically sustainable practices, but the slow adoption of sustainable materials and consumer habits poses challenges. On the other hand, the adoption of circular economy principles, technological advancements, and enabling policies offers potent drivers of economic sustainability. Shifting towards more sustainable fashion involves the concerted effort of all actors in the industry, including brands, consumers and policymakers, to build a more equitable and environmentally responsible sector.

6.13 Non-toxic Coatings and Finishes

With demand for more environmentally friendly, safe, and effective alternatives to the chemical treatments that have defined the textile industry for decades, non-toxic coatings and finishes have emerged as an important part of this field. These developments centre around sustainability, fabric performance, and user safety. Foam-finishing techniques, PFAS-free superhydrophobic coatings, and green antibacterial finishes are some of the approaches that have been attempted. Each provides specific advantages and obstacles, all working together toward the overarching aim of developing sustainability in textile processing.

6.13.1 Foam-Finishing Techniques

Foam finishing is a new, very sustainable way to make padding, reducing substantially the use of water, chemicals and energy. This method also minimises waste and increases sustainable properties while maintaining top-notch performance by applying finishes as foam rather than liquid. As a result, controlled foam application means that there is a lower chemical concentration given that it is evenly distributed in a very thin layer with the minimum amount of chemical needed to be effective, thus this process helps in reducing chemical wastage and minimising the amount of energy used to dry the end product as opposed to conventional cleaning methods. The process has been used for different non-toxic finishes: cross-linkers, oil and water repellents, softeners, and fire retardants, all obtaining similar results to the traditional ones. Cost-effective and eco-friendly, foam-finishing optimizes resource efficiency. This also shows how the tool can be widely adopted with different types and finishes of fabrics.

When properly optimised, foam finishing improves production efficiency and reduces environmental impact. Lower water usage reduces effluent load, while reduced energy consumption reduces operating costs. Hence, as sustainability takes precedence in textile manufacturing, there could be availability of foam-finishing as one of the alternatives that cater to eco-conscious practices while still retaining quality. Its success in implementation proves that greener alternatives can be industry compliant, whilst promoting sustainable production [80].

6.13.2 PFAS-Free Superhydrophobic Coatings

Despite their durable water-repellent (DWR) properties, traditional PFAS-based coatings are associated with serious environmental and health concerns, being persistent and bioaccumulative. To address this, environmentally friendly alternatives, epoxy-free hierarchical nanocoatings that are PFAS-free, are being developed. These new coatings made with silica nanoparticles and long-alkyl-chain silanes create superhydrophobic surfaces without the use of toxic chemicals. These nanocoatings produce a consistent, high-performance water repellency; they are also resistant, even after repeated washing, which makes them the perfect candidates for textiles. Unlike PFAS, they break down harmlessly and are less damaging to the environment. And because their flexible formulation can be tailored to work across a range of fabrics, they provide a sustainable, high-performing solution for waterproofing [104].

Through these nanotechnology-based alternatives, the textile industry can provide eco-safe waterproofing without sacrificing functionality, in keeping with worldwide trends toward safer, more sustainable production [104].

6.13.3 Antibacterial and Antiviral Finishes

Conventional antimicrobial finishes often rely on heavy metals or leaching biocides, posing environmental and health risks. In contrast, next-generation eco-friendly antibacterial finishes provide effective protection without toxicity. Siloxane sulfopropylbetaine (SSPB) offers durable, non-leaching antibacterial activity by chemically bonding to fabrics. It enhances hydrophilicity and strength while remaining skin-safe, making it ideal for medical and everyday textiles. Quaternary ammonium compounds (QACs) deliver broad-spectrum antibacterial and antiviral properties with high durability, resisting degradation from washing and ageing. These coatings are non-irritating and comply with safety standards for skin contact. Both solutions eliminate harmful chemical release, reducing environmental impact while maintaining textile performance. By adopting these advanced finishes, the industry can meet growing demand for sustainable, functional, and skin-friendly antimicrobial textiles [105, 106].

6.13.4 Alternative Water and Oil Repellents

Liquidlike perfluoropolyether (PFPE) brush coatings have recently appealed to many and show the potential to be a non-toxic replacement for short-chain PFAS-based finishes, especially in the contact areas where growing environmental and health concerns are mixed up with the high-performance functionality. These new coatings offer superior repellency to water and oil while maintaining the breathability and comfort of textiles in a variety of applications. In contrast to traditional PFAS that have long environmental persistence and bioaccumulation concerns (e.g., those based on PFOS and PFOA), PFPE brush coatings are engineered for durability, successfully withstanding multiple laundering and abrasion cycles while having a significantly reduced turnover time and environmental footprint. Designed for durability, these coatings effectively address industry standards for performance and sustainability due to their unique liquid-repellent nanostructure. With PFPE-based finishes, textile manufacturers are able to provide the essential protective properties without the use of potentially harmful chemicals, therefore, promoting sustainable innovation and safe use in the end products. This innovation celebrates a major breakthrough towards sustainable functionalities in high-performance textile finishes in accordance with international regulations and consumer demand for eco-friendly solutions [107].

However, now in the quest for sustainable textile finishes, block copolymer brushes without fluorocarbons have been recognised as a novel class of eco-friendly topcoats that can impart both the soil-repelling and soil-releasing behaviours in one coating. This technology removes stubborn environmental pollutants while taking the surfactants out of them, unlike traditional fluorocarbon-based treatments, and allows for the surface behaviour to be switched. Derived from a natural source, their unusual design enables them to switch back and forth between hydrophobic (water-repellent) and hydrophilic (water-attracting) states, meaning that fabrics can repel stains when you're wearing them but then release soil easily in the wash. This active function considerably improves the longevity of textiles while also streamlining their maintenance, minimising the use of harmful detergents. Aside from performance, these coatings conform to principles of green chemistry, using no toxic fluorocarbons and helping to lessen microfiber pollution. They can be appropriate for various applications, including outdoor wear and medical textiles, because of their versatility. This achieves high-tech fabric protection without compromising on sustainability; a new realm of environmentally friendly textile finishes [108].

These developments are great strides, but there are still challenges to address regarding performance, cost, and environmental impact. It is important to understand these aspects for proper transition from conventional to non-toxic finishes and encourage its large-scale implementations in the textile industry. In addition, more studies and development are necessary to enhance the performance and durability of these novel coatings and finishes.

6.14 Key Sustainability Standards for Finishing Processes

From textile sustainability perspectives, these sustainable cross-finishing processes are from the textile, apparel, and industrial sectors, and are critical in minimising the environmental footprint. It is important to use internationally accepted sustainability standards because these processes frequently require significant quantities of water, energy, and chemicals. Important certifications like the Global Organic Textile Standard (GOTS) and OEKO-TEX® STANDARD 100 ensure that finishing treatments are free of toxic substances like formaldehyde and heavy metals. Furthermore, Bluesign® and ZDHC (Zero Discharge of Hazardous) support resource efficiency, smarter chemical selections, compliance with EU Eco-Label and an ISO 14001 overview of environmental management/legislation. For finishing to be truly sustainable, industries need to work on best practices and incorporate water efficiency and energy efficiency measures, which include both low-liquor ratio dyeing and renewable energy. Using non-toxic, bio-based finishes, including enzyme-based treatments and PFC-free water repellents, minimises environmental impact. Various waste reduction tactics like closed-loop water recycling and digital printing, help to reduce ecological footprints even more. Transparency is also key, with brands verifying sustainable practices through Life Cycle Assessments (LCAs) and blockchain traceability. Top players are already looking to these standards to respond to consumer demands and follow regulatory frameworks like the EU Green Deal. Futuristic innovations, like plasma finishing and AI-driven optimisation, hold a future for even greater sustainability. So, there you go! In the end, sustainable finishing is more than a fashionable catchphrase it's a business imperative for enduring ecological and economic vitality.

6.15 Challenges and Future Opportunities

There are a number of reasons why sustainable practices have not been widely adopted by the textile industry. This is due to the high production cost involved with sustainable materials and processes, which is one of the primary factors. Most of the environmentally friendly alternatives available come with high initial costs, making it even more economically unattractive for small and medium enterprises (SMEs). Moreover, in many developing areas, advanced sustainable solutions may not be available, and introducing them into current production systems can be complex and expensive due to technological constraints. The other significant barrier is consumer awareness and demand. In fact, some consumers are still unaware of how devastating textile production is, and sustainable fabric continues to be broadly misunderstood in terms of quality, durability, and price point. Additionally, the industry has been plagued with regulatory and compliance issues, as different regions have varying standards and certifications for sustainability and businesses often find it difficult to comply with environmental regulations in a uniform manner. The complexity of global supply chains is another challenge in the path towards sustainability, considering that sourcing sustainable raw materials at scale is still a major roadblock, as well as maintaining transparency and traceability throughout the production process.

However, there are a number of opportunities for growth and improvement in the sustainable textile industry despite these challenges. Consumer education and behaviour change initiatives can help drive a transition towards sustainable lifestyles. Social media and fashion influencers have a vital role to play in changing the mindset of consumers to make more responsible purchasing decisions. "Brands can help create a culture for sustainable practices by incentivising recycling, upcycling, donating or responsibly disposing of garments," says Yu. To build trust and combat greenwashing, we have to improve transparency and ethical marketing efforts to ensure that consumers make informed purchasing decisions based on verified sustainability claims.

R&D on textile technologies, falling under the umbrella of sustainability, offers significant potential for the industry to transform. Biodegradable and plant-based textiles, lab-based fibres, and alternative dyeing methods are all innovations offering some hope of a better future than our current predicament. Digital transformation through AI, data analytics, and 3D printing can help to optimise sustainable production and minimise waste. By bringing them together, cross-sectoral collaboration among industry, policymakers, and non-governmental organisations (NGOs) can help scale these sustainable solutions and share best practices worldwide.

Sustainability is driven by government support and policies. Looking ahead, government support and policies can push the pace of the transition towards sustainability. Incentives, subsidies, and stricter environmental laws can motivate companies to practice sustainability. As our knowledge evolves, we will implement on-market solutions to circularity, such as rental, resale, and upcycling, that can be profitable and extend the life cycle of any given garment, but simultaneously reduce textile waste. Emerging markets, as well as, targeted consumer segments (e.g., eco-conscious millennials and Gen Z) are also driving demand in sustainable fashion, giving brands the encouragement to evolve their strategies with changing preferences. Moreover, smart & functional textiles embedded with nanotechnology for better functionality with lower environmental effects can also make sustainable fashion more attractive. The pathway to transition the textile industry towards sustainability, a circular economy, and a consumer-centric future involves addressing existing challenges and opportunities for the future.

6.16 Conclusion

This has led to an evolution of the textile industry towards sustainability due to pressure to reduce the impact of manufacturing on the environment, while still creating fabrics that perform and appeal to consumers. Sustainability methods like natural dyes, enzyme-finishing, waterless dyeing and mechanical processes are capable of decreasing resource usage and toxic effluent. Innovative examples like Levi's Water. But opportunities for the future paint a bright picture for the industry. Moreover, growing consumer awareness level, government assistance, and developments of eco-friendly technologies, including nanotechnology and digital technologies, may further promote a shift towards sustainable solutions. Incorporating circular economy principles by embracing repair, resale, and recycling initiatives presents practical solutions to reduce waste and lengthen the lifespan of textiles. Additionally, the adoption of smart textiles and the reach of sustainable practices across budding markets act as growth avenues. Also, Real change will require a collaborative effort between all interested parties along the supply chain from producers and regulators to consumers. The sector's pledge to eco-conscious innovation will ultimately lead to greener, more ethical, and economically viable futures, dovetailing with international environments to solve all of these challenges and promote a circular economy.

References

1. Yılmaz, K., İÖ. Aksu, M. Göçken, and T. Demirdelen. 2024. Sustainable textile manufacturing with revolutionizing textile dyeing: deep learning-based, for energy efficiency and environmental-impact reduction, pioneering green practices for a sustainable future. *Sustainability* 16 (18): 8152. https://doi.org/10.3390/su16188152.
2. Rogale, D., and S. Firšt Rogale. 2024. Energy and environmental aspects of the sustainability of clothing production. *Sustainability* 16(20): 9100. https://doi.org/10.3390/su16209100
3. Wei, Y., Y. Liu, A. Siraj, J. Xu, S. Taneja, and N. Wang. 2024. A strategic framework for sustainable quality management in the waste textile recycling industry within the context of the circular economy. *Business Strategy and the Environment* 33 (8): 9214–9232. https://doi.org/10.1002/bse.3965.
4. Militký, J., D. Křemenáková, M. Venkataraman, and A. P. Aneja. 2024. Towards sustainable textiles for a safer planet: main topics. *Sustainability* 16 (13): 5344. https://doi.org/10.3390/su16135344.

5. Fidan, F. Ş, E. K. Aydoğan, and N. Uzal. 2024. Comprehensive analysis of social subcategories throughout life cycle assessment approach for the textile industry. *International Journal of Life Cycle Assessment*. https://doi.org/10.1007/s11367-024-02340-8.
6. Rubik, F., et al. 2024. Textiles on the path to sustainability and circularity—results of application tests in the business-to-business sector. *Sustainability* 16 (14): 5954. https://doi.org/10.3390/su16145954.
7. Kır, A., E. Ozturk, U. Yetis, and M. Kitis. 2024. Resource utilization in the sub-sectors of the textile industry: Opportunities for sustainability. *Environmental Science and Pollution Research* 31 (17): 25312–25328. https://doi.org/10.1007/s11356-024-32768-2.
8. Rahaman, M. T., A. D. Pranta, M. R. Repon, M. S. Ahmed, and T. Islam. 2024. Green production and consumption of textiles and apparel: Importance, fabrication, challenges and future prospects. *Journal of Open Innovation: Technology, Market, and Complexity* 10 (2): 100280. https://doi.org/10.1016/j.joitmc.2024.100280.
9. Huang, Y., D. R. Eyers, M. Stevenson, and M. Thürer. 2021. Breaking the mould: Achieving high-volume production output with additive manufacturing. *IJOPM* 41 (12): 1844–1851. https://doi.org/10.1108/IJOPM-05-2021-0350.
10. Kim, M., et al. 2024. Reduction of greenhouse gas emissions by optimizing the textile dyeing process using digital twin technology. *Fashion and Textiles* 11 (1): 17. https://doi.org/10.1186/s40691-024-00384-w.
11. Karmakar, S., A. Majumdar, and B. S. Butola. 2024. A sustainable recycling process and its life cycle assessment for valorising post-consumer textile materials for thermal insulation applications. *Waste Management Research* 0734242X241270933. https://doi.org/10.1177/0734242X241270933
12. Hammar, T., D. Peñaloza, and A.-C. Hanning. 2024. Life cycle assessment of a circular textile value chain: The case of a garment made from chemically recycled cotton. *International Journal of Life Cycle Assessment* 29 (10): 1880–1898. https://doi.org/10.1007/s11367-024-02346-2.
13. Wiedemann, S. G., S. J. Clarke, Q. V. Nguyen, Z. X. Cheah, and A. T. Simmons. 2023. Strategies to reduce environmental impacts from textiles: Extending clothing wear life compared to fibre displacement assessed using consequential LCA. *Resources, Conservation and Recycling* 198: 107119. https://doi.org/10.1016/j.resconrec.2023.107119.
14. Choudhury, K., M. Tsianou, and P. Alexandridis. 2024. Recycling of blended fabrics for a circular economy of textiles: Separation of cotton, polyester, and elastane fibers. *Sustainability* 16 (14): 6206. https://doi.org/10.3390/su16146206.
15. Alegbe, E. O., and T. O. Uthman. 2024. A review of history, properties, classification, applications and challenges of natural and synthetic dyes. *Heliyon* 10 (13): e33646. https://doi.org/10.1016/j.heliyon.2024.e33646.
16. Kumasi, G., and G. K. Fobiri. 2022. Synthetic dye application in textiles: a review on the efficacies and toxicities involved. *Textile and Leather Review* 5: 180–198. https://doi.org/10.31881/TLR.2022.22
17. Soni, I., P. Kumar, S. Sharma, and G. Kudur Jayaprakash. 2021. A short review on electrochemical sensing of commercial dyes in real samples using carbon paste electrodes. *Electrochem* 2 (2): 274–294. https://doi.org/10.3390/electrochem2020020
18. Oyeniran, D. O., T. O. Sogbanmu, and T. A. Adesalu. 2021. Antibiotics, algal evaluations and subacute effects of abattoir wastewater on liver function enzymes, genetic and haematologic biomarkers in the freshwater fish, *Clarias gariepinus*. *Ecotoxicology and Environmental Safety* 212: 111982. https://doi.org/10.1016/j.ecoenv.2021.111982.
19. Sojobi, A. O., and T. Zayed. 2022. Impact of sewer overflow on public health: A comprehensive scientometric analysis and systematic review. *Environmental Research* 203: 111609. https://doi.org/10.1016/j.envres.2021.111609.
20. Al-Tohamy, R., et al. 2022. A critical review on the treatment of dye-containing wastewater: Ecotoxicological and health concerns of textile dyes and possible remediation approaches for environmental safety. *Ecotoxicology and Environmental Safety* 231: 113160. https://doi.org/10.1016/j.ecoenv.2021.113160.

21. Calace, N., T. Campisi, A. Iacondini, M. Leoni, B. M. Petronio, and M. Pietroletti. 2005. Metal-contaminated soil remediation by means of paper mill sludges addition: Chemical and ecotoxicological evaluation. *Environmental Pollution* 136 (3): 485–492. https://doi.org/10.1016/j.envpol.2004.12.014.
22. Markandeya, D. M., and S. P. Shukla. 2022. Hazardous consequences of textile mill effluents on soil and their remediation approaches. *Cleaner Engineering and Technology* 7: 100434. https://doi.org/10.1016/j.clet.2022.100434
23. Kant, R. 2012. Textile dyeing industry an environmental hazard. *NS* 4 (01): 22–26. https://doi.org/10.4236/ns.2012.41004
24. Rudnicka-Kępa, P., and A. Zaborska. 2021. Sources, fate and distribution of inorganic contaminants in the Svalbard area, representative of a typical Arctic critical environment—a review. *Environmental Monitoring and Assessment* 193 (11): 724. https://doi.org/10.1007/s10661-021-09305-6.
25. Sanchez, P. S., M. I. Z. Sato, C. M. R. B. Paschoal, M. N. Alves, E. V. Furlan, and M. T. Martins. 1988. Toxicity assessment of industrial effluents from S. Paulo state, Brazil, using short-term microbial assays. *Environmental Toxicology and Water Quality* 3 (1): 55–80. https://doi.org/10.1002/tox.2540030107.
26. Oliveira, D. P., P. A. Carneiro, M. K. Sakagami, M. V. B. Zanoni, and G. A. Umbuzeiro. 2007. Chemical characterization of a dye processing plant effluent—Identification of the mutagenic components. *Mutation Research/Genetic Toxicology and Environmental Mutagenesis* 626 (1–2): 135–142. https://doi.org/10.1016/j.mrgentox.2006.09.008.
27. Bakhshandeh, E., R. Francaviglia, and G. Renzi. 2019. A cost and time-effective method to evaluate soil microbial respiration for soil quality assessment. *Applied Soil Ecology* 140: 121–125. https://doi.org/10.1016/j.apsoil.2019.04.023.
28. Slama, H. B., et al. 2021. Diversity of synthetic dyes from textile industries, discharge impacts and treatment methods. *Applied Sciences* 11 (14): 6255. https://doi.org/10.3390/app11146255.
29. Roy, R., A. Fakhruddin, R. Khatun, M. Islam, M. Ahsan, and A. Neger. 1970. Characterization of textile industrial effluents and its effects on aquatic macrophytes and algae. *Bangladesh Journal of Scientific and Industrial Research* 45 (1): 79–84. https://doi.org/10.3329/bjsir.v45i1.5187.
30. Bilotta, G. S., and R. E. Brazier. 2008. Understanding the influence of suspended solids on water quality and aquatic biota. *Water Research* 42 (12): 2849–2861. https://doi.org/10.1016/j.watres.2008.03.018.
31. Imran, M., D. E. Crowley, A. Khalid, S. Hussain, M. W. Mumtaz, and M. Arshad. 2015. Microbial biotechnology for decolorization of textile wastewaters. *Reviews in Environmental Science & Biotechnology* 14 (1): 73–92. https://doi.org/10.1007/s11157-014-9344-4.
32. Ahmed, K., M. Das, M. Islam, M. S. Akter, S. Islam, and M. A. Al-Mansur 2011. Physico-chemical properties of tannery and textile effluents and surface water of river Buriganga and Karnatoli, Bangladesh.
33. Imtiazuddin, S. M., Majid Mumtaz, and Khalil A. Mallick. 2012. Pollutants of wastewater characteristics in textile industries. *Journal of Basic and Applied Sciences* 8 (2): 554–556. https://doi.org/10.6000/1927-5129.2012.08.02.47
34. Garcia, V. S. G., et al. 2021. Multiple adverse effects of textile effluents and reactive Red 239 dye to aquatic organisms. *Environmental Science and Pollution Research* 28 (44): 63202–63214. https://doi.org/10.1007/s11356-021-15115-7.
35. Cai, H., J. Liang, X. Ning, X. Lai, and Y. Li. 2020. Algal toxicity induced by effluents from textile-dyeing wastewater treatment plants. *Journal of Environmental Sciences* 91: 199–208. https://doi.org/10.1016/j.jes.2020.01.004.
36. Praful, N., and B. K. Pattnaik. 2024. A critical review on electrocoagulation and nanofiltration including hybrid methods for the treatment of Dairy industry wastewater. *Environmental Technology Reviews* 13 (1): 251–268. https://doi.org/10.1080/21622515.2024.2328661.
37. Aragaw, T. A., and F. M. Bogale. 2023. Role of coagulation/flocculation as a pretreatment option to reduce colloidal/bio-colloidal fouling in tertiary filtration of textile wastewater: A

review and future outlooks. *Frontiers in Environmental Science* 11: 1142227. https://doi.org/10.3389/fenvs.2023.1142227.
38. Mishra, V., et al. 2023. Integrating microalgae into textile wastewater treatment processes: Advancements and opportunities. *Journal of Water Process Engineering* 55: 104128. https://doi.org/10.1016/j.jwpe.2023.104128.
39. Shang, S. M. 2013. Process control in dyeing of textiles. In *Process control in textile manufacturing*, 300–338. Elsevier. https://doi.org/10.1533/9780857095633.3.300
40. Haji, A., and M. Naebe. 2020. Cleaner dyeing of textiles using plasma treatment and natural dyes: A review. *Journal of Cleaner Production* 265: 121866. https://doi.org/10.1016/j.jclepro.2020.121866.
41. Park, J., and J. Shore. 2009. Evolution of right-first-time dyeing production. *Coloration Technology* 125 (3): 133–140. https://doi.org/10.1111/j.1478-4408.2009.00194.x.
42. Ammayappan, L., S. Jose, and A. Arputha Raj. 2016. Sustainable production processes in textile dyeing. In *Green fashion. Environmental Footprints and Eco-design of Products and Processes*, eds. S. S. Muthu and M. A. Gardetti, 185–216. Springer Singapore. https://doi.org/10.1007/978-981-10-0111-6_8
43. Behera, M., J. Nayak, S. Banerjee, S. Chakrabortty, and S. K. Tripathy. 2021. A review on the treatment of textile industry waste effluents towards the development of efficient mitigation strategy: An integrated system design approach. *Journal of Environmental Chemical Engineering* 9 (4): 105277. https://doi.org/10.1016/j.jece.2021.105277.
44. Lara, L., I. Cabral, and J. Cunha. 2022. Ecological approaches to textile dyeing: A review. *Sustainability* 14 (14): 8353. https://doi.org/10.3390/su14148353.
45. Arputharaj, A., A. S. M. Raja, and S. Saxena. 2016. Developments in sustainable chemical processing of textiles. In *Green fashion. Environmental footprints and eco-design of products and processes,* ed. S. S. Muthu and M. A. Gardetti, 217–252. Springer Singapore. https://doi.org/10.1007/978-981-10-0111-6_9.
46. Jajpura, L., and R. Nayak. 2020. Ultrasound applications in textiles and apparels. In *Sustainable technologies for fashion and textiles*, 143–161. Elsevier. https://doi.org/10.1016/B978-0-08-102867-4.00007-4
47. Kan, C.-W. 2015. Plasma treatments for sustainable textile processing. In *Sustainable apparel*, 49–118. Elsevier. https://doi.org/10.1016/B978-1-78242-339-3.00003-0
48. Yusuf, M., and M. Shahid. 2022. *Emerging technologies for textile coloration*, 1st ed. CRC Press. https://doi.org/10.1201/9781003140467
49. Muthu, S. S. (eds.). 2015. *Handbook of sustainable apparel production.* CRC Press. https://doi.org/10.1201/b18428
50. Pizzicato, B., S. Pacifico, D. Cayuela, G. Mijas, and M. Riba-Moliner. 2023. Advancements in sustainable natural dyes for textile applications: A review. *Molecules* 28 (16): 5954. https://doi.org/10.3390/molecules28165954.
51. Li, S., B. Mu, X. Wang, and A. Wang. 2021. Recent researches on natural pigments stabilized by clay minerals: A review. *Dyes and Pigments* 190: 109322. https://doi.org/10.1016/j.dyepig.2021.109322.
52. Karadag, R. 2023. Establishing a new international standard for natural dyed textile goods [Natural Organic Dye Standard (NODS)]. *Journal of Natural Fibers* 20 (1): 2162187. https://doi.org/10.1080/15440478.2022.2162187.
53. Saxena, S., and A. S. M. Raja. 2014. Natural dyes: Sources, chemistry, application and sustainability issues. in *Roadmap to sustainable textiles and clothing. Textile science and clothing technology*, ed. S. S. Muthu, 37–80. Springer Singapore. https://doi.org/10.1007/978-981-287-065-0_2
54. Islam, T., K. Md. Rashedul Islam, S. Hossain, M. Abdul Jalil, and M. Mahbubul Bashar. 2024. Understanding the fastness issues of natural dyes. In *Dye chemistry—exploring colour from nature to lab*, ed. B. Kumar. IntechOpen. https://doi.org/10.5772/intechopen.1005363
55. Li, N., Q. Wang, J. Zhou, S. Li, J. Liu, and H. Chen. 2022. Insight into the progress on natural dyes: Sources, structural features, health effects, challenges, and potential. *Molecules* 27 (10): 3291. https://doi.org/10.3390/molecules27103291.

56. Zhang, B., L. Wang, L. Luo, and M. W. King. 2014. Natural dye extracted from Chinese gall—the application of color and antibacterial activity to wool fabric. *Journal of Cleaner Production* 80: 204–210. https://doi.org/10.1016/j.jclepro.2014.05.100.
57. Selvam, R. M., G. Athinarayanan, A. U. R. Nanthini, A. J. A. R. Singh, K. Kalirajan, and P. M. Selvakumar. 2015. Extraction of natural dyes from *Curcuma longa, Trigonella foenum graecum* and *Nerium oleander*, plants and their application in antimicrobial fabric. *Industrial Crops and Products* 70: 84–90. https://doi.org/10.1016/j.indcrop.2015.03.008.
58. Shahmoradi Ghaheh, F., S. M. Mortazavi, F. Alihosseini, A. Fassihi, A. Shams Nateri, and D. Abedi. 2014. Assessment of antibacterial activity of wool fabrics dyed with natural dyes. *Journal of Cleaner Production* 72: 139–145. https://doi.org/10.1016/j.jclepro.2014.02.050
59. Yu, D., Y. Liu, Y. Ren, X. Liu, and H. Qu. 2022. One-pot strategy to simultaneously prepare dyed, flame retardant and UV-resistant silk fabric based on a safflower yellow derivative. *RSC Advances* 12 (40): 26147–26159. https://doi.org/10.1039/D2RA04873F.
60. Hou, X., X. Chen, Y. Cheng, H. Xu, L. Chen, and Y. Yang. 2013. Dyeing and UV-protection properties of water extracts from orange peel. *Journal of Cleaner Production* 52: 410–419. https://doi.org/10.1016/j.jclepro.2013.03.004.
61. Islam, M. S., S. Md. M. Alam, and S. Ahmed. 2020. Attaining optimum values of colourfastness properties of sustainable dyes on cotton fabrics. *FIBRES & TEXTILES in Eastern Europe* 28 (144): 110–117. https://doi.org/10.5604/01.3001.0014.3806
62. Mongkholrattanasit, R., et al. 2015. Ecological dyeing of silk fabric with lac dye by using padding techniques. *The Journal of The Textile Institute* 106 (10): 1106–1114. https://doi.org/10.1080/00405000.2014.976957.
63. Islam, T., M. Ahmed, Md. R. Karim, S. Kaiser, and P. Barua. 2019. Assessment of fastness properties of knitted cotton fabric dyed with natural dyes: a sustainable approach of textile coloration. *JTEFT* 5 (3). https://doi.org/10.15406/jteft.2019.05.00199
64. Ali, S., T. Hussain, and R. Nawaz. 2009. Optimization of alkaline extraction of natural dye from Henna leaves and its dyeing on cotton by exhaust method. *Journal of Cleaner Production* 17 (1): 61–66. https://doi.org/10.1016/j.jclepro.2008.03.002.
65. Das, S., and S. Roy Maulik. 2023. Recent approaches and advancements in natural dyes. In *Natural dyes and sustainability. Sustainable textiles: Production, processing, manufacturing & chemistry*, ed. S. S. Muthu, 63–78. Springer Nature Switzerland. https://doi.org/10.1007/978-3-031-47471-2_4
66. Kumar, D., R. Bhardwaj, S. Jassal, T. Goyal, A. Khullar, and N. Gupta. 2021. Application of enzymes for an eco-friendly approach to textile processing. *Environmental Science and Pollution Research* 30 (28): 71838–71848. https://doi.org/10.1007/s11356-021-16764-4.
67. Yachmenev, V. G., N. R. Bertoniere, and E. J. Blanchard. 2002. Intensification of the bioprocessing of cotton textiles by combined enzyme/ultrasound treatment. *Journal of Chemical Technology & Biotechnology* 77 (5): 559–567. https://doi.org/10.1002/jctb.579.
68. Li, B., et al. 2022. Development of environmentally friendly wool shrink-proof finishing technology based on L-cysteine/protease treatment solution system. *IJMS* 23 (21): 13553. https://doi.org/10.3390/ijms232113553.
69. Wang, L., J. Yao, J. Niu, J. Liu, B. Li, and M. Feng. 2018. Eco-friendly and highly efficient enzyme-based wool shrinkproofing finishing by multiple padding techniques. *Polymers* 10 (11): 1213. https://doi.org/10.3390/polym10111213.
70. Ikbal, Md. S., F. A. Tisha, A. I. Asheque, E. Hasnat, and M. A. Uddin. 2024. Eco-friendly biopolishing of cotton fabric through wasted sugarcane bagasse-derived enzymes. *Heliyon* 10 (4): e26346. https://doi.org/10.1016/j.heliyon.2024.e26346.
71. Alam, A. 2023. Harnessing the power of AI to create intelligent tutoring systems for enhanced classroom experience and improved learning outcomes. In *Intelligent communication technologies and virtual mobile networks. Lecture notes on data engineering and communications technologies*, vol. 171, ed. G. Rajakumar, K.-L. Du, and Á. Rocha, 571–591. Springer Nature Singapore. https://doi.org/10.1007/978-981-99-1767-9_42
72. Ibrahim, N. A., B. M. Eid, and H. A. Amin. 2021. Sustainable textile finishing processes and pollution control based on enzyme technology. In *Green chemistry for sustainable textiles*, 385–415. Elsevier. https://doi.org/10.1016/B978-0-323-85204-3.00018-X

73. Gorjanc, M., et al. 2016. Dyeing of plasma treated cotton and bamboo rayon with *Fallopia japonica* extract. *Cellulose* 23 (3): 2221–2228. https://doi.org/10.1007/s10570-016-0951-9.
74. Correia, J., et al. 2021. Surface functionalization of greige cotton knitted fabric through plasma and cationization for dyeing with reactive and acid dyes. *Cellulose* 28 (15): 9971–9990. https://doi.org/10.1007/s10570-021-04143-8.
75. De Oliveira, C. R. S., P. V. De Oliveira, L. Pellenz, C. R. L. De Aguiar, and A. H. Da Silva Júnior. 2024. Supercritical fluid technology as a sustainable alternative method for textile dyeing: An approach on waste, energy, and CO_2 emission reduction. *Journal of Environmental Sciences* 140: 123–145. https://doi.org/10.1016/j.jes.2023.06.007
76. Van Der Kraan, M., M. V. Fernandez Cid, G. F. Woerlee, W. J. T. Veugelers, and G. J. Witkamp. 2007. Dyeing of natural and synthetic textiles in supercritical carbon dioxide with disperse reactive dyes. *The Journal of Supercritical Fluids* 40 (3): 470–476. https://doi.org/10.1016/j.supflu.2006.07.019
77. Wu, Q. 2024. Application of artificial intelligence-based visual arts pedagogy in traditional painting education. *Applied Mathematics and Nonlinear Sciences* 9 (1): 20242994. https://doi.org/10.2478/amns-2024-2994.
78. Lara, F. 2021. Why a virtual assistant for moral enhancement when we could have a socrates? *Science and Engineering Ethics* 27 (4): 42. https://doi.org/10.1007/s11948-021-00318-5.
79. Wang, M., et al. 2019. Highly efficient treatment of textile dyeing sludge by CO_2 thermal plasma gasification. *Waste Management* 90: 29–36. https://doi.org/10.1016/j.wasman.2019.04.025.
80. Mohsin, M., and S. Sardar. 2020. Development of sustainable and cost efficient textile foam-finishing and its comparison with conventional padding. *Cellulose* 27 (7): 4091–4107. https://doi.org/10.1007/s10570-020-03033-9.
81. Periyasamy, A. P., and S. Periyasami. 2023. Critical review on sustainability in denim: A step toward sustainable production and consumption of denim. *ACS Omega* 8 (5): 4472–4490. https://doi.org/10.1021/acsomega.2c06374.
82. Powar, A. S., A. Perwuelz, N. Behary, L. Hoang, and T. Aussenac. 2020. Application of ozone treatment for the decolorization of the reactive-dyed fabrics in a pilot-scale process—optimization through response surface methodology. *Sustainability* 12 (2): 471. https://doi.org/10.3390/su12020471.
83. Chourasiya, R., S. Pandey, and R. K. Malviya. 2023. Sustainable manufacturing adoption in textile industries: A systematic state-of-art literature review and future research outline. *Sustainable Development* 31 (2): 612–638. https://doi.org/10.1002/sd.2434.
84. Patti, A. 2025. Green advances in wet finishing methods and nanoparticles for daily textiles. *Macromolecular Rapid Communications* 46 (2): 2400636. https://doi.org/10.1002/marc.202400636.
85. Meganathan, M. K., and S. Ramalingam. 2024. Green nanoengineered fabrics: Waste-derived polyphenol-zinc@ silica core-shell reactive Janus nanoparticles for functional fabrics. *ACS Applied Materials & Interfaces* 16 (30): 40004–40017. https://doi.org/10.1021/acsami.4c08268.
86. Kumari, I., S. Kaur, and R. Das. 2024. Green nanomaterials in textile industry. In *Green synthesis of nanomaterials*, 1st edn., ed. A. Chakravarty, P. Singh, S. Ikram, and R. N. Yadava, 114–130. Wiley. https://doi.org/10.1002/9781119900931.ch6
87. Rashid, A., et al. 2024. Sonoprocesses for sustainable textile processing and nanofinishing: A review. *Textile Research Journal* 94 (23–24): 2789–2814. https://doi.org/10.1177/00405175241252562.
88. Eren, S., I. Ozcan, I. Yigit, and H. A. Eren. 2023. Waterless dyeing of polytrimethylene terephthalate and polybutylene terephthalate fabrics via supercritical carbon dioxide. *The Journal of Supercritical Fluids* 201: 106026. https://doi.org/10.1016/j.supflu.2023.106026.
89. Hussain, N., et al. 2021. Introducing deep eutectic solvents as a water-free dyeing medium for poly (1,4-cYclohexane dimethylene isosorbide terephthalate) PICT nanofibers. *Polymers* 13 (16): 2594. https://doi.org/10.3390/polym13162594.

90. Mohsin, M., S. Sardar, M. Hassan, N. Akhtar, A. Hassan, and M. Sufyan. 2020. Novel, sustainable and water efficient nano bubble dyeing of cotton fabric. *Cellulose* 27 (10): 6055–6064. https://doi.org/10.1007/s10570-020-03187-6.
91. Zhao, L., E. Hussam, J.-T. Seong, A. Elshenawy, M. Kamal, and E. Alshawarbeh. 2024. Revolutionizing art education: Integrating AI and multimedia for enhanced appreciation teaching. *Alexandria Engineering Journal* 93: 33–43. https://doi.org/10.1016/j.aej.2024.03.011.
92. Shao, Y., J. Li, and D. Qi. 2024. A salt-free, near zero-discharge, and dyebath-recyclable cold pad dyeing technology of cotton fabric with less water and energy consumption based on cationic pigment hybrid latexes. *Sustainable Materials and Technologies* 40: e00981. https://doi.org/10.1016/j.susmat.2024.e00981.
93. Ahmad, S. F., Mohd. K. Rahmat, M. S. Mubarik, M. M. Alam, and S. I. Hyder. 2021. Artificial intelligence and its role in education. *Sustainability* 13 (22): 12902. https://doi.org/10.3390/su132212902
94. Rehman, M., A. Petrillo, M. Ortíz-Barrios, A. Forcina, I. Baffo, and F. De Felice. 2024. Sustainable fashion: Mapping waste streams and life cycle management. *Journal of Cleaner Production* 444: 141279. https://doi.org/10.1016/j.jclepro.2024.141279.
95. Wood, J., J. Redfern, and J. Verran. 2023. Developing textile sustainability education in the curriculum: Pedagogical approaches to material innovation in fashion. *International Journal of Fashion Design, Technology and Education* 16 (2): 141–151. https://doi.org/10.1080/17543266.2022.2131913.
96. D'Adamo, I., M. Gagliarducci, M. Iannilli, and V. Mangani. 2024. Fashion wears sustainable leather: A social and strategic analysis toward sustainable production and consumption goals. *Sustainability* 16 (22): 9971. https://doi.org/10.3390/su16229971.
97. Balasubramanian, M., and P. Sheykhmaleki. 2024. Comprehending the consumer behavior toward sustainable apparel. *Sustainability* 16 (18): 8026. https://doi.org/10.3390/su16188026.
98. Thorisdottir, T. S., L. Johannsdottir, E. R. G. Pedersen, and K. Niinimäki. 2024. Social, environmental, and economic value in sustainable fashion business models. *Journal of Cleaner Production* 442: 141091. https://doi.org/10.1016/j.jclepro.2024.141091.
99. Wu, B., et al. 2022. Merchandising for sustainable fashion: A systematic literature review. *Sustainability* 14 (20): 13422. https://doi.org/10.3390/su142013422.
100. Köksal, D., and J. Strähle. 2021. Social sustainability in fashion supply chains—understanding social standard implementation failures in Vietnam and Indonesia using agency theory. *Sustainability* 13 (4): 2159. https://doi.org/10.3390/su13042159.
101. Saunders, C., et al. 2024. A social practices approach to encourage sustainable clothing choices. *Sustainability* 16 (3): 1282. https://doi.org/10.3390/su16031282.
102. Gonçalves, A., and C. Silva. 2021. Looking for sustainability scoring in apparel: A review on environmental footprint, social impacts and transparency. *Energies* 14 (11): 3032. https://doi.org/10.3390/en14113032.
103. Puvzhiarasi, M., Ilakiya, K. M. Pachiyappan, R. Divya Raman Sathyam, M. Sharmila. 2024. A review article on the impact and evolution of sustainable brands. *IJFMR* 6 (5): 28354. https://doi.org/10.36948/ijfmr.2024.v06i05.28354
104. Dignes, C., N. J. Nicolas, T. B. H. Schroeder, and J. Aizenberg. 2024. Robust PFAS-free superhydrophobicity exhibited in hierarchically nanostructured coatings on textiles. *Advanced Engineering Materials* 2401736. https://doi.org/10.1002/adem.202401736
105. Chen, S., et al. 2011. Environmentally friendly antibacterial cotton textiles finished with siloxane sulfopropylbetaine. *ACS Applied Materials & Interfaces* 3 (4): 1154–1162. https://doi.org/10.1021/am101275d.
106. Gu, J., et al. 2018. Non-leaching bactericidal cotton fabrics with well-preserved physical properties, no skin irritation and no toxicity. *Cellulose* 25 (9): 5415–5426. https://doi.org/10.1007/s10570-018-1943-8.
107. Shabanian, S., S. K. Lahiri, M. Soltani, and K. Golovin. 2023. Durable water- and oil-repellent textiles without long- or short-chain perfluoroalkylated substances. *Materials Today Chemistry* 34: 101786. https://doi.org/10.1016/j.mtchem.2023.101786.

108. Messerschmidt, M., et al. 2018. Fluorocarbon-free dual-action textile finishes based on covalently attached thermoresponsive block copolymer brush coatings. *ACS Applied Materials & Interfaces* 10 (46): 40088–40099. https://doi.org/10.1021/acsami.8b11448.

Dr. Jyoti Rani is an Assistant Professor Parul Insitute of Design, Parul University, Vadodara-391760, Gujarat, India, with a Ph.D. in Design from Mody University of Science and Technology, India. She holds a Master's degree from Lovely Professional University and a Bachelor's degree from Kurukshetra University (KUK). With over two years of academic and research experience, her expertise spans fashion design, textiles, natural dyeing, consumer behaviour, sustainability, artificial intelligence, and merchandising. Dr. Rani has published 28 research papers in reputed refereed journals and authored five book chapters. She is committed to advancing sustainable practices and technology integration in the fashion and textile industries.

Dr. Sachthivel Santhanam is a Professor and Head of the Department of Textile Technology, Park College of Engineering and Technology, Coimbatore, Tamilnadu, India. He has been a faculty member since 2022. He was the Academic and Research person for several research and community activities during the academic year 2003–2024 in various universities in India and Ethiopia. Sakthivel has completed his Ph.D. at Anna University and his undergraduate and postgraduate studies at the same university in India. His research interests lie in recycling and sustainable product development of textile waste materials, ranging from theory to design and the development of materials to implementation. He has collaborated actively with researchers in several other disciplines of garment Engineering and Apparel waste minimization techniques and technology particularly reusing all the waste materials to minimize the cost-effectiveness to make it value-added new innovative product development from the textile and apparel waste materials; Sakthivel has served on roughly 23 conference and workshop program committees and served as the Program Chair and organizer for several years. He has served on the NBA and NACC as executive Committee leader, the Steering Committee for the AICTE/Anna University Inspection in the year 2015–2022, Textile and Garment technology Curriculum development, and the Technical development Education Board member. Sakthivelis the instructor for a popular TQM on Postgraduate topics in the functional Garment development process. He served many research projects and completed government research projects in India during 2012-2015 and he got the Best Researcher award in the year 2011–2015 is a remarkable achievement in his credits

Dr. Ramratan Guru Assistant Professor Level III, School of Design, Mody University of Science and Technology, Lakshmangarh-332311, Rajasthan, India. Research and Academic Experience: He has more than 10 Years of experience in academics and research. Research Area: His areas of interest include nonwovens, clothing comfort, medical textile, sewing thread/garment, biodegradables, plastic recycling, nanotechnology, sustainable materials and high performance fibers and Fashion Merchandising. Number of Published papers: He has contributed more than 95 Research papers in referred journals and 51 research papers in International Conferences of repute. He has contributed 12 book chapters published by Intech UK, B P International UK, and Woodhead Publications. He has applied for more than T30 Indian/ International Patents and also 06 books published. Appreciate his research work to U.K., Canada, Australia and U.S.A. Government, Europe, Brazil etc.

Raphael Kanyire Seidu (Ph.D.) is a Research Assistant at the School of Fashion and Textiles at The Hong Kong Polytechnic University, Hong Kong. He holds both a Ph.D. in Fashion and Textiles Design from The Hong Kong Polytechnic University and an MFA degree in Textile Design and a BA. Degree in Industrial Art (Textile option) from Kwame Nkrumah University of Science and Technology, Ghana. He was also a Visiting Research Student at Nottingham Trent University (NTU) in the UK. He has several publications in leading journals in textile, design, technology, and material science. His recent paper was adjudged the "Second-Best Paper" presented at the 12th Sustainable Education and Development Research Conference in Somanya, Ghana. His research interest focuses on retro-reflective materials, smart textiles, cloth culture, weaving, and print technologies.

Chapter 7
Biomass Materials from Plants and Animals for Sustainable Textiles

George Kwame Fobiri, Frederick Boamah, Emmanuel Kwame Danso, and Rowena Fatchu Kansanba

Abstract Textile materials obtained from plant and animal sources promise an advancement in green manufacturing in the textile industry, as they are good biomass resources. Recently, a focus on alleviating environmental pollution caused by plant and animal-based materials, especially in the textile industry, has resulted in questions on how these wastes could be significantly processed for sustainable production. Biomass materials are organic substances usually obtained from plants and animals for several industrial applications. Potential biomass materials for sustainable textiles include natural fibres such as cotton, flax, bamboo, hemp, viscose/lyocell from wood pulp, recycled fibres, etc. Also, animal-based materials such as wool, silk, leather, etc., are potential biomass materials. These products are biodegradable and can easily undergo processing to acquire composite materials for sustainable textile production. As resources for textile materials are exhausting, renewable biomass material manufacturing for textiles is an innovative approach to increasing the scope of textile raw materials for sustainable production in the industry. With advanced technologies/methodologies, renewable resources are transformed into valuable products for applications in several sectors. Analysis of these technologies from a textile production point of view is crucial to provide more insight into the appropriate methodologies for utilising plants and animal-based biomass in the textile industry, the challenges involved, and the future directions to help expand the subsector. This chapter presents sustainable approaches to textile production, focusing on employing biomass materials from plants and animals. It provides insight into different production techniques, the challenges involved, and the prospective application of the resulting sustainable textile products. Key findings are presented with directions for future explorations and improvement in the field.

Keywords Sustainable textiles · Plant-based biomass · Animal-based biomass · Eco-friendly textiles · Renewable textile materials

G. K. Fobiri (✉) · F. Boamah · E. K. Danso · R. F. Kansanba
Department of Fashion Design and Textiles Studies, Kumasi Technical University, Kumasi, Ghana
e-mail: kfobiri@gmail.com

S.-X. Jiang et al. (eds.), *Advances in Textile Materials and Processing Techniques for Sustainability*, SDGs and Textiles, https://doi.org/10.1007/978-981-95-0469-5_7

7.1 Introduction

Sustainability as a general concept in the textile industry has developed notably over the past few years [1]. In response to increasing resource scarcity, limited space for waste disposal, and its associated pollution problems in the textile industry, biomass materials have been adopted to provide sustainable solutions by offering renewable resources, reducing waste through circular processes, and mitigating environmental impacts through biodegradable materials [2]. The report "Our Common Future" by the World Commission on Environment and Development (WCED) drew attention to the danger of unchecked patterns in most of the sectors of the world's biggest economies, which are capable of depleting the planet's resources, where sustainable growth is synonymous with and linked to environmental balance. The growing importance of sustainability has increased the organisational complexity of implementing policies, as companies must balance environmental goals with economic viability, supply chain constraints, and adoption of innovative yet costly sustainable materials and technology in the textile industry [3]. Textiles constitute a considerable proportion of the global environmental burden. Fibre production has grown to 100 million tonnes, and it has been estimated that textiles is responsible for about 3% to 7% of global human-caused CO_2 emissions [4, 5]. Buttressing this, it is confirmed that fibre production worldwide has multiplied rapidly since the 1970s, leading to numerous environmental impacts, including air and water pollution, textile waste, and microplastic creation [6]. To address the increasing environmental challenges posed by textile waste, which includes low recycling, high landfill contributions, and greenhouse gas emissions, biomass materials offer a sustainable surrogate. Biomass provides a promising alternative for sustainability in the textile industry as post-consumer textile waste can be sourced from agricultural residues and organic by-products, integrating into circular economy models and reducing the environmental impact of textile disposal [5]. Since the textile production value chain is extensive, negative ecological impacts occur in all stages of fibre production, including agriculture, and industrial synthesis via various processes such as spinning, weaving, sewing, application of colour, waterproofing, flame retardants, etc.

Globally, there are concerns about how plants and ecosystems will respond and adapt to new environmental conditions and climate change. De Freitas, et al. [7] stated that Lignocellulose as a biomass material is receiving attention as a renewable source to reduce dependence on fossil-based energy; however, its utilization in textiles manufacturing faces challenges such as variability in raw material properties, pretreatment processes, high costs of conversion technologies, and limited scalability of production methods. Cellulosic fibre obtained from wood and other organic products is a very versatile renewable material on the planet. Still, the textile industry needs to understand how to use biomass for sustainable production [8]. Biomass is a renewable organic material obtained from plants and animals. It is widely recognised in the energy sector for converting liquid and gaseous fuels. Materials such as wood, wood residues, energy crops, agricultural residues, and organic waste are used to produce biofuels like ethanol, biodiesel, and biogas, contributing to renewable energy production and reducing reliance on fossil fuels in the energy sector [9]. Biomass has been demonstrated to be a pluripotent and powerful alternative to meet the ever-increasing demand for the continuous development of human society and carbon neutralisation. Biomass is cheap, naturally versatile, renewable, and biodegradable [10, 11]. In the textile industry, plant-based fibres, including cotton, jute, hemp, flax, kenaf, and sisal, are regarded as renewable sources for sustainable textile production. Their chemical composition aligns with the growing demand for the textile industry to transition away from synthetic fibre production, contributing to global environmental degradation. Unlike synthetic materials, which are energy-intensive and contribute widely to greenhouse gas emissions, microplastic pollution, and freshwater contamination, these natural fibres are biodegradable and environmentally friendly.

Their cost-effectiveness, recyclability, lightweight properties, and broad availability make bio-based fibres viable alternatives for reducing reliance on non-biodegradable, fossil-based materials [12]. This has placed high demand on textiles produced with biodegradable materials. The global bio-based textile market is predicted to increase significantly to US$ 2,128.1 million by 2033, following its worth of US$ 915.6 million in 2023 and a compound annual growth rate of 8.8% [13]. Also, the World Bio Market Insights (WBMI) predicts that the sustainable textile market is expected to see an increase of 12.5% compound annual growth rate from 2022 to 2030 [14]. Currently, the market is gaining more attention as consumers develop a taste for sustainable textiles. These statistics, however, indicate a bright future for the biomass-based textiles market, which aims to overtake its synthetic counterpart with time. Also, with increased interest of researchers in biomass-based textiles, the market hopes to grow significantly with the influence of the academic community, which is constantly conducting groundbreaking research to discover innovative biomass-based textiles for a sustainable textile market. Figure 7.1 shows the publication trend on biomass-based textiles in the last 10 years (from 2015 to 2025) with data extracted from the Scopus online database in March 2025. A total of 1,859 documents were obtained for the analysis. From the chart, 2015 recorded 76 publications regarding biomass-based textiles, after which an accelerating pattern is observed. After a slight drop of 93 publications in 2017 from 79 in 2016, a significant growth is observed, leading to 363 publications in 2024. As of March 2025, 98 studies have been recorded, indicating a possible rise and subsequent recording of a good number of publications by the end of 2025. Generally, the number of publications has increased over the years, indicating the popularity of biomass material application in textile production.

This focus on biomass materials sets the foundation for exploring innovative approaches to integrate plant-based resources into an eco-friendly textile value chain. In India, there has been a record of burning crop residue responsible for air pollution episodes and public health concerns; greenhouse gas emissions, radiation imbalance and a decline in soil organic productivity. Burning crop residues leads to the loss of organic matter, microbial biomass, and soil biodiversity. It has been estimated that about 75–90 million tonnes of sugarcane, 75 million tonnes of cotton plant, 40 million tonnes of banana pseudostem, 18 million tonnes of coconut and 4 million

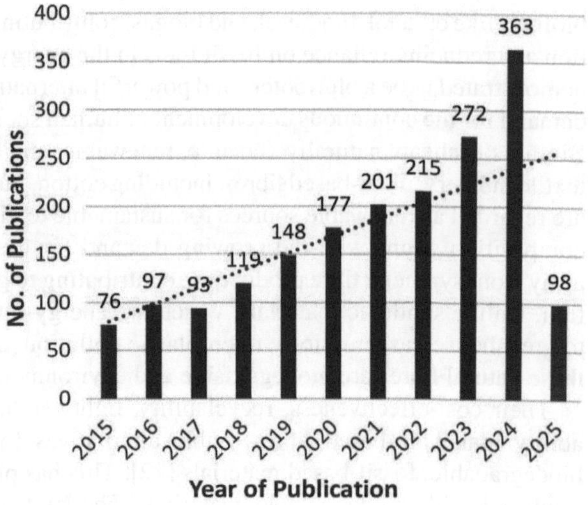

Fig. 7.1 Publication trend on biomass-based textiles

bamboo plants are burnt annually in India. However, this huge amount of lignocellulosic biomass has the potential to generate sustainable and eco-friendly packaging and home textile products. These include fibreboards, particle boards, kraft paper, tissue paper, greaseproof paper, paper carrying bags, eco-friendly cutleries, cushions, sofa covers, curtains, carpets, doormats, table mats, handbags, etc. [15]. There has been a significant development in the utilisation of bio-based textiles in the food and packaging industry [16]. Due to the rising ethical issues on synthetic fibre for various textile products, along with their significant environmental impact, the textile and fashion industry is seeking alternative materials. However, biomass fibres are renewable, biodegradable, and have eco-friendly properties, which make them ideal for addressing environmental challenges in the textile and fashion industry. By applying natural plant and animal resources, the industry has the opportunity to reduce its reliance on non-renewable inputs, minimise pollution, and contribute to a circular economy. This chapter further highlights sustainable biomass resources for textiles, innovative processing techniques for biomass in textiles, and the challenges in biomass utilization for textile manufacturing, and explores the application and impact of biomass-based textiles across industries.

7.2 Sustainable Biomass Resources for Textiles

Material consumption in all sectors of the global economy encompasses renewable biomass and remains the powerhouse of the world's economic growth through production and consumption [17]. Sustainable resourcing and consumption of biomass are essential in the 21st century as there is no negative impact on the ecosystem and the environment across various value chains [18]. However, it is possible to disengage human prosperity from environmental footprint and landfills. It is worth acknowledging that sustainable consumption is attractive yet challenging for the world. The textile framework constitutes renewable biological materials derived from plants and animals that serve as sustainable alternatives to fossil-based sources for fibre production [17]. The plant-based sources include lignocellulose, banana fibre, pineapple leaf fibre, bamboo, hemp, flax and cotton. These plant-based biomass resources are widely used in the textile industry due to their renewability, biodegradability, durability, moisture-wicking, fast-growing, versatility and their negative impact on the environment [19]. Bana residual biomass, for instance, has been identified as a resourceful material for natural fibres for promoting a circular economy [20]. Animal-based biomass sources consist of casein fibre, which is made from milk protein, alpaca, llama fibre, silk and wool fibre. These fibres are renewable, hypoallergenic and offer biodegradable alternatives to synthetic fibres [16]. Biomass materials from plants and animal sources are treated and manipulated in several ways to become useful as textile raw materials [16]. In Fig. 7.2, raw materials from bio-based sources (plants and animals) are presented, showing their characterisation, performance, and application in several sectors, of which the textile industry stands as a key sector. From the Fig. 7.2, biomass materials for textiles are obtained from the seeds, leaves, wood, peels, and roots of plants. Also, shells, bones, wool, and feathers are good sources of biomass materials from animals, as illustrated in Fig. 7.2. These biomass materials are pretreated, manufactured, and characterised to be used for textile products, including fabrics and garments.

Apart from the raw materials indicated in Fig. 7.2, which focuses mainly on animal-based and plant-based biomass materials for textiles, the sources of biomass materials are vast. These could be classified broadly as animal, agricultural, forest, aquatic, and waste biomass. Figure 7.3 shows the classification of fibres with various examples under each category. From the Fig. 7.3, the animal source of biomass is obtained from significant parts (shells, bones, feathers, etc., which are often neglected but have been proven to be good materials for eco-friendly textiles [16]. Agricultural sources, such as leaves, husks, crops, and forest sources of biomass, present similar materials (bark, branches, etc.), which are broadly obtained from plants. Also, materials obtained from wastes (manure, food waste, etc.) and aquatic sources (algae and seaweed) form a significant part of textile materials from biomass. Like plant-based biomass materials, animal-based materials like feathers, hairs, fish skin, etc., are phenomenal.

Feathers as biomass materials from animals are often taken from chickens. It is established that about 6 billion tonnes of chicken feathers are produced out of the poultry business sector, which are mostly discarded in landfills [21]. These have been used to produce biocomposite textiles with recycling concepts in fulfilment of the Sustainable Development Goals (SDG), especially goal 12 (Responsible production and consumption), which seeks to reduce waste. A recent study shows the suitability of feathers as a biomass textile material using the electrospinning process to convert keratin into nanofibers for sustainable textile manufacturing [22]. The study concludes that keratin nanofibres are good materials in biomedicine for tissue engineering and cell regeneration. Also, previous research explores feathers as biomass

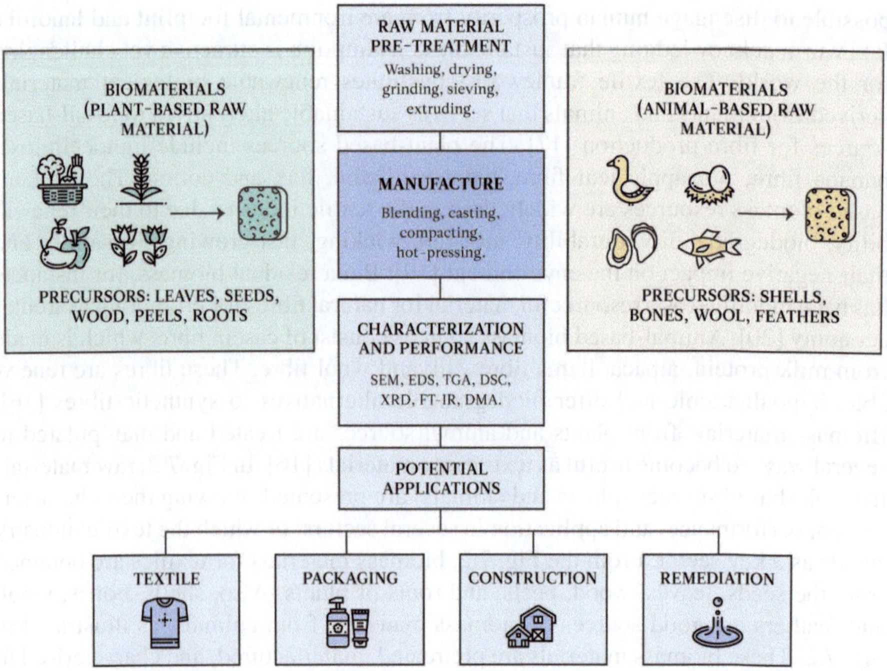

Fig. 7.2 Biomass materials from plant and animal sources for textile production and other potential applications [16]. (This article is an open-access article distributed under the terms and conditions of the Creative Commons Attribution (CC BY) license)

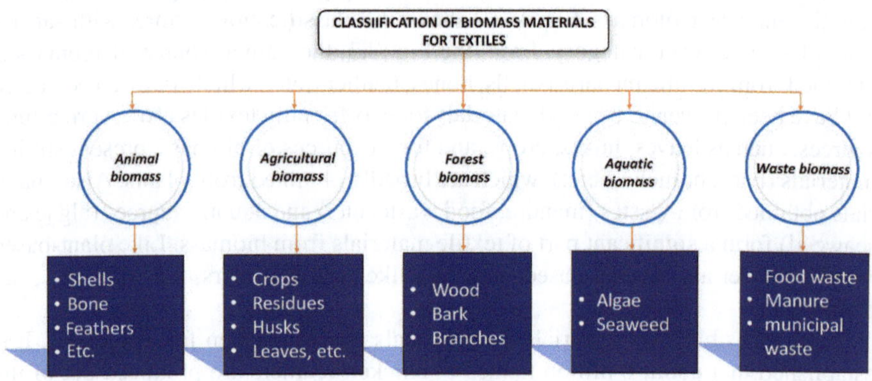

Fig. 7.3 Classification of biomass materials for textiles

materials for nonwoven composite textiles meant for thermal insulation [23]. The study employed the air-laid process to produce composite nonwoven textile from feathers and compared it to commercially available nonwoven thermal insulators. Figure 7.4 shows the feather-fibre fabric, its density, and the SEM image compared to existing materials. In Fig. 7.4a, the density of the feather fibre fabric is observed as 29.8 kg/m^2, which is closely related to that of denim fabric (30.0 kg/m^2) presented in Fig. 7.4c. The SEM image of the feather-fibre textile is shown in Fig. 7.4b. Also, the densities of the other various commercially available thermal insulators used to compare the innovative feather fabric include hemp fabric (83.3 kg/m^2), PET fabric (25.5 kg/m^2), sheep wool fabric (47.6 kg/m^2), and rock wool fabric (110.3 kg/m^2) shown in Fig. 7.4e, g, i, k respectively. The SEM image of the thermal insulators is also shown in Fig. 7.4d (denim fibre fabric), f (hemp fibre fabric), h (PET fabric), j (sheep wool), and l (rock wool). It is reported that the biomass-based textile is suitable for thermal insulation with a low thermal conductivity of 0.033 W/(m K) for the feather-based fabric sample with a 59 kg/m^3 density [23].

Also, Hou et al. [24] explored aquatic biomass for innovative textile material using fish scales as a contribution to the fulfilment of the SDGs. The collagen modification technology for polyester was focused on in the study to obtain skin-friendly, soft, and good moisture regain properties. The process involved the acquisition of recycled tilapia fish scales, washing, freeze-drying, smashing, treating with enzymes to acquire short amino acids, and filtering to obtain the collagen peptides. Figure 7.5a shows the

Fig. 7.4 Feather-based non-woven fabric and its SEM image compared to commercially available non-woven thermal insulators and their SEM images [23]. (This is an open-access article distributed under the terms of the Creative Commons CC BY license)

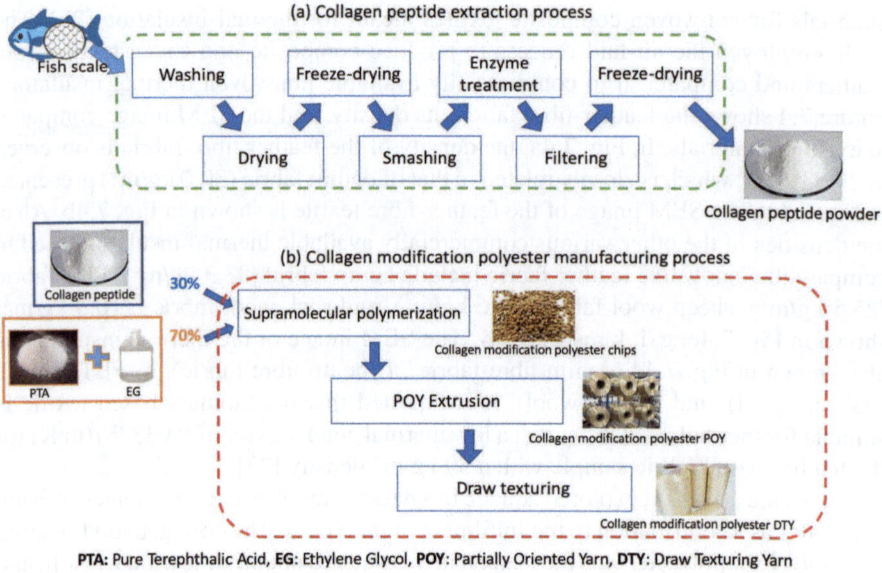

Fig. 7.5 Schematic illustration of collagen modification of polyester with aquatic biomass (fish scales). **a** The extraction process for collagen peptide powder from tilapia scales biomass. **b** Manufacturing collagen-modified polyester through super molecular polymerisation, extrusion, and drawing/spinning [24]. (This article is available under the Creative Commons CC-BY-NC-ND license)

extraction of collagen from fish biomass for textile modification. After the extraction, the novel polyester fabric was obtained through the process, including polymerisation where the extracted collagen was combined with benzenedicarboxylic acid (terephthalic acid or isophthalic acid) and ethylene glycol, along with catalysts (Sb_2O_3 and TiO_2). The mixture was then taken through the super molecular polymerisation process in which polyester chips modified with collagen are formed. The fibre extrusion is performed after the super molecular polymerisation process to first obtain a partially oriented polyester yarn, which is later textured to obtain a draw textured yarn with improved characteristics. It is worth noting that polyester yarns are synthetic but assume new properties after being treated with the biomass material obtained from fish scales. Figure 7.5b also shows the schematic illustration of the process to obtain the collagen-modified polyester yarns.

Sheep wool is a good biomass material from animal sources for polymer composites. They are normally used as reinforcements to enhance the properties of textiles in material engineering. They possess high aspect ratio, surface morphology, and chemical composition, which helps to improve the mechanical properties of their composites [25]. Biomass materials from non-wood sources are also demonstrated to be a sustainable resource for textile material (man-made cellulose fibres) production for effective use in clothing manufacturing [26]. These materials include feedstocks with cellulose properties used significantly for textiles through processes like

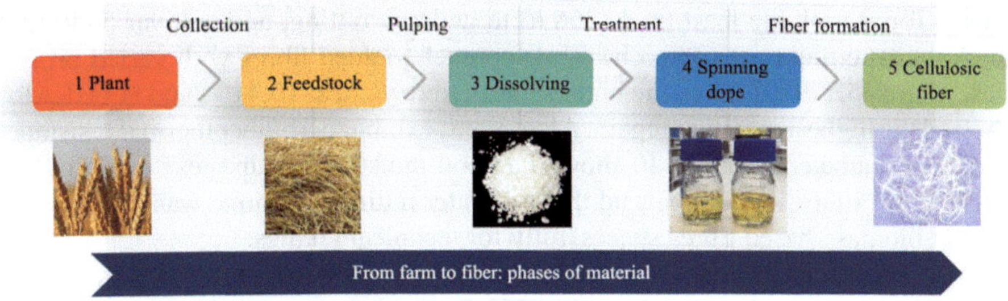

Fig. 7.6 The process involved in converting non-wood biomass to man-made cellulosic fibres [27]. (This article is available under the Creative Commons CC-BY-NC-ND 4.0 license)

pulping, purification, dissolution, doping, spinning, and yarn production for textiles and clothing [26, 27]. Figure 7.6 shows the conversion of plant biomass from non-wood sources for man-made cellulose fibre production. Other emerging and innovative biomass resources have been identified with biodegradable and antimicrobial properties. This includes mycelium, seaweed, algae, soy protein fibre, and cattail fibre. Research reveals the utilisation of plant and animal biomass to produce biofuels, biomaterials, and biochemicals to replace petroleum-based energy, synthetic textile materials, and other unsustainable chemical substances toward a sustainable and effective circular economy [19, 28]. Biomass resources enhance growth and environmental degradation, are renewable and biodegradable, reduce dependence on fossil fuels, support sustainable development by utilising agricultural and industrial by-products, promote resource efficiency, align with circular economic principles and support the growth of countries in economic transitions [9]. It also has the advantage of being able to be stored and used only when demand justifies it [29].

7.3 Innovative Processing Techniques for Biomass in Textiles

Biomass materials from diverse sources are treated through innovative techniques to make them suitable for textile products. With sustainability being the focus of biomass-based textile production, the innovative approaches not only yield textiles with enhanced performance and functionality but also contribute to environmental sustainability. To begin with, canola biomass has been proven to be an appropriate source of bio-based textile for technical applications in an attempt to obtain an alternative material for cotton for technical textiles [30]. It is established that the plant has a cellulose content of 61.3%, making its biomass suitable for innovative fibre production [30, 31]. Shuvo et al. [30] explored the water retting approach as an innovative technique of extracting bast fibre (canola biomass) from four different cultivars (HYHEAR 1, Topas, 5440, and 45H29). In this approach, the stems of the cultivars were retted with water at room temperature for a duration range of 266–380.3 h, where Topas took the least time (266 h) to undergo retting, and cultivar 5440 took the longest retting time (380.3 h). The approach yielded fibres with varied properties, where 45H29 exhibited the highest mean fibre yield at 10.41%, and HYHEAR 1 showed the highest moisture regain. Also, Topas exhibited a higher thermal resistance and fibre diameter, while 5440 showed a good moisture regain, next to HYHEAR 1. From the study, it could be said that the water retting technique was employed to extract biomass-based fibres successfully for technical textiles.

Also, the acetate process is adjudged an innovative technique for processing non-woody biomass into functional textiles [26]. Soto-Salcido et al. [32] used novel fractionation and acetylation methods to produce biomass-based textile from birch wood and agave bagasse biomass. The process involved pretreatment with natural deep eutectic solvent (choline chloride and lactic acid) and bleaching. It is observed from the study that the fractionation technique could be used effectively in birch wood and agave bagasse biomass processing for textiles with the help of natural deep eutectic solvent. Also, acetylation was employed to convert the fractionated cellulose into acetates. Acetylation involves the substitution of acetyl groups with hydroxyl groups [32]. Acetylation has also been adopted with electrospinning to produce biomass-based textile materials from sugarcane bagasse [33]. In the exploration, cellulose materials were obtained from the sugarcane bagasse, acetylated, electrospun to form acetate cellulose fibre, deacetylated, and cationized to form a bio-based material for textiles. It should be said that, apart from the acetate method of producing innovative biomass-based textiles, Larrota et al. [33] identified electrospinning and cationization as innovative techniques for obtaining functional bio-based materials for textiles. With the electrospinning approach, cost-effective, versatile, and functional fibres are produced using a high-voltage electric field [34]. The process yields varied sizes of fibres with large surface area and high porosity, as there is precise control over the process regarding their morphologies and dimensions [35]. Larrota et al. [33] found that the electrospinning technique works well when converting biomass materials (sugarcane bagasse) into textiles, as it yielded fibres of 261–4452 nm with tunable porosity of 20.8–35.5% and high thermal stability. Cationization is also an innovative technique for improving the dyeability of biomass-based cellulose textiles [33]. It helps to improve colour fixation while maintaining the morphological characteristics of the fibre.

Other innovative techniques are explored to demonstrate how biomass materials from non-wood sources, including feedstocks could be prepared through conventional technologies for textile materials like man-made cellulose fibres [27]. A typical example of this conventional technology is illustrated in Fig. 7.7, which demonstrates the procedure for dissolving pulp. In this process, wood chips are cooked and transformed into pulp through a defibrating process, washing and bleaching are done, and made to dry, then cut into sheets [27, 36].

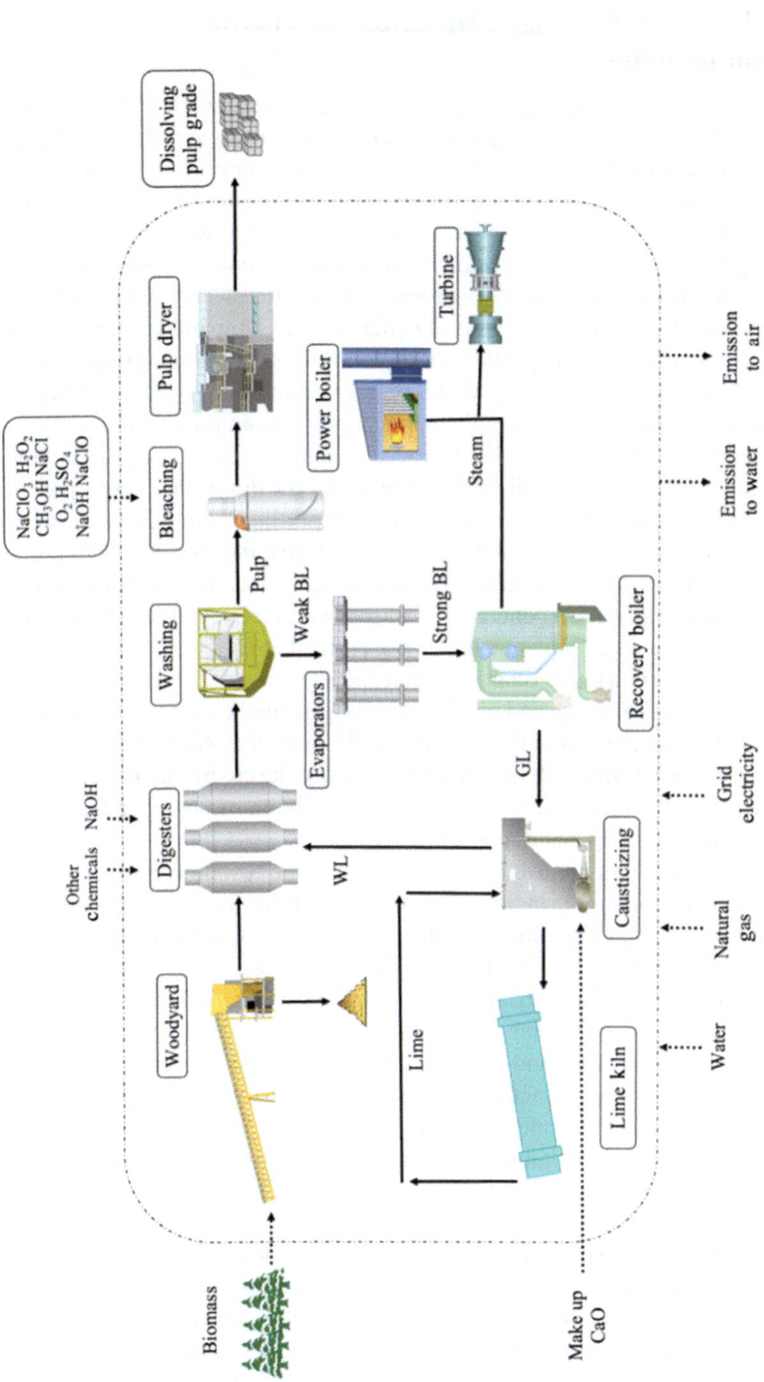

Fig. 7.7 Conventional technique for pulp dissolving into textile materials [27] (This article is available under the Creative Commons CC-BY-NC-ND 4.0 license)

7.4 Challenges in Biomass Utilisation for Textile Manufacturing

Despite the enormous efforts made by researchers in various organisations and countries, the utilisation of solid organic textile waste as raw material for the manufacturing of useful products is not being widely used as an alternative in the textile industry [37]. Biomass production mostly relies on agricultural and textile waste, making it highly susceptible to weather conditions, economic fluctuations, and production rates. These dependencies can result in raw material shortages, disrupting the consistent supply needed for textile manufacturing. Comparably, other renewable sources are also influenced by atmospheric conditions. Biomass supply is further constrained by its reliance on seasonal and regional availability [29, 38]. One key challenge the textile industry faces in the application of biomass in some regions is the exportation of wood pellets to other European countries for higher prices, which has significantly declined the availability of biomass for domestic energy production in the textile industry. The economic gains in exporting biomass products have limited their use within the textiles industry, which undermines the efforts to reduce energy dependence on external sources. It has been estimated that around 4 billion tons of agricultural residue are generated worldwide annually, and nearly a quarter of this is burned in the field. This practice is widespread in Brazil, China, and India. These countries also contribute to air pollution and greenhouse gas emissions globally [39]. The agricultural waste and residues are the principal element in the biomass conversion process, where they can be transformed into sustainable fibres, yarns, and fabrics. This innovative approach not only helps reduce the environmental impact of traditional textile production but also supports the creation of eco-friendly textiles such as biodegradable fabrics and non-toxic dyes.

The textile industry can contribute to waste reduction, resource conversion, and promoting circular economy principles that will benefit local communities and the environment by applying agricultural waste. There are other unlisted challenges in extracting fibres from biomass. For instance, ‘hemp’, a bast fibre, used in making varieties of commercial and industrial products such as ropes, clothing, shoes, papers, and bioplastics, often requires labour-intensive and energy-consuming processes such as retting, decortication, and scutching to separate the fibres from the plant stalks. These processes can sometimes be inefficient, costly, and environmentally challenging, rendering large-scale production and adoption difficult for the textile industry [40–42]. The hemp plants’ stalks often contain significant concentrations of earth-alkali salts and metals, including sodium (Na), magnesium (Mg), potassium (K), and calcium (Ca). These contaminants pose considerable health and safety risks during the production process [43]. In some circumstances, the oxidation states of transition metals can catalyse explosive radical decomposition reactions when using solvents like N-methymorpholine-N-oxide (NMMO) for fibre processing [26, 44]. Such risks portray the environmental and economic trade-offs associated with utilising hemp biomass in production, specifically when derived from phytostimulation efforts. However, hemp plants hold a reputable record in the paper industry as compared to wood fibre due to their faster growth rate, higher cellulose content, and lower environmental impact, making hemp a more sustainable alternative for producing high-quality paper products [42]. Furthermore, converting biomass fibre into value-added sustainable products requires knowledge of product design, raw material manufacturing, processing, checking market dynamics, supply chain logistics, and integrating eco-friendly technologies to ensure minimal environmental impact and maximum efficiency throughout the production lifecycle [45].

7.5 Applications and Impact of Biomass-Based Textiles Across Industries

Using biomass as an alternative raw material is essential for understanding different types and scenarios for applicability in various areas that vary with its origin, characteristics, composition, and disposal [46]. Biomass-based textiles have been utilized in construction, energy, environmental protection, sports, agriculture, health, etc. to alleviate environmental pollution and resource shortages [47–50]. Additionally, its use has helped reduce the high cost of producing carbon fibres for diverse applications [51]. Research reveals that bio-based textiles have emerged as the preferred textile choice over traditional textiles due to their renewable characteristics, tensile strength, durability, and resistance to some environmental factors such as moisture and ultraviolet exposure [52]. Bio-based construction materials possess sustainable properties and positively impact social well-being [47]. Flax, bamboo, hemp, sisal, kenaf, etc. have been combined to create strong and lightweight composite materials for structural and non-structural applications [53]. The development of concrete and cement alternatives that use bio-based aggregate to reduce the carbon footprint of concrete production is evident [54]. Bio-based textiles and fibres epitomise an innovative shift in the textile industry, contributing sustainable alternatives to conventional fibres derived from resource-intensive agricultural practices. Smart wearables (flexible conductive fabrics) have been manufactured from biomass materials in the clothing manufacturing sector [55]. This was achieved with soy-based waterborne polyurethane, carboxylated carbon nanotubes, and collagen fiber nonwovens. It is reported that the biomass-based textile has excellent hydrophobic, friction-electric, and electrothermal conversion properties which greatly impact its multifunctionality as a conductive textile for smart wearables. The use of these wearables in the clothing industry adds to the technological advancements sprouting from the industry. The ability of textiles produced from biomass materials to resist flame and combat ultraviolet (UV) rays makes them applicable in protective clothing manufacturing. An experimental study to understand the innovative modification of lyocell fabric to withstand unfriendly weather conditions revealed the suitability of biomass materials in producing a gallic acid-based intumescent flame retardant and UV protective finish [56].

It is established that biomass-based textile applications in the health sector have seen significant improvement as they are used for antimicrobial wound dressing purposes, artificial blood vessels, tissue design, etc. [25]. Again, bio-based textiles derived from microbial biomass known as bacterial cellulose, are promising materials in industrial applications. Bacterial cellulose is another form of biomass material with features similar to vegetable cellulose with the only difference found in their appearance and water content and used as raw material for sustainable textile production [46]. It is said that bacterial cellulose has exceptional properties such as pure cellulose that is free of lignin, hemicellulose, or pectin making it preferred over cellulose from vegetables as no preparatory processes are needed to get rid of the unwanted agents, hence avoiding harmful chemicals [57–59]. Textiles from this biomass material are used in the medical field, food industry, cosmetics, etc. Figure 7.8 shows the industrial application of biomass-based textiles based on bacterial cellulose. The European Union has targeted biomass-derived textiles as a medium to ensure a good environmental impact [46]. As a strategy put in place by the European Commission in 2022, textiles derived from biomass materials must greatly and positively impact the environment in 2030 with features such as their recyclability, containing more recycled fibres, hazard-free production processes, and material used [60].

7.6 Key Findings

Findings indicate that biomass resources enhance growth and reduce environmental degradation, are renewable and biodegradable, reduce dependence on fossil fuels, support sustainable development by utilising agricultural and industrial by-products, promote resource efficiency, align with circular economic principles, and support the growth of countries in economic transitions [9]. It also has the advantage of being able to be stored and used only when demand justifies it [29]. The study also revealed that using biomass-based textiles has increased sustainable production and application of eco-friendly materials in various industries. Biomass-based textiles have been utilised in construction, energy, sports, agriculture, environmental protection, etc., to mitigate environmental pollution and resource shortages. They are also significant materials in the clothing industry for smart and sustainable wearables, cosmetics, and the medical field [46, 61]. It is revealed that the biomass-based textile market is gaining more attention as consumers develop a taste for sustainable textiles. The global bio-based textile market is predicted to increase significantly to US$ 2,128.1 million by 2033, following its worth of US$ 915.6 million in 2023 and a compound annual growth rate of 8.8% [13]. Also, the World Bio Market Insights (WBMI) predicts that the sustainable textile market is expected to see an increase of 12.5% compound annual growth rate from 2022 to 2030 [14].

Notwithstanding the popularity received by biomass-based textiles recently, findings revealed challenges in their manufacturing. It was found that biomass-based textile production mostly relies on agricultural and textile waste, making it highly susceptible to weather conditions, economic fluctuations, and production rates. The

Fig. 7.8 Application of biomass-based textiles (bacterial cellulose) across industries [61]. (This is an open-access article under the terms of the Creative Commons Attribution License)

dependence on plants and animals for raw materials leads to a raw material shortage, limiting production efficiency. Regarding raw material acquisition for production, the study found sources of biomass materials for textiles to include plant sources (seeds, leaves, wood, orange waste films, etc.) and animal sources (sheep wool, fish scales, feathers, bones, shells). Conventional techniques are also employed to convert raw materials into textile materials. For instance, pulp dissolving technology is an appropriate technique for producing textiles from cellulose materials [27]. The innovative processing techniques for biomass in textiles encompass a range of approaches, including the water retting approach to bast fibre (canola biomass) extraction [30], acetylation and fractionation [26, 32], and electrospinning and cationization [33]. It is worth noting that biomass materials from diverse sources (mostly plant and animal sources) are good alternatives and promise sustainable textile production and consumption.

7.7 Future Research Studies

Findings indicate prospects of yarns obtained from biomass materials in flexible conductive fabrics. Future studies should focus on the scalable production of fully biomass-based conductive yarns to improve wearable textile electronics, sustainable production, and consumption. For instance, technologies like acetylation and fractionation could be used to produce fully bio-based textiles from biomass materials for consumption in the market al.so, modification of textiles with fish biomass through collagen extraction and extrusion technologies as described by Hou et al. [24] would help improve yarns that fall within the synthetics but are valuable in textile production. This hybridisation process ensures sustainable production rather than relying fully on synthetic materials. A focus should also be placed on improving the functional properties of biomass-based textiles, such as their UV protection and flame retardancy. This would leave industries with several options and opportunities to explore biomass-based textiles in their production processes. Feathers have also been identified as good biomass materials for thermal insulation. With the abundance of feathers in the poultry business sector, future studies could venture into the scalable production of high-performance insulation textile materials from feather biomass. In this case, nonwoven fabrics could be produced using the air laid technology, which has been validated by previous research [23] as an appropriate method of manufacturing feather-based textile for thermal insulation.

7.8 Conclusion

The chapter highlights sustainable approaches to textile production, focusing on employing biomass materials from plants and animals. Innovative materials from plants (seeds, leaves, wood, and orange waste films) and animals (sheep wool, feathers, fish scales, bones, and shells) were identified as raw materials for biomass-based textiles. Also, innovative processing techniques for biomass in textiles include the water retting approach to bast fibre (canola biomass) extraction, acetylation, fractionation, electrospinning, and cationization. Biomass-based textiles are used in various industries, including health, construction, food, environmental protection, etc. As protective clothes, biomass-based modified textiles are used as protective clothing to resist UV and perform as flame retardants. It could be said that the emphasis on utilising renewable resources and minimising waste is essential for shaping the future of textile production. As the industry evolves, collaborative efforts between researchers, designers, and industry stakeholders will be crucial in driving innovation and fostering a more sustainable textile ecosystem.

References

1. Prieto-Jiménez, E., L. López-Catalán, B. López-Catalán, and G. Domínguez-Fernández. 2021. Sustainable development goals and education: A bibliometric mapping analysis. *Sustainability* 13: 2126. https://doi.org/10.3390/su13042126.
2. Keßler, L., S. A. Matlin, and K. Kümmerer. 2021. The contribution of material circularity to sustainability—Recycling and reuse of textiles. *Current Opinion in Green and Sustainable Chemistry* 32: 100535. https://doi.org/10.1016/j.cogsc.2021.100535.
3. Larsson, J., and L. Larsson. 2020. Integration, application and importance of collaboration in sustainable project management. *Sustainability* 12: 585. https://doi.org/10.3390/su12020585.
4. Laitala, K., I. Klepp, and B. Henry. 2018. Does use matter? Comparison of environmental impacts of clothing based on fiber type. *Sustainability* 10: 2524. https://doi.org/10.3390/su10072524.
5. Vera, R. E., F. Zambrano, A. Suarez, A. Pifano, R. Marquez, M. Farrell, et al. 2022. Transforming textile wastes into biobased building blocks via enzymatic hydrolysis: A review of key challenges and opportunities. *Cleaner and Circular Bioeconomy* 3: 100026. https://doi.org/10.1016/j.clcb.2022.100026.
6. Lawson, L., L. M. Degenstein, B. Bates, W. Chute, D. King, and P. I. Dolez. 2022. Cellulose textiles from hemp biomass: Opportunities and challenges. *Sustainability* 14: 15337. https://doi.org/10.3390/su142215337.
7. de Freitas, E., J. Salgado, R. Alnoch, A. Contato, E. Habermann, M. Michelin, et al. 2021. Challenges of biomass utilization for bioenergy in a climate change scenario. *Biology (Basel)* 10: 1277. https://doi.org/10.3390/biology10121277.
8. Konstantinavičienė, J., and V. Vitunskienė. 2023. Definition and classification of potential of forest wood biomass in terms of sustainable development: A review. *Sustainability* 15: 9311. https://doi.org/10.3390/su15129311.
9. Scarlat, N., J.-F. Dallemand, F. Monforti-Ferrario, and V. Nita. 2015. The role of biomass and bioenergy in a future bioeconomy: Policies and facts. *Environment and Behaviour* 15: 3–34. https://doi.org/10.1016/j.envdev.2015.03.006.
10. Wang, R., Y. Feng, D. Li, K. Li, and Y. Yan. 2024. Towards the sustainable production of biomass-derived materials with smart functionality: A tutorial review. *Green Chemistry* 26: 9075–9103. https://doi.org/10.1039/D4GC01771D.
11. Dukare, A., R. Yadav, S. Kautkar, P. Kuppusamy, K. Sharma, A. Shaikh, et al. 2024. Waste to wealth: Microbial-based sustainable valorization of cotton biomass, processing waste and by-products for bioenergy and other value-added products to promote circular economy. *Waste Management Bulletin* 2: 262–280. https://doi.org/10.1016/j.wmb.2024.11.011.

12. Parvin, M., M. Shadhin, M. Dulal, and M. Rahman. 2025. Harnessing Cattail biomass for sustainable fibers and engineered bioproducts: A review. *Global Challenges* 9. https://doi.org/10.1002/gch2.202400183.
13. Fact.MR. 2002. Bioplastic textiles market outlook (2023 to 2033).
14. WBMI. 2022. The biggest bio-textile investments of 2022 (So Far).
15. Rathinamoorthy, R., T. Sharmila Bharathi, M. Snehaa, and C. Swetha. 2023. Mycelium as sustainable textile material—Review on recent research and future prospective. *International Journal of Clothing Science and Technology* 35: 454–476. https://doi.org/10.1108/IJCST-01-2022-0003.
16. Sandoval, S. S., A. Amenábar, I. Toledo, N. Silva, and P. Contreras. 2024. Advances in the sustainable development of biobased materials using plant and animal waste as raw materials: A review. *Sustainability* 16: 1073. https://doi.org/10.3390/su16031073.
17. Destek, M. A., S. A. Sarkodie, and E. F. Asamoah. 2021. Does biomass energy drive environmental sustainability? An SDG perspective for top five biomass consuming countries. *Biomass and Bioenergy* 149: 106076. https://doi.org/10.1016/j.biombioe.2021.106076.
18. Sadhukhan, J., E. Martinez-Hernandez, R. J. Murphy, D. K. S. Ng, M. H. Hassim, K. Siew Ng, et al. 2018. Role of bioenergy, biorefinery and bioeconomy in sustainable development: Strategic pathways for Malaysia. *Renewable and Sustainable Energy Reviews* 81: 1966–1987. https://doi.org/10.1016/j.rser.2017.06.007.
19. Zhu, J. Y., U. P. Agarwal, P. N. Ciesielski, M. E. Himmel, R. Gao, Y. Deng, et al. 2021. Towards sustainable production and utilization of plant-biomass-based nanomaterials: A review and analysis of recent developments. *Biotechnology for Biofuels* 14: 114. https://doi.org/10.1186/s13068-021-01963-5.
20. Jayaprakash, K., A. Osama, R. Rajagopal, B. Goyette, and O. P. Karthikeyan. 2022. Agriculture waste biomass repurposed into natural fibers: A circular bioeconomy perspective. *Bioengineering* 9: 296. https://doi.org/10.3390/bioengineering9070296.
21. Kavitha, S. 2024. Transforming chicken feathers into sustainable textile solutions. Fibre2Fashion. https://www.fibre2fashion.com/industry-article/10032/transforming-chicken-feathers-into-sustainable-textile-solutions. Accessed 1 Apr 2025.
22. de Oliveira, F.G., A.A. Passos, and M.G. Mothé. 2025. Keratin nanofiber: Sustainable innovation in the nonwoven textile sector to mitigate the generation of microplastic. *Biomass Conversion and Biorefinery*.
23. Dieckmann, E., R. Onsiong, B. Nagy, L. Sheldrick, and C. Cheeseman. 2021. Valorization of waste feathers in the production of new thermal insulation materials. *Waste Biomass Valorization* 12: 1119–1131. https://doi.org/10.1007/s12649-020-01007-3.
24. Hou, E.-J., C.-S. Huang, Y.-C. Lee, and H.-T. Chu. 2022. Upcycled aquaculture waste as textile ingredient for promoting circular economy. *Sustainable Materials and Technologies* 31: e00336. https://doi.org/10.1016/j.susmat.2021.e00336.
25. Kulkarni, M. B., V. Gavande, P. A. Mahanwar, A. R. Shah, R. K. Shuib, A. M. Khare, et al. 2024. Review on biomass sheep wool–based polymer composites. *Biomass Convers Biorefin* 14: 30961–30982. https://doi.org/10.1007/s13399-023-04912-4.
26. Frazier, R. M., M. Lendewig, R. E. Vera, K. A. Vivas, N. Forfora, I. Azuaje, et al. 2024. Textiles from non-wood feedstocks: Challenges and opportunities of current and emerging fiber spinning technologies. *Journal of Bioresources and Bioproducts* 9: 410–432. https://doi.org/10.1016/j.jobab.2024.07.002.
27. Frazier, R. M., K. A. Vivas, I. Azuaje, R. Vera, A. Pifano, N. Forfora, et al. 2024. Beyond cotton and polyester: An evaluation of emerging feedstocks and conversion methods for the future of fashion industry. *Journal of Bioresources and Bioproducts* 9: 130–159. https://doi.org/10.1016/j.jobab.2024.01.001.
28. Munawar, M. A., A. H. Khoja, S. R. Naqvi, M. T. Mehran, M. Hassan, R. Liaquat, et al. 2021. Challenges and opportunities in biomass ash management and its utilization in novel applications. *Renewable and Sustainable Energy Reviews* 150: 111451. https://doi.org/10.1016/j.rser.2021.111451.

29. Nunes, L.J.R., R. Godina, and J.C. de O. Matias. 2019. Technological Innovation in biomass energy for the sustainable growth of textile industry. *Sustainability* 11: 528. https://doi.org/10.3390/su11020528.
30. Shuvo, I. I., M. Rahman, T. Vahora, J. Morrison, S. DuCharme, and L.-P. Choo-Smith. 2020. Producing light-weight bast fibers from canola biomass for technical textiles. *Textile Research Journal* 90: 1311–1325. https://doi.org/10.1177/0040517519886636.
31. Tofanica, B. M., E. Cappelletto, D. Gavrilescu, and K. Mueller. 2011. Properties of Rapeseed (*Brassica napus*) stalks fibers. *Journal of Natural Fibers* 8: 241–262. https://doi.org/10.1080/15440478.2011.626189.
32. Soto-Salcido, L. A., I. Anugwom, L. Ballinas-Casarrubias, M. Mänttäri, and M. Kallioinen. 2020. NADES-based fractionation of biomass to produce raw material for the preparation of cellulose acetates. *Cellulose* 27: 6831–6848. https://doi.org/10.1007/s10570-020-03251-1.
33. Larrota, O. A. F., R. Vera-Graziano, A. López-Córdoba, and E. Y. Gómez-Pachón. 2021. Electrospun ultrafine cationic cellulose fibers produced from sugarcane bagasse for potential textile applications. *Polymers (Basel)* 13: 3927. https://doi.org/10.3390/polym13223927.
34. Li, T., M. Sun, and S. Wu. 2022. State-of-the-art review of electrospun gelatin-based nanofiber dressings for wound healing applications. *Nanomaterials* 12: 784. https://doi.org/10.3390/nano12050784.
35. Bonakdar, M. A., and D. Rodrigue. 2024. Electrospinning: Processes, structures, and materials. *Macromol* 4: 58–103. https://doi.org/10.3390/macromol4010004.
36. Chen, C., C. Duan, J. Li, Y. Liu, X. Ma, L. Zheng, et al. 2016. Cellulose (dissolving pulp) manufacturing processes and properties: A mini-review. *BioResources* 11: 5553–5564.
37. Ruiz, B., E. Fuente, A. Pérez, L. Taboada-Ruiz, J. M. Sanz, L. F. Calvo, et al. 2023. Employment of conventional and flash pyrolysis for biomass wastes from the textile industry with sustainable prospects. *Journal of Analytical and Applied Pyrolysis* 169: 105864. https://doi.org/10.1016/j.jaap.2023.105864.
38. Guilhermino, A., G. Lourinho, P. Brito, and N. Almeida. 2018. Assessment of the use of forest biomass residues for bioenergy in Alto Alentejo, Portugal: Logistics, economic and financial perspectives. *Waste Biomass Valorization* 9: 739–753. https://doi.org/10.1007/s12649-017-9830-3.
39. Yadav, M., N. Singh, Annu, S.A. Khan, C.J. Raorane, and D.K. Shin. 2024 Recent advances in utilizing lignocellulosic biomass materials as adsorbents for textile dye removal: A comprehensive review. *Polymers (Basel)* 16: 2417. https://doi.org/10.3390/polym16172417.
40. Kraszkiewicz, A., M. Kachel, S. Parafiniuk, G. Zając, I. Niedziółka, and M. Sprawka. 2019. Assessment of the possibility of using hemp biomass (*Cannabis Sativa L.*) for energy purposes: A case study. *Applied Sciences* 9: 4437. https://doi.org/10.3390/app9204437.
41. Sieracka, D., J. Frankowski, S. Wacławek, and W. Czekała. 2023. Hemp biomass as a raw material for sustainable development. *Applied Sciences* 13: 9733. https://doi.org/10.3390/app13179733.
42. Chaowana, P., W. Hnoocham, S. Chaiprapat, P. Yimlamai, K. Chitbanyong, K. Wanitpinyo, et al. 2024. Utilization of hemp stalk as a potential resource for bioenergy. *Materials Science for Energy Technologies* 7: 19–28. https://doi.org/10.1016/j.mset.2023.07.001.
43. Zafeiraki, E., K.M. Kasiotis, P. Nisianakis, and K. Machera. 2021. Macro and trace elements in hemp (*Cannabis sativa L.*) cultivated in Greece: Risk assessment of toxic elements. *Frontiers in Chemistry* 9. https://doi.org/10.3389/fchem.2021.654308.
44. Rosenau, T., A. Potthast, H. Hettegger, M. Bacher, M. Opietnik, T. Röder, et al. 2021. On the role of N-methylmorpholine-N-oxide (NMMO) in the generation of elemental transition metal precipitates in cellulosic materials. *Cellulose* 28: 10143–10161. https://doi.org/10.1007/s10570-021-04181-2.
45. Adam, R., A. Pollex, T. Zeng, C. Kirsten, L. Röver, F. Berger, et al. 2023. Systematic homogenization of heterogenous biomass batches—Industrial-scale production of solid biofuels in two case studies. *Biomass and Bioenergy* 173: 106808. https://doi.org/10.1016/j.biombioe.2023.106808.

46. Dias, J. C., S. Marques, P. C. Branco, T. Rodrigues, C. A. V. Torres, F. Freitas, et al. 2025. Biopolymers derived from forest biomass for the sustainable textile industry. *Forests* 16: 163. https://doi.org/10.3390/f16010163.
47. Yadav, M., and M. Agarwal. 2021. Biobased building materials for sustainable future: An overview. *Materials Today: Proceedings* 43: 2895–2902. https://doi.org/10.1016/j.matpr.2021.01.165.
48. Miller, S. A. 2018. Natural fiber textile reinforced bio-based composites: Mechanical properties, creep, and environmental impacts. *Journal of Cleaner Production* 198: 612–623. https://doi.org/10.1016/j.jclepro.2018.07.038.
49. Sharma, N., B. Allardyce, R. Rajkhowa, A. Adholeya, and R.A. Agrawal. 2022. Substantial role of agro-textiles in agricultural applications. *Frontiers in Plant Science* 13. https://doi.org/10.3389/fpls.2022.895740.
50. Hassan, M. M., and C. M. Carr. 2021. Biomass-derived porous carbonaceous materials and their composites as adsorbents for cationic and anionic dyes: A review. *Chemosphere* 265: 129087. https://doi.org/10.1016/j.chemosphere.2020.129087.
51. Wu, Y., X. Gao, T. T. Nguyen, J. Wu, M. Guo, W. Liu, et al. 2022. Green and low-cost natural lignocellulosic biomass-based carbon fibers—Processing, properties, and applications in sports equipment: A review. *Polymers (Basel)* 14: 2591. https://doi.org/10.3390/polym14132591.
52. Shadhin, M., M. Rahman, R. Jayaraman, Y. Chen, D. Mann, and W. Zhong. 2023. Natural biomass & waste biomass fibers—Structures, environmental footprints, sustainability, degumming methods, & surface modifications. *Industrial Crops and Products* 204: 117252. https://doi.org/10.1016/j.indcrop.2023.117252.
53. Elfaleh, I., F. Abbassi, M. Habibi, F. Ahmad, M. Guedri, M. Nasri, et al. 2023. A comprehensive review of natural fibers and their composites: An eco-friendly alternative to conventional materials. *Results in Engineering* 19: 101271. https://doi.org/10.1016/j.rineng.2023.101271.
54. Rabbat, C., S. Awad, A. Villot, D. Rollet, and Y. Andrès. 2022. Sustainability of biomass-based insulation materials in buildings: Current status in France, end-of-life projections and energy recovery potentials. *Renewable and Sustainable Energy Reviews* 156: 111962. https://doi.org/10.1016/j.rser.2021.111962.
55. Feng, J., X. Chen, S. Han, H. Liu, P. Zhang, and M. Li. 2025. Biomass-derived multifunctional conductive fabrics with aluminum ion coordination: Integrating hydrophobic triboelectric and electrothermal conversion properties. *Journal of Colloid and Interface Science* 683: 101–110. https://doi.org/10.1016/j.jcis.2024.12.066.
56. Zhao, J., L. Jiang, C. Zuo, W. Tan, Y. Ren, and X. Liu. 2024. Biomass-inspired fabrication of eco-friendly, durable flame retardant and ultraviolet resistant lyocell fabric. *International Journal of Biological Macromolecules* 279: 135332. https://doi.org/10.1016/j.ijbiomac.2024.135332.
57. Felgueiras, C., N.G. Azoia, C. Gonçalves, M. Gama, and F. Dourado. 2021. Trends on the cellulose-based textiles: Raw materials and technologies. *Frontiers in Bioengineering and Biotechnology* 9. https://doi.org/10.3389/fbioe.2021.608826.
58. Reiniati, I., A. N. Hrymak, and A. Margaritis. 2017. Recent developments in the production and applications of bacterial cellulose fibers and nanocrystals. *Critical Reviews in Biotechnology* 37: 510–524. https://doi.org/10.1080/07388551.2016.1189871.
59. Singh, A., A. Kaur, A.K. Patra, and R. Mahajan. 2018. A sustainable and green process for scouring of cotton fabrics using xylano-pectinolytic synergism: Switching from noxious chemicals to eco-friendly catalysts. *3 Biotech* 8: 184. https://doi.org/10.1007/s13205-018-1193-3.
60. European Commission. n.d. EU strategy for sustainable and circular textiles. https://EnvironmentEcEuropaEu/Strategy/Textiles-Strategy_en. Accessed 18 Feb 2025.
61. Girard, V., J. Chaussé, and P. Vermette. 2024 Bacterial cellulose: A comprehensive review. *Journal of Applied Polymer Science* 141. https://doi.org/10.1002/app.55163.

George Kwame Fobiri is an Assistant Lecturer at the Department of Fashion Design and Textiles Studies at Kumasi Technical University. He holds a bachelor's degree in Industrial Art (Textiles) and a master of Fine Art (MFA) in Textile Design from the Department of Industrial Art, Kwame Nkrumah University of Science and Technology (KNUST). He is currently pursuing a Ph.D. in Textile Design Technology at KNUST. His research interests cover textiles, fashion, computer-aided design (CAD), e-textiles, smart textiles, textile printing, textile dyeing, etc.

Frederick Boamah holds a Ph.D. from the Department of Industrial Art, Kwame Nkrumah University of Science and Technology (KNUST). He also holds a B.A. Industrial Art-Textiles, MBA logistics and Supply chain management and MFA Textiles Design from the same university. His research borders on Textiles as a tool in solving social problems, investigation into indigenous fibres, African textile designs, supply chain risk management and nature-inspired textile designs. He is currently working on the internal structures of natural objects as inspiration for African Textile Print designs.

Emmanuel Kwame Danso is a researcher and a talented fashion/textile designer with a Diploma and a Bachelor of Technology in Fashion Design and Textiles from Kumasi Technical University. He has honed his skills in creative design and technical expertise. Awards and recognition include the Best in Textiles Award from Kumasi Technical University's Fashion Department. Drawing inspiration from nature and the environment, Emmanuel's designs blend functionality with sustainability. At the periphery of his professional interests, he explores diverse subjects through research and analysis. However, his research interest covers textile design, fashion design, sustainability, etc. Aspiring to become a world-class education specialist, Emmanuel strives for excellence in fashion, textiles, and academic research. His unique blend of creativity, technical skill, and passion for learning makes him a compelling voice in his field.

Rowena Fatchu Kansanba is a lecturer at the Department of Fashion Design and Textile Studies of the Kumasi Technical University in Ghana. She is currently a Ph.D. student at the Kwame Nkrumah University of Science and Technology (KNUST), offering a Ph.D. in Fashion Design Technology. Her research is in technology adoption, and she has an interest in exploring technology adoption in the textiles and apparel industry. She has published papers in the Scientific African journal and the Journal of Arts and Humanities.

Chapter 8
Biodegradable Materials: The Future of Sustainable Fabrics

Bismark Sarkodie, Benjamin Tawiah, and Haijin Liu

Abstract Sustainability has been a buzzword in many aspects of operation. In the context of environmental sustainability, biodegradable materials have been identified as one of the primary pillars that are going to safeguard the lives of the current and future generations from the hazards presented by non-biodegradable materials. As a result, the investigation and use of biodegradable materials in a wide range of industries is seen as important and necessary. The vogue of fast fashion and consumer cravings has exacerbated environmental threats posed by the textile and garment industry. The textile industry is among the world's largest suppliers of garbage, with the majority of materials being synthetic and non-biodegradable. A thorough investigation into the development and application of biodegradable materials is required to protect the environment from the harmful impacts of large non-biodegradable materials and to provide more sustainable textiles for various purposes. This chapter discusses the rationale behind replacing non-biodegradable materials with biodegradable materials, the classification of biodegradable materials (both natural and synthetically generated biodegradable materials), the biodegradation mechanism, literature on biodegradable fibers/fabrics, major challenges, and future opportunities.

Keywords Biodegradation · Environment · Fiber · Fabric · Microbes · Sustainability

---

B. Sarkodie (✉) · B. Tawiah
Department of Industrial Art (Textiles), Kwame Nkrumah University of Science and Technology, Kumasi, Ghana
e-mail: bismarksarkodie@yahoo.com; bismark.sarkodie@knust.edu.gh

B. Sarkodie · H. Liu
School of Environment, Key Laboratory for Yellow River and Huaihe River Water Environment and Pollution Control, Ministry of Education, Henan Key Laboratory for Environmental Pollution Control, Henan Normal University, Xinxiang, PR China

S.-X. Jiang et al. (eds.), *Advances in Textile Materials and Processing Techniques for Sustainability*, SDGs and Textiles, https://doi.org/10.1007/978-981-95-0469-5_8

8.1 Introduction

Environmental sustainability is on the current global agenda, with the goal of eliminating the serious and terrible hazards caused by pollution for mankind and the environment. A variety of diseases are linked to people's exposure to air, water, and solid pollutants in the surroundings. The debate and discourse on environmental and human existence have revealed various preconceptions that economic and environmental interests are in conflict. However, the current study argues that this dichotomy between growth and conservation is not only unwarranted but actually harmful to both aims. Reaching the goal of sustainability will rely on our capacity to protect both societies and sound ecology [1]. There are countless efforts throughout the globe to phase out the use of non-biodegradable materials in ecosystems. The concept of sustainability stretches beyond almost 30 years to the new mission established by the International Union for Conservation of Nature in 1969. It was a central issue of the United Nations Conference on the Human Environment in Stockholm in 1972. The concept was created expressly to argue that economic progress and industrial development could be achieved without causing adverse ecological effects [2]. The United Nations issued an advisory on June 2, 2022, that the consequences of humanity's ecological footprint are becoming increasingly unsustainable; therefore, the heads of state assembled in Stockholm reiterated their dedication to urgent and coordinated efforts to combat the global turmoil and build a brighter future for all humanity [3]. Throughout these decades, the concept of sustainable growth has expanded. The Brundtland Report identifies sustainable growth as "development which addresses current demands devoid of jeopardizing the capacity of posterity to satisfy their needs" [2]. These non-biodegradable compounds have a large spectrum of applications, particularly in textiles. The current pre- and post-fabrication processes endanger humans and the environment, particularly in their application. Textile items often last less than five years. The present business-as-usual strategy is expected to result in the buildup of more than 4.5 gigatons of textile waste by 2040. Existing textile waste processing procedures, such as incineration and recycling, can only handle limited volumes of trash; hence the vast majority of post-consumer textile waste ends up in landfills. Following an estimation by the United States Environmental Protection Agency in 2018, the United States alone generates around 17.03 million tons of textile waste each year. Out of the waste generated annually, 2.51, 3.22, and 11.3 million tons (66.35%) are recycled, charred, and discarded in trash dumps, respectively [4]. This trend has highlighted an urgency to safeguard humans and ecosystems via the widespread introduction, investigation, and adoption of biodegradable materials in the textile industry.

This chapter discusses biodegradable materials and classifications of biodegradable materials, with examples that include both natural and synthetic origins. The chapter highlights biodegradable materials that find application in the textile industry. In addition, the process of biodegradation as well as literature on biodegradable fibers/fabrics are covered. Within the context of textiles, the challenges that arise in the production and use of biodegradable materials have also been examined, as well as the possibilities for the future.

8.2 Why Biodegradable Materials?

In recent years, a large proportion of synthetic polymers utilized as raw ingredients in fabrics are non-biodegradable since they are petrol-based. "Many people succumb to buying seasonal trends that are then thrown away within a couple of months, and it's just not sustainable," says Nadya Hutagalung, an iconic figure across Asia [5]. The widespread usage of non-biodegradable textiles has an influence on the environment [6]. Figure 8.1 depicts the primary data on worldwide textile waste trends from the production to the disposal stage. It is apparent that billions of textiles are manufactured annually, resulting in around ninety-two million tons of garbage. About 60% of the billions of textiles used each year are produced with non-biodegradable materials (plastic), resulting in about forty-two million tons of textile trash. Surprisingly, approximately one percent of textile waste gets recycled into fresh clothing. The rising concern about the avoidance of ecological destruction, in conjunction with the necessity for eco-friendly materials, has brought the spotlight on materials derived from sustainable natural sources and biodegradable plant/animal components. In today's world, customers are more concerned not only with the effect of materials on their bodies or the small radius in which they find themselves, but also with the financial and ecological effects of what they buy. Furthermore, media attention and commercial promotion have increased customer expectations about material origins, manufacturing, and disposal after use. Moreover, there has been the implementation of government rules on a plethora of businesses to help address the environmental effect of products [7]. A novice might question why synthetic non-biodegradable materials are used in the textile industry since they are not biodegradable. The distinct structure and enthralling properties of non-biodegradable materials, such as stability and strength, render them useful in a myriad of applications. Unfortunately, the key features of non-biodegradable materials make them difficult to decompose and hence persistent in the environment. Furthermore, the production of certain polymeric materials necessitates the use of harmful substances or generates harmful byproducts [8]. Based on the aforementioned issues, the disposal of most synthetic non-biodegradable textile wastes is a serious environmental problem that we face today, namely the alarming risk of global warming caused by carbon emissions from the combustion of non-biodegradable conventional polymers such as polyethylene, polypropylene, and polyvinyl chloride. "At the launch of the United Nations Sustainable Fashion Alliance, it was released that researchers are engineering intriguing fibers that are sustainable, have low water impact, and low impact on the environment where they are produced" [1]. In general, humanity cannot afford to continually commit themselves to the substantial consumption of non-biodegradable materials; so, the flag of biodegradable materials and innovations must be robustly erected [6].

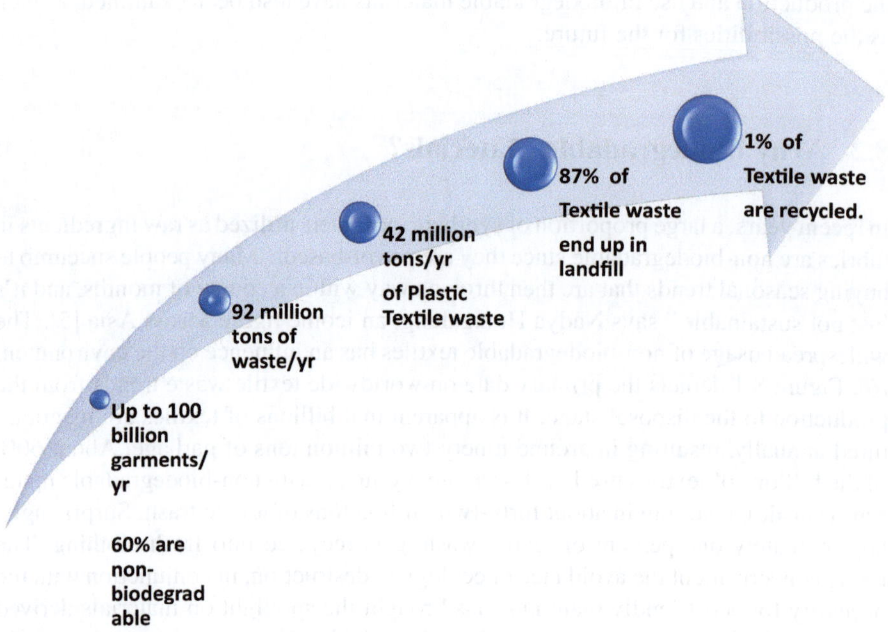

Fig. 8.1 Key statistics of waste in the textile sector—from production to the disposal stage (Data source: https://theroundup.org/textile-waste-statistics/)

8.3 Biodegradable Materials

The word "degradation" can be defined as the process by which substances are converted into simpler compounds by chemical or microbiological processes using fungus or bacteria, etc. Premising on the definition given for "degradation", the word "biodegradation" can simply be explained as the process by which microbial organism break down or transform the structure of materials introduced into the environment into simpler compounds. The action of biodegradation is mostly necessary at the post-consumer stage to ensure green environment. Therefore, biodegradable polymers have gained popularity as an alternative to traditional polymers used in fabric manufacture. Biodegradable polymers are those that can be spontaneously decomposed by microbes into beneficial compounds like compost and CH_4. Biodegradable polymers (biopolymers) are materials that can operate for a brief time prior to their disintegration into simple matter, takes a monitored process. They may be made from various wastes or bioresources, such as food, animal, and agro-waste, along with other materials such as starch and cellulose. Biopolymers made from sustainable sources are frequently less expensive than those made from microbial supplies; hence producers prioritize engineering biopolymers from sustainable sources. The use of biodegradable polymers has many environmental benefits, including raw material regeneration, biodegradation, and lower carbon dioxide emissions. It is critical to understand that the biodegradation process is determined by the content of the material. The polymer's shape, structure, chemical treatment, and molecular weight, among others, all have an influence on the manner in which it breaks down. Integration of polymeric materials has been used to enhance the biological decomposition of non-biodegradable materials. In most cases, surface modification of natural biodegradable materials is done to enhance compatibility with non-biodegradable material prior to blending. This strategy can be said to be a route for controlling non-biodegradable elements in fabric manufacturing to reduce the threats posed on the ecology.

8.3.1 *Classification of Biodegradable Materials*

Classifications of biodegradable materials can best be deduced from the polymeric substances used. There are a great number of biopolymers that have been widely used in a wide variety of fields of life for the purpose of manufacturing a vast range of items that are employed in both domestic and commercial settings. Natural and synthetic biodegradable polymers are the two primary types of biodegradable polymers that are shown in Fig. 8.2. The latter is engineered in a laboratory, while the former is highly organic material that predominantly originates from nature (renewable or biological sources such as animal, plant, marine, and microbial sources). Among the natural biodegradable polymers, there is a category of natural biodegradable polyesters that are produced by microorganisms [6]. Some examples of these polyesters are polyhydroxyalkanoate and poly-3-hydroxybutyrate. The production of synthetic biodegradable polymers is accomplished using chemical approaches [6]. To narrow biodegradable materials to fibers, Bao et al. categorize them into three: natural biobased fibers, synthetic biobased fibers, and regenerated biobased fibers (as shown in Fig. 8.3).

8.3.1.1 Natural Origin

Fibers derived from natural sources constitute the majority of biodegradable materials. As shown in Fig. 8.4, natural fibers may also be categorized according to their sources, which can be either plant, animal, or mineral. In most cases, the structure of the fiber, which is predominantly composed of cellulose and protein in both plant and animal fibers, makes them susceptible to biodegradation. Although it is classified as a natural fiber, asbestos is very non-biodegradable due to its composition, which consists of silicate chains that are connected with magnesium, iron, calcium, aluminum, and sodium, or trace elements. This section will briefly delve into a few of the plant and animal fibers that are regularly used, with a particular emphasis on their composition and applications.

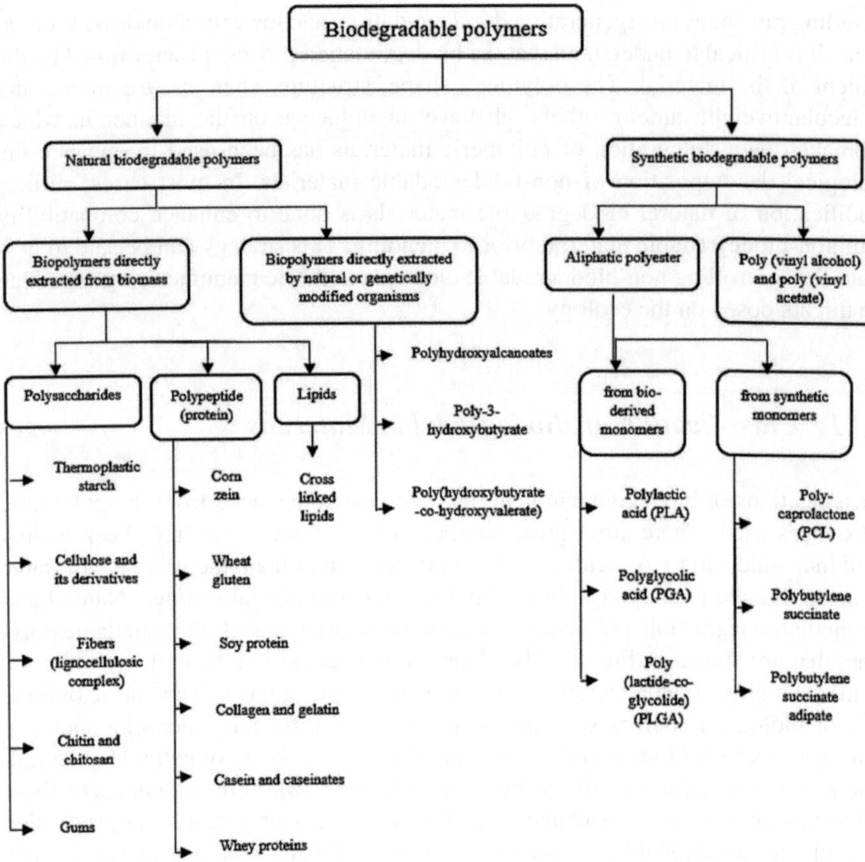

Fig. 8.2 Schematic shows biodegradable polymers classification based on their origin and method of production. *Reprinted with permission from* [6], 2022. An open-access article by Springer Nature

Plant fibers

Plant fibers, among the most extensively used natural fibers, may be categorized into four groups based on their tissue origin in the plant: (1) seed hair, such as cotton and kapok. (2) leaves or fruit envelopes, such as sisal and coconut; (3) stems (or bast), such as hemp, flax, and ramie; or (4) roots, such as sugar beet and carrot [11], among others. Cellulose is undoubtedly one of the top materials utilized in the textile industry [12]. Cellulose fibers, which are prevalent in a range of textile goods, have been viewed as environmentally friendly fibers since microorganism can swiftly degrade them once buried [12]. Given the variations in chemical and physical features, not all cellulose fibers exhibit the same biodegradable propensity. This is due to the variances in their characteristics: the chemical compositions, crystallinity, degree of polymerization, and manufacturing procedures of cotton, linen, rayon, and acetate are all distinct from one another. In addition, the non-cellulose content and the

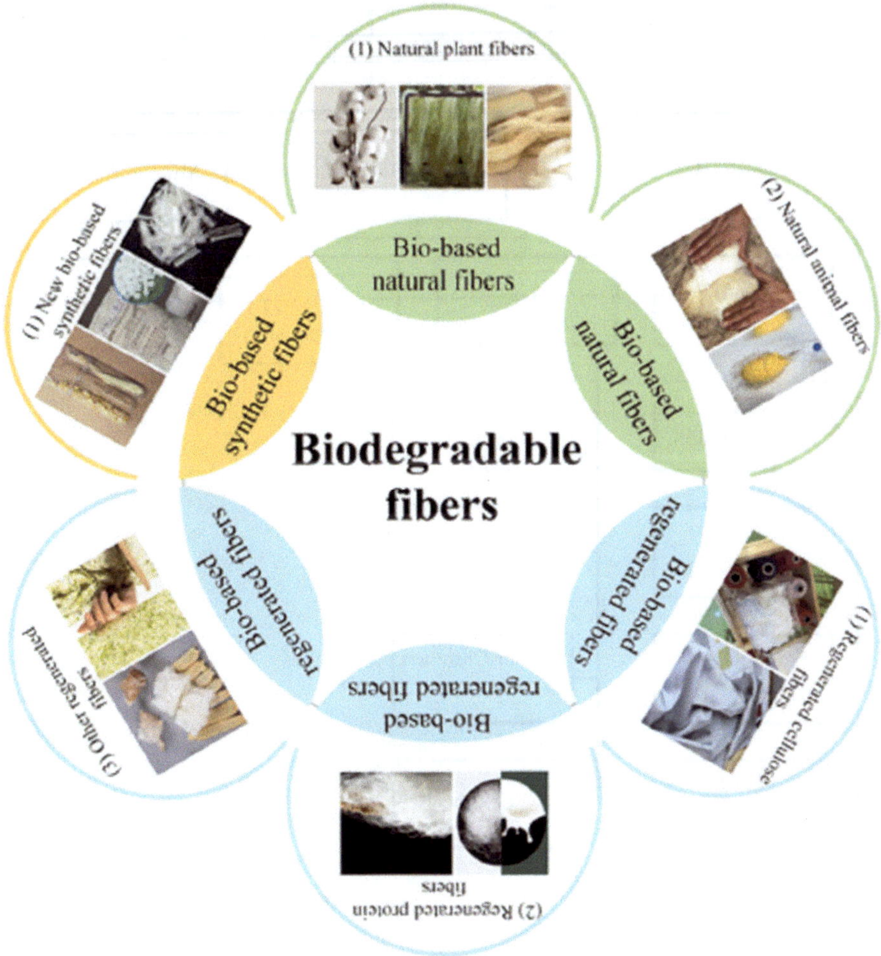

Fig. 8.3 Classification of biodegradable fiber materials in textile and apparel. *Reprinted with permission from* [9], 2024. Taylor and Francis

composition of each unique fibrous substance are distinct from one another. When all of these variables are taken into consideration, the breakdown behavior of cellulose fibers would be affected, which would result in varying degrees of biodegradability [13].

Cotton fiber: Approximately 38% of the fiber that is consumed globally is cotton. Although it is cultivated as a perennial plant in its natural state, the cotton plant is produced as an annual plant for commercial purposes [14]. Cotton fibers are long, thin tubes that span virtually the whole length of the fiber and include a lumen, which is a central channel that runs through the fiber. Cotton fiber is composed of approximately 90–95% cellulose, which is a plant substance that is composed

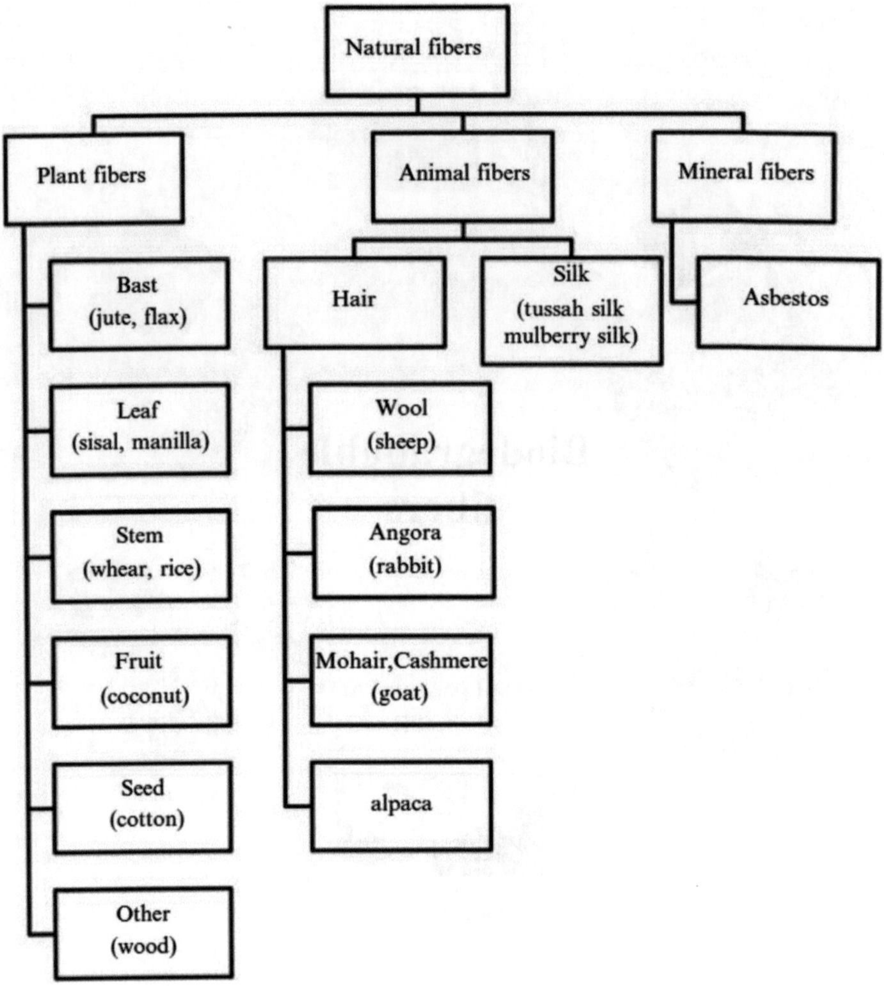

Fig. 8.4 Classification of natural fibers. *Reprinted with permission from* [10]

of carbohydrates. Additionally, cotton fiber consists of additional elements such as pectin, waxes, organic acids, and inorganic matter. The fibers are constructed in layers that are concentric, and the cuticle layer of the fiber contains pectin and wax that are separated from other fibers. The structure has a lengthy linear molecular chain consisting of more than 10,000 cellulosic units, that is, glucose units connected by β-1 → 4 linkage, and are kept together by strong intermolecular interactions. As a consequence of this, it is rather robust within the parameters of its use. Crystalline cellulose fibrils make up the primary layer of the fiber, which is the layer that extends the highest around the circumference of the fiber. There are three distinct layers that make up the innermost layer, which is the layer that contains the bulk of the cellulose located inside the fibers. These layers are made up of parallel fibrils that are densely packed and spirally wounded [15]. The long length-to-diameter ratio of cotton fiber offers a great deal of flexibility and functionality for long-term use. The vast majority of cotton fibers are used in the production of traditional fibrous goods, primarily clothing. This is as a result of their inherent elegance and aesthetic qualities that they possess. Towels, sheets, pillows, bedspreads, tablecloths, and upholstery are among the items that are often made from cotton fibers [12, 16, 17]. Cotton fibers are also frequently utilised for interior and domestic products.

Flax fiber: The inner bark of the stem of the flax plant is the source of flax fiber, which qualifies it as a bast fiber. It is more pliable, flexible, and durable than cotton; however, it has fewer elastic properties. The length of flax fiber ranges from 10 to 100 cm, while its diameter ranges from 40 to 80 μm. Flax Fiber is considerably the earliest textile fiber on record for humankind. Throughout the ages, it has been used in the manufacturing of linen fabric. Flax is a plant that is farmed for its fibers and seeds in many different regions of the globe [18]. Flax fibers are made up of roughly forty constituent fibrils that are held together by a matrix that is composed of lignin and hemicellulose. A little over 70 weight percent of the bulk fiber is made up of cellulose, 18–19 weight percent of hemicellulose, 2–2.5 weight percent of lignin, and 1.5–2 weight percent of wax. Crystalline and amorphous forms of cellulose make up the vast bulk of the cellulose that is found within the elementary fiber [19]. Suturing thread, sewing thread, luxury bed linen, kitchen towels, napkins, runners, and upholstery are all examples of applications for linen [20, 21]. Linen is also used in the production of clothes and tableware.

Hemp fiber: The hemp plant, also known as "cannabis sativa," is the source of hemp, which is a bast fiber that is comparable to jute and is characterised by its fineness, light colour, and glossy appearance [22, 23]. There are four primary components that make up hemp fiber: cellulose (53–91%), hemicellulose (4–18%), lignin (1–21%), and pectin (1–17%). The precise content of hemp fiber may change depending on the cultivar of hemp and the circumstances in which it is grown [11]. Hemp has a phenomenal potential for applications in a variety of directions. Currently, hemp fiber is being used for a wide variety of textile applications, including nonwovens, wovens, and knit configurations, all of which are areas in which its inherent natural physical qualities show remarkable potential. Hemp is also being studied for its use as a fiber reinforcement in novel composite applications [24]. These applications are particularly interesting because the sustainability, fiber strength, elongation, and potential economic advantages of using hemp. The following are some examples of apparel that are often created using hemp: dresses, skirts, trousers, jackets, hoodies, and clothing for children. Specifically, due to its resistance to wear and tear, it is also quite popular in the production of T-shirts. Others include upholstery, as well as the production of raw materials for technological goods [21, 25].

Animal origin

There are a number of fibers that originates from animals. They are relatively more expensive than plant fibers and mostly obtained from the skin of animals, except for silk which is spun by organisms. Animal fibers such as silk and wool have been discussed below.

Silk fiber: As the only natural fiber that can be found in filament form, silk is an animal fiber that is generated by silkworms for the purpose of constructing their cocoons and webs. Silk fiber is semicrystalline and is made up of two primary components: a fiber protein known as fibroin and a gummy substance known as sericin, which is responsible for holding the fibers together. The large amounts of glycine, alanine, and serine in silk fibroins are the essential feature of the kind of amino acid. The silkworm species Bombyx mori and Anaphe pernyi contain around 80–85% of the total amino acids [26]. Generally, silk fibroin accounts for 75–83% of the chemical constitution, sericin accounts for 17–25%, waxes account for around 1.5%, and miscellaneous components account for approximately 1.0% by weight [27]. There are a great number of insects that are capable of producing silk; however, the commercial silk industry exclusively makes use of the filament that is generated by the larvae of the caterpillar of the domesticated Bombyx mori moth and some other genera belonging to the same species [27]. Silk is widely used in the production of high-end textiles owing to the impressive qualities it has, such as its sensual handling, comfort, softness, and ability to absorb water. In recent years, silk has gained novel functions in healthcare owing to its amazing physical properties and biocompatibility, as well as its high degradability [28]. Silk is also very friendly with living organisms. Raw silk is used in the production of garments such as shirts, suits, ties, blouses, lingerie, pajamas, and jackets. Hand-spun mulberry silk is utilized in the production of comforters and sleeping bags. Mulberry silk is used to create a wide range of other types of fabric materials, including dupions, plain silk, deluxe, satin, chiffon, chinnons, crepe, and brocades.

Wool fiber: For a very long time, wool fiber has been widely used as a textile fabric owing to its exceptional moisture and warmth properties. The architecture of the wool fiber is hierarchical, and it is a member of the family of proteins and keratins [29]. Protein known as keratin, which is the main component of wool, is characterized by a polypeptide linkage that is accompanied by amino acid side chains. In addition to having a helical chain structure, keratin also has robust hydrogen bonding [30]. Wool is the animal fiber, and it has characteristics that are distinct from those of hair and fur. Wool fiber, also called crimpy fiber, grows in staples or clusters. It is more particularly hydrophilic and hygroscopic than other textile fibers. Staples of wool are held together by a combination of lanolin and grease [31]. The fabrication of blankets, clothing, and carpets is done using wool fibers. They also have applications as noise-absorbent materials in various technologies [29].

8.3.1.2 Biodegradable Regenerated Fibers

Natural plants and animal materials are processed into fibers with chemicals are known as biodegradable regenerated fibers. Regenerated cellulose fibers are produced by chemically treating and mechanically processing natural cellulose sources, namely, cotton staple, wood, and bagasse. These cellulosic materials undergo treatment with chemicals. Given the various processing methods that are used, regenerated

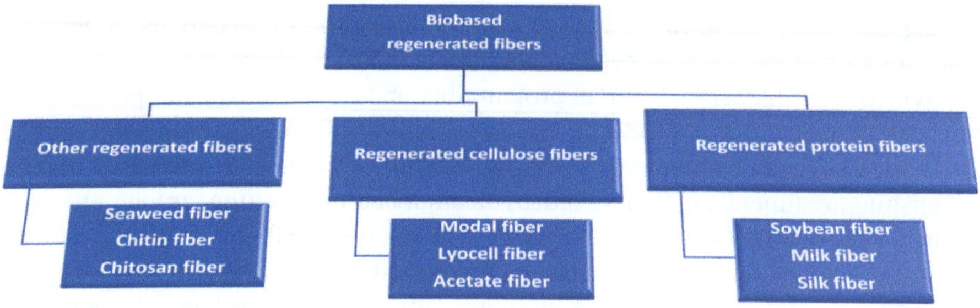

Fig. 8.5 Classification of biodegradable regenerated fibers

cellulose fibers may be classified as either common viscose fibers or novel regenerated cellulose fibers. In order to produce ordinary viscose fibers, cotton staples are used in a procedure known as carbon disulfide that somewhat hazardous to human health and severely harmful to the environment. Novel regenerated cellulose fibers, on the other hand, are mostly green fibers that are produced using novel raw materials or procedures. Some examples of this category of fibers are modal fibers, lyocell fibers, and acetate fibers [32]. Among the several types of protein fibers that are identified under regenerated fibers are silk protein fibers, milk protein fibers, soybean protein fibers, and others. Chitin fibers, seaweed fibers, and other types of fibers are examples of biodegradable regenerated fibers (Fig. 8.5).

Lyocell fiber: Lyocell fibers are a kind of cellulose fiber that are classified as belonging to the third generation. It is manufactured by a method known as N-Methyl morpholine oxide solvent spinning with wood pulp. Lyocell fiber has a cross-section that is either oval or nearly circular, which is a characteristic of its architectural structure. In the course of the wet spinning process, a skin–core-like structure emerges. Lyocell fiber has a breaking strength that ranges between 4 and 4.2 cN/dtex, which gives it a physical property. Lyocell fiber has a moisture recovery of around 1–13% according to traditional methods. The fiber has remarkable mechanical and physical qualities. Good dry and wet strength, high tensile modulus, durability, and a small margin of shrinkage are all characteristics of this fiber. Lyocell fibers are commonly used in knitted sweaters and are particularly ideal for worsted knitted and woven textiles. The fiber surface does not generate microfibers; hence, enabling it to preserve its unique qualities.

Acetate fiber: Acetate fiber is obtained by spinning a solution of cellulose pulp into fibers using wet or dry spinning methods. They are categorized into diacetate and triacetate fibers based on the number of acetyl groups replacing hydroxyl groups in glucose units of cellulose. Acetate fibers are mostly characterized by smooth and high uniformity along their length with obvious grooves. The cross section has a cloverleaf shape. The development of the application in the field of textile and apparel has advanced significantly, leading to its application in the production of satin used in dresses, women's garments, undergarments, etc. Nonetheless, its application in the arena of textiles and fashion has been narrowed due to the poor strength and dimensional stability of acetate fiber.

Regenerated protein fiber: It is generated from protein precursors either from dairy milk or plants such as peanuts and maize, among others. As a result, it may be divided into regenerated animal protein fiber and regenerated plant protein fiber. Soybean protein fiber is produced by wet spinning of biotechnologically synthesized oiled soybean residue precursor into bundles. The bundles undergo a number of finishing treatments, including aldehydization and heat setting, among others, to create a high-quality fiber for textile usage. The fiber is mostly made up of polyvinyl alcohol and soy protein bicomponents. It is appropriate for high-quality knitted undergarments, sweaters, and women's clothing. However, the mechanical features of regenerated protein fibers derived from plants are significantly inferior compared to wool fibers. Additionally, the strength of the fiber in a wet environment is drastically low, with poor heat stability and higher susceptibility to wrinkling and fluffing.

8.3.1.3 Biopolyesters

Biopolyesters are a new main class of biodegradable polymer obtained or engineered from biological monomers like polylactic acid, polyhydroxybutyrate, and polycaprolactone. These biopolymers can be categorized into bacterial biopolyesters and synthetic biopolyesters.

Bacterial biopolyesters: Polyhydroxyalkanoates (PHAs) are biopolymers derived from bacteria, including poly (3-hydroxybutyric acid) and other similar aliphatic polyesters (e.g., poly(3-hydroxybutyrate-co-3-hydroxyhexanoate)). Microbes may generate PHAs by enzymes in vivo, producing a real natural polyester suitable for cell storage. Approximately ninety-one (91) distinct polyhydroxyalkanoic acid elements under PHAs have been reported, and their total count is constantly increasing. A limitless quantity of PHA copolymers may be produced by combining monomers in different fashions. PHAs, which are biodegradable in a variety of conditions, are an appealing alternative to existing disposable polymers that are undesirable for reintroduction into the production line owing to their low standard. With their consequential biodegradable character, PHAs develop a recycling path that aids in the reduction of harmful effects on the environment. The functioning and manufacturing processes vary depending on the kind of PHA [33]. The main issue with some microbes that yield polyesters is their poor tensile qualities, limiting their application in the garment industry. Nevertheless, due to their promising application in non-woven textiles, they are now replacing conventional petroleum-based polymers in a variety of sectors. Considering the insatiable quest of the world to provide ecologically safe materials, scientists have been investigating various means of altering and enhancing polyesters with other polymers, resulting in different composite materials with higher quality [33].

Synthetic biopolyesters: Polycaprolactone, polylactic acid (PLA), and poly (butylene adipate-co-terephthalate [PBAT]) are viable substitutes for certain traditional synthetic fibers to reduce textile waste. PLA remains one of the particularly intriguing polymers with a renewable, organic source and a strong track record in the garment sector. PLA is made by fermenting lactic acid and then polymerizing via chemical means. PLA and PBAT are biodegradable, enhancing the sustainable qualities of natural fibers while also mimicking the functionality of petroleum-based textiles. The PBAT polymer has mechanical qualities similar to polypropylene, enabling its application in the rope industry and garment linings. Additionally, it serves as a replacement for polyethylene geotextiles and organic waste sacks [4]. It is frequently found in wrapping products, dinnerware, and textiles. It performs the requisite functionality and may decompose in the industrial waste recycling process [4].

Despite the existence of numerous manufacturing techniques, some bio-based polymers remain non-biodegradable. Only PLA and PHAs are entirely biodegradable and have a complete carbon cycle, similar to cellulosic materials. Nonetheless, unlike PHAs, PLA is biodegradable in specific conditions but not in saltwater, hindering its application in combating waste released into the environment. Migrating to bio-based biodegradable polymers ensures the realization of a greener choice by employing an upcycling method, in which the yield of biodegradation is used as the input for the same polymer. Furthermore, poly(butylene adipate-co-terephthalate), another type of petroleum-based polymer, is biodegradable, but owing to their manufacturing technique, they do not follow a circular economy pattern. Inasmuch as biodegradable plastics have the tendency to be reused, most of them have short life cycles as a result of the reduced and inferior properties of the biodegraded plastics [33].

8.4 Biodegradation Mechanism

The biodegradation of polymers in the environment involves two basic processes: oxidation and hydrolysis. Free radicals produced from oxygen and ozone in the atmosphere disrupt the polymer's bonds, resulting in oxidative breakdown. In hydrolysis, the water content in the atmosphere assists in the breakdown of polymer connections such as covalent bonds. Microbes initiate the degradation of textile polymers through aerobic or anaerobic digestion, contingent upon the presence of oxygen in the surrounding environment, after they have been decomposed into smaller fragments through oxidation and/or hydrolysis (as illustrated in Fig. 8.6). During aerobic biodegradation, microorganisms utilize oxygen as an electron acceptor to acquire energy for the degradation of polymers [4]. Enzymes from the cell and other factors such as oxidation, light irradiation, and hydrolysis break down long-chain polymers into shorter chains. The biological means of polymer decomposition process is used to calcify oligomers, which are bioassimilated by microorganisms in the second phase. Degradation may be aerobic or anaerobic: aerobic decomposition takes place in the midst of oxygen, producing CO_2, H_2O, biomass, residue, etc., and anaerobic decomposition happens in an atmosphere devoid of oxygen and produces CO_2, H_2O, biomass, residue, etc.

Siracusa et al. [34] provide more insight on the breakdown of polymer backbone into its constituent parts polymers in the presence of microorganisms (such as

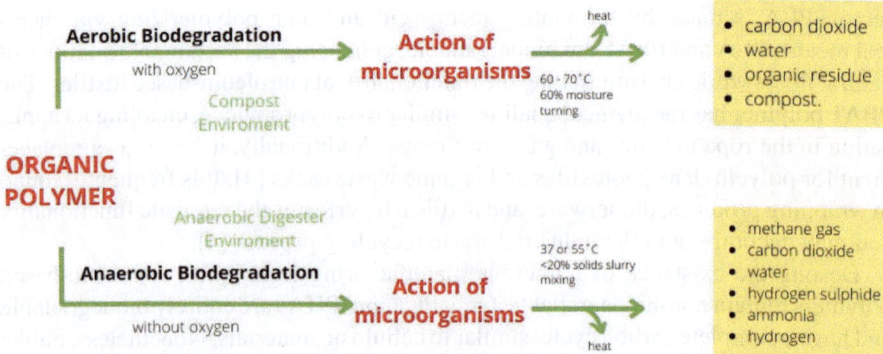

Fig. 8.6 Aerobic and anaerobic degradation pathways for organic polymers. *Reprinted with permission from* [7], 2022. An open access article by MDPI

bacteria and fungi). This process happens in a variety of contexts since the microorganisms involved vary and are potent in the soil. Additionally, polymer properties such as chain mobility, crystallinity, molecular weight, functional groups, the type of substituents in the polymer structure, and the presence of plasticizers and/or additives may all have a substantial effect on the degradation process. The degradation process consists of two key phases: (i) the severing of polymer chains, in which hydrolysis and oxidation, along with extracellular enzymes, are primarily responsible for the breakdown of internal linkages and terminal polymer molecules; and (ii) mineralization, which occurs within the cell and is responsible for microbial degradation. During the degradation process, polymers are reduced into monomers, which are subsequently mineralized. Microorganisms attack the ends of macromolecules. Given that the number of ends in these polymers is inversely proportional to their molecular weight. The aforementioned two-step biodegradation process is particularly noticeable in produced biodegradable polymers such as polylactic acid [4, 9].

8.5 Fabrication Strategies for Biodegradable Textiles

Similar to other textile fibers, yarns and fabrics, there are several strategies used in the production of biodegradable fibers, yarns and fabrics. Natural biodegradable fibers obtained from natural source are transformed into yarns by spinning and then into fabrics by weaving, knitting, etc. Non-woven fabrication techniques involving web formation (air laid, dry-laid, and spun laid) and web bonding (Mechanical methods: needle punching, hydro-entanglement; Thermal methods: calendaring, air bonding, sonic bonding; Chemical methods: Impregnation, foam coating, spray coating, print bonding), melt, dry and wet spinning, etc.

8.6 Literature on Biodegradable Assessment of Fabrics

The need for a critical consideration of the interaction of waste in the environment has demanded in-depth research into the biodegradability of fabrics. Consequently, there have been a series of attempts to examine the biodegradability of materials. Generally, the biodegradability of fabrics can be examined via determination of the amount of CO_2 released, the tensile strength of fabric over delay time in water or soil, the weight loss of the fabric, and the amorphous/crystallinity of the fabric, since degradation has significant effect on the aforementioned features of fabrics [13]. Optical microscopy examines the surface changes caused by biodegradation after soil burial tests. The first indicator of textile biodegradation is the appearance change, which can be viewed with the naked eye (macroscopic scale). Usually, the colors of the textile specimens' surfaces turned brown and black due to the action of microorganisms. The physical changes and microbial colonization in fibers could be detected using microscopy techniques [10]. Textile fibers or films biodegradation process causes complete or partial decomposition of the substrate, resulting in an overall weight loss. Indeed, the most used method for detecting the biodegradation of textile fibers or films is to measure substrate weight loss [10, 35]. Another method (AATCC Soil Burial Method 30–1993) is based on measuring the sample's tensile strength. During the degradation time, water was supplied at a regular interval to maintain the moisture regain of soil at $25 \pm 5\%$. After degradation, samples were rinsed with distilled water and dried, and the tensile strength was measured. The evaluation of biodegradability is from the decreasing tensile strength rate. XRD is also used to investigate changes in the specimen crystallinity and internal structure resulting from degradation [36]. The initial reaction during biodegradation of fiber may produce minor deviations in its structure or composition. Fourier transform infrared spectroscopy (FTIR) is often used to detect chemical changes in the fiber [35]. In brief, the release of a high amount of CO_2, reduction in tensile strength, loss in weight, reduction in crystallinity, formation of brownish appearance and disintegration of materials during the degradation process, are indicators of good biodegradation of material. In this section, more pictographs demonstrating the degradation of materials with concomitant weight loss over a period of time have been focused on to give quick-to-understand illustrations. Park et al. conducted a study on the biodegradability of cellulose fabrics, viz., cotton, rayon, and acetate fabrics.

A study on cotton, rayon, acetate, and linen fibers was done by Park et al. The cotton, rayon, and acetate establish changes in the crystallinities and the internal structures as a result of degradation in soil and activated sewage sludge. The biodegradability decreased in the following order: rayon > cotton $\gg$ acetate [13]. Figure 8.7 shows the XRD patterns of linen and cotton fabrics. Crystallinity reduces over time due to the attack and breakdown of molecular structure by microorganisms. Rayon fibers, which have a low crystallinity and a low degree of orientation, showed the highest biodegradability in most cases. On the other hand, in spite of the low crystallinity of acetate fibers, it exhibited very low biodegradability, probably because of the presence of hydrophobic groups in its structure. Linen showed an inconsistent

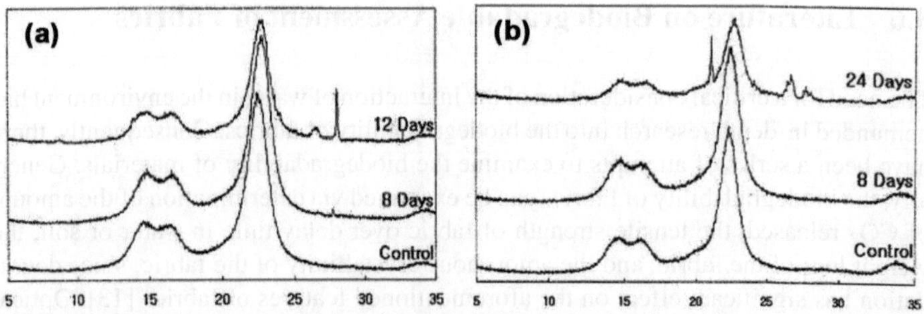

Fig. 8.7 XRD pattern of **a** linen and **b** cotton [13] (The XRD curves according to soil burial time lapse. It was observed that the major crystalline peak, which appeared around 2 theta = 22.8 by cellulose crystalline plane (002), rapidly lowered in 12 days. In the biodegradation mechanism, the disordered non-crystalline region in the internal cellulose fiber would be the first target for soil microorganisms. After the amorphous region has been decomposed, microorganisms will attack the crystalline part, which leads to a decrease in crystallinity). *Reprinted with permission from* [13]

behavior in that it had the highest biodegradability in the soil burial test, but a lower biodegradability than that of cotton in the activated sewage sludge test. This implies that the environment and the nature of the material could influence the biodegradation efficiency of a material.

An investigation into the degradability of 3D-printed PLA biopolymer, natural cotton, and synthetic polyester textile fibers was done by Natani Aparecida do Bem et al. [37]. The comparison was carried out with samples of materials degraded in soil and exposed to the weather for 120 days (Fig. 8.8). Significant results were obtained for mass loss, as follows: 13.4% PLA; 8.9% cotton/flat, and 3.84% polyester/flat. As for the loss of area, the results were 46.5% for PLA; 15.4% for cotton/knit; and 6.25% for polyester/knit. Based on their study, it was inferred that the composition and material construction of the analyzed materials is one of the factors that can determine the period of degradation, since natural fiber fabrics present faster decomposition due to the presence of microorganisms.

In view of biodegradation, knitted fabric appears more unsteady than plain fabric. The bonds disintegrate far more rapidly, leading to a distinct deterioration path for flat, knit, and non-woven textiles [37]. Milošević et al. investigated the biological breakdown of cotton and cotton/PET textiles treated with Ag/TiO_2 particles in the ground [38]. The findings indicate that Ag/TiO_2 particles retarded the breakdown of cotton and cotton/PET textiles. Soil burial tests over many days revealed that a lower concentration of Ag nanoparticles in blended fabric enhances biodegradability. Significant breakdown of the cotton fibers in both blended and unblended fabrics were released. In general, cotton-PET fibers showed insignificant change throughout their decomposition process, as shown in Fig. 8.9. Polyester, a synthetic polymer, has low biodegradation efficiency. Hence, it hinders the significant biodegradability of cotton in cotton/polyester fibers.

Few research investigations on the biodegradability of functional fabrics have been performed. Such includes that of Tomsic et al., who examined the effect

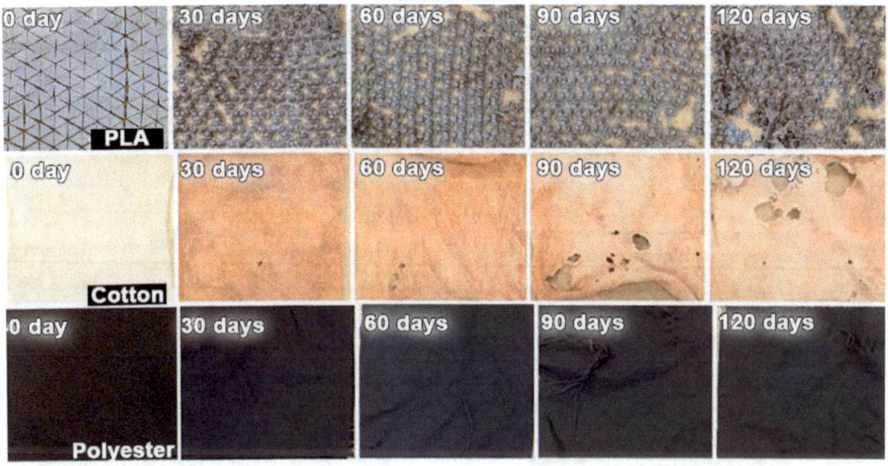

Fig. 8.8 Degradation process of 3D-printed rigid polylactic acid biopolymer, knitted cotton and polyester samples. Reprinted with permission from [37], 2022. An open access article by RBCIAMB

of antimicrobial finishes on the rate of biological degradation of pure cotton and cotton-polyester textiles [39]. Polyester fibers resisted biodegradation, similar to the report of Milošević et al. [38]. The addition of a high level of antimicrobial silver component significantly retarded biological degradation of cellulose fibers in cotton and cotton/polyester textiles. Unlike silver, the propyldimethyloctadecyl (3-trimethoxysilyl) ammonium chloride antimicrobial finish on cotton and cotton/polyester textiles did not hinder the biodegradation process amid extended durations of soil burial. The study displays the potential effect of finishes on biodegradability of fabrics. Therefore, more investigations regarding this area are required to provide a comprehensive knowledge of the biological decomposition of fabrics, especially after use.

Lyocell is a biodegradable fiber that, by the action of living organisms secreted by enzymes, is broken down into water and carbon dioxide. Figure 8.10 illustrates the significant degradation of lyocell in 4 and 7 days. The roughness of the surface results from the breaking action of enzymes released by living organisms. Furthermore, the lyocell fibers entirely degraded after 42 days in a static aerated compost pile and wholly degraded within a week in a sewage farm anaerobic digester, where the residence cycle is about 3 weeks. Lyocell degrades entirely within 84 days in soil burial tests (BS 6085/AATCC 30) [10, 40].

Table 8.1 shows polymers used in textiles and their biodegradability performance. Undoubtedly, synthetic polymers such as PET have significantly poorer biodegradability than biopolymers. There exists a difference in the biodegradability of the biopolymers due to their composition, degree of polymerization, and degradation conditions such as pH, the medium of degradation, and others. Boyandin and

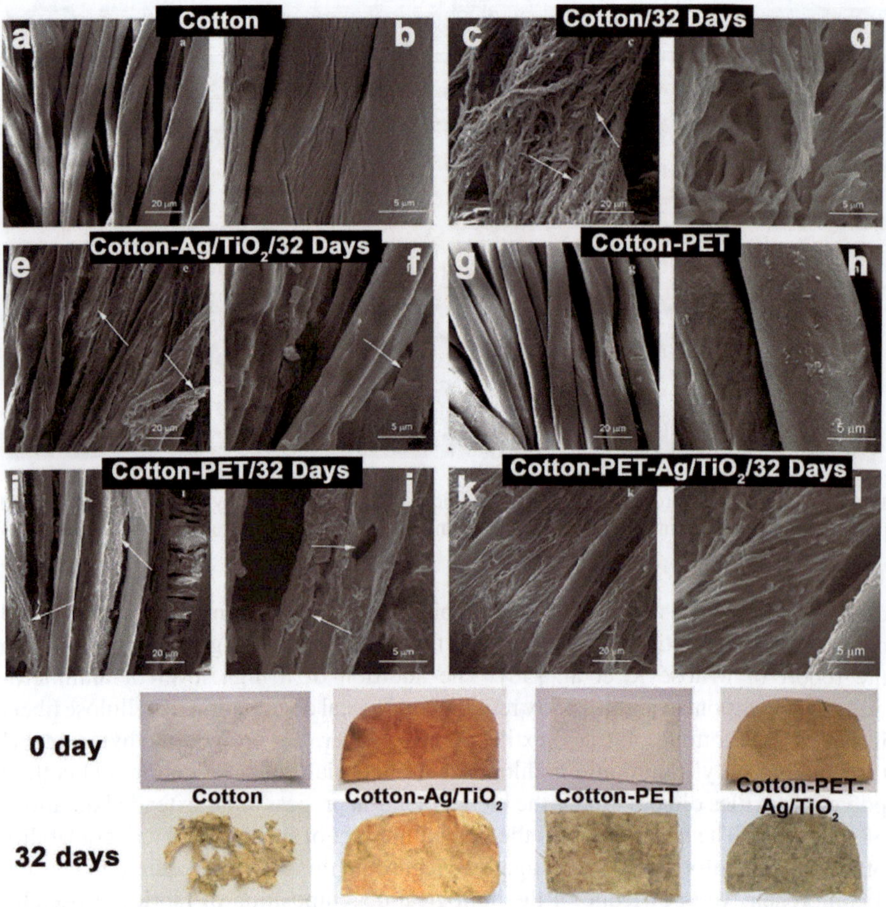

Fig. 8.9 Microscopic pictures of (**a** and **b**) pristine cotton fabric, (**c** and **d**) cotton fabric after one month of soil burial testing, (**e** and **f**) Ag/TiO_2 coated cotton fabric after one month of soil burial testing, (**g** and **h**) pristine cotton/PET fabric, (**i** and **j**) cotton/PET fabric after one month of soil burial testing, and (**k** and **l**) Ag/TiO_2 coated cotton/PET fabric after one month of soil burial testing [38]. The photographs of fabric disintegration after certain time intervals of soil burial testing: cotton fabric, Ag/TiO_2 coated cotton fabric, cotton/PET fabric, and Ag/TiO_2 coated cotton/PET fabric. *Reprinted with permission from* [38], 2017. Elsevier

coworkers reveal that the type and condition of soil can influence the biodegradability of the same fabric [42]. This is buttressed by the difference in the degradation percentage and time of similar polymeric fabrics such as PLA, etc., reported in Table 8.1.

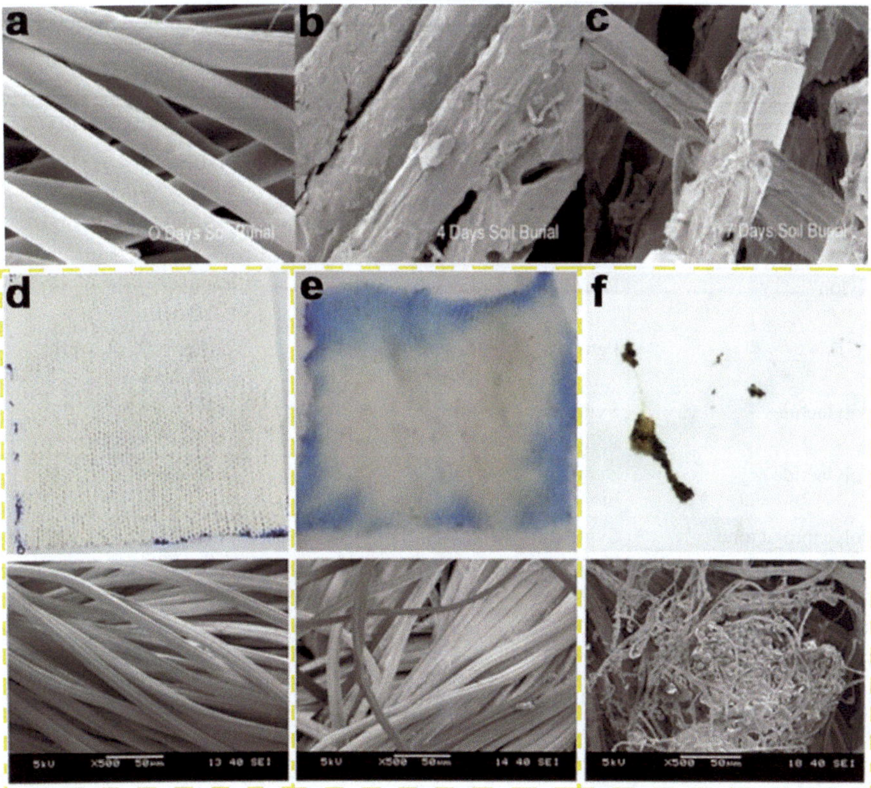

Fig. 8.10 Biodegradability of lyocell fibers after soil burial periods **a** control lyocell, **b** 4 days and **c** 7 days. *Reprinted with permission from* [40], 2005. Elsevier, **d** control viscose fabric, **e** viscose fabric in 1 month and **f** viscose fabric in 4 months burial. *Reprinted with permission from* [41], 2019. An open access article from FIBRES and TEXTILES in Eastern Europe

8.7 Challenges and Future Outlook

The quest for sustainability of the environment for the sake of the present and future is undebatable. Many different routes have been attempted, whilst others are still in the research stream, to drastically crack down on the threat posed by post-consumer textile waste on the environment. With so many challenges in the field of textile biodegradation lie several opportunities for improving processes, policies, and regulations. They may be categorized into pre- and post-consumer challenges. One of the main reasons for the preference of synthetic non-biodegradable fibers in fabric production pre- and during consumer consumption is their good tensile properties, easy-to-process, and economically benign factors. Biodegradable materials, even though they are environmentally friendly, find limited application (especially in less humid and non-microorganism-culturing conditions) relative to synthetic non-biodegradable materials. This has encouraged the blending and/or the spinning of

Table 8.1 Biodegradability of polymers used in textile fabrics

Polymer	Source/ category	Percentage degradation	Medium	Timeline	Refs.
Cotton	Plant	25–100%	Soil	In 25–90 days	[43]
Wool	Animal	43%	Soil	After 168 days	[44]
Flax	Plant	More than 70%	Soil	In 180 days	[45]
Nylon	Petrol	Less than 1%	Soil	After 270 days	[44]
PET	Petrol	Less than 1%	Soil	After 700 days	[46]
Polylactide	Synthetic biopolymer	60–100%	Soil	In 60–120 days	[47]
Polylactide	Synthetic biopolymer	75%	Soil	620 days	[48]
Polycaprolactone	Synthetic biopolymer	70–100%	Soil	In 60–100 days	[46]
Poly (butylene adipate-co-terephthalate)	Synthetic biopolymer	30–67%	Soil	42–47 days	[46]
Polyhydroxybutyrate	Bacteria	30–95%	Soil	25–60 days	[46]
Ramie	Animal	42%		168 days	[45]
PE	Petrol	Less than 1%	Soil	300 days	[42]
Polyhydroxybutyrate	Bacteria	50–100%	Soil	300 days	[42]
Polycaprolactone	Synthetic biopolymer	About 100%	Liquid	140 days	[49]
40% PCL/60% cellulose acetate	Synthetic biopolymer	61.2%	Liquid	140 days	[49]
LDPE	Petrol	Less than 1%	Liquid	140 days	[49]
Polyacrylonitrile	Petrol	Less than 1%	Soil	120 days	[10]
Lyocell	Regenerated cellulose	100%	Soil and sewage farm anaerobic digester	42 days and 7 days	[10, 40]
Poly (butylene succinate)	Synthetic biopolymer	About 20%	Soil	70 days	[49]

biodegradable fibers with non-biodegradable fibers. However, this strategy reduces the biodegradability of the product. This calls for extensive studies on green strategies to improve the properties of biodegradable materials to ensure environmentally benign pre- and post-consumer textiles.

Most studies on the biodegradability of fabrics focus on pristine fabrics/fibers. However, the presence of chemical additives in the form of dyes, pigments, and finishing treatments at the manufacturing stage can significantly affect post-consumer fabric biodegradation process. Primarily, fabrics are treated with several chemical finishes to enhance their function, aesthetics, comfort and durability. The use of such treatments is quite huge in the case of high-performance functional textiles for various industrial applications. For instance, antimicrobial finishes for medical textile applications are often coated with biocidal nanoparticles, which is a hydrophobic barrier on the fiber surface that prevents the growth of microbes, thereby prolonging the biodegradation process. Also, lack of moisture absorption in such fabrics is a major factor hindering biodegradation. Similarly, flame-retardant finishes have a negative impact on the biodegradation of the treated textile as well.

Although the textile sector is not the only source of polyfluoroalkyl substances (PFAS), it appears to have a substantial role. As a result, green chemistry-based remedies have become crucial to substitute the existing water and oil repellent treatments in the textile sector. Water repellency can be accomplished using different coatings, but satisfactory oil repellency can only be achieved with fluorochemical treatments that emit PFAS. Landfill leachate is the main carrier of PFAS owing to the disposal of PFAS-containing consumer items (particularly textiles), which eventually pollutes nearby soil and water systems [50]. The persistent disparity in the functionality and biodegradability of textiles is unavoidable. Identifying and controlling this discrepancy is a significant task in current textile development. Furthermore, functional textiles processed with chemicals can generate hazardous compounds, which may destroy microorganisms in the soil alongside releasing toxins into the environment. The inclusion of these finishing agents impedes the biodegradation mechanism. Currently, obtaining a superior or even equal performance is a challenge, so, more funding and studies into developing novel textiles that have outstanding eco-friendliness are required.

Biodegradable plastics often have drawbacks, notably weak mechanical qualities, and cannot be blended with an array of polymeric materials without compromising their functions. The commercial worth of biodegradable plastics is based primarily on how well they biodegrade when discarded. In view of that, evaluating the biodegradability of polymer remains necessary to guarantee that no chemical interactions interfere with the biodegradation rate. PLA and PHB, the two most common bio-based and biodegradable polymers, are exceedingly fragile when compared to commercial and technical plastics. Despite certain PHAs exhibiting more ductility as their chain length rises, their cost exceeds that of regular plastics and other bio-based biodegradable polymers. Furthermore, the inclusion of petrol-based biodegradable polymers is a possible option to preserve the biodegradability of PHAs, the resultant composites are less ecologically friendly owing to the use of petroleum [33]. Natural fibers offer a solution for lowering the cost of PHAs while improving their best qualities, such as modulus and biodegradability. Yet, the impact of natural fibers, formulation, and compatibilization processes on biodegradation remains unknown.

8.8 Conclusion

Biodegradable materials are considered to be enthralling materials that spear toward reducing carbon footprint to ensure the sustainability of our environment. The textile industry, which contributes massively to the economy of countries and the world, has concomitant devastating output regarding the handling and degradation of its post-consumer waste. Numerous research endeavors into the development and modification of their raw materials to achieve top-quality biodegradable materials for textile products have been an undeniable hotspot. Primarily, natural biodegradable polymers call for their extensive application in textiles, however, they do face some downsides, such as poor tensile strength, elongation and susceptibility to microorganisms during application. Synthetic biodegradable polymers used in the textile field have also demonstrated positive signs; however, most do not have better biodegradability than natural biodegradable polymers. Biodegradation occurs due to the monomerization of polymer by microorganisms. As buttressed by literature, different variables, viz., chemical constitution, orientation, water-holding capacity, crystallinity, extent of polymerization, presence of dyes, chemical finishes, and even the environmental condition, whether in different sewage or soil, also contribute to a textile material's ability to biodegrade. Due to so many variables in play, it is important that the biodegradation claims are made and understood carefully. These factors draw attention to the need for extensive research into the effect of the variables, especially finishes, whether hydrophobic or hydrophilic, to help in the proper disposal and biodegradation of textile waste to achieve environmental sustainability.

References

1. Perspectives: The Science of Sustainability. The nature conservancy. 2018 [cited 2024 22/10/2024]. https://www.nature.org/en-us/what-we-do/our-insights/perspectives/the-science-of-sustainability/#:~:text=Achieving%20a%20sustainable%20future%20will%20be%20dependent%20on,human%20communities%20and%20abundant%20and%20healthy%20natural%20ecosystems.
2. Adams, W.M. 2006. *The future of sustainability: Re-thinking environment and development in the twenty-first century.*
3. Sachs, J., et al. 2002 *Sustainable development report 2022.* Cambridge University Press.
4. Mehta, S. 2023. Biodegradable textile polymers: A review of current scenario and future opportunities. *Environmental Technology Reviews* 12 (1): 441–457.
5. UN Alliance For Sustainable Fashion addresses damage of 'fast fashion'. The UN environment programme. https://www.unep.org/news-and-stories/press-release/un-alliance-sustainable-fashion-addresses-damage-fast-fashion).

6. Samir, A., et al. 2022. Recent advances in biodegradable polymers for sustainable applications. *NPJ Materials Degradation* 6 (1): 68.
7. Kopitar, D., et al. 2022 Biodegradable nonwoven agrotextile and films—A review 14 (11): 2272.
8. Var, C., and S. Palamutcu. 2024. Man-made bio-based and biodegradable fibers for textile applications. In *Sustainable manufacturing practices in the textiles and fashion sector*, ed. S. S. Muthu, 229–280. Cham: Springer Nature Switzerland.
9. Bao, H., et al. 2024. A systematic review of biodegradable materials in the textile and apparel industry. *The Journal of The Textile Institute* 115 (7): 1173–1192.
10. Nofal, R. M. 2023. Biodegradable textiles, recycling, and sustainability achievement. In *Handbook of biodegradable materials*, ed. G. A. M. Ali and A. S. H. Makhlouf, 1449–1485. Cham: Springer International Publishing.
11. Liu, M., et al. 2017. Targeted pre-treatment of hemp bast fibres for optimal performance in biocomposite materials: A review. *Industrial Crops and Products* 108: 660–683.
12. Koh, J. 2011. 3–Dyeing of cellulosic fibres. In *Handbook of textile and industrial dyeing*, ed. M.Clark, 129–146. Woodhead Publishing.
13. Park, C.H., Y.K. Kang, and S.S.J.J.o.A.P.S. Im. 2004. Biodegradability of cellulose fabrics 94: 248–253.
14. Johansson, K. 2007. 10–Plasma modification of natural cellulosic fibres. In *Plasma technologies for textiles*, ed. R.Shishoo, 247–281. Woodhead Publishing.
15. Mahmoud Zaghloul, M.Y., M.M. Yousry Zaghloul, and M.M. Yousry Zaghloul. 2021. Developments in polyester composite materials—An in-depth review on natural fibres and nano fillers. *Composite Structures* 278: 114698.
16. Elmogahzy, Y.E. 2020. 8—Fibers. In *Engineering Textiles,* 2nd ed, ed. Y.E. Elmogahzy, 191–222. Woodhead Publishing.
17. French, A. D., and H. J. Kim. 2018. Cotton fiber structure. In *Cotton fiber: Physics, chemistry and biology*, ed. D. D. Fang, 13–39. Cham: Springer International Publishing.
18. Hejazi, S. M., et al. 2012. A simple review of soil reinforcement by using natural and synthetic fibers. *Construction and Building Materials* 30: 100–116.
19. 1—Natural fibers and their composites. In *Tribology of natural fiber polymer composites*, ed. N. Chand, M. Fahim, 1–58. Woodhead Publishing.
20. Samanta, K.K., S. Basak, and S.J.S.F.f.F.I.V. Chattopadhyay. 2016. *Potential of ligno-cellulosic and protein fibres in sustainable fashion,* 61–109.
21. Muzyczek, M. 2020. The use of flax and hemp for textile applications. In *Handbook of natural fibres*, 147–167. Elsevier.
22. Mini, K.M., D. Sathyan, and K. Jayanarayanan. 2022. 15—Biofiber composites in building and construction. In *Advances in bio-based fiber,* ed. S.M. Rangappa, et al., 335–365. Woodhead Publishing.
23. Goswami, K.J.A.I.C.M. 2018. *Developments in handmade carpets: design and manufacture*, 269–350.
24. Kaufmann, J. 2020. *Hemp: A reintroduction to one of the original textile inputs*. Textile World 2020 [cited 2024 22/10/2024]. https://www.textileworld.com/textile-world/features/2020/12/hemp-a-reintroduction-to-one-of-the-original-textile-inputs/.
25. Zimniewska, M. 2022. Hemp fibre properties and processing target textile: A review 15 (5): 1901.
26. Chu, C.C. 2013. 11—Materials for absorbable and nonabsorbable surgical sutures. In *Biotextiles as medical implants*, ed. M.W. King, B.S. Gupta, and R. Guidoin, 275–334. Woodhead Publishing.
27. Babu, K.M. 2015. Chapter 3—Natural textile fibres: animal and silk fibres. In *Textiles and Fashion*, ed. R. Sinclair, 57–78. Woodhead Publishing.
28. Haji, A., and M. Naebe. 2020. Cleaner dyeing of textiles using plasma treatment and natural dyes: A review. *Journal of Cleaner Production* 265: 121866.
29. Madhu, P., et al. 2022. 1—Introduction to bio-based fibers and their composites. In *Advances in bio-based fiber*, ed. S.M. Rangappa, et al., 1–20. Woodhead Publishing.

30. El Mogahzy, Y.E. 2009. 8—Structure, characteristics and types of fiber for textile product design. In *Engineering Textiles*, ed. Y.E. El Mogahzy, 208–239. Woodhead Publishing.
31. Ngo, T.-D. 2022. Chapter 8—Nanotechnological intervention in the wool composites. In *Wool fiber reinforced polymer composites*, ed. S. Thomas, and S. Jose, 141–154. Woodhead Publishing.
32. Özçelık Kayseri, G., F. Bozdoğan, and L. Hes. 2010. Textile and apparel 20 (3): 208–212.
33. Meereboer, K. W., M. Misra, and A. K. Mohanty. 2020. Review of recent advances in the biodegradability of polyhydroxyalkanoate (PHA) bioplastics and their composites. *Green Chemistry* 22 (17): 5519–5558.
34. Siracusa, V. 2019. Microbial degradation of synthetic biopolymers waste 11 (6): 1066.
35. AK, P.J.B., and S. Fibres. 2005. *Microbial processes in the degradation of fibers*, p. 1.
36. Chidambareswaran, P.K., et al. 1987. Analysis of some textile blends using their X-ray diffraction patterns 57 (3): 167–171
37. do Bem, N.A., et al. 2022. 3D-printed polylactic acid biopolymer and textile fibers: Comparing the degradation process 57 (2): 302–319.
38. Milošević, M., et al. 2017. Biodegradation of cotton and cotton/polyester fabrics impregnated with Ag/TiO_2 nanoparticles in soil. *Carbohydrate Polymers* 158: 77–84.
39. Tomšič, B., et al. 2011. Influence of antimicrobial finishes on the biodeterioration of cotton and cotton/polyester fabrics: Leaching versus bio-barrier formation. *Polymer Degradation and Stability* 96 (7): 1286–1296.
40. White, P.A., et al. 2005. *5—Lyocell fibres.*
41. Sülar, V., G.J.F. Devrim, and T.I.E. Europe. 2019. *Biodegradation behaviour of different textile fibres: Visual, morphological, structural properties and soil analyses.*
42. Boyandin, A. N., et al. 2013. Microbial degradation of polyhydroxyalkanoates in tropical soils. *International Biodeterioration and Biodegradation* 83: 77–84.
43. Ansari, I., G. East, and D.J.J.o.t.T.I. Johnson. 2003. Structure–property relationships in natural cellulosic fibres: Part IV: Biodegradability of textile fibres 94 (1–2): 16–36.
44. Zumstein, M.T., et al. 2018. Biodegradation of synthetic polymers in soils: Tracking carbon into CO_2 and microbial biomass 4 (7): eaas9024.
45. Warnock, M., et al. 2011. Soil burial effects on biodegradation and properties of three cellulosic fabrics. 11 (1).
46. Gómez, E.F., F.C.J.P.D. Michel Jr. and Stability. 2013. Biodegradability of conventional and bio-based plastics and natural fiber composites during composting, anaerobic digestion and long-term soil incubation 98 (12): 2583–2591.
47. Karamanlioglu, M., et al. 2017. Abiotic and biotic environmental degradation of the bioplastic polymer poly (lactic acid): A review 137: 122–130.
48. Urayama, H., T. Kanamori, and Y. Kimura. 2002. Properties and biodegradability of polymer blends of Poly(L-lactide)s with different optical purity of the lactate units 287 (2): 116–121
49. Calil, M. R., et al. 2006. Comparison of the biodegradation of poly(ε-caprolactone), cellulose acetate and their blends by the sturm test and selected cultured fungi. *Polymer Testing* 25 (5): 597–604.
50. Wei, Z., T. Xu, and D. Zhao. 2019. Treatment of per- and polyfluoroalkyl substances in landfill leachate: Status, chemistry and prospects. *Environmental Science: Water Research and Technology* 5 (11): 1814–1835.

Bismark Sarkodie is a lecturer in Kwame Nkrumah University of Science and Technology who embarked on his post-doctoral research in Advanced Fiber Materials Engineering Research Center of Anhui Province, School of Textile and Clothing, Anhui Polytechnic University, China, and School of Environment, Henan Normal University, China. He obtained his doctorate in Material Science and Engineering from East China University of Science and Technology and M.Eng. Textile Engineering from Jiangnan University in the year 2022 and 2018, respectively. His scientific interests include polymer nanomaterials, metal and metal-oxide nanomaterials and their application in the field of textiles and catalysis.

Benjamin Tawiah holds a B.A. in Textiles and a Master of Engineering from Kwame Nkrumah University of Science and Technology, Ghana, and Jiangnan University, China, respectively. He also has a Postgraduate Diploma in Management Information Systems from GIMPA and a PhD in Textile Technology (Chemistry) from Hong Kong Polytechnic University. Previously, he was a postdoctoral fellow at Hong Kong Polytechnic University and City University of Hong Kong. Currently, he lectures at Kwame Nkrumah University of Science and Technology, focusing on polymer nanomaterial composites, flame retardant polymer composites, and second-generation batteries for advanced grid energy storage.

Haijin Liu earned her bachelor's degree in Environmental Protection in 1992, followed by a master's degree in Analytical Chemistry in 2003, and a Ph.D. in Environmental Science in 2010. She is currently a professor at the School of Chemistry and Environmental Science at Henan Normal University in Xinxiang, China. Her research focus is on the wastewater treatment: photocatalytic degradation of perfluorooctanoic acid and reduction of chromium ion in water. She has a significant academic presence with over 121 published works and a notable citation index, indicating her influence in the field.

Chapter 9
Alternative Cellulosic Sources and Processing Techniques for Sustainable Textiles

Emmanuel Abankwah Ofori, Phoebe Naa Afaaley Sackeyfio, Hannah Owusu Ansah, Mutala Jentina Nakpan, Benjamin Tawiah, Joseph Nii Offei Dodoo, and Richard Acquaye

Abstract The textile industry faces increasing pressure to adopt more sustainable practices and innovative processing techniques to reduce environmental impacts associated with conventional textile production. Cellulose, a naturally abundant polymer, is traditionally sourced from cotton and wood pulp; however, these sources often face challenges like high water usage, deforestation, and pesticide dependency. This has influenced the shift in material sourcing to environmentally benign alternatives, with the tendency to decrease reliance on resource-intensive crops and promote a closed-loop cycle. Hence, this review paper provides a comprehensive overview of alternative cellulosic materials as promising solutions for sustainable textile production. Additionally, advanced processing techniques, including enzymatic treatment, electrospinning and wet spinning, are investigated to improve fibre quality, reduce resource consumption, and minimise chemical waste. Furthermore, the economic viability and life cycle assessments (LCA) are discussed to provide a holistic view of

E. A. Ofori (✉) · H. Owusu Ansah · M. J. Nakpan
Hubei Key Laboratory of Energy Storage and Power Battery, Shiyan Key Laboratory of Quantum Information and Precision Optics, Shiyan, China
e-mail: oforiemmanuelabankwah@gmail.com

E. A. Ofori · H. Owusu Ansah
School of Automotive Materials, Hubei University of Automotive Technology, Shiyan, China

P. N. A. Sackeyfio · B. Tawiah (✉)
Department of Industrial Art (Textiles), Kwame Nkrumah University of Science and Technology, Kumasi, Ghana
e-mail: benjamin.tawiah@knust.edu.gh

B. Tawiah
School of Fashion and Textiles, The Hong Kong Polytechnic University, Hong Kong SAR, China

J. N. O. Dodoo
Fashion Department, Tema Technical Institute, Accra, Ghana

R. Acquaye
Department of Textile Design and Technology, Takoradi Technical University, Takoradi, Ghana
e-mail: richard.acquaye@ttu.edu.gh

S.-X. Jiang et al. (eds.), *Advances in Textile Materials and Processing Techniques for Sustainability*, SDGs and Textiles, https://doi.org/10.1007/978-981-95-0469-5_9

sustainability impacts and their integration into existing textile supply chains. In addition, the remaining challenges associated with cost-effectiveness and scaling up for mass production are addressed for industrial implementation, adoption, and promotion of the circular economy. Importantly, the chapter contributes to the growing body of knowledge by contextualising recent advancements in alternative cellulose sourcing and processing within the broader landscape of sustainable textile innovation. It underscores the potential of these materials and technologies to support a circular economy, with practical implications for both research and industry adoption.

Keywords Sustainable textiles · Life cycle assessments · Ethical sourcing · Circular economy

9.1 Introduction

The textile industry is a global enterprise that employs tens of millions of people and supplies the fundamental needs of the people [1]. For the majority of countries, it has remained a major source of revenue [2], as it is the second most essential human need after food and water [3]. With an estimated value of USD 176 billion, China was the largest exporter of all textiles as of 2021. Similarly, the European Union came in second with an export rate of USD 151 billion, followed by Bangladesh, Vietnam, and Turkey, who made out the top five [4]. According to current projections, the global apparel market is expected to produce $1.5 trillion in 2021 and $2 trillion in 2025 [5]. Notwithstanding its economic significance, the textile industry has recently drawn criticism for its environmental impact because of the use of synthetic materials, resource depletion, and excessive water consumption [6, 7]. Hence, sustainability has become a key driver of innovation and policy, necessitating the demand for cleaner production processes and eco-friendly materials.

Cellulose, a naturally occurring polysaccharide, is one of the many raw materials used in the textile industry and has become essential to the industry's sustainable growth. Cellulose-based fibres, which are traditionally derived from plants like cotton, flax, and hemp, are essential to the development of environmentally conscious textiles because of their comfort, biodegradability, and renewability [8, 9]. Crucially, cotton continues to be the most popular natural fabric because of its softness, breathability, and suitability for a variety of applications. Its sustainability is further enhanced by its biodegradability, renewability, and optimisation to lessen environmental burdens, especially through integrated pest management, organic farming, and water-efficient irrigation techniques [6]. This is because 20,000 L of water, 24% insecticides, and 11% pesticides are needed to produce one kilogram of cotton, which is equivalent to one t-shirt and one pair of trousers [10]. In a similar vein, linen, a bast fibre derived from flax, has seen a renaissance in recent years as an environmentally friendly alternative because of its exceptional strength, biodegradability, exceptional breathability, great moisture wicking, long shelf life, and low water usage [11–14]. Their cultivation can also promote crop rotation and enhance soil health, which is consistent with the ideas of regenerative farming. Because it is durable, biodegradable, requires fewer chemicals, and improves soil health, hemp, another type of cellulose, is a sustainable choice for textiles [15, 16].

Notwithstanding the advantages of conventional cellulose and its role in sustainable practices, new developments have led to the optimisation and investigation of substitute sources, and advancements in cellulose chemistry have increased the industry's potential for sustainable textiles. Tencel, for example, is biodegradable [17], uses less water and chemicals, and is sourced sustainably from wood pulp, typically from eucalyptus, beech, or spruce trees [18, 19]. Agricultural residues such as banana stems, pineapple leaves, and wheat straw have also garnered increasing interest. The potential of agricultural waste fibres is enormous because of their strength, affordability, sustainability, durability, and environmental friendliness [20, 21]. These non-traditional sources provide distinctive fabric qualities that either match or exceed those of traditional textiles while also reusing biomass to lessen the environmental impact. In addition to reducing agricultural waste, banana and pineapple leaf fibre (Piña) have special mechanical and aesthetic qualities that make them excellent for textiles. However, one of the amazing properties of bamboo as a natural fibre is that it completely biodegrades in soil without generating dangerous pollutants like methane [22]. Durability, lustre, porosity, uniformity, robustness, and fineness are its distinguishing characteristics. These qualities are essential for turning fibres into yarns, but they also have an impact on the texture, look, breathability, and longevity of the finished product [23].

In contrast, the extraction, modification, and use of cellulose have been completely transformed by advances in cellulose chemistry. Ionic liquids and deep eutectic solvents, for example, have become more environmentally friendly than conventional solvents for cellulose regeneration and dissolution, reducing energy use and hazardous waste [24–26]. Furthermore, because of its high surface area, mechanical strength, and capacity to build robust networks, nanocellulose, including cellulose nanocrystals (CNCs) and cellulose nanofibrils (CNFs), has paved the way for new developments in functional textiles [27]. As the need for more environmentally friendly textiles grows, these substitutes are attracting recognition for their ability to improve the textile supply chain's resilience and diversity in addition to their environmental benefits [28]. The description of SDG number 9 emphasises that "sustainable industry, innovation, and infrastructure" seeks to find long-term solutions to environmental and economic issues. Furthermore, SDG 12 promotes responsible consumption and is strongly tied to sustainable methods of production and consumption [29].

In light of this, the chapter aims to investigate alternative cellulosic materials and their processing techniques for sustainable textiles, highlighting their environmental benefits and economic implications. It also discusses how these alternatives contribute to circularity in the textile industry, aiming to reduce waste, energy consumption, and overall environmental burden.

9.2 Sources of Cellulose

Cellulose is one of the most prevalent biopolymers found in nature. It can be generated from dried hemp (57%), dry woods (40–50%), and cotton fibre (90%) [30]. Given their highly organised regions and well-defined structures, the cellulose fibrils have a high structural crystallinity. However, some areas are less organised and amorphous, with low structural crystallinity [31]. As shown, Fig. 9.1a [32] illustrates how this low-ordered region is more accessible for hydrogen bonding and other available chemicals. On a chemical level, cellulose makes up the majority of all vegetable fibres, but they also contain different proportions of hemicellulose (see Fig. 9.1b), pectin (see Fig. 9.1c), wax and fat (see Fig. 9.1d), and lignin (see Fig. 9.1e) [9, 33–35]. The many forms of cellulose exhibit notable differences in shape and mechanical strength, although they have comparable molecular structures and chemical compositions. This makes cellulose a viable material for its distinct uses, particularly in the textile sector. Nonetheless, Fig. 9.1f depicts a flowchart of fibre classifications, whereas Fig. 9.1g lists some natural fibre resources [36] that are used to make cellulose.

Cotton

Cotton fibres are the purest form of natural cellulose, as 90% of it is glucose only [38]. Due to their hollow structure, these fibres are often referred to as breathable and absorbent fibres. Morphologically, cotton fibre has a bean-like cross-section that turns round when it absorbs water. The natural ability of cotton fibres to retain water is 24–27 times their weight [39]. Cotton fibres typically have a micronaire value of 3.5–5.5 and a length of 20–32 mm [40]. To protect the environment and soil fertility, organic cotton is now grown without the use of chemicals or fertilisers.

Hemp

The Cannabis sativa plant species, which is mostly grown for industrial purposes, is the source of hemp fibre. Hemp fibres are mostly grown in the United States, Canada, France, and China. This class of cellulose material is a fast-growing annual herbaceous plant with inexpensive, high-quality natural bast fibres growing taller than three meters. The desirable qualities of hemp, including its durability, stiffness, and lightweight properties, make it useful for numerous applications [41]. Textiles, construction, paper, composites, and animal bedding are the main industries that use hemp fibres. Compared to cotton, hemp is a coarser, darker-coloured fibre that is more difficult to bleach. Additionally, it has a cylindrical shape with joints and cracks on the surface [41, 42].

Jute

The plants of the genus Corchorus, which has about 100 species, are the source of jute, the second most valuable and affordable natural fibre after cotton [43]. When compared to other bast fibres, jute fibre has the highest manufacturing volume. Jute is valued for its substantial environmental advantages as well as its exceptional tensile

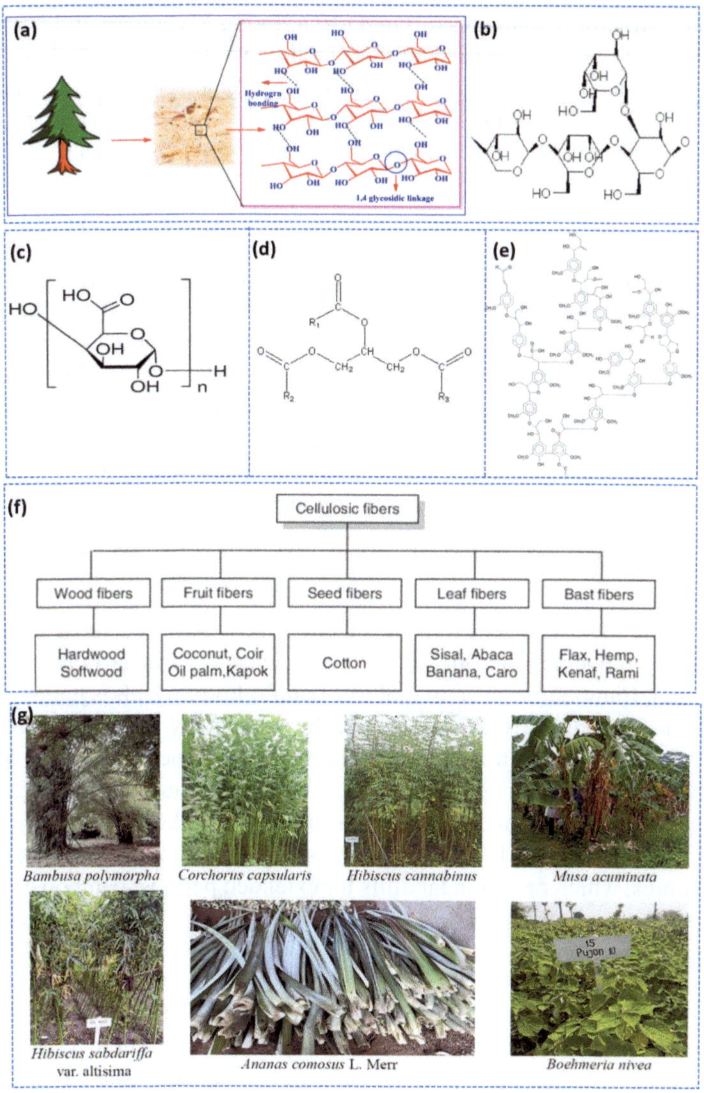

Fig. 9.1 **a** Source and structure of cellulose with an explanation of hydrogen bonding between different chains of the cellulose molecule [32]. This is an open access article distributed under the Creative Commons CC BY license **b** hemicellulose **c** pectin **d** lignin **e** waxes and fats [34]. This is an open access article distributed under the Creative Commons CC BY license **f** Flowchart of fibre classifications. Reproduced with permission from Ref. [37], Copyright 2008 Sage Publications **g** Some natural fibre resources for cellulose generation [36]. This is an open access article distributed under the Creative Commons CC BY license

strength and structural soundness. Bangladesh, India, Myanmar, and China are the top producers of jute fibre because of their favourable development conditions [44]. The jute plant can reach heights of 2.5–5 m and fibre lengths of 1–3.5 m. The majority of jute is composed of cellulose (45–71.5%), lignin (12–26%), and hemicellulose (13.6–21%) [45]. Importantly, a single jute fibre is often polygonal in shape with round corners.

Flax/Linen

The flax plant (Linum usitatissimum) is the source of linen, another cellulosic fibre among the oldest textiles known to exist [46]. Made up primarily of cellulose, hemicellulose, lignin, and other hydrophilic materials, flax is a hardy crop that can grow in a range of temperatures, particularly temperate ones. It is frequently cultivated in areas with limited water supplies. Because it uses less water, fertiliser, and pesticide than cotton, linen is a more environmentally friendly material. Additionally, because of its exceptional breathability, high moisture wicking, biodegradability, and long shelf life, linen is a popular option among customers who care about the environment [13, 14].

Bamboo

Bamboo fabric, also known as virgin bamboo, possesses the remarkable quality of being fully biodegradable in soil without releasing harmful pollutants such as methane [22]. Bamboo is the source of this natural fibre, which is hailed as an eco-friendly and sustainable textile resource for the present day. It is distinguished by its durability, lustre, porosity, homogeneity, resilience, and fineness. These characteristics affect the final product's texture, look, breathability, and lifespan in addition to being necessary for the conversion of fibres into yarns [23]. Consequently, garments crafted from pure bamboo have an insignificant impact on the environment. Unlike synthetic fibres, which persist in landfills for extended periods, bamboo clothing can be composted organically [10].

Agricultural residues

Around the world, agricultural waste fibres are utilised as building materials, decorative products, and a variety of raw materials. They are also of significant economic and cultural importance. Because of their durability, affordability, sustainability, strength, and environmental friendliness, agricultural waste fibres have great promise [20, 21]. Additionally, agricultural residues are viable sources for cellulosic fibres. Materials such as rice husk, açai, wheat straw, corn stalks, and sugarcane bagasse, which are often treated as waste products, contain valuable cellulose that is extracted and processed into textile fibres [47–49]. These cellulosic materials can be found in abundance in fruit processing waste, such as orange peels, banana stems, pineapple leaves (Piñatex), and food residues for textile and clothing manufacturing.

9.2.1 *Extraction and Processing Techniques*

Depending on the type of fibre and intended end use, several procedures are used in the production of textile fibres to convert raw materials into usable fibres for fabric manufacturing. Extraction is a term that is used to define the transfer of any compound from its original state such as solid or liquid to another phase. It can be understood as the separation of the compound from a matrix [50]. The extraction process often entails separating fibres from their parent materials, whereas processing techniques specify how these fibres are ready for knitting, weaving, and other fabric manufacturing techniques. This section of the chapter, which is discussed separately below, examines the methods for extracting and processing different types of cellulosic fibres.

9.2.2 *Mechanical Processing*

Mechanical techniques are successful and efficient because they extract fibre bundles using physical forces rather than dangerous chemicals. This procedure lessens the ecological imprint and is more economical and ecologically friendly [51]. For processing plant fibres, hammer mills, breaking and scutching equipment, and manual stripping are the most widely used mechanical methods [52]. However, not all types of plant fibre can be processed mechanically, especially those that are more lignified or less flexible, and the yield could be reduced [53]. Innovative methods, including the use of nanowhiskers and nanofibers, have been investigated in addition to conventional mechanical procedures. In their discussion of the extraction of cellulose nanofibers and nanocrystals, Benito-González et al. [54] emphasised the significance of mechanical procedures in creating high-quality nanostructured cellulose from a variety of biomass sources, such as rice straw and Posidonia oceanica waste. These methods enhance the biodegradability of the finished textiles while also lessening their negative effects on the environment.

9.2.3 *Chemical Processing*

Textiles can be made more valuable by adding useful features and improving their appearance by chemical processing [55]. It is frequently performed in aqueous media and necessitates the employment of numerous chemicals, many of which are not biodegradable, and a lot of water. Unused chemicals are subsequently dumped into process water, resulting in effluent that needs expensive treatment to comply with environmental regulations [56]. Since different treatments require high temperatures and the particular input requirements vary depending on the material and gear used, the manufacture of textiles is incredibly energy-intensive. The sustainability of the textile processing sector depends on addressing the issues of high energy consumption, soft water requirements, and aqueous effluent discharge [57].

The mitigation of these has therefore been the main focus of current developments. The textile sector faces major environmental difficulties related to solid waste management, gaseous emissions, and water pollution. Over time, they deteriorate and leak into water sources, endangering broader environmental conservation initiatives. Numerous novel approaches are being researched to meet these challenges. Wet processing reuses mixed textile effluent after it has been decomposed using high-energy gamma radiation [58]. Membrane systems are utilised to recover valuable chemicals and water, and ion exchange and pH regulation aid in the purification of water. Similarly, biological approaches help with wastewater reclamation, and ultrafiltration technologies improve purification.

9.2.4 Enzyme-Based Processing

Enzymes are biocatalysts produced by living cells through biochemical processes, mainly cell metabolism [59]. Enzymes from natural sources have been utilised since ancient times to make foods like cheese, sourdough, wine, vinegar, beer, and indigo. Fermentation methods have improved during the past century, especially for producing purified enzymes on a large scale [60]. Recombinant gene technology has improved the processes used to produce enzymes. Most commercial enzymes hydrolyse natural substances to break them down. Textiles and leather processing are two of the many sectors that use enzymes. These consist of β-mannanase, cellulase, glucoamylase, amylase, phytase, xylanase, protease, and others [61, 62]. Enzymes have greatly improved product quality and decreased environmental impact in textile processing.

9.2.5 Emerging Technologies

The textile industry is adopting new technologies to reduce its environmental impact and promote sustainability, with the primary objectives being to use less water and energy, use fewer chemicals, and use eco-friendly materials. Advanced separation techniques such as steam explosion and ultrasonic treatment have shown promise in efficiently breaking down plant materials to liberate their cellulose component [63]. Similar to this, Radhakrishnan [64] believes that biotechnology, a recent spur, has witnessed quick advancements in genetic engineering techniques, which opens the door to the creation of organisms that maximize the production of known or unknown metabolites of commercial significance as well as the transfer of genetic material (genes) from one organism to another. The harsh chemical procedures used in textile preparation and finishing are being replaced by biotechnology, namely, designed enzymes and microbial treatments [65–67]. To effectively break down cellulose, remove impurities, and change the properties of fibre, these biochemicals require significantly less energy and water.

Additionally, the use of plasma technology, a dry, chemical-free method for surface modification of textiles, is growing in popularity [68]. It improves fabric properties such as water repellency and dye absorption without the need for traditional wet processing methods, and ultrasonic energy is being used for more efficient dyeing and finishing, reducing water and chemical usage by up to 40% [69, 70]. Another innovation is supercritical CO_2 dyeing, which uses pressurised carbon dioxide as a dyeing medium instead of water, recovering 95% of the dye and preventing wastewater outflow. Furthermore, the introduction of nanotechnology ten years ago led to significant advancements and improvements in this area of textile technology.

With the application of NT, fabric finishing has taken on new forms and demonstrated significant promise for breakthroughs [71]. The fabric industry can now provide apparel with remarkable quality, elasticity, and durability that is maintained after washing and wearing, thanks to the development of nanotechnology [72]. Large corporations have been driven into the textile industry by the unique and desirable qualities of nanomaterials and the associated financial rewards. Nanomaterials' high surface-to-volume ratio preserves their high surface energy, which improves their affinity for textiles and contributes to long-term durability. Furthermore, a thin coating and layering of nanoparticles on the fabric gives it a smooth, breathable texture [73].

9.3 Fabrication Technologies and Fibre Properties

The increasing demand for sustainable and high-performance materials in the textile industry has led to significant advancements in the production of cellulose-based fibres in recent years. The morphological, mechanical, and functional qualities of the fibres produced are greatly influenced by the fabrication process, which also affects their suitability for a variety of applications. Each production process has its advantages, from the durable and multifunctional fibres formed by wet spinning to the ultrafine nanofibers created by electrospinning. Hence, the manufacturing technologies for cellulose-based fibres and the relationship between fabrication and fibre qualities are discussed below.

9.3.1 Electrospinning

Since the late 1990s, electrospinning has emerged as a sophisticated, cutting-edge, and adaptable process due to its simplicity in producing nanofibers with diameters ranging from 2 nm (nm) to a few micrometres using a variety of materials [74–77]. An electric field is used in the novel process of electrospinning to create ultrafine fibres [78]. The concentration of the polymer solution, the spinning distance, the applied voltage, and the volume flow rate are the four electrospinning process factors that most affect the shape of electrospun fibres, according to Nurwaha et al. [79]. The electrospinning parameters and their impact on fibre morphology are shown in Table 9.1.

As seen in Fig. 9.2a, b, this method generally uses a straightforward configuration consisting of four main parts: a fibre collector, a syringe pump, a spinneret (a syringe with a blunt-tip needle), and a high-voltage power source [90]. With the use of a syringe pump, a polymer is dissolved in a highly volatile solvent and then released from a syringe at a steady, controlled rate. Positive and negative charges split apart inside the liquid due to the electrostatic field, and charges that have the same sign as the needles polarity move toward the surface to create a charged polymer droplet at the needle's tip [91]. The strength of the electrostatic field increases the surface charge density of the droplet, increasing the mutual charge repulsion on the liquid's surface. Consequently, the droplet's surface area increases, and decreases repulsion. A Taylor cone replaces the drop shape. Nonetheless, a reliable electrospinning method that controls nanofiber diameter and form requires a strong Taylor cone [92]. However,

Table 9.1 Electrospinning parameters and their effects on fibre morphology

	Parameters	Parameter values	Morphological effect	Refs.
Solution parameters	Increased concentration	10% w/v	Fibre diameter, beads disappear	[80]
	Increased viscosity	530000 MWs	Fibre diameter, beads disappear	[81]
	Low viscosity	180000 MWs	Discontinuation of filament formation, formation of beads	[81, 82]
	High molecular weight	10–15 wt.%	Increases fibre diameter and uniformity, where beads disappear	[83]
Process parameters	Increased voltage	12–18kv	Affects the force of the fluid jet, which favours fibre formation and diameter	[84]
	The distance between the needle and the collector	20 cm	Shrinks the beads and changes the diameter of the fibre	[85]
	Flow rate	0.5–1.5 mL/hr	Affects fibre diameter based on increased variation	[86]
Ambient parameters	Temperature	283, 293, and 303	Fibre diameter	[87]
	Humidity	50% RH	Controversial effect on fibre diameter	[88, 89]

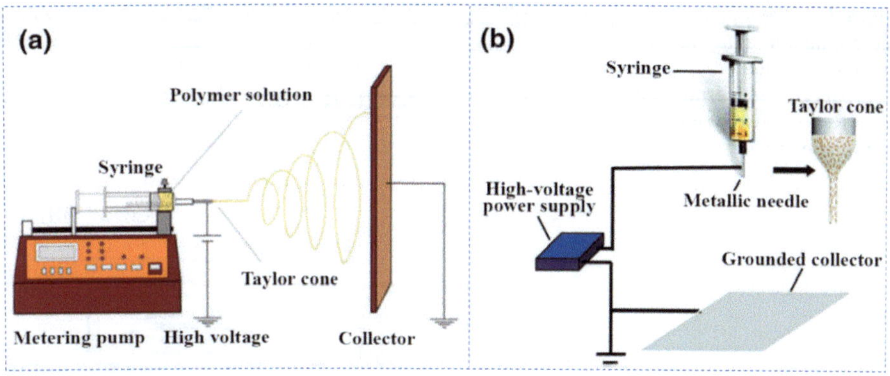

Fig. 9.2 Diagram of **a** electrospinning process. Reproduced with permission from [90] Copyright 2016 Elsevier Ltd. **b** the vertical setup [94]. This is an open access article distributed under the Creative Commons CC BY license

non-uniform nanofibers or beads could be produced by unstable or uneven Taylor cones. Additionally, Zhao et al. [93] demonstrated a direct correlation between the diameter of the generated nanofibers and the height of the Taylor cone, which was supported by both theoretical and experimental evidence. Ultimately, surface tension is overcome by electrostatic repulsion, creating a jet that swiftly approaches the collection. The solvent evaporates as the polymer solution elongates and whips as the jet approaches the collection. When these procedures are combined, incredibly small fibres are produced.

9.3.2 *Wet Spinning*

The wet spinning method is the most popular technique for renewing cellulose fibres because it combines rheological and diffusional behaviours, where fibre orientation determines both structure development and mechanical qualities like high tensile strength, modulus, fibre diameter, porosity, and surface roughness. The formation of a continuous fibre without breaking and the elastic behaviour of the polymer solution during spinning are caused by the rheological component, sometimes referred to as the viscoelastic property of the fibres [95]. The process of wet spinning typically involves four main steps: dissolution, extrusion, coagulation, and collection. Typically, a spinneret is used to extrude a spinning dope (a polymer soaked in an appropriate solvent) into a coagulation bath that contains either a "poor solvent," a "non-solvent," or both [96, 97], as seen in Fig. 9.3. Due to solvent/nonsolvent exchange-induced phase inversion, the filament's polymer coagulates and solidifies in contact with the bath [98]. The continuous fibres then go through a sequence of cleaning and heated rollers (drying drawing rollers) before being shaped and gathered on a rotating spool. For instance, Björquist et al. [19] investigated the textile qualities

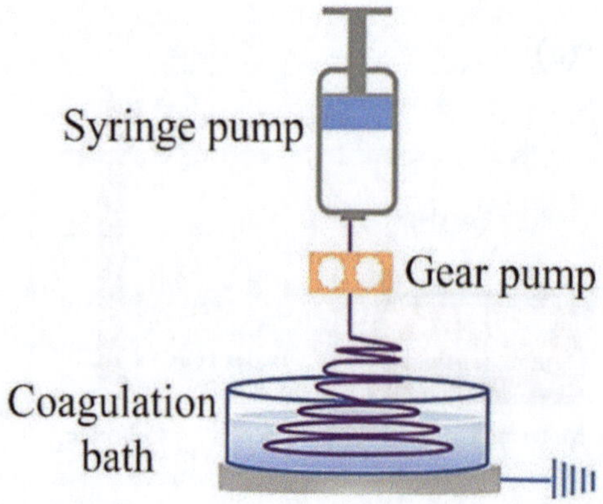

Fig. 9.3 Diagrammatic representation of the wet spinning technique. Reproduced with permission from Ref. [97] Copyright 2014, Elsevier

of lyocell fibres derived from cotton waste pulp, suggesting that waste materials can be effectively converted into high-quality dissolving pulp suitable for wet spinning.

9.4 Functionalization Methods, Functional Finishes and Applications

Alternative cellulose functionalization for textile-based materials has emerged as a promising avenue for producing green textiles with improved qualities and functionalities for a range of uses (see Fig. 9.4). There are two general methods for introducing these functional chemicals onto textiles: ex-situ (indirect incorporation) and in situ (direct incorporation) strategies. The preparation part of the ex-situ method entails synthesising the materials separately, and the uploading step comprises transferring the as-synthesised ingredients onto the textiles. The materials are created directly in the textile matrix during the in-situ process, though. It is significant to note that any of the aforementioned methods can be used to blend cellulosic fabrics with metal oxides, metal nanoparticles, and metal–organic frameworks (MOFs) [99]. Notably, the numerous functionalization methods are categorised into three main groups: chemical functionalization, physical functionalization, and biological functionalization techniques. These are treated free-standing below.

9.4.1 Chemical Functionalization

Chemical modification involves altering the cellulose structure by introducing functional groups to improve properties such as moisture absorption, antimicrobial

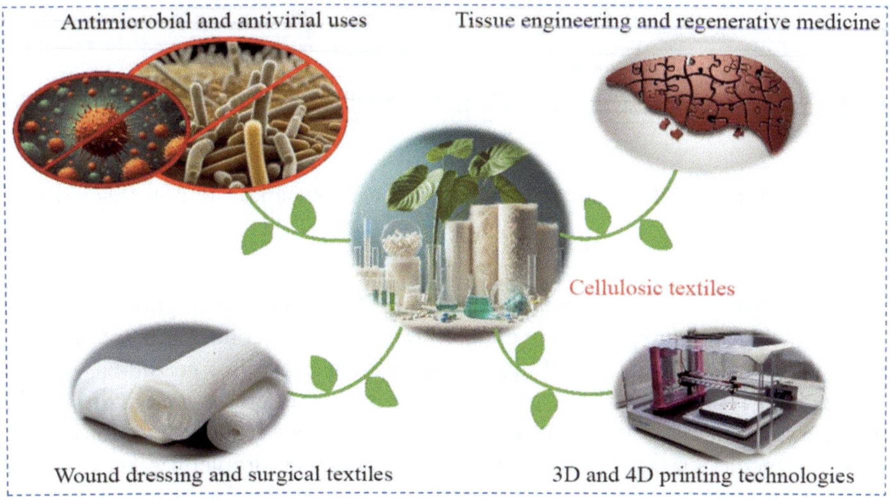

Fig. 9.4 Functionalization and application of cellulosic textiles [100]. This is an open access article distributed under the Creative Commons CC BY license

activity, UV protection, or flame resistance [101]. The most common chemical modifications include acylation and esterification, where the structure of the textile material is altered using acetic and lauric acid by the introduction of ester groups. Both long and short-chain cellulose esters are synthesised via this approach to enhance the mechanical properties of cellulose materials [102, 103]. Recent studies have reviewed various synthesis pathways for these unconventional cellulose derivatives, utilising different solvents and new activating agents to enhance functionalization, offering opportunities for designing advanced cellulose materials with tailored properties [104]. Similarly, grafting and crosslinking as chemical functionalization methods are promising approaches offering the industry a bright future. Grafting involves the covalent bonding of functional molecules or polymers onto textile fibre or fabric surfaces [105], rendering them multi-functional in satisfying performance requirements [106]. Additionally, the emergence of nanomaterials, including silver (Ag NP) and zinc oxide (ZnO NPs) nanoparticles offers antimicrobial and UV-blocking properties and, thus, finds application in medical textiles and outdoor clothing [107, 108].

9.4.2 Physical Functionalization

Physical modifications enhance the mechanical properties such as tensile strength, elasticity, and surface texture by changing the cellulose fibres physical structure. Among these are plasma therapy and electrospinning. Electrospinning is a voltage-driven method that creates fibres from a polymer solution and is controlled by

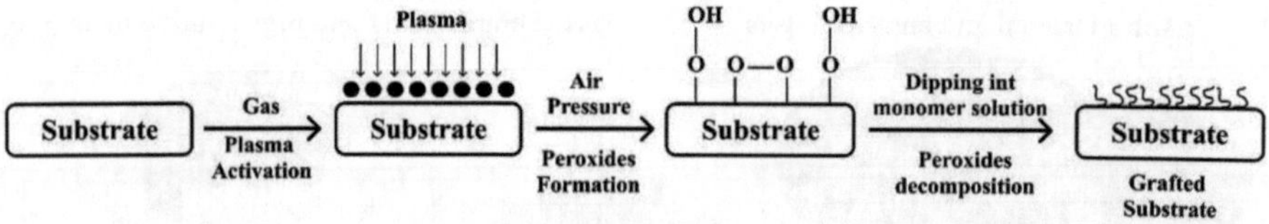

Fig. 9.5 Mechanism of plasma-induced grafting. An open access article distributed under the CC-BY License [119]

the electro-hydrodynamic phenomenon [109]. The simplest configuration for this method includes a pump, a high-voltage power source, a collector, a solution in a reservoir, usually a syringe, and tipped with a blunt needle for needle-based electrospraying [110, 111]. Electrostatic spinning is used to create nanofibers from a polymer solution that contains the desired spinning dope (spinning solution) [112].

At the needle, known as the Taylor cone tip, the liquid droplet becomes charged and stretched into a cone form as the voltage rises [113]. The apex of the cone releases a jet of the polymer solution, which speeds in the direction of the collector and, following solvent evaporation, eventually concentrates into solid fibres [114]. Numerous methods, such as melt-spinning techniques and single-needle, multi-needle, coaxial, and needleless methods (such as wire and revolving drums), have been used in the electrospinning of nanofibers [77, 115]. However, using reactive species, plasma treatment (see Fig. 9.5) improves the fibers physicochemical and biological characteristics [116–119]. Power, pressure, gas flow rate, and exposure time are only a few of the many variables that affect the treatment's highly adjustable effects and morphology, coatings, and dyeability.

9.4.3 Biological Functionalization

The biological functionalization method involves the interaction between biological agents, such as microbes and fungi, or processes that change the properties of the fibres, improving performance, sustainability, and functionality. Common bacteria used are the acetobacter species, which causes fermentation or the use of Trametes versicolor fungal networks. These influence properties such as crystallinity, water retention, and enhanced antimicrobial and antifungal properties. Nonetheless, bacterial cellulose can be functionalized with natural dyes, making it useful for sustainable fashion applications [120].

9.4.4 Dyeability

Due to the significant volumes of water, energy, and chemicals used in the wet treatment process, the dyeing and finishing process is a hotspot in the textile chain, resulting in remarkable greenhouse gas emissions (more than 3.3 billion metric tons annually) [121]. Researchers are increasingly focusing on natural dyes as more sustainable substitutes for synthetic ones to meet environmental requirements. Similarly, using cellulose-producing bacteria to produce self-pigmented textiles is one of the green methods for dying environmentally friendly fabrics [122]. Interestingly, alternative cellulose textiles like bamboo, lyocell, and modal tend to have strong dyeability, meaning they can effectively absorb and retain dye. However, depending on the intended colour and degree of sustainability, the particular dyeing method and type of colour may need to be modified. Walker et al. [122], for instance, genetically modified Komagataeibacter rhaeticus to produce a bacterial strain that produces self-pigmenting BC to dye fabrics with eumelanin sustainably. In another study, Periyasamy et al. [123] observed green dyeing using natural material such as syzygium cumini fruit explored as an environmentally friendly alternative dye for cellulose fibres (see Fig. 9.6). Likewise, Gan et al. [124] investigated the dyeing capabilities of cellulose powder derived from leftover cotton fibres, which were then dyed to serve as a novel class of printing pigments. Black carrot extract has been used as a natural dye in cellulose-based flax and cotton studies, along with sodium alginate and gall oak as bio-mordants [125].

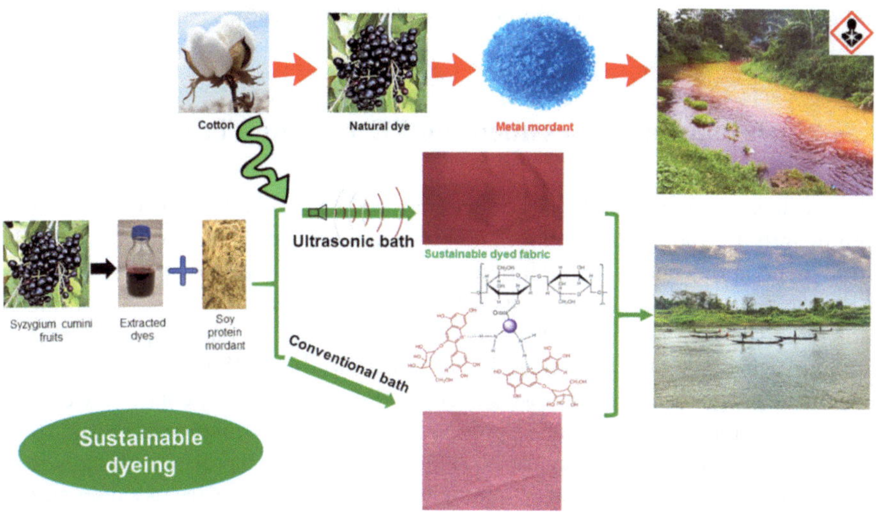

Fig. 9.6 Natural dyeing of cellulose fibres using syzygium cumini fruit extracts [123]. This is an open access article distributed under the terms of the Creative Commons CC BY license

9.4.5 *Flame-Retardant and Antimicrobial Finishes*

Biodegradable textiles made from the most prevalent biopolymer, cellulose, have drawn a lot of attention from scientists. This can be attributed to the chemical modification of cellulose, the incorporation of multifunctional groups, and its renewable and biodegradable properties [126]. Flame retardancy, antimicrobial/bacterial coatings, and self-cleaning properties are examples of these functional finishes. To prevent flammable gases from escaping, the flame-retardant treatments change the thermal breakdown process, create a protective layer of char, or encourage the development of a non-flammable barrier (Fig. 9.7a). Because they lower the danger of fire-related accidents and property damage, flame-retardant fabrics are crucial for improving safety in a variety of applications, such as home furnishings, public transportation interiors, and gear for firemen and industrial personnel. For instance, Bai et al. [127] surface-treated lyocell fibres with phosphorus compounds and determined the respective effects on the flame-retardant properties of the fibres. The flame-retardant properties of the lyocell fibres increased with the increase in di-ammonium-hydrogen phosphate (DAHP) content, forming a dense and thick char layer. Additionally, materials like quaternary ammonium compounds (QACs) [128], guanidine polymers [129], chitosan [130], silver nanoparticles (Ag NPs) [131], titanium dioxide nanoparticles (TiO_2 NPs) [132], and zinc oxide nanoparticles (ZnONPs) [133] are usually applied for the development of antimicrobial functions in textiles. By inhibiting the growth of bacteria and fungi, antimicrobial agents can prevent the production of volatile compounds that contribute to unpleasant smells [134, 135] as shown in Fig. 9.7b. Because of these qualities, antimicrobial finishes are utilised in many different fields, such as the biomedical field for wound care, tissue regeneration/substitution, drug delivery systems, biosensors, and cancer diagnosis; and the paper and textile industries for fibre composites and coatings [136].

9.5 Sustainability Assessment and Economic Viability

The textile sector is one of the largest and most influential in the world, contributing significantly to trade, job creation, and economic growth. With several issues like waste generation, chemical pollution, and water consumption, it is also one of the most resource-intensive and environmentally harmful industries. According to McKinsey's research from 2020, consumers' decisions to purchase mass-market clothing will be greatly influenced by sustainability by 2025 [138]. As consumer awareness and governance frameworks surrounding sustainability continue to evolve, the textile sector is facing mounting pressure to lessen its environmental impact while maintaining its financial sustainability. This sector's sustained prosperity hinges on its twin focus on sustainability and economic feasibility, which are discussed separately below.

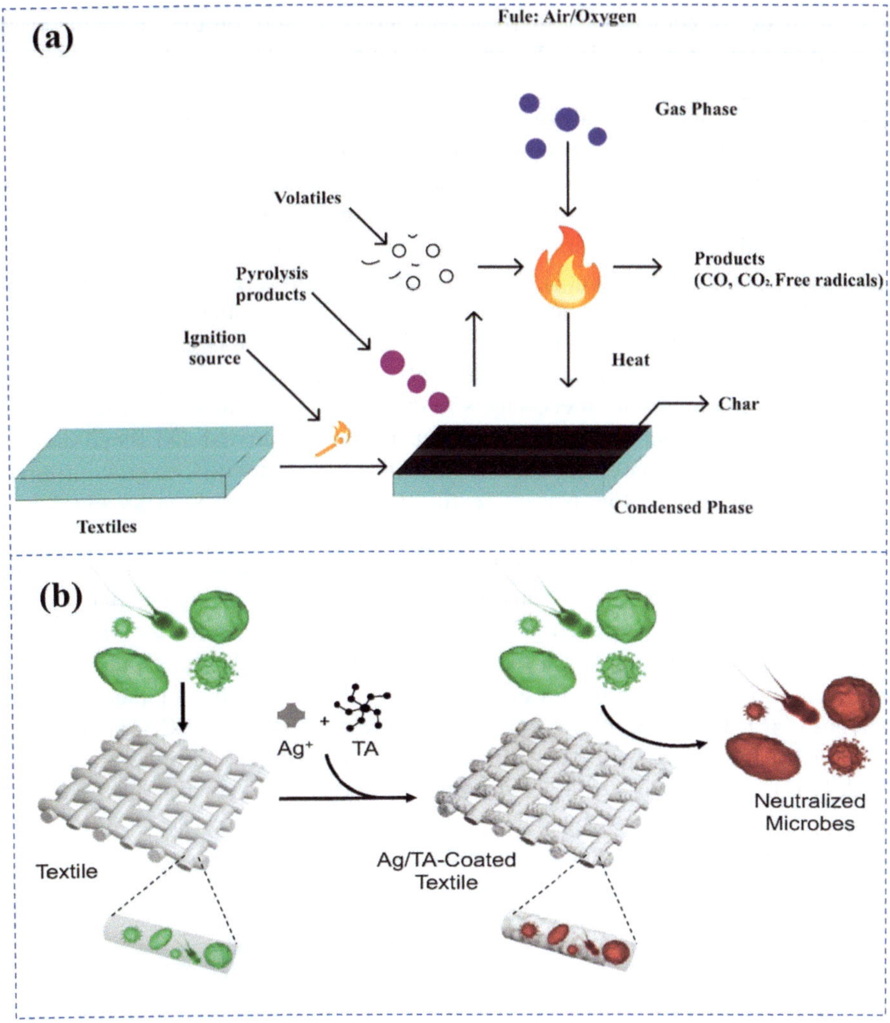

Fig. 9.7 **a** The mechanism of flame-retardant textiles. Reproduced with permission from Ref. [137]. This is an open access article distributed under the terms of the Creative Commons NC-ND BY license **b** Schematic illustration of the formation of colourless Ag/TA coating on textiles and their interactions with microbes. While pathogenic and odour-causing microbes can survive on uncoated textiles, Ag/TA-coated textiles rapidly neutralise a wide range of microbes [135]. This is an open access article distributed under the terms of the Creative Commons CC BY license

9.5.1 Environmental Impact Metrics

The textile industry's extensive use of chemicals, energy, and water has a major negative influence on the environment. With 195 nations pledging to keep the temperature increase to less than 2 °C [139], international measures such as the United Nations

Paris Climate Agreement have therefore put more pressure on the sector to lessen its environmental impact [118]. A significant initiative that tackles the issue of environmental degradation and encourages more nations and areas to go on the path of sustainable development to safeguard the environment is green labelling as a standard [140]. It also helps consumers comprehend a product's sustainability characteristics and certifications, as well as helping businesses sustainably acquire raw resources. According to the Encyclopaedia of Agriculture and Food Systems, an "ecolabel" within a certain product or service category indicates a product or service's shown preference for the environment [141]. Approximately 100 ecolabels covering various stages of the industrial life cycle are used by the textile sector both domestically and internationally. Oeko-Tex Standard 100 (see Fig. 9.8a), Global Organic Textile Standard (GOTS, Fig. 9.8b), Organic Exchange Standard, and Organic Content Standard are the most commonly used eco-labels in the global textile industry. These labels ensure that the product has undergone extensive environmental and social assessments; their Australian counterpart, BMP Certified Cotton, focuses on growing environmentally friendly cotton; Nature Textile Best, and the Blue Sign Standard (see Fig. 9.8c) [142–144].

The Oeko-Tex Standard 100 is the most representative ecological barrier to commerce in textile products and resources, ensuring that neither human health nor the environment is endangered by textile products [145]. A standard framework that outlines the four stages of a life cycle assessment has also been published by the International Organisation for Standardisation (ISO) and other regulatory organisations [146]. The ideas and structure of Life Cycle Assessment (LCA) are covered by ISO 14040, even though ISO 14044 outlines the criteria and requirements for LCA [147]. Additionally, using environmentally friendly production methods like digital printing or waterless dyeing reduces chemical and water waste [148, 149], all of which promote a circular economy and a sustainable environment.

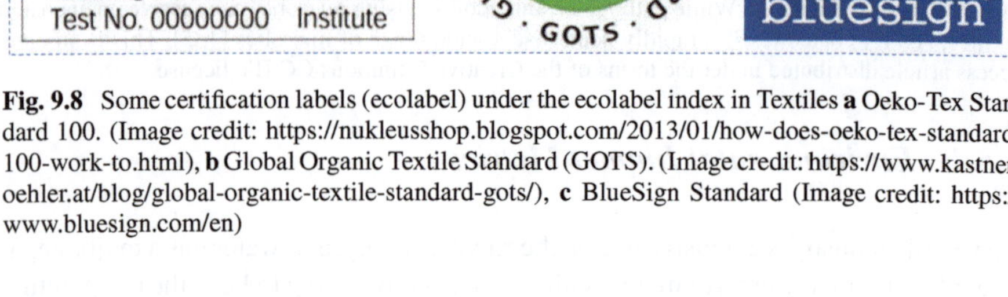

Fig. 9.8 Some certification labels (ecolabel) under the ecolabel index in Textiles **a** Oeko-Tex Standard 100. (Image credit: https://nukleusshop.blogspot.com/2013/01/how-does-oeko-tex-standard-100-work-to.html), **b** Global Organic Textile Standard (GOTS). (Image credit: https://www.kastner-oehler.at/blog/global-organic-textile-standard-gots/), **c** BlueSign Standard (Image credit: https://www.bluesign.com/en)

9.5.2 Social and Economic Implications

The growing demand for sustainable materials has led to an increased interest in alternative cellulose sources such as agricultural residues, including sugarcane bagasse, rice husks, and fast-growing plants, including bamboo and hemp, as substitutes for traditional wood pulp. These alternatives offer significant environmental benefits and present transformative impacts across social and economic domains. As a result, businesses are starting to view sustainability as a crucial strategic concern [150].

For instance, the use of alternative sources of cellulose, including hemp, flax, organic cotton, low-chemical or chemical-free cotton, lyocell (Tencel®), and bamboo, leads to improvements in the environment and sustainability initiatives [151]. Utilisation of locally sourced feedstock allows for complete traceability of fibre from feedstock source to end-use application. Alternative sources of cellulose also promote whole plant utilisation, which supports a circular economy as well as the overall bioeconomy [36, 152]. Furthermore, the adoption of alternative cellulosic fibres is associated with increased efficiency and cost-effectiveness, especially when economies of scale are achieved. Levi Strauss & Co. emphasises that scaling up production of sustainable fibres can lead to more affordable options, which in turn can facilitate broader adoption and positive economic impacts. Similarly, Patagonia, a leader in sustainability, for instance, balances cost and environmental impact by reducing environmental impact through innovative materials and technology. Moreover, the conversion of waste into construction materials exemplifies how sustainable fibre utilisation can generate economic efficiency benefits, contributing to both cost savings and social sustainability by promoting resource reuse [153].

9.6 Conclusion and Future Perspectives

Exploring alternative cellulosic sources and innovative processing techniques represents a crucial frontier in sustainable textile development. As the global textile industry grapples with environmental challenges and resource scarcity, the emergence of novel cellulosic materials from agricultural residues offers promising pathways toward a more sustainable future. Alternative cellulosic fibres, such as those derived from bamboo, hemp, and banana plants, offer several advantages over conventional cotton. These fibres are typically more resistant to pests, require fewer chemical inputs, and can be grown in regions with lower water availability, making them more eco-friendly and resource-efficient. Additionally, the use of agricultural residues, like wheat straw, corn stalks, and rice husks, can provide a valuable second-life use for materials that would otherwise be discarded as waste. This not only contributes to reducing agricultural waste but also promotes the circular economy model by recycling plant-based resources into high-value textile products. Furthermore, this provides new economic opportunities for farmers and rural communities. Additionally, advancements in biotechnologies and bioengineered materials have enabled the development of eco-friendly textile finishes, such as natural flame retardants, water repellents, and anti-microbial treatments, all without compromising on performance or durability.

Despite these promising developments, several challenges need to be addressed in the future. The scalability of alternative cellulosic fibres and processing techniques remains a significant concern, especially with the need to meet the mass demands of the global textile market. Hence, the need for collaboration and shared innovative platforms between agricultural producers, textile manufacturers, and research institutions to accelerate the transition from prototypes to market-ready products. Furthermore, while some alternative sources like bamboo have gained attention, standardisation of quality metrics and processing protocols will be crucial for industry-wide adoption. Additionally, policies promoting circular economy principles and restricting environmental impacts could accelerate the adoption of alternative cellulosic materials, whereas government support for research and development in this field could catalyse technological breakthroughs. Nonetheless, the integration of artificial intelligence and machine learning could optimise processing parameters and predict fibre properties, streamlining the development of new cellulosic materials.

With continued innovation and interdisciplinary research, these materials could become a cornerstone of sustainable textiles, offering consumers eco-friendly alternatives without sacrificing performance. The future of textiles lies in the balance of environmental stewardship and technological innovation, where sustainable sourcing and processing practices go hand in hand with the need for high-quality, functional, and affordable products.

References

1. Harsanto, B., et al. 2023 Sustainability innovation in the textile industry: A systematic review. *Sustainability* 15. https://doi.org/10.3390/su15021549.
2. Memon, H., et al. 2022. Sustainable textile marketing—Editorial. *Sustainability* 14 (19): 11860.
3. Madhav, S., et al. 2018. A review of textile industry: Wet processing, environmental impacts, and effluent treatment methods. *Environmental Quality Management* 27 (3): 31–41.
4. Smith, P. 2022. *Leading exporting countries of textile worldwide by value in 2021.* [August 12, 2023] https://www.statista.com/statistics/1054452/textile-apparel-goods-leading-exporters-worldwide/.
5. Seidu, R.K., B. Eghan, and R. Acquaye. 2023. A review of circular fashion and bio-based materials in the fashion industry. *Circular Economy and Sustainability.*
6. Rahaman, M. T., and M. S. H. Khan. 2025. Green merchandising of textiles and apparel in a circular economy: Recent trends, framework, challenges and future prospects towards sustainability. *Journal of Open Innovation: Technology, Market, and Complexity* 11 (1): 100457.
7. Niinimäki, K., et al. 2020. The environmental price of fast fashion. *Nature Reviews Earth & Environment* 1 (4): 189–200.
8. Novaković, M., et al. 2020. Development of comfortable and eco-friendly cellulose based textiles with improved sustainability. *Journal of Cleaner Production* 267: 122154.

9. Thyavihalli Girijappa, Y.G., et al. 2019. Natural fibers as sustainable and renewable resource for development of eco-friendly composites: A comprehensive review. *Frontiers in Materials* 6.
10. Plakantonaki, S., et al. 2023. A review of sustainability standards and ecolabeling in the textile industry. *Sustainability* 15. https://doi.org/10.3390/su151511589.
11. Winqvist, C., J. Ahnström, and J. Bengtsson. 2012. Effects of organic farming on biodiversity and ecosystem services: Taking landscape complexity into account. *Annals of the New York Academy of Sciences* 1249 (1): 191–203.
12. Ramesh, M. 2019. Flax (*Linum usitatissimum L.*) fibre reinforced polymer composite materials: A review on preparation, properties and prospects. *Progress in Materials Science* 102: 109–166.
13. Gębarowski, T., I. Jęśkowiak, and B. Wiatrak.2022. Investigation of the properties of linen fibers and dressings. *International Journal of Molecular Sciences* 23 (18).
14. Ibrahim, N. A., A. R. Zaher, and H. M. El-Hennawi. 2024. Antimicrobial printed linen fabric by using brewer's yeast enzyme. *Beni-Suef University Journal of Basic and Applied Sciences* 13 (1): 76.
15. Duque Schumacher, A.G., S. Pequito, and J. Pazour. 2020. Industrial hemp fiber: A sustainable and economical alternative to cotton. *Journal of Cleaner Production* 268: 122180.
16. Zampori, L., G. Dotelli, and V. Vernelli. 2013. Life cycle assessment of hemp cultivation and use of hemp-based thermal insulator materials in buildings. *Environmental Science & Technology* 47 (13): 7413–7420.
17. Parajuli, P., et al. 2021. 4 - Regenerated cellulose in textiles: rayon, lyocell, modal and other fibres. In *Fundamentals of natural fibres and textiles*, ed. M.I.H. Mondal, 87–110. Woodhead Publishing.
18. Alam, M. N., and L. P. Christopher. 2017. A novel, cost-effective and eco-friendly method for preparation of textile fibers from cellulosic pulps. *Carbohydrate Polymers* 173: 253–258.
19. Björquist, S., et al. 2017. Textile qualities of regenerated cellulose fibers from cotton waste pulp. *Textile Research Journal* 88 (21): 2485–2492.
20. Dungani, R., et al. 2016. Agricultural waste fibers towards sustainability and advanced utilization: A review. *Asian Journal of Plant Sciences* 15: 42–55.
21. Owonubi, S.J., et al. 2021. Non-woody biomass as sources of nanocellulose particles: A review of extraction procedures. *Frontiers in Energy Research* 9.
22. Amjad, A. I. 2024. Bamboo fibre: A sustainable solution for textile manufacturing. *Advances in Bamboo Science* 7: 100088.
23. Harwood, J., and R. Harwood. 2012. 12—Testing of natural textile fibres. In *Handbook of natural fibres*, ed. R.M. Kozłowski, 345–390. Woodhead Publishing.
24. Alqahtani, A. S. 2024. Indisputable roles of different ionic liquids, deep eutectic solvents and nanomaterials in green chemistry for sustainable organic synthesis. *Journal of Molecular Liquids* 399: 124469.
25. Rachiero, G.P., P. Berton, and J. Shamshina. 2022. Deep eutectic solvents: alternative solvents for biomass-based waste valorization. *Molecules* 27. https://doi.org/10.3390/molecules27196606.
26. Zhu, J., et al. 2025. Recent progress on the dissolution of cellulose in deep eutectic solvents. *Industrial Crops and Products* 228: 120844.
27. Rostamabadi, H., et al. 2024. Cellulose nanofibers, nanocrystals, and bacterial nanocellulose: Fabrication, characterization, and their most recent applications. *Future Postharvest and Food* 1 (1): 5–33.
28. Akbari, M., et al. 2017. Emperical social network analysis in sustainable supply chain in Vietnam. In *Proceedings of the 31st Annual Australian & New Zealand Academy of Management Conference (ANZAM)*.
29. Ofori, E. A., et al. 2025. A review on upcycling as a strategic marketing tool for sustainability in the fashion industry. *Journal of Responsible Production and Consumption* 2 (1): 1–24.
30. Bakri, M. K. B., M. R. Rahman, and F. I. Chowdhury. 2022. Sources of cellulose. In *Fundamentals and recent advances in nanocomposites based on polymers and nanocellulose*, 1–18. Elsevier.

31. Tanpichai, S., et al. 2022. Review of the recent developments in all-cellulose nanocomposites: Properties and applications. *Carbohydrate polymers* 286: 119192.
32. Zhang, Z., et al. 2023. Cellulose/inorganic nanoparticles-based nano-biocomposite for abatement of water and wastewater pollutants. *Chemosphere* 313: 137483.
33. Karimah, A., et al. 2021. A review on natural fibers for development of eco-friendly bio-composite: Characteristics, and utilizations. *Journal of Materials Research and Technology* 13: 2442–2458.
34. Romanowska, B., W. Różańska, and M. Zimniewska. 2024. The influence of the chemical composition of flax and hemp fibers on the value of surface free energy. *Materials* 17. https://doi.org/10.3390/ma17051104.
35. Mohammadi, M., I.M. Ridzwan, and M.T.H. Sultan. 2024. Exploring chemical and physical advancements in surface modification techniques of natural fiber reinforced composite: A comprehensive review. *Journal of Natural Fibers* 21 (1): 2408633.
36. Karimah, A., et al. 2021. A comprehensive review on natural fibers: technological and socio-economical aspects. *Polymers* 13. https://doi.org/10.3390/polym13244280.
37. Nourbakhsh, A., et al. 2008. Effect of a novel coupling agent, polybutadiene isocyanate, on mechanical properties of wood-fiber polypropylene composites. *Journal of Reinforced Plastics and Composites* 27 (16–17): 1679–1687.
38. Chopra, L., and Manikanika. 2022. Extraction of cellulosic fibers from the natural resources: A short review. *Materials Today: Proceedings* 48: 1265–1270.
39. Ravandi, S. H., and M. Valizadeh. 2011. Properties of fibers and fabrics that contribute to human comfort. In *Improving comfort in clothing*, 61–78. Elsevier.
40. Fang, D.D. 2018. *Cotton fiber: Physics, chemistry and biology*. Springer.
41. Morin-Crini, N., et al. 2019. Hemp-based adsorbents for sequestration of metals: A review. *Environmental Chemistry Letters* 17: 393–408.
42. Bouloc, P., S. Allegret, and L. Arnaud. 2013. *Hemp: Industrial production and uses*. CABI.
43. Samanta, K. K., S. Basak, and S. Chattopadhyay. 2016. Potential of ligno-cellulosic and protein fibres in sustainable fashion. *Sustainable Fibres for Fashion Industry:* 2: 61–109.
44. Faruk, O., et al. 2012. Biocomposites reinforced with natural fibers: 2000–2010. *Progress in polymer science* 37 (11): 1552–1596.
45. Laraba, S. R., et al. 2022. Development of sandwich using low-cost natural fibers: Alfa-Epoxy composite core and jute/metallic mesh-Epoxy hybrid skin composite. *Industrial Crops and Products* 184: 115093.
46. Akin, D. E. 2013. Linen most useful: Perspectives on structure, chemistry, and enzymes for retting flax. *ISRN Biotechnol* 2013: 186534.
47. Ansari, M. M., et al. 2024. Nanocellulose derived from agricultural biowaste by-products–Sustainable synthesis, biocompatibility, biomedical applications, and future perspectives: A review. *Carbohydrate Polymer Technologies and Applications* 8: 100529.
48. Razali, N.A., et al. 2022. Comparative study on extraction of cellulose fiber from rice straw waste from chemo-mechanical and pulping method. *Polymers* 14. https://doi.org/10.3390/polym14030387.
49. Siqueira, M. U., et al. 2022. Brazilian agro-industrial wastes as potential textile and other raw materials: A sustainable approach. *Materials Circular Economy* 4 (1): 9.
50. Roy, S., and J.-W. Rhim. 2021. Fabrication of cellulose nanofiber-based functional color indicator film incorporated with shikonin extracted from Lithospermum erythrorhizon root. *Food Hydrocolloids* 114: 106566.
51. Khan, A., et al. 2023. An examination of cutting-edge developments in bamboo-PLA composite research: A comprehensive review. *Renewable and Sustainable Energy Reviews* 188: 113832.
52. Song, Y., et al. 2024. Separation process of plant fibers for textile and composite application: A review of recent advances. *Journal of Natural Fibers* 21.
53. Vazquez, A., et al. 2015. Extraction and production of cellulose nanofibers. In *Handbook of Polymer Nanocomposites. Processing, Performance and Application: Volume C: Polymer Nanocomposites of Cellulose Nanoparticles*, eds. J.K. Pandey, et al., 81–118. Springer Berlin Heidelberg: Berlin, Heidelberg.

54. Benito-González, I., et al. 2019. Cellulose nanocrystal-based films produced by more sustainable extraction protocols from Posidonia oceanica waste biomass. *Cellulose* 26 (13): 8007–8024.
55. Arputharaj, A., R. Asm, and S. Saxena. 2016. *Developments in Sustainable Chemical Processing of Textiles*, 217–252.
56. Silva, J.A. 2023. Wastewater treatment and reuse for sustainable water resources management: A systematic literature review. *Sustainability* 15. https://doi.org/10.3390/su151410940.
57. Saxena, S., R. Asm, and A. Arputharaj. 2017. *Challenges in Sustainable Wet Processing of Textiles*, 43–79.
58. Bhuiyan, M. A. R., et al. 2016. Scope of reusing and recycling the textile wastewater after treatment with gamma radiation. *Journal of Cleaner Production* 112: 3063–3071.
59. Robinson, P. K. 2015. Enzymes: Principles and biotechnological applications. *Essays in Biochemistry* 59: 1–41.
60. Voigt, C. A., S. Kauffman, and Z. G. Wang. 2000. Rational evolutionary design: The theory of in vitro protein evolution. *Advances in Protein Chemistry* 55: 79–160.
61. Araújo, R., M. Casal, and A. Cavaco-Paulo. 2008. Application of enzymes for textile fibres processing. *Biocatalysis and Biotransformation* 26 (5): 332–349.
62. Koh, J., and S.M. Mamun Kabir. 2021. Sustainable textile processing by enzyme applications. In *Biodegradation Technology of Organic and Inorganic Pollutants*, eds. K. Ferreira Mendes, R. Nogueira de Sousa, and K. Cabral Mielke. IntechOpen: Rijeka.
63. Wang, C., et al. 2023. The principle of steam explosion technology and its application in food processing by-products. *Foods* 12 (17).
64. Radhakrishnan, S. 2014. *1. Book chapter: Roadmap to sustainable textiles and clothing ISBN 978-981-287-064-3* https://doi.org/10.1007/978-981-287-065-0 *Springer Science+ Business Media, Singapore 2014 (*www.springer.com*) Chapter 9: Application of biotechnology in the processing of textile fabrics; Pg. 277–325*, 277–325.
65. Sen, A., et al. 2021. Biotechnological applications of microbial enzymes to replace chemicals in the textile industry-A review 82: 68–73.
66. Guebitz, G. 2001. Biotechnology in the textile industry—perspectives for the new millennium. *Journal of biotechnology* 89: 89–90.
67. Eid, B., and N. Ibrahim. 2020. Recent developments in sustainable finishing of cellulosic textiles employing biotechnology. *Journal of Cleaner Production* 284: 124701.
68. Šimor, M., et al. 2010. The influence of surface DBD plasma treatment on the adhesion of coatings to high-tech textiles. *Journal of Adhesion Science and Technology* 24: 77–97.
69. Rehman, F., et al. 2024. Ultrasonic assisted enhancement in natural dye extraction from eucalyptus for process optimization. *Journal of Natural Fibers* 21 (1): 2386596.
70. Periyasamy, A. P., and S. Periyasami. 2023. Critical review on sustainability in denim: A step toward sustainable production and consumption of denim. *ACS Omega* 8 (5): 4472–4490.
71. Omietimi, H., et al. 2023. An overview of nanotechnology and its application. *E3S Web of Conferences* 391.
72. Krifa, M., and C. Prichard. 2020. Nanotechnology in textile and apparel research—An overview of technologies and processes. *The Journal of The Textile Institute* 111: 1–16.
73. Brown, P.J., and K. Stevens. 2007. *Nanofibers and nanotechnology in textiles*, 1–528.
74. Govind Kumar, S., and J. Nirmala Rachel. 2022. Electrospinning: The technique and applications. In *Recent developments in nanofibers research*, eds. K. Maaz and C. Samson Jerold Samuel, Ch. 1. IntechOpen: Rijeka.
75. Wang, B., et al. 2023. Assembling nanocelluloses into fibrous materials and their emerging applications. *Carbohydrate Polymers* 299: 120008.
76. Xue, J., et al. 2019. Electrospinning and electrospun nanofibers: Methods, materials, and applications. *Chemical Reviews* 119 (8): 5298–5415.
77. Zahir, A., et al. 2002. Biomaterials for medical and healthcare products. In *Medical textiles from natural resources*, ed. E.M.I.H. Mondal, 43–86. Woodhead Publisher.
78. Baji, A., et al. 2010. Electrospinning of polymer nanofibers: Effects on oriented morphology, structures and tensile properties. *Composites Science and Technology* 70 (5): 703–718.

79. Nurwaha, D., W. Han, and X. Wang. 2013. Effects of processing parameters on electrospun fiber morphology. *Journal of the Textile Institute* 104.
80. Bosworth, L., and S. Downes. 2012. Acetone, a sustainable solvent for electrospinning Poly(ε-Caprolactone) fibres: Effect of varying parameters and solution concentrations on fibre diameter. *Journal of Polymers and the Environment* 20.
81. Zaarour, B., L. Zhu, and X. Jin. 2019. Controlling the surface structure, mechanical properties, crystallinity, and piezoelectric properties of electrospun PVDF nanofibers by maneuvering molecular weight. *Soft Materials* 17: 1–9.
82. Korotcenkov, G. 2021. Electrospun metal oxide nanofibers and their conductometric gas sensor application. Part 1: Nanofibers and features of their forming. Nanomaterials 11. https://doi.org/10.3390/nano11061544.
83. Mwiiri, F.K., and R. Daniels. 2020. Influence of PVA molecular weight and concentration on electrospinnability of Birch Bark extract-loaded nanofibrous Scaffolds intended for enhanced wound healing. *Molecules* 25. https://doi.org/10.3390/molecules25204799.
84. Chowdhury, M., and G. Stylios. 2010. Effect of experimental parameters on the morphology of electrospun Nylon 6 fibres. *International Journal of Basic & Applied Sciences* 10.
85. Hekmati, A., et al. 2013. Effect of needle length, electrospinning distance, and solution concentration on morphological properties of polyamide-6 electrospun nanowebs. *Textile Research Journal* 83: 1452–1466.
86. Zargham, S., et al. 2012. The effect of flow rate on morphology and deposition area of electrospun nylon 6 nanofiber. *Journal of Engineered Fibers and Fabrics* 7: 42–49.
87. De Vrieze, S., et al. 2009. The effect of temperature and humidity on electrospinning. *Journal of Materials Science* 44 (5): 1357–1362.
88. Szewczyk, P. K., and U. Stachewicz. 2020. The impact of relative humidity on electrospun polymer fibers: From structural changes to fiber morphology. *Advances in Colloid and Interface Science* 286: 102315.
89. Cheng, M., et al. 2017. Use of electrospinning to directly fabricate three-dimensional nanofiber stacks of cellulose acetate under high relative humidity condition. *Cellulose* 24 (1): 219–229.
90. Riazi, K., et al. 2016. Polystyrene comb architectures as model systems for the optimized solution electrospinning of branched polymers. *Polymer* 104: 240–250.
91. Subbiah, T., et al. 2005. Electrospinning of nanofibers. *Journal of Applied Polymer Science* 96: 557–569.
92. Ballengee, J., and P. Pintauro. 2011. Morphological control of electrospun nafion nanofiber mats. *Journal of The Electrochemical Society* 158: B568–B572.
93. Zhao, K., et al. 2019. From Taylor cone to solid nanofiber in tri-axial electrospinning: Size relationships. *Results in Physics* 15: 102770.
94. Chinnappan, B.A., et al. 2022. Electrospinning of biomedical nanofibers/nanomembranes: Effects of process parameters. *Polymers* 14. https://doi.org/10.3390/polym14183719.
95. Pereira, C., et al. 2024. Sustainable and naturally derived wet spun fibers: A systematic literature review. *Fibers* 12. https://doi.org/10.3390/fib12090075.
96. Rohani Shirvan, A., A. Nouri, and A. Sutti. 2022. A perspective on the wet spinning process and its advancements in biomedical sciences. *European Polymer Journal* 181: 111681.
97. Imura, Y., R.M.C. Hogan, and M. Jaffe. 2014. *Dry spinning of synthetic polymer fibers*, 187–202.
98. Skjåk-Bræk, G., and K. Draget. 2012. Alginates: Properties and Applications. *Polymer Science: A Comprehensive Reference, 10 Volume Set* 10: 213–220.
99. Emam, H. E. 2019. Generic strategies for functionalization of cellulosic textiles with metal salts. *Cellulose* 26 (3): 1431–1447.
100. Nocca, G., et al. 2023. Cellulosic textiles—An appealing trend for different pharmaceutical applications. *Pharmaceutics* 15. https://doi.org/10.3390/pharmaceutics15122738.
101. Altynov, Y., et al. 2025. Nanocellulose hydrogels from agricultural wastes: methods, properties, and application prospects. *Nanoscale*.
102. Tanaka, S., T. Iwata, and M. Iji. 2017. Long/short chain mixed cellulose esters: Effects of long acyl chain structures on mechanical and thermal properties. *ACS Sustainable Chemistry & Engineering* 5 (2): 1485–1493.

103. Wang, Y., et al. 2018. Functional nanomaterials through esterification of cellulose: A review of chemistry and application. *Cellulose* 25 (7): 3703–3731.
104. Heinze, T., and T. Liebert. 2001. Unconventional methods in cellulose functionalization. *Progress in Polymer Science* 26 (9): 1689–1762.
105. Silva, A. C., et al. 2021. Modification of textiles for functional applications. In *Fundamentals of natural fibres and textiles*, 303–365. Elsevier.
106. Joshi, M., and A. Bhattacharyya. 2011. Nanotechnology–a new route to high-performance functional textiles. *Textile Progress* 43 (3): 155–233.
107. Kalia, S., et al. 2011. Cellulose-based bio-and nanocomposites: A review. *International Journal of Polymer Science* 2011 (1): 837875.
108. Hassan, T., et al. 2021. Functional nanocomposites and their potential applications: A review. *Journal of Polymer Research* 28 (2): 36.
109. Das, S., et al. 2021. *Techniques and methods used for the fabrication of bionanocomposites*, 17–43.
110. Moradi, S., et al. 2020. Fabrication of electrospun sensor based on a synthesized component doped into PAN (polyacrylonitrile) nanofibers for electrochemical detection of zearalenone mycotoxin in foods simulant. *Sensing and Bio-Sensing Research* 28: 100321.
111. Ameer, J.M., A.K. Pr, and N. Kasoju. 2019. Strategies to tune electrospun Scaffold porosity for effective cell response in tissue engineering. *Journal of Functional Biomaterials* 10. https://doi.org/10.3390/jfb10030030.
112. Nandhini, J., et al. 2024. Pitolisant nanofibers: A promising frontier in drug delivery for narcolepsy-formulation, optimisation, and characterization insights. *Biomedical Technology* 6: 9–16.
113. Sinha Ray, S. 2019. *Bio-Waste Based Nanofiber Materials.*
114. Shang, L., et al. 2019. Spinning and applications of bioinspired fiber systems. *ACS Nano* 13 (3): 2749–2772.
115. Partheniadis, I., et al. 2020. A mini-review: Needleless electrospinning of nanofibers for pharmaceutical and biomedical applications. *Processes* 8. https://doi.org/10.3390/pr8060673.
116. Bhatt, P., et al. 2023. Plasma modification techniques for natural polymer-based drug delivery systems. *Pharmaceutics* 15. https://doi.org/10.3390/pharmaceutics15082066.
117. Bie, P., et al. 2016. Structural characteristics and rheological properties of plasma-treated starch. *Innovative Food Science & Emerging Technologies* 34: 196–204.
118. Pan, Y., D.-W. Sun, and Z. Han. 2017. Applications of electromagnetic fields for nonthermal inactivation of microorganisms in foods: An overview. *Trends in Food Science & Technology* 64: 13–22.
119. Karam, L., et al. 2013. Study of surface interactions between peptides, materials and bacteria for setting up antimicrobial surfaces and active food packaging. *J. Mater. Environ. Sci* 4 (5): 798–821.
120. He, X., et al. 2020. Novel bacterial cellulose membrane biosynthesized by a new and highly efficient producer Komagataeibacter rhaeticus TJPU03. *Carbohydrate Research* 493: 108030.
121. Fashion, M. 2018. *Environmental impact of the global apparel and footwear industries study.* Full report and methodological considerations. Quantis.
122. Walker, K.T., et al. 2024. Self-pigmenting textiles grown from cellulose-producing bacteria with engineered tyrosinase expression. *Nature Biotechnology* 1–10.
123. Periyasamy, A. P. 2022. Natural dyeing of cellulose fibers using syzygium cumini fruit extracts and a bio-mordant: A step toward sustainable dyeing. *Sustainable Materials and Technologies* 33: e00472.
124. Gan, L., et al. 2019. Dyeing and characterization of cellulose powder developed from waste cotton. *Polymers (Basel)* 11 (12).
125. Pars, A., and R. Karadag. 2024. Sustainable bio-dyeing of cellulosic-based fabrics with anthocyanins from black carrot (*Daucus carota L.*). *Fibres & Textiles in Eastern Europe* 32 (3): 50–56.
126. Islam, M. S., and T. G. van de Ven. 2021. Cotton-based flame-retardant textiles: A review. *BioResources* 16 (2): 4354.

127. Bai, B. C., et al. 2014. Improved flame-retardant properties of lyocell fiber achieved by phosphorus compound. *Materials Letters* 135: 226–228.
128. Asri, L. A., et al. 2014. A shape-adaptive, antibacterial-coating of immobilized quaternary-ammonium compounds tethered on hyperbranched polyurea and its mechanism of action. *Advanced Functional Materials* 24 (3): 346–355.
129. Han, H., et al. 2019. Inherent guanidine nanogels with durable antibacterial and bacterially antiadhesive properties. *Advanced Functional Materials* 29 (12): 1806594.
130. Verlee, A., S. Mincke, and C. V. Stevens. 2017. Recent developments in antibacterial and antifungal chitosan and its derivatives. *Carbohydrate polymers* 164: 268–283.
131. Eckhardt, S., et al. 2013. Nanobio silver: Its interactions with peptides and bacteria, and its uses in medicine. *Chemical reviews* 113 (7): 4708–4754.
132. Liu, Y., et al. 2015. Bioinspired, stimuli-responsive, multifunctional superhydrophobic surface with directional wetting, adhesion, and transport of water. *Advanced Functional Materials* 25 (31): 5047–5056.
133. Qi, K., et al. 2017. Review on the improvement of the photocatalytic and antibacterial activities of ZnO. *Journal of Alloys and Compounds* 727: 792–820.
134. Gulati, R., S. Sharma, and R. K. Sharma. 2022. Antimicrobial textile: Recent developments and functional perspective. *Polymer Bulletin* 79 (8): 5747–5771.
135. Richardson, J. J., et al. 2022. Rapid assembly of colorless antimicrobial and anti-odor coatings from polyphenols and silver. *Scientific Reports* 12 (1): 2071.
136. Felgueiras, C., et al. 2021. Trends on the cellulose-based textiles: Raw materials and technologies. *Frontiers in Bioengineering and Biotechnology* 9.
137. Ghosh, J., et al. 2025. Grafting modification for textile functionalization: Innovations and applications. *Discover Applied Sciences* 7 (1): 49.
138. Balchandani, A., et al. 2020. *The state of fashion 2021: In search of promise in perilous times*, p. 8–118. McKinsey: Atlanta, GA, USA.
139. Abbate, S., et al. 2024. Sustainability trends and gaps in the textile, apparel and fashion industries. *Environment, Development and Sustainability* 26 (2): 2837–2864.
140. Zhang, Q., J. Mi, and H. Shen. 2019. Green labeling and sustainable development. In *Encyclopedia of Sustainability in Higher Education*, ed. W. Leal Filho, 1–7. Springer International Publishing: Cham.
141. Müller, K., et al. 2014. *Advances in Pesticide Risk Reduction*, 17–34.
142. Hayat, N., A. Hussain, and H. D. Lohano. 2020. Eco-labeling and sustainability: A case of textile industry in Pakistan. *Journal of Cleaner Production* 252: 119807.
143. Velasco-Molpeceres, A., et al. 2023. Slow fashion as a communication strategy of fashion brands on instagram. *Sustainability* 15. https://doi.org/10.3390/su15010423.
144. Diekel, F., et al. 2021. Life cycle based comparison of textile ecolabels. *Sustainability* 13. https://doi.org/10.3390/su13041751.
145. Xu, J., Y. Liu, and L. Yang. 2018. A comparative study of the role of China and India in sustainable textile competition in the U.S. market under green trade barriers. *Sustainability* 10. https://doi.org/10.3390/su10051348.
146. Gonzalez, V., X. Lou, and T. Chi, 2023. Evaluating environmental impact of natural and synthetic fibers: A life cycle assessment approach. *Sustainability* 15. https://doi.org/10.3390/su15097670.
147. Finkbeiner, M., et al. 2006. The new international standards for life cycle assessment: ISO 14040 and ISO 14044. *The International Journal of Life Cycle Assessment* 11 (2): 80–85.
148. Eren, H. A., et al. 2020. Sustainable textile processing with zero water utilization using super critical carbon dioxide technology. In *Sustainability in the textile and apparel industries: Production process sustainability*, ed. S. S. Muthu and M. A. Gardetti, 179–196. Cham: Springer International Publishing.
149. Galarraga Gallastegui, I. 2002. The use of eco-labels: A review of the literature. *European Environment* 12 (6): 316–331.
150. Morais, D. O. C., and B. S. Silvestre. 2018. Advancing social sustainability in supply chain management: Lessons from multiple case studies in an emerging economy. *Journal of Cleaner Production* 199: 222–235.

151. Kwon, S.-H. 2017. Green can be the new white for wedding dresses. *The Design Journal* 20 (5): 595–616.
152. Lawson, L., et al. 2023. Processes and challenges for the manufacturing of lyocell fibres with alternative agricultural feedstocks. *Applied Sciences* 13. https://doi.org/10.3390/app132312759.
153. Samir, A., et al. 2022. Recent advances in biodegradable polymers for sustainable applications. *NPJ Materials Degradation* 6 (1): 68.

Emmanuel Abankwah Ofori is a graduate student and researcher with a degree in B.Sc. Textile Design and Technology from the Industrial Art Department of Kwame Nkrumah University of Science and Technology. Having worked under the tutelage of Dr. Benjamin Tawiah, Emmanuel's research interest focuses on sustainable environmental practices, textile science, polymers, and energy efficiency. Recently, he co-authored a paper that was adjudged the "Second-Best Paper" presented at the 12th Sustainable Education and Development Research Conference in Somanya, Ghana. Emmanuel is dedicated to fostering the next generation of environmental and scientific leaders as a passionate textile scientist and educator.

Phoebe Naa Afaaley Sackeyfio is a researcher and a graduate of the Kwame Nkrumah University of Science and Technology, Kumasi. She was in the Department of Industrial Art where she studied B.Sc. Textile Design and Technology. Her interest is in the research fields of textile design, digital media, and environmental sustainability. Reading, listening to music, designing, and watching movies are hobbies she enjoys.

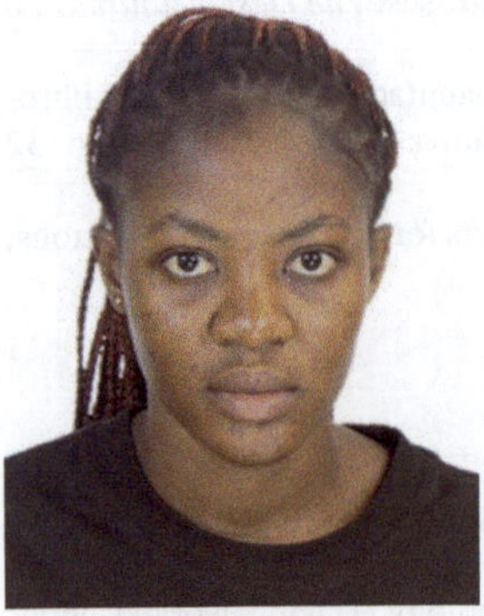

Hannah Owusu Ansah is a Materials Physics and Chemistry student at Hubei University of Automotive Technology (HUAT). She has a solid background in physics, which forms the foundation of her research in Material Science. Her academic interests focus on single-molecular electronics as building blocks for electronic circuits, next-generation energy storage systems, and semiconductor nanocrystals for their unique optical and electrical properties, with potential applications in sensors, smart textiles, and photovoltaics. Her work represents an important intersection of physics, chemistry, and engineering.

Mutala Jentina Nakpan earned his Bachelor of Science in Textile Design and Technology from Kwame Nkrumah University of Science and Technology (KNUST) in Ghana. Currently, he is a master's student at the School of Material Science and Engineering, Hubei University of Automotive Technology (HUAT) in China, specializing in Material Science and Engineering. His research focuses on advancing sustainable solutions in material science, with interests in eco-friendly dyes that provide flame-retardant properties, the development of electrospun nanofibers, and the exploration of piezoelectric composites. He is committed to advancing the field through impactful research that bridges science, design, and sustainability.

Benjamin Tawiah (Ph.D.) received his B.A. Industrial Art (Textiles Option) and MEng in Textile Engineering from the Kwame Nkrumah University of Science and Technology, Ghana, and Jiangnan University, China, respectively. Prior to his Master of Engineering degree, he earned a postgraduate Diploma in Management Information Systems from the Ghana Institute of Management and Public Administration (GIMPA) and then studied for his Ph.D. at the Hong Kong Polytechnic University where he also worked as a postdoctoral Fellow. He also worked as a research associate at the City University of Hong Kong. He is a lecturer at the Kwame Nkrumah University of Science and Technology. His research interest focuses on flame retardant polymer composites, advanced energy electrochemical energy storage systems, and flexible fiber-based current collectors for zinc ion batteries.

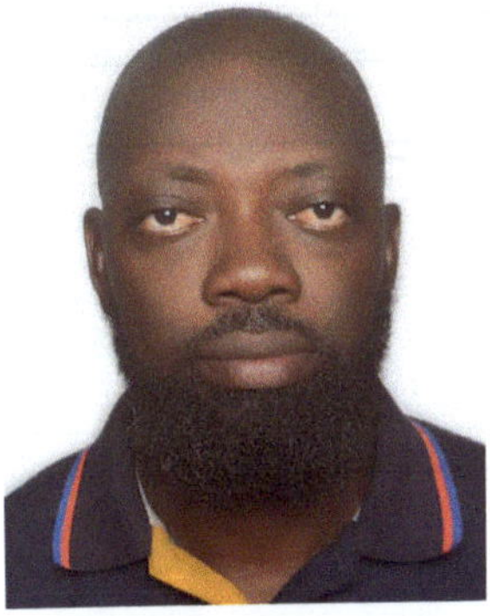

Joseph Nii Offei Dodoo is a dedicated facilitator at Tema Technical Institute, where he has served for 14 years. He holds a Master of Philosophy in Fashion Design Technology from Kwame Nkrumah University of Science and Technology, alongside a Bachelor of Science in Fashion Design and Textiles Education, a Higher National Diploma in Commercial Arts (Textile's option), and a Diploma in Education. His research interests focus on sustainable textile production practices and sustainable apparel product development. Joseph is also a member of the Institute of Textile and Fashion Professionals-Ghana (ITFP-GH).

Richard Acquaye (Ph.D.) is a Professor at the Department of Textile Design and Technology and the Dean of the School of Graduate Studies, Takoradi Technical University, Takoradi – Ghana. His scholarly pursuits are within textile science and design. His expertise encompasses exploring Computer-Aided Design and Computer-Aided Manufacturing (CAD/CAM) application in textiles. His research extends to contemporary textile practices, exploring sustainable solutions such as recycling and upcycling of fabrics. His impact resonates beyond the confines of academia, as evidenced by his publication record in high-impact journals, serving as a source of inspiration and knowledge dissemination for fellow researchers and practitioners worldwide.

Chapter 10
Bio-Manufacturing and Processing of Textiles for Sustainability

Chanchal Kumar Kundu, Mehedi Hasan Chaion, and Bin Fei

Abstract Textile processing and manufacturing is a chemical and energy-intensive industry and is one of the most significant contributors to polluting the environment by various means and poses threats to living beings. The rapidly changing fashion behavior and unscrupulous consumption by the end users are making this situation further worsen. Thus, to address this catastrophe, all stakeholders of a textile supply chain should come up with sustainable practices in every stage of their activities so that they can grow as a sustainable industry. Meanwhile, textile industries around the globe have already started adopting numerous innovative technologies and sustainable practices while putting focus on implementing ethical manufacturing processes to keep energy consumption, resource depletion, and waste generation as low as possible. Among them, bio-processing and manufacturing of textile goods have become a premier choice within the textile industry, which effectively reduces carbon footprint, replaces traditional hazardous chemicals, and ultimately helps to diversify the supply chain and move forward for a cleaner manufacturing process. Herein, this chapter will showcase such novel practices and manufacturing techniques while critically explaining the ongoing challenges and shedding light on future opportunities.

Keywords Textile processing · Textile manufacturing · Sustainable textiles · Bio-processing · Green supply chain

C. K. Kundu (✉) · M. H. Chaion
Department of Textile Engineering, Jashore University of Science and Technology, Jashore, Bangladesh
e-mail: ck.kundu@just.edu.bd

C. K. Kundu · B. Fei (✉)
School of Fashion and Textiles, The Hong Kong Polytechnic University, Kowloon, Hong Kong, P. R. China
e-mail: bin.fei@polyu.edu.hk

S.-X. Jiang et al. (eds.), *Advances in Textile Materials and Processing Techniques for Sustainability*, SDGs and Textiles, https://doi.org/10.1007/978-981-95-0469-5_10

10.1 Introduction

Textile manufacturing is a pivotal and dynamic stage in a typical textile supply chain, encompassing a diverse range of operations [1]. Traditionally, textile manufacturing is energy- and water intensive and threatens the environment due to its huge chemical consumption and subsequent generation of enormous waste. However, with time, innovations are taking place to knock off the traditional manufacturing of textiles, just putting sustainability at its core to face the challenges of environmental concerns arising from such manufacturing techniques. Among them, bio-based processing and manufacturing of textile goods have attained deep attention within industries, owing to their looming benefits for the environment and all stakeholders of a typical textile supply chain. A distinctive supply chain commonly adopted in textile industries is depicted in Fig. 10.1.

Bio-manufacturing and processing are incredibly necessary to meet the upcoming challenges of the Industry 4.0 revolution associated with several key factors depicted in Fig. 10.2. For example, traditional textile manufacturing processes frequently utilize harmful chemicals originating from crude oil. Meanwhile, bio-manufacturing can minimize the dependence on synthetic dyes and chemicals by employing natural dyes and enzymes, which are less toxic and biodegradable. This shift will not only reduce environmental pollution but also lessen the existing dependency on fossil fuel-based chemicals and dyes. Consequently, such manufacturing and processing usually generate a lower carbon footprint compared to traditional manufacturing techniques. In addition, bio-manufacturing processes generally require less energy that can further decrease greenhouse gas emissions associated with textile production and can lead to conserving the planet from the consequences of climate change due to global warming. Such manufacturing and processing can also foster a circular economy by utilizing waste to make a new product, minimizing waste, and maximizing resource efficiency. This type of manufacturing also can help to comply with the stringent laws and policies set by the environmental and governmental policymakers to go green and to be sustainable in the manufacturing and processing of

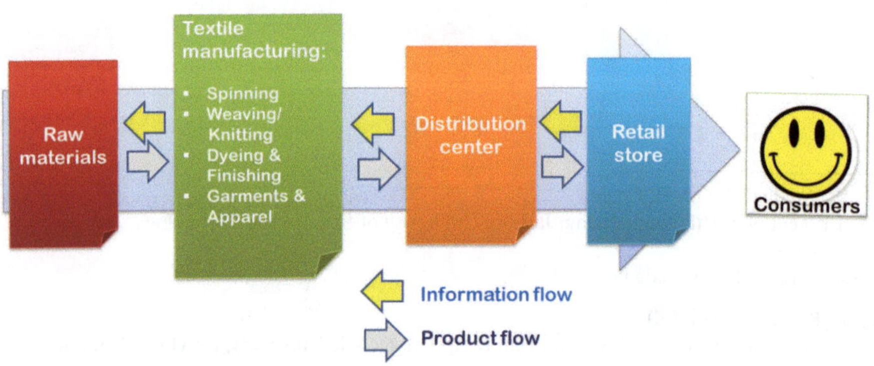

Fig. 10.1 A typical textile supply chain

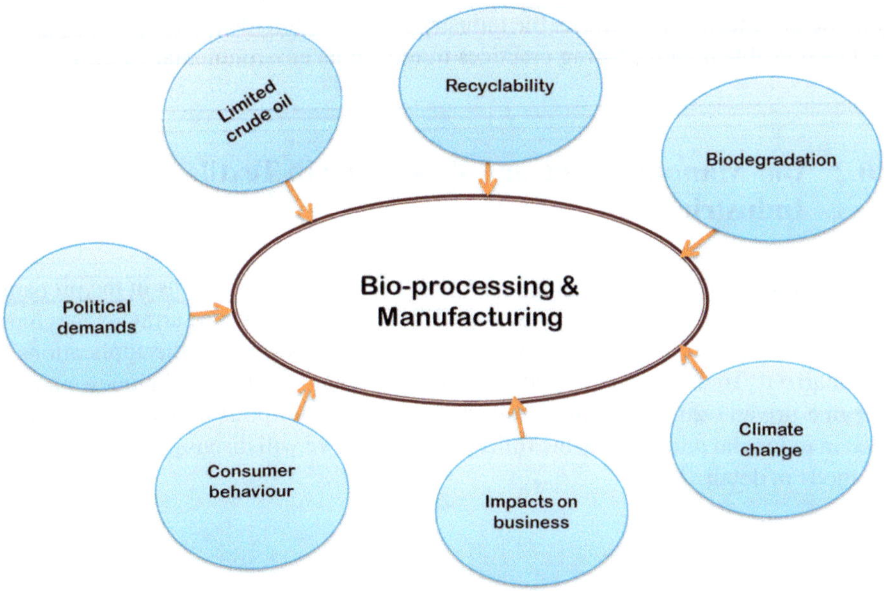

Fig. 10.2 Factors affecting the creation of a bio-based processing and manufacturing technique

textile goods and products. Nonetheless, this can further generate immense opportunities for innovation, which may have a huge impact on business through creating differentiate brands in the market, appealing to eco-conscious consumers. This will not only bolster brand reputation but also create new avenues for revenue generation.

Meanwhile, the market for biomanufacturing and processing of textiles has been experiencing significant growth in the recent time, driven by increasing consumer demand for sustainable and eco-friendly products. A recent statistic indicates that the global sustainable textile market is projected to reach over $100 billion by 2025, with a compound annual growth rate (CAGR) of approximately 9% from 2020 to 2025 [2]. Simultaneously, current trends show a rising preference for innovations in biodegradable textiles and eco-friendly processing of textiles. Additionally, the circular economy model is gaining traction, with brands increasingly adopting recycling and upcycling initiatives to reduce waste [3]. This shift reflects a broader commitment to sustainability within the textile industry, influencing both consumer choices and corporate strategies.

Herein, this chapter explores the innovative practices and technologies in biomanufacturing and processing textiles that aim to enhance sustainability within the textile industry. The chapter also discusses various sustainable techniques, such as the use of bio-derived compounds like enzymes, biomolecules, biopigments, eco-friendly dyes, and waste reduction strategies. By focusing on these advancements, the chapter aims to inform all stakeholders of a textile supply chain about the potential for creating a more sustainable textile industry that can minimize ecological footprints while meeting demands of smart consumers. The overall aim of this chapter is to provide insights into how the textile industry can transition towards more sustainable and responsible manufacturing practices to reduce its environmental footprint.

10.2 Bio-Manufacturing and Processing in Textile Industries

Biomanufacturing and processing in textiles utilize living organisms in the preparation and production of textile goods, primarily leveraging a diverse array of enzymes. In Table 10.1, the role of some widely used enzymes in various textile applications is summarized. To date, several bioprocessing and manufacturing techniques, namely enzyme-driven sizing, desizing, scouring, bleaching, bio-polishing, etc., were realized in industrial practices. In the following section, we will discuss such application methods in detail.

Table 10.1 List of enzymes used in different textile preparation processes

No	Name of enzyme	Working parameters	Specific applications	References
01	α-amylase	• Temperature: not mentioned • pH: 6.0	• Used for removing sizing material from cotton surfaces	[4]
02	Endoglucanase (EGPh) from *Pyrococcus horikoshii*, a hyperthermostable enzyme	• Temperature: >97 °C, • pH: Optimum: 5.4–6.0	• Suitable for industrial cellulose hydrolysis • Enhances softness and appearance of cotton through biopolishing • Can be used in high-temperature desizing processes (>70 °C) for combining desizing and biopolishing in one step	[5]

(continued)

Table 10.1 (continued)

No	Name of enzyme	Working parameters	Specific applications	References
03	• Cellulase • Pectinase • Lipase • Glucose oxidase	Cellulase, Pectinase, Lipase (for Bioscouring): • Temperature: 55 °C • Time: 40 min • Cellulase: 1.79 U/g cotton • Pectinase: 0.118 U/g cotton • Lipase: 0.434 U/g cotton • pH: 6.5 (acetate buffer) • Other conditions: bath ratio of 1:8, with surface active agent and sequestrant added Glucose oxidase (for hydrogen peroxide generation in bleaching): • Temperature: 35 °C • pH: 5.5	Cellulase, Pectinase, and Lipase: • Effectively remove non-cellulosic impurities from knitted cotton fabrics Glucose oxidase: • Generates hydrogen peroxide, acting as a bleaching agent • Eliminates natural pigments, enhancing fabric whiteness	[6]
04	Laccase	• Concentration: 10 g/L of protein (from Trametes villosa) • Temperature: 60 °C • pH: 5 (acetate buffer)	• Used as a pre-treatment for cotton fabrics to enhance conventional peroxide bleaching • Improved fabric whiteness comparable to two consecutive peroxide bleaches • Reduced hydrogen peroxide usage by up to 50%	[7]
05	Laccase	• Concentration: 0.125 g/L • Temperature: 50 °C • pH: 5 (acetate buffer)	• Used as a pre-treatment to enhance bleaching effect on cotton fabrics • Increased fabric whiteness by 1.67 Berger units compared to conventional peroxide bleaching • Significantly reduced hydrogen peroxide usage, lowering both chemical and energy consumption	[8]

(continued)

Table 10.1 (continued)

No	Name of enzyme	Working parameters	Specific applications	References
06	Laccase	• Concentration: 2% owf (on weight of fabric) • Temperature: 60 °C (for laccase pre-treatment) • pH: 5 (acetate buffer)	• Used as a pre-treatment to enhance bleaching of linen fabrics • Combination with hydrogen peroxide enables effective oxidative bleaching	[9]
07	Glucose oxidase	• Concentration: 10 g/L • pH: 5.1 • Temperature: 35 °C • Reaction time: 60 min for glucose oxidation	• Generates hydrogen peroxide, which is activated with TAED to form peracetic acid (PAA) • Enables effective low-temperature (55–60 °C) enzymatic bleaching of cotton fabrics	[10]
08	Alkaline pectinase	• Concentration: 2 g/L • Temperature: 55 °C • pH: 8.5 • Time: 60 min	• Used to remove impurities such as pectin, waxes, and non-cellulosic materials from cotton fabrics	[11]
09	Amyloglucosidase	• Temperature: 55 °C (for enzymatic desizing) • pH: 5.0 • Reaction time: 60 min	• Primarily used for desizing cotton fabrics	[12]
10	Acid cellulase	• Enzyme dosage: 20 g/L • Temperature: 50 °C for batching • pH: 5 (used for cellulase treatment)	• Used for biofinishing and biopolishing cellulose-based fabrics, particularly cotton	[13]
11	Cellulase	• pH: 4.8 • Temperature: 55 °C • Time: 1 h	• Used for biopolishing cotton fabrics by removing protruding fibres	[14]
12	Cellulase	• Conc.: 0.5, 1.25 and 2% • Time: 30, 45 and 60 min for acid enzyme and 60, 75 and 90 min for neutral cellulase	• Used for creating a fading look to the denim garment also helps to attain a softer feel through removing hairiness	[15]

10.2.1 Sizing with Enzymes

Sizing is an important process applicable to warp yarns in imparting some novel properties like higher strength, smoothness, abrasion resistance, etc., facilitating weaving. However, traditional techniques of the sizing process raise sustainability concerns due to their high water and energy consumption, wastewater generation, and the potential toxicity and non-biodegradability of sizing chemicals. Conventional sizing agents, such as starch and its derivatives, poly(vinyl alcohol) (PVA), carboxymethylcellulose (CMC), acrylic, and wax, offer varying performance and environmental impacts. Among them, starch is found to be less harmful to the environment, though its sizing performance is poor due to its limited durability connected to its readily water soluble behavior, lack of strong film-forming behavior, temperature sensitivity, etc. Meanwhile, starch derivatives and polyacrylates provide better sizing performance but have higher costs and unreliable environmental effects. PVA, on the other hand, demonstrates excellent sizing performance but poses a high risk of water pollution [16].

Thus, to overcome the aforementioned drawbacks of conventional sizing techniques and to promote sustainability, the textile industry is adopting eco-friendly sizing and desizing practices. Key approaches include innovative sizing technologies, optimized green recipes, reuse of chemicals, clean/dry technologies, and the use of bio-derived sizing agents. For example, sizing agents, namely wheat gluten, corn gluten, dried grains, soy protein, keratin, collagen, casein, chitosan, and alginate, derived from protein and polysaccharide origins, have emerged as promising alternatives. These bio-based sizing agents offer improved adhesion, film-forming ability, abrasion resistance, and biodegradability over traditional sizing agents [17–19]. Reddy et al. [20] found that corn distillers dried grains (DDGS) exhibited superior abrasion resistance compared to polyvinyl alcohol (PVA) when used as a sizing agent on cotton, polyester, and polyester/cotton blend fabrics and claimed a promising alternative to conventional sizing agents. In another work, Chen et al. [21] found that soy protein-based sizing agents outperformed conventional polyvinyl alcohol (PVA) sizing in terms of enhancing yarn strength and abrasion resistance. In a recent work, Rafikov et al. [22] developed a collagen-based sizing system that enriched the microstructure and morphology of cellulose fibres, leading to reduced sizing agent consumption. Obtained results indicated an increase in yarn breaking load by 25–35% and a reduction in yarn breaking elongation by 15–20%.

10.2.2 Desizing with Enzymes

Desizing is a textile processing technique carried out to remove sizing materials applied on warp yarns during the weaving process. These size materials need to be removed during wet processing steps, such as dyeing and finishing. Enzymatic desizing utilizes enzymes, typically amylases or cellulases, to break down and degrade the sizing agents. Of them, amylase enzyme is well-recognized as a desizing agent in the textile industry, particularly within enzymatic desizing processes or biodesizing [23, 24]. Amylase enzyme is classified based on their biological sources, namely β-amylases (saccharogenic) and α-amylases (dextrinogenic), with the latter acting more rapidly. Alpha-amylases exhibit catalytic activity at various sites on starchy substrates, making them particularly suitable for textile desizing [25–27].

These amylase enzymes promote the hydrolytic breakdown of water-insoluble starch into water-soluble products, allowing for their easy removal from cotton in the following steps [4, 28].

The enzymatic desizing usually takes place at a temperature of 70 °C and higher, with even greater temperatures often being preferable [28]. Standard amylases have a temperature range of 25–55 °C, while medium-range amylases work at 50–95 °C, and some selected amylases can perform at temperatures exceeding 70 °C [5]. Meanwhile, thermally-stable amylases exhibit efficacy above 95 °C, ensuring continuous pre-treatment during desizing [24]. Additionally, amylases can be modified to enhance their performance. High-temperature amylases offer distinct advantages over their low-temperature counterparts for textile desizing. Desizing with α-amylases that operate at lower temperatures (20–60 °C) can be a slow and time-consuming process [28]. In contrast, amylases that function optimally at higher temperatures (95–100 °C or higher) can significantly expedite the desizing operation. These thermostable amylases need to be deployed in a steam chamber to enable a fully continuous desizing process [5]. This continuous processing approach facilitated by high-temperature amylases provides efficiency improvements over the slower, more intermittent low-temperature desizing methods. The enhanced catalytic activity and thermal stability of the high-temperature amylases make them a more suitable choice for optimizing industrial-scale textile desizing operations.

The enzymatic desizing process yields numerous simple sugar compounds as an end product during the hydrolysis process, and that happens in three main stages like impregnation, incubation, and after-wash. Initially, during pre-wash non-starch, water-soluble additives are removed from the fibre surface and thus facilitate interaction between amylase enzyme and starch materials. The impregnation stage is carried out at a temperature above 70 °C, and the textile substrate will be incubated for a longer period, allowing the fabric to soak in the amylase solution. The incubation period typically lasts from 2 to 16 h, depending on factors such as stability of the enzyme, reaction activity with temperature and pH, as well as the type of the size material and textile substrate. Finally, textile substrates undergo frequent washing in alkaline liquor at temperatures above 80 °C, followed by a neutral wash.

Meanwhile, Choudhury [24] proposed an additional incorporation of enzymes such as lipases along with amylases to enhance the desizing process through synergistic effects. These lipases are particularly effective at breaking down hydrophobic components like fat-based lubricants available in size bath, and ultimately, these lubricants can be removed during the sizing process. Otherwise, they can be challenging to eliminate during the scouring process. This improved desizing method is beneficial as it can reduce processing time and uniformly yield high-value products [25]. So, this bio-desizing process offers an effective method for eliminating all undesirable sizing materials from textile substrates, potentially enhancing desizing performance compared to conventional washing or alkaline treatments.

10.2.3 Bio-Scouring

Bio-scouring typically refers to the scouring process that utilizes enzymes, emerging as a viable alternative to traditional chemical-based methods and is considered a more environmentally friendly option. This is due to the benefits of enzymatic bio-scouring, which include a broader pH range, reduced operating temperatures, and the elimination of harsh chemicals typically used in textile processing [29]. For instance, the bio-scouring of cotton normally employs a combination of cellulase and pectinase enzymes. Pectinase acts to disrupt the cotton cuticle structure by breaking down pectin, the material that binds the cuticle to the main body of the cotton fibre. Meanwhile, cellulase enzymes help remove the cuticle by digesting the cellulose found in the primary wall immediately beneath it. In comparison to traditional alkaline scouring, bio-scouring presents several benefits. It results in lower Biological Oxygen Demand (BOD), Chemical Oxygen Demand (COD), and Total Dissolved Solids (TDS) in wastewater, along with creating a milder alkaline environment. Furthermore, bio-scouring minimizes cotton weight loss, making it advantageous for knitting applications. The enzymatic approach is also selective, focusing on pectin and waxes rather than cellulose, which helps to reduce damage to the cotton fibres. Additionally, bio-scouring gives the fabric a softer feel, contrasting with the harsher hand associated with alkaline scouring. Overall, enzymatic bioscouring is an eco-friendly alternative that reduces health risks for operators by eliminating exposure to harsh chemicals [30].

For example, Pušić et al. [31] experimented with the enzymatic bioscouring of cotton textile in the presence of acid and neutral pectinases while they also compared the outcome with the traditional alkali scouring process. From the experimental results, it was found that the use of acid and neutral pectinases for bioscouring was much more effective than traditional alkali scouring. They also claimed that the use of neutral pectinase was advantageous, as it eliminates the need for neutralization of the wastewater. Meanwhile, in an earlier application [32], researchers included a biodegradable macromolecule, namely β-cyclodextrin in a co-application with pectinase to get a better scouring effect for a cotton textile substrate. Herein, they claimed that the β-cyclodextrin acted as a scouring agent in a way to remove hydrophobic waxes present in the cuticle of cotton to further facilitate the action of pectinase enzyme, where they found that the β-CD treatment followed by pectinase would offer a much better scouring effect.

Additionally, bio-scouring offers significant benefits for cotton blend textiles with silk, wool, and cashmere, as a severe alkaline condition commonly employed in traditional scouring processes can adversely affect the quality of these delicate fibres. Traditional alkaline scouring often leads to damage, including weakening of the fibres and alterations in their texture and appearance, which diminishes the overall quality of the blended fabric. In contrast, bio-scouring utilizes enzymatic methods that operate under milder conditions, reducing the risk of harm to these sensitive fibres while effectively removing impurities. This gentler approach preserves the integrity and luxurious characteristics of silk, wool, and cashmere, ensuring that the final product maintains its desired quality and performance. Thus, bio-scouring stands out as an ideal choice for those working with blends that includes these exquisite fibres, providing an effective and quality-preserving alternative to traditional methods [25].

10.2.4 Bio-Bleaching

Bio-bleaching in textiles is a cutting-edge process that employs biological agents, primarily enzymes, to eliminate color from fabrics, especially cotton. This technique presents a sustainable alternative to conventional bleaching methods, which often rely on harsh chemicals that can be detrimental to both the environment and the fibres. Specifically, bio-bleaching commonly utilizes enzymes like laccases and peroxidases, which effectively break down colorants in the fabric while preserving its structural integrity. These enzymes function under milder conditions, enhancing the energy efficiency of the process. By utilizing enzymes, bio-bleaching also reduces the production of toxic waste and minimizes water pollution compared to traditional bleaching practices. This eco-friendly approach resonates with the increasing demand for sustainable practices within the textile industry. Moreover, unlike traditional bleaching agents that can weaken fibres, bio-bleaching helps maintain the strength and quality of the fabric, resulting in a softer feel and improved overall performance [33].

In a recent study, Reis et al. [6] investigated the bleaching of cotton knit fabric using hydrogen peroxide that was enzymatically produced by glucose oxidase (EC 1.1.3.4) during the oxidation of glucose. The whiteness index of the fabric treated with this enzymatic bleaching method was measured at 52 ± 1 Berger degree, exhibiting better hydrophilic behavior, with a capillarity rate of 5.2 cm in 10 min. The research findings suggest that this enzymatic process could serve as a viable alternative for the textile preparation industry. Notably, it significantly reduces water consumption, with usage being four times lower compared to conventional bleaching methods. Before this, numerous research teams explored the potential of bio-bleaching, focusing particularly on the use of enzymes through various innovative methods. For instance, Tzanov et al. [7] revealed that employing a combination of the laccase-hydrogen peroxide method along with ultrasound energy significantly enhances the bleaching of cotton. This process was carried out under mild conditions, specifically at a pH of 5 and a temperature of 60 °C for 30 min. The results indicated not only improved bleaching efficiency but also an enhancement in the overall quality of the fabric, successfully achieving the desired level of whiteness. The improved effectiveness of bleaching when ultrasound energy is applied can be largely attributed to the enhanced activity of the enzyme as well as the improved diffusion of the enzyme to the surface of the substrate. The ultrasonic waves create microscopic cavitation bubbles, which can increase the local concentration of the enzymes and facilitate better interaction with the cotton fibres [8, 9]. Moreover, research has demonstrated that the enzyme glucose oxidase has significant potential for generating hydrogen peroxide from glucose during the bleaching process for cotton. This enzymatic generation of hydrogen peroxide offers a more environmentally friendly approach to bleaching, eliminating the need for externally supplied hydrogen peroxide while simultaneously enhancing the efficiency of the bleaching process. This innovative combination of enzymatic action and synergy with ultrasound energy marks a promising advancement in the textile bleaching industry, highlighting the shift toward more sustainable and effective practices.

However, it is possible to conduct pre-treatment processes such as desizing, scouring, and bleaching simultaneously, which can streamline production and enhance efficiency in textile processing. For instance, Mojsov [10] demonstrated this approach by using sized cotton fabric to undergo enzymatic treatments that combined desizing, scouring, and bleaching into a single process. In this study, a specialized enzyme mixture was employed, consisting of starch-degrading enzymes, pectinases, and glucose oxidases (GOxs). The inclusion of GOxs was particularly important as it facilitated the generation of hydrogen peroxide (HP) during the process.

In the past decade, various researchers [11, 12, 28] have explored the benefits of performing desizing and bleaching processes in a single bath or applying bleaching in conjunction with a reused desizing bath. Their findings indicate that these combined methods lead to a significantly greater degree of whiteness in the fabric, as well as enhancements in mechanical properties such as tensile strength and tear strength. This integrated approach not only improves the aesthetic quality of the textiles but also contributes to their durability.

It has been established that the enzyme glucose oxidase can generate hydrogen peroxide under a milder acidic to neutral condition at a moderate temperature range. This feature offers a promising avenue for achieving bleaching with minimal environmental impact. However, traditional bleaching methods often require higher temperatures, typically ranging from 80 to 90 °C, along with an alkaline pH of around 11.0 to effectively reach the desired bleaching outcomes [34, 35]. The challenge lies in reconciling the optimal conditions for enzyme activity with the requirements for effective bleaching, prompting further research into enzyme combinations and process optimizations that can facilitate efficient bleaching while maintaining the integrity of the fibres. By addressing these challenges, researchers aim to advance the development of eco-friendly bleaching techniques that contribute to the sustainable processing of textiles.

10.2.5 Bio-Dyeing

The conventional textile dyeing process, which relies on synthetic organic dyes, poses significant environmental risks, requiring effective wastewater treatment to reduce its impact. In light of these issues, the textile industry has developed a range of eco-friendly dyeing methods as alternatives to traditional practices, focusing on more sustainable and environmentally responsible approaches [36]. In this transformative journey of textile dyeing, the bio-dyeing process has progressed by incorporating innovative strategies, including the use of bio-pigments, biodegradable dyes, and both natural and fermentation-based dyeing techniques [37]. For example, microbial pigments have emerged as a demandable choice among biogenic dyes, largely due to their wide color range, excellent efficiency, and rapid production, and thus become suitable for industrial applications [38]. Several studies have already revealed the efficacy of biopigments as a textile dye, which are available from microorganisms like bacteria [39–41], fungi [42], yeasts [43], and microalgae [44]. These bio-based dyes play a crucial role in decreasing water pollution and lessening the reliance on harmful chemicals typically associated with conventional dyeing methods. By leveraging the natural capabilities of these microorganisms, the textile industry can move towards more sustainable practices that not only meet the demand for color but also contribute positively to environmental preservation. The biotechnological production of Indigo dye is another example that exemplifies a significant shift towards sustainability in the textile industry. Historically, Indigo Blue derived from plant sources has been utilized for dyeing purposes since around 4000 BCE in various parts of the world, including India, Egypt, Mesopotamia, and China. However, the arrival of the petrochemical industry marked a turning point, as synthetic Indigo began to be manufactured from petroleum-based aniline. This process, while economically advantageous, often involves environmentally detrimental methods and hazardous chemicals, leading to significant ecological concerns. In recent years, groundbreaking research has illuminated the possibility of producing bio-indigo through innovative biotechnological approaches involving engineered microorganisms. This new method presents a more sustainable alternative to traditional Indigo production, mitigating the environmental impact associated with synthetic processes. Specifically, this innovative technique utilizes recombinant Escherichia coli cells that are genetically modified to contain a unique gene encoding flavin-containing monooxygenase (FMO). This enzyme catalyzes the conversion of tryptophan into indole derivatives, which are then further processed to produce bio-indigo. The potential for this sustainable method to replace conventional Indigo dye synthesis offers exciting prospects for the textile industry while preserving the integrity of the environment [45].

Recently, the Colorifix® process has appeared as a groundbreaking choice for textile dyeing, representing an innovative dyeing method [46]. This sustainable approach uses microbial biopigments that are cultivated on-site, using renewable feedstock as a growth medium. Through a fermentation process, engineered microorganisms generate the desired colorants, and the biopigment-rich recombinant cells can be directly introduced into conventional dyeing equipment. This integration not only eliminates the necessity for additional chemicals and specialized machinery but also streamlines the dyeing operation and significantly reduces the environmental impact compared to traditional textile coloration techniques. Meanwhile, a recent study investigated the potential of beetroot as a natural dye source for nylon 6 fabric [47]. The research examined the effectiveness of four natural mordants- pomegranate peel, walnut husk, pistachio hull, and Terminalia Chebula in enhancing color fastness properties, comparing them with traditional metal mordants. The findings revealed that pre-mordanting with these natural mordants resulted in fabrics exhibiting a range of color strengths, with walnut husk displaying the highest intensity. All bio-mordants demonstrated excellent washing fastness, while Terminalia Chebula notably stood out for its exceptional light fastness. Furthermore, the washing fastness of the bio-mordants was found to be comparable to, and in some instances superior to, that of metal mordants.

Natural colorants can be derived not only from plant sources but also from animals, insects, algae, fungi, and minerals, providing vibrant hues for textile materials [48, 49]. For dyeing processes to be classified as true bio-dyeing, it is essential to use a natural or bio-based mordant in conjunction with the natural colorants. The inclusion of a natural or bio-based mordant is critical, as it facilitates the effective fixation and application of the natural colorants to textile materials.

Bio-mordants present eco-friendly, abundant, and non-toxic alternatives to traditional metallic mordants, often sourced from high-tannin plants that can serve as both dyes and mordants due to their tannin content and chemical properties [50, 51]. Common bio-mordants like pomegranate, turmeric, henna, acacia, gallnut, neem, etc., have been applied across a variety of textile substrates [52, 53]. For instance, Shahmoradi et al. [54] examined the natural dyeing of cotton fabric with Hibiscus sabdariffa, utilizing various bio-mordants such as tannic acid, pine cone, lemon peel, and sodium alginate. The resulting bio-mordanted fabrics displayed exceptional colorfast properties, including light fastness (grades 5–6), wash fastness (grades 4–5), and dry/wet rub fastness (grades 4–5), outperforming those dyed without mordants. Another study by Zhang et al. [53] used Eriobotrya japonica fruit seeds to dye wool fabric, employing Chinese tallow, Folium artemisiae argyi, and Cinnamomum camphora as bio-mordants, with Chinese tallow achieving the highest color strength at 9.73. The dyed fabrics showed wash fastness (grades 4–5), rubbing fastness (grades 4–5), and light fastness (grades 5–6). Additionally, Amin et al. [55] utilized cochineal insects as a natural dye and applied henna leaves, acacia bark, and pomegranate peel as bio-mordants for silk fabric. Pomegranate peels produced the highest color strength (K/S) values, reaching 9.2 at a 3% concentration during pre-mordanting and 10.75 at a 1% concentration for post-mordanting. Rani et al. [56] explored the dyeing of various protein fabrics, including wool, silk, soya, and milk, using papaya leaves as a natural dye and orange peel, pomegranate peel, amla, and harda as bio-mordants.

The application of bio-dyeing with natural colorants and bio-mordants has also been expanded to various synthetic textile substrates. For instance, Rehman et al. [57] utilized natural dyes derived from saffron and safflower to dye nylon fabric. They incorporated turmeric and pomegranate as bio-mordants, employing both pre-mordanting and post-mordanting methods. The optimal results were found using a 7% pomegranate extract in the post-mordanting approach and a 7% turmeric extract in the pre-mordanting method, both of which demonstrated increased color strength (K/S) values.

Research has shown that these bio-pigments and natural dyes offer a variety of functional benefits, including UV protection, antimicrobial activity, insect repellency, antioxidant properties, deodorizing capabilities, and anti-fungal effects, making them highly advantageous for textile applications [48, 53, 58–64]. For example, natural colorants, such as those derived from plants like Rheum emodi, Gardenia jasminoides, Lonicera periclymenum, Eucalyptus globulus, Citrus sinensis, and Pterocarya fraxinifolia, have been identified as possessing beneficial attributes for skin protection. These natural dyes are rich in bioactive compounds, including flavonoids, curcumin, and anthraquinone-based structures, which are capable of absorbing harmful ultraviolet rays. Numerous studies have documented the efficacy of these natural dyes in promoting skin health, indicating their potential as a functional enhancement in textiles used for clothing, health, and wellness applications [65–70]. Ultimately, the inherent functionalities of these natural, biologically-derived colorants unlock opportunities for developing innovative textile materials that claim improved performance characteristics, specifically designed to cater to the diverse needs of various industries and consumer markets.

10.2.6 Bio-Finishing

10.2.6.1 Bio-Polishing

Biopolishing is an advanced finishing technique in the textile industry that significantly elevates the quality of fabric, particularly in cellulose-based knitted materials. By employing enzymatic treatments, this process effectively minimizes the tendency of fabrics to pill and become fuzzy, which can detract from their overall aesthetic and tactile appeal. During biopolishing, specific enzymes are utilized to target and break down protruding fibres and slubs that are often present on the surface of knitted fabrics. This enzymatic action not only removes these imperfections but also has a profound impact on the fabric's texture. As a result, the finished textile exhibits a markedly reduced pilling propensity, leading to a smoother, softer hand feel that is more pleasant for the wearer. The permanent effects of biopolishing enhance the fabric's appearance, giving it a clean and polished look that is desirable in various applications, from casual apparel to high-end fashion [71].

Cellulase enzymes play a pivotal role in the biopolishing process of textiles, particularly due to their unique specificity for cellulose substrates and their ability to hydrolyze cellulosic micro-fibrils. This hydrolytic action is central to enhancing the fabric's quality by effectively targeting and breaking down the protruding fibres that contribute to pilling and fuzziness. One of the key advantages of cellulase enzymes is their optimal kinetics; they act slowly and steadily, allowing for precise control over the finishing process without adversely affecting the integrity or overall quality of the fabric. These enzymes are particularly suited for biopolishing applications in the textile industry because they strike a delicate balance between effectiveness and fabric preservation. Unlike harsher treatments, cellulase enzymes gently refine the surface of the fabric, leading to a smoother finish while maintaining its structural characteristics [13, 72–74].

Sources of cellulase enzymes are diverse, encompassing a range of organisms, including various microbes and fungi, as well as certain animals and plants. These enzymes can be obtained through two primary fermentation methods: immersed fermentation and solid-state fermentation. Immersed fermentation typically involves cultivating microorganisms in a liquid medium, allowing for efficient enzyme production. In contrast, solid-state fermentation leverages microorganisms growing on solid substrates, often using agricultural byproducts, which can be a more sustainable and eco-friendly approach [75]. In a recent study [14], a sustainable and cost-effective method for textile biopolishing by extracting cellulase enzymes from waste sugarcane bagasse was reported. The extraction was performed in an aqueous medium, and the biopolishing process was optimized by varying pH and time. The optimal conditions were determined to be 1 h at pH 4.8 and 55 °C using the solution fermented for three days, successfully removing nearly 100% of protruding fibres of cotton knit fabric. Throughout the process, fabric weight decreased by up to 5.26% and strength by up to 10.54%, confirming the effectiveness of the enzymatic treatment. A comparative analysis with commercial cellulase samples demonstrated similar biopolishing effectiveness, validating the viability of this approach.

10.2.6.2 Bio-Based Garment Washing

Bio-based washing of ready-made garments has significantly expanded the scope of denim finishing techniques. This innovative approach utilizes specific enzymes to enhance the quality and variety of finishes that can be applied to denim fabrics. This enzymatic action leads to a range of desirable effects, such as softening the fabric, enhancing color vibrancy, and creating unique texture variations. By leveraging these bio-based washing methods, manufacturers can achieve finishes with a minimal impact on physical properties of textiles that were previously difficult or impossible to attain through conventional chemical treatments. Moreover, this eco-friendly alternative not only improves the aesthetic appeal of denim garments but also aligns with the growing demand for sustainable practices in the fashion industry. The use of enzymes reduces the reliance on harsh chemicals, thereby minimizing environmental impact and promoting a more sustainable production process.

Traditionally, pumice-stone-aided finishing is carried out to remove surface dyes from a denim garment to give it a specific faded look through mechanical abrasion. This process often involves the use of pumice stones treated with sodium hypochlorite or potassium permanganate. While stone washing with pumice stones can create the desired aesthetic for denim fabric, it comes with several drawbacks. These include issues such as back-staining, potential damage to garments along hems and seams, significant environmental impacts, and the requirement for large quantities of pumice stones [76, 77]. To overcome these above-mentioned drawbacks of stonewashing by pumice stones, the enzyme is used as an alternative method for achieving a stone-washed effect, beginning in the 1980s [78, 79]. Some of the common enzymes used in denim finishing are amylase, cellulase, and laccase, and among them, cellulase is extensively used for giving a worn or faded look to the denim garment, which is usually done by pumice stone in a traditional wash process. Meanwhile, cellulase enzyme-driven treatment helps to eliminate hairiness from the denim surface, thus offering a smoother and softer feel [15]. Meanwhile, amylase enzymes are used in removing warp sizes from the denim garment. Typically, low-temperature amylase enzymes operated at 30–50 °C and pH 5.5–7.5 are considered in denim garment desizing. Laccase enzyme is particularly considered in denim bleaching, specifically decoloration of indigo vat dye. This laccase enzyme performs an oxidation reaction to split indigo dye chromophore into colorless isatin molecules [80].

10.2.6.3 Bio-Based Functional Finishing

The use of bio-based and biodegradable materials in crafting functional finishes onto the textile surfaces has also been garnered extensively in recent times. Bio-based finishes are increasingly recognized for their potential to enhance performance while also being environmentally friendly and safer for human health. These cutting-edge materials support global sustainability initiatives by leveraging renewable resources. The common sources of biomolecules encompass a diverse range of natural entities, including plants, microorganisms, and animals, which present a compelling alternative to traditional synthetic materials. Nanocellulose, lignin, chitosan, alginate, gelatin, collagen, keratin, and polylactic acid (PLA) are some of the notable examples. Each of them possesses distinct functional properties that can be tailored to various applications. They offer benefits such as anti-microbial, flame-retardant, UV protective, and antioxidant capabilities, making them highly valuable across multiple industries. Moreover, such bio-derived compounds are finding critical applications in advanced medical fields such as tissue engineering, drug delivery systems, and wound healing solutions. By substituting synthetic polymers with biopolymers, industries can make significant strides toward sustainability while capitalizing on their functional advantages [81].

For example, a type of functional finishing was applied to the cotton textile substrate using chitosan, a natural polymer, and onion-peel extract. Obtained results demonstrated that the chitosan-onion composite imparted robust antioxidant activity and effective antibacterial performance against E. coli. The synergistic interaction between chitosan and the active phytochemicals in the onion extract was identified as the key driver for the composite's ability to inhibit bacterial growth [82]. In a different application, Kundu et al. [83] used all green polyelectrolytes, such as chitosan (CS), phytic acid (PA), and oxidized sodium alginate (OSA), to improve the flame retardant properties of the PA66 textiles. Chitosan, derived from natural sources such as crustacean shells, is known for its excellent biocompatibility and biodegradability, while phytic acid acts as a natural flame retardant due to its phosphorus content. Oxidized sodium alginate, a derivative of natural alginate, contributes additional functional properties, including forming a protective char layer upon exposure to heat. Such finishes not only gain improved flame resistance but also adhere to environmental sustainability goals by leveraging bio-derived materials. This method showcases the potential of combining green chemistry with textile technology, paving the way for the development of high-performance, eco-friendly fabrics suitable for a variety of applications where safety and environmental impact are of paramount importance.

In particular, chitosan biopolymer has exhibited immense potential in imparting assorted functionalities across the types of textile substrates. For example, chitosan-sericin composite finishing can effectively enhance the antioxidant properties of cotton fabrics, allowing them to be considered for medical uses [84]. It has been experimented to cater some other properties like wrinkle resistance [85, 86], water and oil repellency [87], antistatic properties, etc., to textile substrates [88], contributing to broaden its application areas and market demand. In a recent application, chitosan is combined with silica moieties (i.e., SiO_2) originating from rice husk waste to impart hydrophobic properties to textiles [89]. Studies have revealed that chitosan-based functional finishing can provide long-lasting antibacterial effects, offer durable functionalities, thus supporting sustainable consumption practices within the textile supply chain by reducing waste and resource depletion [90–92].

10.2.7 Textile Effluent Treatment by Enzymes

The wastewater or effluents released from textile processing industries are one of the most significant pollutants due to their substantial volume and toxicity [93]. This textile wastewater typically contains 10 to 200 mg/L of dyes, along with various organic chemicals, inorganic compounds, and additives. Alarmingly, even after treatment, around 90% of these dyes are still released into water bodies without any chemical alteration. Meanwhile, biodegradation of these dyes is difficult because of their xenobiotic properties, aromatic structures, high thermal stability, and photostability, which pose serious threats to aquatic life and to the environment [94, 95]. This dye-mediated water pollution from the textile industry has escalated in recent times due to the harmful effects of synthetic dyes and their byproducts, which can be mutagenic and carcinogenic. This underscores the pressing need for eco-friendly and economical methods to treat wastewater, utilizing physical, chemical, or biological approaches. Herein, bioremediation offers a viable solution for addressing textile dye pollution occurring in water bodies and soil [96]. Bioremediation utilizes living organisms, such as bacteria, fungi, algae, and plants, to transform textile dyes into non-toxic end products [97]. This process relies on enzymes, which are proteins that act as a biocatalyst to accelerate the degradation of dye pollutants. These enzymes are encoded by genes and facilitate reactions, including the degradation of textile dyes through either degradation or biotransformation mechanisms [98]. Enzymes offer advantages over traditional physicochemical methods, as they typically do not produce harmful by-products and can be easily digested by native microorganisms post-treatment, making enzymatic wastewater treatment an ecologically viable option. Apart from environmental benefits, enzymes show immense potential in several ways; like they can be operated in milder conditions and a lower temperature range along with lower chemical inputs than conventional methods, which can lessen energy demand and conserve resources. Meanwhile, enzymes can be selectively used and designed to target specific compounds in textile wastewater, leading to more efficient and selective treatment. However, these enzymes usually exhibit numerous challenges from cost and scalability perspectives as the production and purification of enzymes is usually expensive and also challenging for the scale-up application in industries. In addition, careful selection and optimization is required for complex textile effluents containing diverse pollutants as enzymes are subjected to work on specific substrates. Evermore, operation parameters need to be properly controlled as enzymes are sensitive to several factors like pH, temperature, and the presence of inhibitors, which can limit their operational lifetime and performance. Key enzymes involved in the bioremediation of textile dyes include lignin peroxidase, laccase, azo reductase, manganese peroxidase, reductase, veratryl alcohol oxidase, tyrosinase, riboflavin reductase and others [96].

Apart from enzymes, biopolymers like chitosan have also shown potential in wastewater treatment applications and are regarded as a sustainable choice in wastewater management in textile manufacturing [99–101]. It functions as a natural adsorbent to remove toxic dyes, heavy metals, and organic pollutants, thereby refining effluent quality and reducing sludge generation [102, 103].

10.3 Bio-Manufacturing and Processing in a Commercial Scale

The bio-based processing of textiles has already been materialized on a commercial scale, where a varied class of enzymes is available in different trade names and ready to perform in industrial application methods. For example, IFF Textile solutions have introduced enzyme-based bio-processing agents suitable and effective for desizing, scouring, bleaching, color modification, etc., of textile goods in industrial applications. Optisize® and PrimaGreen® are two examples of α-amylase enzymes developed by IFF, which are capable of breaking down and solubilizing starch present in the sizing of woven fabrics. These α-amylase enzymes, Optisize® and PrimaGreen®, operate effectively across a broad range of pH and temperature conditions, making them suitable for the diverse desizing processes employed in the textile industry. Additionally, PrimaGreen® EcoScour from IFF is capable of bioscouring textiles at a low to medium temperature and a neutral pH, leading to savings in both water and energy usage. Furthermore, PrimaGreen® EcoScour enzyme enhances fabric quality by providing a softer feel, reducing fabric weight loss, and improving dyeability in comparison to traditional scouring agents. They have also launched PrimaGreen® EcoWhite enzyme for the bleaching treatment of textile goods, which operates effectively at medium temperatures and neutral pH, promoting sustainability in the bleaching process. PrimaGreen® EcoWhite has proven to be a successful and competitive alternative to traditional bleaching methods, considering both quality and environmental factors. IFF's IndiAge®, PrimaGreen, and Primafast enzyme solutions for stonewashing are highly effective in denim and garment finishing, providing desirable stonewashed appearances on denim [104].

10.4 Challenges and Future Direction

The biomanufacturing and processing of textiles aimed at sustainability encounter a myriad of challenges that need to be navigated to unlock their transformative potential in the industry. A primary challenge is the limited scalability of bio-based materials, which often struggle to match the production volume and cost-effectiveness of conventional textiles derived from synthetic fibres. This is compounded by the need for advanced biotechnological processes to enhance the efficiency and sustainability of production methods, such as fermentation or enzyme-based treatments, which are still in the nascent stages of development. Furthermore, there is a critical need for stringent quality control measures, as bio-based textiles can exhibit significant variability in performance attributes, such as durability, comfort, and aesthetic appeal, depending on their source and processing methods. The environmental impact of sourcing raw materials also raises questions, particularly concerning land use and resource consumption, necessitating a cautious approach that respects indigenous communities' rights, avoids exploitation, and promotes biodiversity conservation. Sourcing renewable and biodegradable raw materials at scale remains a significant hurdle, as securing a consistent supply and maintaining quality standards can be complex. Bio-manufacturing processes require specific equipment and technical know-how, which can result in higher costs compared to traditional technologies. Additionally, the shift towards sustainable textile production through bio-manufacturing and eco-friendly processing methods raises important social and ethical considerations. Ensuring fair labor practices and safe working conditions for textile workers becomes increasingly crucial as new technologies and biomaterials are adopted. The use of biotechnology also raises concerns about potential health risks and environmental impacts, necessitating transparent communication and rigorous safety assessments. Furthermore, bio-textile products must be accessible and affordable for diverse consumers, promoting social inclusion and equitable distribution of sustainable benefits. The use of hazardous chemicals in conventional textile processing also necessitates the development of eco-friendly alternatives and closed-loop systems to minimize environmental pollution and worker exposure. Addressing these chemical-related challenges will be crucial for creating truly sustainable textile products. Additionally, fostering consumer awareness and changing buying behavior towards sustainable textiles is paramount.

Future directions in this domain appear promising, focusing on innovation and collaboration. One potential avenue is the integration of synthetic biology and biotechnology in developing engineered fibres that exhibit enhanced properties, such as biodegradability, moisture retention, and antimicrobial features. Additionally, scaling up the production of bio-based dyes and finishing agents could dramatically reduce the harmful chemicals currently used in textile processing, resulting in a safer and more sustainable production cycle. Embracing circular economy principles is another pivotal direction, which entails designing textiles with their end-of-life in mind, promoting recycling, and fostering the use of post-consumer waste as raw materials for new textile products. This would not only minimize waste but also reduce dependence on virgin resources. Future studies should also include effective strategies for educating consumers, promoting transparency in supply chains, and incentivizing the adoption of bio-based and recyclable textile options. Ultimately, achieving these advancements will require robust partnerships between manufacturers, researchers, policy-makers, and consumers to create a cohesive framework that supports sustainable practices and drives the widespread adoption of bio-manufactured textiles in the global market.

By tackling these multifaceted challenges through continued research, technological advancements, and collaborative efforts across the textile value chain, the industry can accelerate the transition towards a more sustainable and responsible future for bio-manufacturing and textile processing.

10.5 Conclusions

The textile industry is undergoing a shift from a traditional to a more sustainable paradigm. Herein, biomanufacturing and processing can play a pivotal role in a collective transformation toward sustainability across the supply chain. This chapter highlighted advancements in bio-processing and innovative manufacturing techniques that can significantly reduce the industry's carbon footprint and reliance on hazardous chemicals. This change is relatively more prominent and visible in the preparatory processes of textile substrates, especially in desizing, scouring, and bleaching. In line, such practices are also gaining much attention in treatment processes like dyeing and finishing with time, even these practices are extended in the denim washing and finishing processes and in the treatment of textile wastewater. Enzymes derived from various sources are the main players in this field, while bio-derived compounds, especially biopolymers originating from living organisms also contribute effectively in achieving the much-needed functionalities. These approaches not only promote environmental preservation but also cater to a growing base of eco-conscious consumers. However, challenges are therein, particularly in technology adoption, investment, and reluctance to change and their industrial viability and commercialization. It has been realized that this journey toward sustainability is imperative for the industry's future. By embracing sustainable practices, the textile sector can lead efforts to combat climate change and protect biodiversity. Collaboration among manufacturers, consumers, policymakers, and researchers will be vital to implement these practices successfully. Prioritizing sustainability will enable the textile industry to address critical environmental issues and foster a healthier, more equitable future for all and will help to create a resilient industry capable of meeting modern consumer demands while enhancing brand reputation.

References

1. Kundu, C. K., and R. Rajesh. 2010. *Analysis of challenges in existing textile retail business for implementing sustainable resilient supply chain*. Master in Applied Textile Management, School of Textiles, University of Borås, Swedish School of Textiles. https://doi.org/10.13140/RG.2.2.18498.49607.
2. Textile Market Size, Share & Growth Analysis Report. 2030.
3. Looking back, looking ahead: Textile industry trends-Specialty Fabrics Review.
4. Opwis, K., T. Mayer-Gall, E. Schollmeyer, C. Dammer, T. Titscher, A. Nickisch-Hartfiel, O. Grün, C. Spurk, C. Schloderer, A. Köppe, and C. Dörfler. 2010. Generation of methane from textile desizing liquors. *Engineering in Life Sciences* 10: 293–296.
5. Ando, S., H. Ishida, Y. Kosugi, and K. Ishikawa. 2002. Hyperthermostable endoglucanase from pyrococcus horikoshii. *Applied and Environment Microbiology* 68: 430–433. https://doi.org/10.1128/AEM.68.1.430-433.2002.
6. Reis, C. Z., O. Fogolari, D. Oliveira, S. M. De Arruda Guelli Ulson De Souza, and A. A. U. De Souza. Bioscouring and bleaching of knitted cotton fabrics in one-step process using enzymatically generated hydrogen peroxide. *The Canadian Journal of Chemical Engineering* 95: 2048–55. https://doi.org/10.1002/cjce.22891.
7. Tzanov, T., C. Basto, G. M. Gübitz, and A. Cavaco-Paulo. 2003. Laccases to improve the whiteness in a conventional bleaching of cotton. *Macromolecular Materials and Engineering* 288: 807–810. https://doi.org/10.1002/mame.200300100.
8. Basto, C., T. Tzanov, and A. Cavaco-Paulo. 2007. Combined ultrasound-laccase assisted bleaching of cotton. *Ultrasonics Sonochemistry* 14: 350–354. https://doi.org/10.1016/j.ultsonch.2006.07.006.
9. Abou-Okeil, A., A. El-Shafie, and M. M. El Zawahry. 2010. Ecofriendly laccase–hydrogen peroxide/ultrasound-assisted bleaching of linen fabrics and its influence on dyeing efficiency. *Ultrasonics Sonochemistry* 17: 383–390. https://doi.org/10.1016/j.ultsonch.2009.08.007.
10. Mojsov, K. 2019. Enzymatic desizing, bioscouring and enzymatic bleaching of cotton fabric with glucose oxidase. *The Journal of The Textile Institute* 110: 1032–1041. https://doi.org/10.1080/00405000.2018.1535240.
11. Zhou, C., M. Li, R. Garcia, A. Crawford, K. Beck, D. Hinks, et al. 2012. Time-of-flight-secondary ion mass spectrometry method development for high-sensitivity analysis of acid dyes in nylon fibers. *Analytical Chemistry* 84: 10085–10090. https://doi.org/10.1021/ac3025569.
12. Hebeish, A., M. Hashem, N. Shaker, M. Ramadan, B. El-Sadek, and M. A. Hady. 2009. New development for combined bioscouring and bleaching of cotton-based fabrics. *Carbohydrate Polymers* 78: 961–972. https://doi.org/10.1016/j.carbpol.2009.07.019.
13. Das, S. C., M. A. Uddin. 2022. BUTEXDCE2022C01: a qualitative testing method for assessing enzymatic biopolishing effect on textile substrate. In *1st international conference on textile science & engineering BUTEX*, 77–85, Dhaka.
14. Campos, R., A. Kandelbauer, K. H. Robra, A. Cavaco-Paulo, and G. M. Gübitz. 2001. Indigo degradation with purified laccases from Trametes hirsuta and Sclerotium rolfsii. *Journal of Biotechnology* 89: 131–139. https://doi.org/10.1016/S0168-1656(01)00303-0.
15. Ul Islam, S., B. S. Butola, and M. Srivastava. 2023. Synthesis of novel bio-composite material for functional finishing of cellulosic textile substrate. *Journal of Natural Fibers* 20: 2226884. https://doi.org/10.1080/15440478.2023.2226884.
16. Var, C., and S. Palamutcu. 2024. Sustainable approaches in textile-sizing process. In *Sustainable manufacturing practices in the textiles and fashion sector*, ed. Muthu, S.S, 55–74, Springer Nature Switzerland, Cham. https://doi.org/10.1007/978-3-031-51362-6_3.
17. Xu, X., K. Song, B. Xing, W. Hu, Q. Ke, and Y. Zhao. 2019. Thermal-tenacity-enhanced and biodegradable textile sizes from cellulose nanocrystals reinforced soy protein for effective yarn coating. *Industrial Crops and Products* 140: 111701. https://doi.org/10.1016/j.indcrop.2019.111701.

18. Yang, M., H. Xu, X. Hou, J. Zhang, and Y. Yang. 2017. Biodegradable sizing agents from soy protein via controlled hydrolysis and dis-entanglement for remediation of textile effluents. *Journal of Environmental Management* 188: 26–31. https://doi.org/10.1016/j.jenvman.2016.11.066.
19. Zhao, Y., Y. Zhao, H. Xu, and Y. Yang. 2015. A sustainable slashing industry using biodegradable sizes from modified soy protein to replace petro-based poly(Vinyl Alcohol). *Environmental Science and Technology* 49: 2391–2397. https://doi.org/10.1021/es504988w.
20. Reddy, N., Y. Zhang, and Y. Yang. 2013. Corn distillers dried grains as sustainable and environmentally friendly warp sizing agents. *ACS Sustainable Chem Eng* 1: 1564–1571. https://doi.org/10.1021/sc4002017.
21. Chen, L., N. Reddy, and Y. Yang. 2013. Soy proteins as environmentally friendly sizing agents to replace poly(vinyl alcohol). *Environmental Science and Pollution Research* 20: 6085–6095. https://doi.org/10.1007/s11356-013-1601-5.
22. Rafikov, A. S., M. S. Khakimova, D. A. Q. Fayzullayeva, and A. F. Reyimov. 2020. Microstructure, morphology and strength of cotton yarns sized by collagen solution. *Cellulose* 27: 10369–10384. https://doi.org/10.1007/s10570-020-03450-w.
23. Araújo, R., M. Casal, and A. Cavaco-Paulo. 2008. Application of enzymes for textile fibres processing. *Biocatalysis and Biotransformation* 26: 332–349. https://doi.org/10.1080/10242420802390457.
24. Muthu, S. S. (eds.). 2014. *Roadmap to sustainable textiles and clothing: eco-friendly raw materials, technologies, and processing methods*. Singapore: Springer Singapore. https://doi.org/10.1007/978-981-287-065-0.
25. Madhu, A., and J. N. Chakraborty. 2017. Developments in application of enzymes for textile processing. *Journal of Cleaner Production* 145: 114–133. https://doi.org/10.1016/j.jclepro.2017.01.013.
26. Hossain, S., and K. Uddin. 2011. Comparative analysis between conventional pretreatment and bio-preparation. *International Journal of Engineering and Technology* 11: 16–21.
27. Mojsov, K. 2011. Application of enzymes in the textile industry; a review. In *Proceedings of the II international congress 'engineering, ecology and materials in the processing industry'*, 9–11th March, 230–239, Jahorina, Bosnia and Herzegovina.
28. Shahid, M., F. Mohammad, G. Chen, R.-C. Tang, and T. Xing. 2016. Enzymatic processing of natural fibres: white biotechnology for sustainable development. *Green Chemistry* 18: 2256–2281. https://doi.org/10.1039/C6GC00201C.
29. Rahman, M., M. M. Billah, D. Hack-Polay, and A. Alam. 2020. The use of biotechnologies in textile processing and environmental sustainability: an emerging market context. *Technological Forecasting and Social Change* 159: 120204. https://doi.org/10.1016/j.techfore.2020.120204.
30. Mojsov, K. 2014. Trends in bio-processing of textiles: a review. *SavrTechnology* 3: 135–138. https://doi.org/10.5937/savteh1402135M.
31. Pušić, T., A. Tarbuk, and T. Dekanić. 2015. Bio-innovation in cotton fabric scouring-acid and neutral pectinases. *FIBRES & TEXTILES in Eastern Europe* 1(109): 98–103.
32. Wang, Q., X. Fan, J. Yuan, P. Wang, L. Cui, and J. Chen. 2008. Optimization of two-step cotton scouring with β-Cyclodextrin and alkaline pectinase. *Engineering in Life Sciences* 8: 339–343. https://doi.org/10.1002/elsc.200700043.
33. Biobased Solutions for the Textiles Industry-IFF Bioscience. https://bioscience.iff.com/solutions/textile-processing.
34. Anis, P., A. Davulcu, and H. A. Eren. 2009. Enzymatic pre-treatment of cotton Part 2: peroxide generation in desizing liquor and bleaching. *Fibres & Textiles in Eastern Europe* 17: 87–90.
35. Farooq, A., S. Ali, N. Abbas, G. A. Fatima, and M. A. Ashraf. 2013. Comparative performance evaluation of conventional bleaching and enzymatic bleaching with glucose oxidase on knitted cotton fabric. *Journal of Cleaner Production* 42: 167–171. https://doi.org/10.1016/j.jclepro.2012.10.021.
36. Ibrahim, N. A., K. EL-Badry, B. M. Eid, and T. M. Hassan. 2011. A new approach for biofinishing of cellulose-containing fabrics using acid cellulases. *Carbohydrate Polymers* 83:116–21. https://doi.org/10.1016/j.carbpol.2010.07.025.

37. Ikbal, Md. S., F. A. Tisha, A. I. Asheque, E. Hasnat, and M. A. Uddin. 2024. Eco-friendly biopolishing of cotton fabric through wasted sugarcane bagasse-derived enzymes. *Heliyon* 10: e26346. https://doi.org/10.1016/j.heliyon.2024.e26346.
38. Patra, A. K., A. Madhu, and N. Bala. 2018. Enzyme washing of indigo and sulphur dyed denim. *Fash Text* 5: 3. https://doi.org/10.1186/s40691-017-0126-9.
39. Kundu, C. K., Md. T. Hossen, and R. Saha. 2022. Coloration with nanoparticles: scope for developing simultaneous colouring and functional properties onto textile surfaces—a short review. *Coloration Technology* 138: 443–55. https://doi.org/10.1111/cote.12621.
40. Alves, M. B. N., A. M. S. Jorge, and J. F. B. Pereira. 2024. The biotechnology revolution in textile dyeing. *Trends in Biotechnology* 42: 1211–1214. https://doi.org/10.1016/j.tibtech.2024.02.010.
41. Fried, R., I. Oprea, K. Fleck, and F. Rudroff. 2022. Biogenic colourants in the textile industry–a promising and sustainable alternative to synthetic dyes. *Green Chemistry* 24: 13–35. https://doi.org/10.1039/D1GC02968A.
42. Morales-Oyervides, L., J. Oliveira, M. Sousa-Gallagher, A. Méndez-Zavala, and J. C. Montañez. 2017. Assessment of the dyeing properties of the pigments produced by Talaromyces spp. *JoF* 3: 38. https://doi.org/10.3390/jof3030038.
43. Mussagy, C. U., P. G. P. Silva, C. F. Amantino, J. F. M. Burkert, F. L. Primo, A. Pessoa, et al. 2022. Production of natural astaxanthin by Phaffia rhodozyma and its potential application in textile dyeing. *Biochemical Engineering Journal* 187: 108658. https://doi.org/10.1016/j.bej.2022.108658.
44. Azeem, M., N. Iqbal, R. A. Mir, S. Adeel, F. Batool, A. A. Khan, et al. 2019. Harnessing natural colorants from algal species for fabric dyeing: A sustainable eco-friendly approach for textile processing. *Journal of Applied Phycology* 31: 3941–3948. https://doi.org/10.1007/s10811-019-01848-z.
45. Hsu, T. M., D. H. Welner, Z. N. Russ, B. Cervantes, R. L. Prathuri, P. D. Adams, et al. 2018. Employing a biochemical protecting group for a sustainable indigo dyeing strategy. *Nature Chemical Biology* 14: 256–261. https://doi.org/10.1038/nchembio.2552.
46. Nugent, D. G. H., et al. 2023. Colorifix Ltd. Method of dyeing fabric using microorganisms, US11781265B2.
47. Shahmoradi Ghaheh, F., A. Haji, E. Daneshvar, and M. R. Repon. 2024. Bio dyeing of nylon 6 fabrics using beetroot and natural mordants. *ChemistrySelect* 9: e202402783. https://doi.org/10.1002/slct.202402783.
48. Shahid, M., Mohammad F. Shahid-ul-Islam. 2015. Recent advancements in natural dye applications: a review. *Journal of Cleaner Production* 53: 310–31. https://doi.org/10.1016/j.jclepro.2013.03.031.
49. Kasiri, M. B., and S. Safapour. 2014. Natural dyes and antimicrobials for green treatment of textiles. *Environmental Chemistry Letters* 12: 1–13. https://doi.org/10.1007/s10311-013-0426-2.
50. Erdem İşmal, Ö., L. Yıldırım, and E. Özdoğan. 2014. Use of almond shell extracts plus biomordants as effective textile dye. *Journal of Cleaner Production* 70: 61–67. https://doi.org/10.1016/j.jclepro.2014.01.055.
51. Doty, K., S. Haar, and J. Kim. 2016. Black walnut, Osage orange and eastern redcedar sawmill waste as natural dyes: effect of aluminum mordant on color parameters. *Fash Text* 3: 22. https://doi.org/10.1186/s40691-016-0074-9.
52. Habib, N., S. Adeel, F. Ali, N. Amin, and S. R. Khan. 2021. Environmental friendly sustainable application of plant-based mordants for cotton dyeing using Arjun bark-based natural colorant. *Environmental Science and Pollution Research* 28: 54041–54047. https://doi.org/10.1007/s11356-021-14536-8.
53. Zhang, Y., Q. Zhou, L. J. Rather, and Q. Li. 2021. Agricultural waste of Eriobotrya japonica L. (Loquat) seeds and flora leaves as source of natural dye and bio-mordant for coloration and bio-functional finishing of wool textile. *Industrial Crops and* Products 169: 113633. https://doi.org/10.1016/j.indcrop.2021.113633.

54. Shahmoradi Ghaheh, F., M. K. Moghaddam, and M. Tehrani. 2021. Comparison of the effect of metal mordants and bio-mordants on the colorimetric and antibacterial properties of natural dyes on cotton fabric. *Coloration Technology* 137: 689–698. https://doi.org/10.1111/cote.12569.
55. Amin, N., F. Rehman, S. Adeel, T. Ahamd, M. Muneer, and A. Haji. 2020. Sustainable application of cochineal-based anthraquinone dye for the coloration of bio-mordanted silk fabric. *Environmental Science and Pollution Research* 27: 6851–60. https://doi.org/10.1007/s11356-019-06868-3.
56. Rani, N., L. Jajpura, and B. S. Butola. 2020. Ecological dyeing of protein fabrics with Carica papaya L. Leaf natural extract in the presence of bio-mordants as an alternative copartner to metal mordants. *Journal of The Institution of Engineers (India): Series E* 101: 19–31. https://doi.org/10.1007/s40034-020-00158-1.
57. Rehman, F. U., S. Adeel, W. Haddar, R. Bibi, M. Azeem, R. Mia, et al. 2022. Microwave-assisted exploration of yellow natural dyes for nylon fabric. *Sustainability* 14: 5599. https://doi.org/10.3390/su14095599.
58. Yusuf, M., A. Ahmad, M. Shahid, M. I. Khan, S. A. Khan, N. Manzoor, et al. 2012. Assessment of colorimetric, antibacterial and antifungal properties of woollen yarn dyed with the extract of the leaves of henna (Lawsonia inermis). *Journal of Cleaner Production* 27: 42–50. https://doi.org/10.1016/j.jclepro.2012.01.005.
59. Baaka, N., M. Ben Ticha, W. Haddar, M. T. P. Amorim, and M. F. Mhenni. 2018. Upgrading of UV protection properties of several textile fabrics by their dyeing with grape pomace colorants. *Fibers Polym* 19: 307–312. https://doi.org/10.1007/s12221-018-7327-0.
60. Baseri, S. 2020. Eco-friendly production of anti-UV and antibacterial cotton fabrics via waste products. *Cellulose* 27: 10407–10423. https://doi.org/10.1007/s10570-020-03471-5.
61. Grifoni, D., G. Roscigno, E. D. Falco, A. Vece, F. Camilli, F. Sabatini, et al. 2020. Evaluation of dyeing and UV protective properties on hemp fabric of aqueous extracts from vegetal matrices of different origin. *Fibers Polym* 21: 1750–1759. https://doi.org/10.1007/s12221-020-1045-8.
62. Karabulut, K., and R. Atav. 2020. Dyeing of cotton fabric with natural dyes without mordant usage part I: determining the most suitable dye plants for dyeing and UV protective functionalization. *Fibers Polym* 21: 1773–1782. https://doi.org/10.1007/s12221-020-9365-2.
63. Rather, L. J., Q. Zhou, A. Ali, Q. M. R. Haque, and Q. Li. 2021. Valorization of agro-industrial waste from peanuts for sustainable natural dye production: focus on adsorption mechanisms, ultraviolet protection, and antimicrobial properties of dyed wool fabric. *ACS Food Science & Technology* 1: 427–42. https://doi.org/10.1021/acsfoodscitech.1c00005.
64. Abate, M. T., Y. Zhou, J. Guan, G. Chen, A. Ferri, and V. Nierstrasz. 2020. Colouration and bio-activation of polyester fabric with curcumin in supercritical CO2: Part II–effect of dye concentration on the colour and functional properties. *The Journal of Supercritical Fluids* 157: 104703. https://doi.org/10.1016/j.supflu.2019.104703.
65. Grifoni, D., L. Bacci, G. Zipoli, L. Albanese, and F. Sabatini. 2011. The role of natural dyes in the UV protection of fabrics made of vegetable fibres. *Dyes and Pigments* 91: 279–285. https://doi.org/10.1016/j.dyepig.2011.04.006.
66. Mongkholrattanasit, R., J. Kryštůfek, J. Wiener, and M. Viková. 2011. UV protection properties of silk fabric dyed with eucalyptus leaf extract. *Journal of the Textile Institute* 102: 272–279. https://doi.org/10.1080/00405001003722369.
67. Rattanaphol, M., K. Jiří, W. Jakub, and V. Martina. 2011. Dyeing, fastness, and UV protection properties of silk and wool fabrics dyed with eucalyptus leaf extract by the exhaustion process. *Fibres & Textile in Eastern Europe* 19: 94–99.
68. Chattopadhyay, S. N., N. C. Pan, A. K. Roy, S. Saxena, and A. Khan. 2013. Development of natural dyed jute fabric with improved colour yield and UV protection characteristics. *Journal of The Textile Institute* 104: 808–818. https://doi.org/10.1080/00405000.2012.758352.
69. Ebrahimi, I., and Gashti M. Parvinzadeh. 2015. Extraction of juglone from Pterocarya fraxinifolia leaves for dyeing, anti-fungal finishing, and solar UV protection of wool. *Coloration Technology* 131: 451–457. https://doi.org/10.1111/cote.12180.

70. Bonet-Aracil, M. Á., P. Díaz-García, E. Bou-Belda, N. Sebastiá, A. Montoro, and R. Rodrigo. 2016. UV protection from cotton fabrics dyed with different tea extracts. *Dyes and Pigments* 134: 448–452. https://doi.org/10.1016/j.dyepig.2016.07.045.
71. https://textilelearner.net/biopolishing-in-textile-process-advantages/
72. Shabbir, M., and J. N. Sheikh. 2020. Introduction to textiles and finishing materials, 1st ed. In *Frontiers of textile materials*, ed. Shabbir, M., S. Ahmed, J. N. Sheikh, 1–11, Wiley. https://doi.org/10.1002/9781119620396.ch1.
73. Hoque, M. T., N. Mazumder, M. T. Islam. 2021. Enzymatic wet processing, 1st ed. In *Sustainable practices in the textile industry*, ed. Rather, L.J., M. Shabbir, A. Haji, 87–110, Wiley. https://doi.org/10.1002/9781119818915.ch4.
74. Zhang, X., Y. P. Zhang. 2013. Cellulases: characteristics, sources, production, and applications, 1st ed. In *Bioprocessing technologies in biorefinery for sustainable production of fuels, chemicals, and polymers*, ed. Yang, S., H. A. El-Enshasy, N. Thongchul, 131–46, Wiley. https://doi.org/10.1002/9781118642047.ch8.
75. Pedersen, A. H., and P. T. N. N. Schneider. 1998. US Pat. 5795855 A. US-Patent.
76. Rehman, M. M. A., and M. A. Imran. 2014. Revolution of biotechnology in finishing sector of textile. *Revolution* 6(2). Chemistry and Materials Research www.iiste.org. ISSN 2224-3224 (Print) ISSN 2225-0956 (Online) 6(2).
77. Parthiban, M., M. R. Srikrishnan, and P. Kandhavadivu (eds.). 2018. *Sustainability in fashion and apparels*. WPI Publishing. https://doi.org/10.1201/9781351185875.
78. Colomera, A., and H. Kuilderd. 2015. *Biotechnological washing of denim jeans*. Denim, Elsevier, 357–403. https://doi.org/10.1016/B978-0-85709-843-6.00012-3.
79. Daget, T. M., B. B. Kassie, and D. F. Tassew. 2025. A shift from synthetic to bio-based polymer for functionalization of textile materials: a review. *International Journal of Biological Macromolecules* 306: 141637. https://doi.org/10.1016/j.ijbiomac.2025.141637.
80. Kumar Kundu, C., W. Wang, S. Zhou, X. Wang, H. Sheng, Y. Pan, et al. 2017. A green approach to constructing multilayered nanocoating for flame retardant treatment of polyamide 66 fabric from chitosan and sodium alginate. *Carbohydrate Polymers* 166: 131–138. https://doi.org/10.1016/j.carbpol.2017.02.084.
81. Mondal, M. I. H., S. C. Sarker, F. Ahmed, M. N. Pervez, and J. Saha. 2024. Fabrication of sustainable functional cotton fabric with silk sericin and chitosan for protective textiles. *Heliyon* 10: e39250. https://doi.org/10.1016/j.heliyon.2024.e39250.
82. Teli, M. D., J. Sheikh, and P. Bhavsar. 2013. Multifunctional finishing of cotton using chitosan extracted from bio-waste. *International Journal of Biological Macromolecules* 54: 125–130. https://doi.org/10.1016/j.ijbiomac.2012.12.007.
83. Xiao, L. 2018. Chitosan application in textile processing. *CTFTTE* 4. https://doi.org/10.19080/CTFTTE.2018.04.555635.
84. Ivanova, N. A., and A. B. Philipchenko. 2012. Superhydrophobic chitosan-based coatings for textile processing. *Applied Surface Science* 263: 783–787. https://doi.org/10.1016/j.apsusc.2012.09.173.
85. Abdel-Halim, E. S., F. A. Abdel-Mohdy, S. S. Al-Deyab, and M. H. El-Newehy. 2010. Chitosan and monochlorotriazinyl-β-cyclodextrin finishes improve antistatic properties of cotton/polyester blend and polyester fabrics. *Carbohydrate Polymers* 82: 202–208. https://doi.org/10.1016/j.carbpol.2010.04.077.
86. Ratnawulan, Rahmadhani D, Y. K. D. Zulkifli, E. Frida, and A. Taufiq. 2024. Hydrophobic and antibacterial properties of textiles using nanocomposite chitosan and SiO2 from rice husk ash as-coating. *South African Journal of Chemical Engineering* 48: 366–74. https://doi.org/10.1016/j.sajce.2024.03.010.
87. Stella, F., S. Fraterrigo Garofalo, N. Cavallini, D. Fino, and F. A. Deorsola. 2024. Closing the loop: analysis of biotechnological processes for sustainable valorisation of textile waste from the fast fashion industry. *Sustainable Chemistry and Pharmacy* 38: 101481. https://doi.org/10.1016/j.scp.2024.101481.
88. Maliki, S., G. Sharma, A. Kumar, M. Moral-Zamorano, O. Moradi, J. Baselga, et al. 2022. Chitosan as a tool for sustainable development: a mini review. *Polymers* 14: 1475. https://doi.org/10.3390/polym14071475.

89. Elamri, A., K. Zdiri, M. Hamdaoui, and O. Harzallah. 2023. Chitosan: a biopolymer for textile processes and products. *Textile Research Journal* 93: 1456–1484. https://doi.org/10.1177/00405175221127315.
90. Gita, S., S. P. Shukla, N. Saharan, P. Prakash, and G. Deshmukhe. 2019. Toxic effects of selected textile dyes on elemental composition, photosynthetic pigments, protein content and growth of a freshwater chlorophycean alga Chlorella vulgaris. *Bulletin of Environmental Contamination and Toxicology* 102: 795–801. https://doi.org/10.1007/s00128-019-02599-w].
91. Jamee, R., and R. Siddique. 2019. Biodegradation of synthetic dyes of textile effluent by microorganisms: an environmentally and economically sustainable approach. *European Journal of Microbiology and Immunology* 9: 114–118. https://doi.org/10.1556/1886.2019.00018.
92. Sharma, J., S. Sharma, and V. Soni. 2021. Classification, and impact of synthetic textile dyes on aquatic flora: a review. *Regional Studies in Marine Science.* 45: 101802. https://doi.org/10.1016/j.rsma.2021.101802.
93. Inamdar, S., K. Vanarase, and S. Inamdar. 2023. Enzyme engineering for advanced treatment of textile wastewater. In *Current developments in bioengineering and biotechnology*. Elsevier, 679–710. https://doi.org/10.1016/B978-0-323-91235-8.00010-3.
94. Vikrant, K., B. S. Giri, N. Raza, K. Roy, K.-H. Kim, B. N. Rai, et al. 2018. Recent advancements in bioremediation of dye: current status and challenges. *Bioresource Technology* 253: 355–367. https://doi.org/10.1016/j.biortech.2018.01.029.
95. Mishra, S., and A. Maiti. 2019. Applicability of enzymes produced from different biotic species for biodegradation of textile dyes. *Clean Technologies and Environmental Policy* 21: 763–781. https://doi.org/10.1007/s10098-019-01681-5.
96. Qamar, S. A., M. Ashiq, M. Jahangeer, A. Riasat, and M. Bilal. 2020. Chitosan-based hybrid materials as adsorbents for textile dyes–a review. *Case Studies in Chemical and Environmental Engineering* 2: 100021. https://doi.org/10.1016/j.cscee.2020.100021.
97. Vakili, M., M. Rafatullah, B. Salamatinia, A. Z. Abdullah, M. H. Ibrahim, K. B. Tan, et al. 2014. Application of chitosan and its derivatives as adsorbents for dye removal from water and wastewater: a review. *Carbohydrate Polymers* 113: 115–130. https://doi.org/10.1016/j.carbpol.2014.07.007.
98. Kyzas, G. Z., D. N. Bikiaris, and A. C. Mitropoulos. 2017. Chitosan adsorbents for dye removal: a review: Chitosan adsorbents for dye removal. *Polymer International* 66: 1800–1811. https://doi.org/10.1002/pi.5467.
99. Yan, B., Y. Dai, L. Xin, M. Li, H. Zhang, H. Long, et al. 2024. Research progress in the degradation of printing and dyeing wastewater using chitosan based composite photocatalytic materials. *International Journal of Biological Macromolecules* 263: 130082. https://doi.org/10.1016/j.ijbiomac.2024.130082.
100. De Medeiros, A. D. M., C. J. G. Da Silva Junior, J. D. P. De Amorim, H. A. Do Nascimento, A. Converti, A. F. D. S. Costa, et al. 2021. Biocellulose for treatment of wastewaters generated by energy consuming industries: a review. *Energies* 14: 5066. https://doi.org/10.3390/en14165066.
101. https://bioscience.iff.com/solutions/textile-processing
102. Ren, Y., J. Gong, R. Fu, Z. Li, Z. Yu, J. Lou, et al. 2017. Dyeing and functional properties of polyester fabric dyed with prodigiosins nanomicelles produced by microbial fermentation. *Journal of Cleaner Production* 148: 375–385. https://doi.org/10.1016/j.jclepro.2017.01.168.
103. Kanelli, M., M. Mandic, M. Kalakona, S. Vasilakos, D. Kekos, J. Nikodinovic-Runic, et al. 2018. Microbial production of violacein and process optimization for dyeing polyamide fabrics with acquired antimicrobial properties. *Frontiers in Microbiology* 9: 1495. https://doi.org/10.3389/fmicb.2018.01495.
104. DeBritto, S., T. D. Gajbar, P. Satapute, L. Sundaram, R. Y. Lakshmikantha, S. Jogaiah, et al. 2020. Isolation and characterization of nutrient dependent pyocyanin from Pseudomonas aeruginosa and its dye and agrochemical properties. *Science and Reports* 10: 1542. https://doi.org/10.1038/s41598-020-58335-6.

Chapter 11
Energy Efficiency in Textile Production: Strategies and Solutions

Raphael Kanyire Seidu, George Kwame Fobiri, Edem Kwami Buami, Jyoti Rani, Benjamin Tawiah, and Shouxiang Jiang

Abstract The textile industry is a major contributor to the carbon footprint and consumption of high energy during manufacturing processes, thus leading to rising energy costs. This limits efforts to meet the global climatic goals and minimise any environmental impacts, hence efforts to adopt sustainable practices. Energy efficiency, however, poses a critical solution and strategy to ensure economic gains, limit environmental impacts and enhance effective energy usage in the textile industry. This is because, at the production stages of the textile industry, high energy consumption is experienced. In this chapter, we discuss the strategies and solutions for energy efficiency in textile production in three (3) key areas: strategies for energy efficiency, technological innovations and benefits of technological innovations for energy efficiency in textile production processes. The barriers to energy efficiency of textile production are further discussed. Even though there are numerous benefits of practising energy efficiency, there are some challenges and Regulatory Issues that hinder the full realisation of this approach to lower carbon emissions, limit environmental

R. K. Seidu (✉) · B. Tawiah · S. Jiang (✉)
School of Fashion and Textiles, The Hong Kong Polytechnic University, Kowloon, Hong Kong SAR, China
e-mail: seiduraphael@gmail.com; raphael-kanyire.seidu@connect.polyu.hk

S. Jiang
e-mail: kinor.j@polyu.edu.hk

G. K. Fobiri
Department of Fashion Design and Textiles Studies, Kumasi Technical University, Kumasi, Ghana

E. K. Buami
Department of Industrial Art, Ho Technical University, Ho, Ghana

J. Rani
Parul Institute of Design, Parul University, Vadodara, Gujarat, India

B. Tawiah
Department of Industrial Art, Kwame Nkrumah University of Science and Technology, Kumasi, Ghana

S. Jiang
Research Centre of Textiles for Future Fashion, The Hong Kong Polytechnic University, Kowloon, Hong Kong SAR, China

S.-X. Jiang et al. (eds.), *Advances in Textile Materials and Processing Techniques for Sustainability*, SDGs and Textiles, https://doi.org/10.1007/978-981-95-0469-5_11

impacts and high energy consumption. The chapter lastly concludes by providing promising areas for future research opportunities for sustainable energy efficiency in the textile industry.

Keywords Energy efficiency · Technological innovations · Barriers and strategies · Textile industry

11.1 Introduction

The textile industry has played a significant role in revenue generation, provision of jobs and improving the overall growth of an economy. There is, however, utilisation of large-scale equipment, consumption of large volumes of chemicals and water in the production processes. This requires the supply and utilisation of energy to power the production plants, hence making the industry energy-intensive. This makes the textile industry a major consumer of energy [1]. As a cost component and process input in the production process, thermal energy and electricity are the two forms of energy used in the textile industry [1]. The consumption of higher energy in production is required to respond to the growth of fast fashion. This makes energy an indispensable input that is needed in the production chain of the textile industry [2]. Alternatively, there are higher carbon footprints based on the type of energy consumed during the production chain. For example, top textile-producing countries such as Bangladesh, China and India heavily rely on electricity powered by fossil fuels, which contributes to their carbon footprint [3]. As more energy is consumed, more carbon footprint is produced, contributing to about 10% (approximately 1.2b tonnes of CO_2) of greenhouse gas emissions worldwide [4]. The contribution to greenhouse gas emissions by the textile industry is shown in Fig. 11.1. This implies that conventional energy sources have led to higher emissions and energy costs, hence, advancement in energy efficiency is necessary for a cost-effective and sustainable industry.

Energy efficiency in the textile production chain is imperative due to the rapid concerns of energy costs, resource depletion and climate change. This has raised awareness for improved energy efficiency and reduced energy demands [2]. It further forms a greater contributor to reduced air pollution and conservation of "primary energy resources" [1], hence making the applications of cleaner production and adoption of energy efficiency for the textile industry necessary. For example, the utilisation of solar energy is regarded as an alternative source of energy for the textile industry [5, 6]. This form of renewable energy source contributes to limiting the environmental impacts from the industry. As such, adopting effective innovations and practices that will lead to reducing energy consumption and carbon emissions without compromising the production process and the eventual quality of products is needed in the textile industry. Hence, optimising energy use at every stage from fibre production to finishing is important to reduce any high operational costs and waste. This further implies that textile manufacturers can significantly reduce their energy use while maintaining good standards of production. Some effective energy

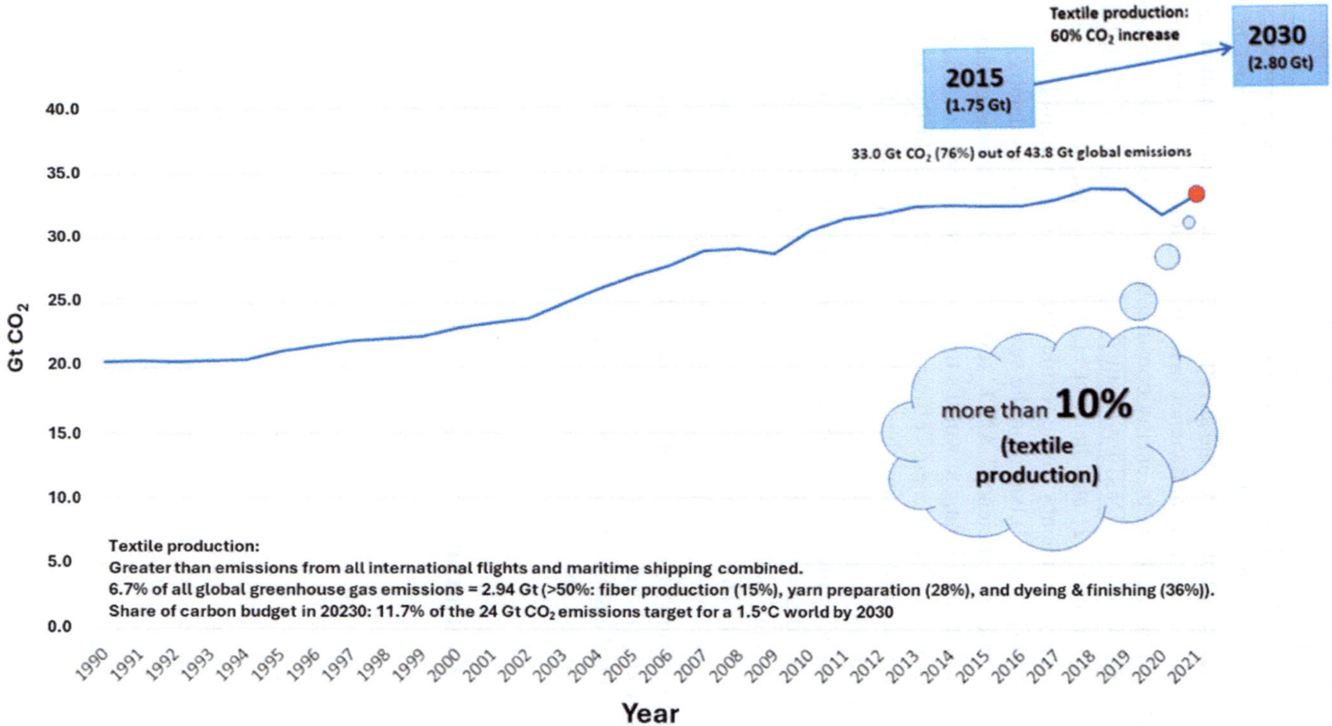

Fig. 11.1 The contribution to greenhouse gas emissions by the textile industry as compared to global energy production [3]. (This is an open-access article distributed under the terms of the creative commons attribution license (CC BY))

efficiency strategies in the textile industry include the upgrade of technology, fine-tuning of equipment and high-capacity utilisation [7]. Energy efficiency is affected by the technologies applied in the production process [8].

Aside from the benefits of energy efficiency, there are several challenges, such as the maintenance of advanced technologies, the need for skilled personnel, higher costs of energy-efficient technologies and government support or collaborations associated with the utilisation of energy efficiency in the textile industry. To further understand the challenges and effects of energy consumption from the textile industry, several studies have been conducted by Rasul [9], Palanichamy et al. [10], Tang et al. [11], Pardo Martínez [12] and Jiang et al. [13], amongst others. For effective strategies on energy efficiency in the textile industry, publications in this area have fairly increased (Fig. 11.2) based on the search on the Scopus database using the search blocks "energy efficiency" AND "textile industry". It was evident that researchers from India and China, the world's top textile-producing countries, have contributed to the subject area of energy efficiency in the textile industry. Advancement in these areas is critical to ensure the optimisation of the processes for effective machinery replacement [14]. Moreover, enhancing energy efficiency in the textile industry can ensure economic benefits, minimise environmental effects, and decrease carbon emissions. It is thus imperative for effective cooperation, participation, knowledge improvements and awareness creation in the textile industry for optimisation of energy saving [15]. This chapter aims to outline strategies and solutions for improving energy efficiency in textile production. Additionally, it addresses various barriers, challenges, and regulatory concerns linked to energy efficiency within the industry. By grasping these issues, policymakers, stakeholders, and manufacturers can promote essential energy-efficient strategies and practices, paving the way for a greener and more sustainable future in the textile sector.

11.2 Energy Consumption in the Textile Industry

The consumption of energy in textile production cannot be overemphasised since processes such as bleaching, washing, curing, dyeing, and drying require a lot of energy or electricity to power the various equipment [1, 12]. Alternatively, the processing of fibre types will determine the energy required and consumed [16] and even the volume of emissions generated. For example, in the production of a cotton t-shirt, 2.1 kg CO_2e is produced, as compared to a higher 5.5 kg CO_2e for a polyester t-shirt [17]. This implies that there is a higher carbon footprint during the production of synthesis as compared to cotton [3]. In addition to fibre types, length in textile facilities, process structure, and machine properties influence energy consumption in the textile industry.

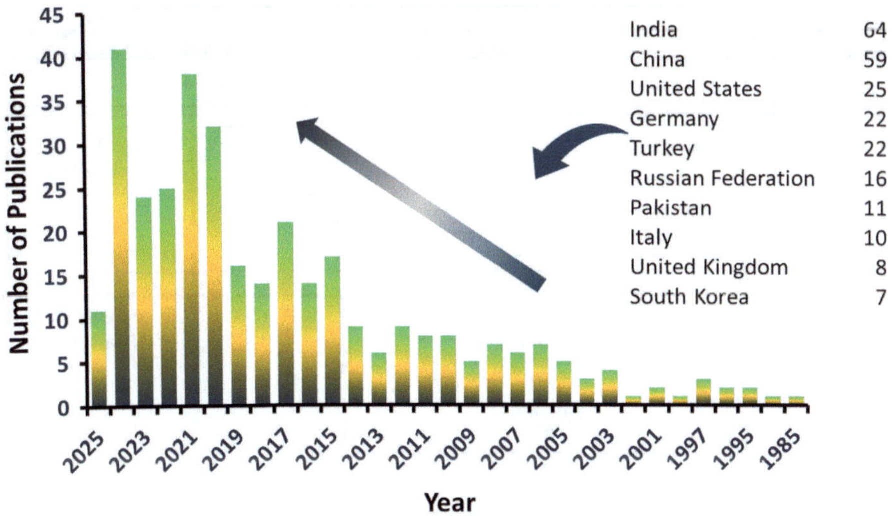

Fig. 11.2 Publication trends in energy efficiency in the textile industry

11.2.1 Energy Use in Different Stages of Textile Production

The textile industry has several production stages with large machinery that consumes high energy, impacting environmental sustainability and costs. The understanding of this energy use at each stage provides an opportunity for improvements in efficiency and savings. Furthermore, mechanisation in these stages has resulted to increase in energy intensity [18]. As such, when there is an increase in consumption and energy, the cost of production also increases [15]. However, the type of energy influences the type of machinery used. For example, generated steam from natural gas, coal, LPG and oil is used in finishing processes, scouring or bleaching, washing and dyeing, in the form of thermal energy [19]. These processes use temperatures ranging from 60 to 120°C [15]. Others in the form of electricity are used to power office equipment at the offices, lighting, control systems for temperature and cooling, and for machinery [15]. Figure 11.3 provides a comprehensive outline of the key processes in the manufacturing activities of the textile industry.

Key manufacturing processes like spinning, wet processing, weaving and clothing manufacture utilise large machinery that consumes high energy. Lower energy and source of energy (either heat energy or electricity) are required for some processes as compared to others [20]. For example, electric energy is consumed to ensure the conversion of fibres into yarns through the spinning process. The level of energy consumed is largely dependent on several factors, such as the characteristics of the raw materials, spinning machine types, yarn properties and spinning system [21]. The spinning machinery consumes the majority of the energy in the spinning process mill [22]. Furthermore, others like the HVAC (heating, ventilation and air conditioning) systems consume about 16% of the electricity in the spinning mill

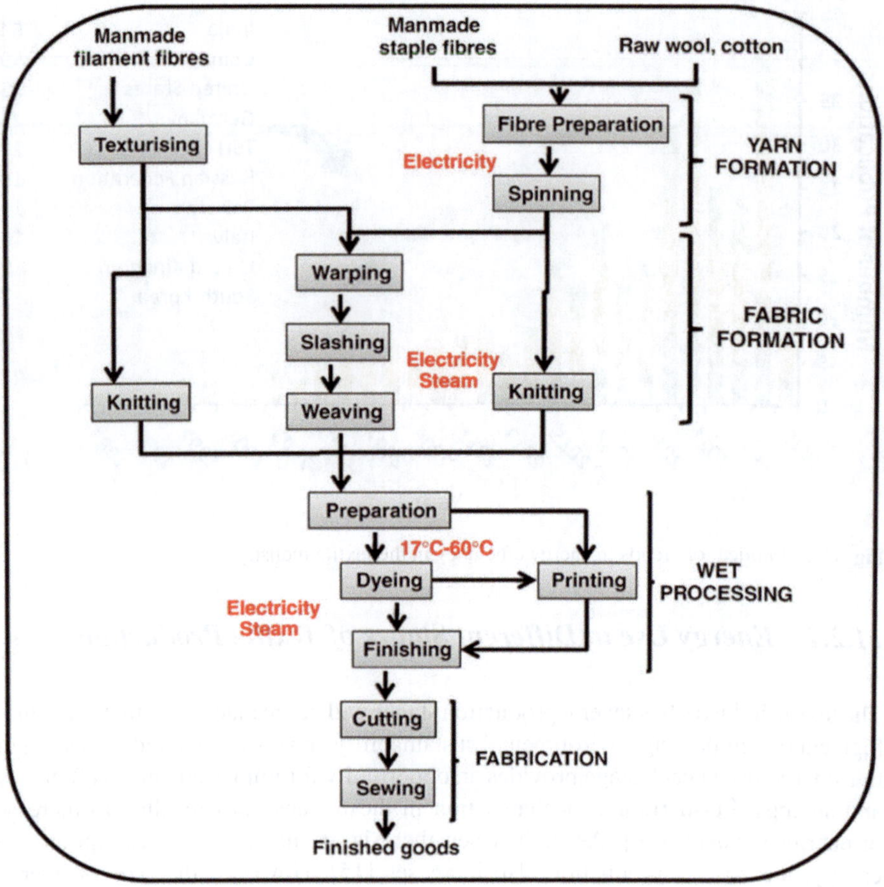

Fig. 11.3 Key processes in the manufacturing activities of the textile industry. Reprinted with permission from Ref. [12] Copyright 2010 Elsevier

[20]. For wet processing, hot air, steam, and hot water are consumed in processes such as drying-fixation, post treatments, dyeing, bleaching and pretreatment [20]. Electric energy is, however, consumed by the mechanical moving parts of the wet processing machinery. The same electric energy is consumed in the stages of warp preparation (sizing and warping). For example, indirect steam, which uses electricity aids in yarn sizing. Additionally, electric energy is consumed by the weaving process and the conditions in the controlled climatic room. It is also important to note that the technical parameters and the fabric structure further influence the energy consumed from the weaving process [20]. The climatic conditions are controlled to limit any flammability of fibres such as cotton lint when stuck in the moving parts of the weaving machines [21]. In clothing manufacture, electric energy is consumed by sewing, cutting and laying up, while hot air or steam is used in the ironing stage.

11.2.2 Factors Influencing Energy Consumption

Aside from the significant role the textile sector plays in the economic development of countries globally [23], the operations of the industry consume huge energy. The latter is used to power several machines and equipment used in production and the offices, respectively. Energy consumption of energy in the textile industry is influenced by several factors, which contribute to its global carbon footprint. Key areas identified by Li and Lin [24] include the impact of technological advancements, the role of foreign trade in promoting efficiency, the influence of energy prices, and the effectiveness of various policies aimed at reducing energy consumption, such as optimisation and upgrading of production structure, which will in turn result in notable increases in the industry's overall usage of energy [25]. By examining these aspects, we can better assess the current challenges and opportunities for enhancing energy efficiency within the textile manufacturing chain.

Technological progress is a pivotal factor in enhancing energy efficiency within the textile industry, as it directly influences production methods and energy consumption patterns. Advancements in technology not only streamline manufacturing processes but also facilitate the adoption of energy-saving practices. For instance, innovations such as automated machinery and advanced textile processing techniques can significantly reduce energy inputs while maintaining or improving product quality. Lin et al. [26] further underscore the relationship between production technology and energy efficiency, suggesting that the integration of state-of-the-art equipment and processes can lead to substantial energy savings. For example, modern dyeing technologies, such as digital printing and waterless dyeing, are designed to minimise water and energy usage compared to traditional methods [1]. These technologies not only reduce operational costs but also lessen the environmental impact associated with high energy consumption [27].

Moreover, foreign trade plays a crucial role in promoting technological progress within the textile sector. Enhanced exposure to global markets encourages local manufacturers to adopt and adapt new technologies to remain competitive. This technical exchange often results in the importation of more efficient production technologies and practices from countries with advanced textile industries [28]. As a result, the technical effects of export trade become evident, as manufacturers may seek to meet international standards that prioritise energy efficiency. Also, in the textile industry, the prices of energy influence energy efficiency in the textile industry, where there is more efficient energy use due to higher energy prices [29]. Furthermore, rising energy prices in primary energy sources such as coal and electricity impact the operational costs of companies in the industry, a situation that could influence them to reduce consumption and enhance significant efficiency in the use of energy [28].

Policy interventions significantly influence the industrial transitions towards sustainable work operations by promoting energy efficiency and supporting the adoption of these technologies in the sector [30]. Several regulatory policies and standards have been implemented to ensure limitations in the energy use for fibre and textile production. For example, in the textile mill sector, Khattab et al. [31] reported on the adoption of the EU IPPC directive, which aims at ensuring a significant reduction in the use of energy and water. The elimination of energy-intensive equipment and the promotion of energy-efficient technologies can significantly lower energy consumption and intensity in the textile industry [23].

11.3 Strategies for Energy Efficiency for Textile Production

Energy costs represent a significant portion of operational expenses in the textile industry. In response, firms in the textile industry have adopted key strategies such as cleaner technologies to reduce greenhouse gas emissions and energy consumption [32], even though some are cost-effective [1]. Key strategies for reducing energy intensity and emissions include the adoption of energy-efficient machinery, the integration of renewable energy sources, and the implementation of smart manufacturing technologies. Investing in energy-efficient machinery is a fundamental strategy. Upgrading to modern, energy-efficient machines can significantly reduce energy consumption. For instance, high-efficiency motors and automated systems can lead to substantial energy savings. Additionally, regular maintenance of existing equipment ensures optimal performance and prevents energy loss due to inefficiencies.

Integrating renewable energy sources into textile production represents a transformative strategy that addresses both economic and environmental challenges faced by the industry. As highlighted by Yoon et al. [30], the adoption of renewable energy, such as solar, wind, and biomass, can significantly diminish dependence on conventional fossil fuels, which are not only finite but also subject to volatile market prices and increasing regulatory scrutiny regarding carbon emissions. Again, the installation of solar panels is particularly advantageous for textile manufacturers, as these systems can be deployed on rooftops or unused land, allowing textile companies to harness sunlight for their energy needs [33]. There are a total of 1914 solar panels, each with a capacity of 545 watts, one transformer and eight inverters. These panels are installed on the roof of the textile factory.

This not only reduces energy costs by providing a stable, predictable energy source but also minimises the carbon footprint associated with traditional energy consumption [34]. For instance, solar energy can power various operations within textile facilities, such as spinning, weaving, and dyeing, which are typically energy-intensive processes. Similarly, biomass energy offers a sustainable alternative by utilising organic materials such as agricultural residues and by-products from textile production as fuel. This approach not only mitigates waste but also creates a closed-loop system where energy is derived from materials that would otherwise contribute to environmental degradation. By converting these materials into energy, companies can lower their overall energy costs while promoting sustainability within their operations. The financial benefits of adopting renewable energy solutions extend beyond immediate cost savings. Ozturk et al. [1] illustrate that companies investing in renewable technologies often see long-term financial returns through reduced energy

Table 11.1 Overview of factors influencing energy efficiency in textile production

Factors influencing energy efficiency	Energy consumption parameters	Strategies to ensure energy efficiency	Refs
Type of machinery and technology used	Energy consumption by process: 34% spinning, 23% weaving, 38% chemical processing, 5% miscellaneous	Upgrade to energy-efficient machinery	[1]
Upgrade to energy-efficient machinery	Dominant energy sources: fuel for wet processing, electricity for yarn spinning	Optimise processes and reduce steps	[36]
Adoption of renewable energy sources	Potential for reduced reliance on conventional energy	Implement solar, wind, or biomass energy solutions	[1, 30]
Technological advancements	Efficiency of heating and cooling processes	Use heat recovery systems and innovative drying technologies	[1, 27]

bills, lower operational costs, and potential incentives or subsidies from governments aimed at promoting green energy.

The implementation of smart manufacturing technologies, such as IoT devices and data analytics, can enhance energy management [35]. These technologies ensure the optimisation of machines for production and the monitoring of energy consumption in real-time. Although IoT adoption remains limited worldwide, some manufacturers have implemented these systems to manage and track energy consumption [35]. In the manufacturing sector, IoT utilises wireless devices like sensor networks to collect real-time data from the shop floor, including machine status, inventory levels, shipment progress, and energy consumption. Consequently, several smart meters and sensors, both wired and wireless, are available in the market. Table 11.1 outlines factors that affect energy efficiency in textile production. In addition, maintenance practices, employee training and awareness and fuel selection and combustion efficiency are related factors that influence energy efficiency.

11.4 Technological Innovations for Energy Efficiency in Textile Production Processes

Several attempts have been made to ensure efficient energy applications in the textile industry for sustainable and economical production practices. To begin with, innovative energy conservation strategies have been adopted to improve motor efficiency, ensure machine modification, and adjust processes after an energy audit in Southern Tamil Nadu of India [37]. This led to the energy reservation of 56,053 kWh per month. The study reports technological innovations, including installing IE3 motors in ring frame machines for high efficiency, adjusting spindle speed for reduced power consumption, replacing traditional aluminium fans with energy-saving ones, installing solenoid valves to ensure proper control of suction and reusing compressor exhaust heat to avoid heating lamps in speed frame machines. Biomass materials have also been used as a sustainable and innovative technology to reduce energy consumption and enhance efficiency in the textile industry [38]. The study found that biomass could be used to replace traditional energy sources like propane gas, naphtha, and natural gas in the textile industry to achieve about 35% reduction in the cost of energy as a result of steam generation. Moreover, a sustainable energy consumption model has been designed to help manage energy usage in the textile industry through a fully intuitionistic fuzzy optimisation approach [39]. From the study, an energy consumption assessment was done in various stages of the textile production line in Pakistan, including spinning, weaving, rewinding, sizing, and dyeing. The Fully Intuitionistic Fuzzy Linear Programming model was then developed to help energy allocation for an enhanced and energy-saving operation. It is reported that the model led to a lower energy cost of PKR 5,927,615 per month compared to traditional approaches.

Also, solar energy generation is proposed as a sustainable method compared to using non-renewable energy sources for the operation of the textile industry [39]. Prior to this, previous studies had proposed using solar energy generation as a technological innovation to help efficient energy usage in the textile industry, especially for process heating in cotton-based textile industries [40], boiler feed water heating and textile dyeing process [6], and an energy-efficient dyeing process with solar energy where an installed solar system is reported to have led to 60% energy savings in summer and 25% in winter [33]. Moreover, Internet of Things (IoT) technologies have been used to achieve sustainable and energy-efficient operations in the textile industry. It is reported that IoT-driven operations ensure real-time monitoring of the manufacturing environment and lead to compliance of regulatory standards for efficient, sustainable usage of resources, including energy in the textile industry [41]. Additionally, an innovative approach has been adopted in an integrated woolen textile facility to ensure efficient energy consumption in the textile industry, as reported in a previous study [1]. The innovation involves using a developed systematic decision-making model to identify 13 appropriate energy-saving techniques out of 85 potentials with techno-economic feasibility, potential savings, and environmental benefits as a benchmark. The selected techniques are presented in Table 11.2 with their respective details. With these techniques, the study reports a significant reduction of 8–27%, 12–28%, and 23–45% in electricity, thermal energy, and air emissions, respectively.

11.4.1 Benefits of Technological Innovations for Energy Efficiency

Technological innovation, such as IoT sensors, leads to operational monitoring in the textile industry for significant improvement in sustainable usage of resources

Table 11.2 Energy efficiency techniques employed in the textile industry and the outcome

S/N	Energy efficiency technique	Details and outcome of the technique regarding energy efficiency
1	Process optimisation in the finishing unit	The technique involves avoiding irrelevant stenter processes to save electricity and thermal energy of 8–12% and 12–32% respectively
2	Monitoring of inputs-outputs in terms of quality and quantity	This involves tracking chemicals, energy, raw materials, water, and emissions through a systematic approach to significantly reduce 1–5% energy consumption in the industry
3	Installation of an energy monitoring system	Tracking and optimising energy use in the industry through the installation of computerised systems. This reduces up to about 28% of energy consumption
4	Control of recirculation air in stenters	Incoming air in stenters is preheated using air emissions to reduce thermal losses, improve energy efficiency, and save 0.4–1% of thermal energy
5	Collection of cooling water and waste heat recovery	The technique involves recovering heat from cooling processes by installing closed-loop cooling water systems. This leads to 0.4–3% thermal energy savings
6	Waste heat recovery from hot wastewater streams	Reducing about 1–3.5% of energy consumption by separating hot wastewater streams and using heat exchangers to recover energy
7	Installation of a cogeneration system	Generating electricity and heat from a single fuel source using a combined heat and power system to reduce 8–15% of energy consumption
8	Preheating the boiler feed water and combustion air	Preheating feed water and combustion air with waste heat recovered from boiler exhaust gases to improve boiler efficiency by 3–6%
9	Changing the feeding technique in radio frequency dryers	This technique reduces drying period and electricity consumption by 0.3–1% through a modification of fibre feeding so that it would spread out the fibres instead of being fed in bulk
10	Positioning the compressed air system on the north side	The technique involves reducing compressor energy consumption by 3–12% by moving the air compressor intake to the north side to lower inlet air temperature
11	Establishment of a compressor monitoring and control system	Installation of a control system to enhance compressor load and levels of pressure to save 4–6% of electricity
12	Modification of the HVAC system	The use of variable speed drives and enhancing fan motors to cut HVAC electricity consumption by 6–9%
13	Filtering of air emissions with a wet scrubber system	Capturing and neutralising air pollutants with an installed wet scrubber system. This helps to reduce 30–50% of emissions

and compliance with regulatory standards [41]. IoT has been identified as a prospect for energy efficiency in industries by increasing the share of renewable energy and minimising the environmental impacts of energy usage [42]. Also, it is established that using appropriate technological innovations for efficient usage of energy in the textile industry leads to a significant reduction of 8–27%, 12–28%, and 23–45% in electricity, thermal energy, and air emissions, respectively [1]. This indicates the beneficial roles of adaptive techniques such as process optimisation in the finishing unit, installation of a cogeneration system, installation of an energy monitoring system, preheating the boiler feed water and combustion air, etc., as proposed by Ozturk et al. [1] in a study to implement innovative energy-saving techniques in the textile industry. Again, the installation of solar systems in textile industries yields about 60% energy savings compared to traditional non-renewable energy usage, especially in the summer [33], indicating the energy-saving and cost-effective nature of the solar technology. Solar energy systems adoption in the textile industry again leads to a reduction in carbon emissions, which is non-avoidable with fossil fuels. A recent study has also indicated that implementing energy-saving strategies using frameworks like Minimum Energy Performance Standards leads to about 20% energy saving in European Union countries [43]. Again, studies have posited that adopting renewable energy resources in manufacturing industries is estimated to yield about 28 million jobs globally by 2050 [44]. This indicates the promising future of technological innovations such as green/renewable energy implementation and the benefits it renders to industries which adopt it. As the world seeks to adopt green operations and maintain ethical standards in all manufacturing sectors [45, 46], technological innovation to ensure efficient energy usage in industries like the textile sector is highly encouraged for economical and sustainable manufacturing through energy-saving strategies.

11.5 Barriers to Energy Efficiency of Textile Production

The high-energy nature of the industry and its heavy dependence on fossil fuels have led to carbon emissions while its use of water in production processes has led to water stress and pollution in several regions [47]. Energy use, water use and pollution make the textile industry one of the worst environmental offenders. The fabrication of textile products requires energy-consuming and water-intensive processes such as scouring, washing, bleaching, sizing, dyeing, and finishing, all of which result in extensive wastewater generation comprising diverse chemical compositions. The textile industry is a fossil fuel-intensive field, especially in countries where most of the energy is coal-based [48, 49]. Based on these impacts, ensuring energy efficiency in the textile production processes is important to optimise energy use and reduce carbon emissions. However, several barriers to energy efficiency in textile production have been identified.

The textile industry is facing several challenges to improve its energy efficiency and lessen its environmental footprint. This is a significant challenge with the industry's reliance on energy-intensive technologies, like high-temperature dyeing and finishing processes that are hard to replace with more efficient alternatives [7, 50]. Another hurdle is the industry's fragmented structure, where numerous small- and medium-sized enterprises frequently lack the capital and know-how to invest in energy-efficient solutions [51]. The significant initial investment required for upgrading equipment and systems may also be a disincentive for textile manufacturers to implement sustainable practices. The absence of regulatory frameworks and enforcement mechanisms in some areas can further lead to an uneven playing field, enabling producers less committed to the environment to outcompete those who have made investments in sustainable technologies [52].

It is also worth noting that due to globalisation, the global supply chains of the textile industry make the process even more complicated, as production units are often placed in other countries with soft environmental policies, making it more challenging to work on energy efficiency and reduce environmental impacts across the entire value chain [53, 54]. To overcome these barriers and to encourage energy efficiency textile industry, many strategies have been proposed. Such strategies involve the adoption of new technologies, including eco-friendly natural dyes and water-saving dyeing methods, and promoting stronger energy-efficient policies and regulations. That is what suits individual consumers to a more significant extent [55].

In the textile sector, the adoption of energy-efficient technologies/practices faces several challenges. This is primarily due to the persistence of energy-intensive technologies within the industry, such as high-temperature dyeing and finishing processes, making it difficult to implement more efficient alternatives. The diversity within the sector, in terms of the number of small- and medium-sized enterprises, also restricts the availability of resources and expertise to invest in cutting-edge energy-efficient technologies [56]. Upgrades to equipment and infrastructure require a very high upfront cost, which makes adopting sustainable practices difficult for textile manufacturers. Furthermore, inconsistent regulations and enforcement across different regions lead to an uneven competitive landscape where environmentally conscious producers may be undercut by those with lax sustainability practices. The global complexity of the textile industry's supply chains aggravates how the industry struggles to improve the energy efficiency and environmental impact of the value chain more broadly, as it tends to outsource production to regions with weaker regulations.

11.5.1 Studies on Energy Efficiency in Textile Production

They have made some patents for all confer methods that are not discussed here for textile production attempts to innovate them for better energy efficiency and environmental effects. One avenue that shows promise is the development of biodegradable natural dyes, which may be used as an alternative to energy-intensive man-made dyes [57]. Dyes, from natural sources, such as food, wood, and agricultural waste, are being extracted at a lower cost and possibly with a lower impact on the environment [58].

Computer-aided design systems like Pro-Cad and WKCAD have revolutionised the textile industry by automating and facilitating the drafting and production of complex Jacquard patterns, leading to optimised and innovative design in textiles [59]. Another area of research focuses on emerging technologies for energy and water efficiency in textile manufacturing. This includes the use of membrane bioreactors, which can significantly reduce water consumption and wastewater pollution, as well as the development of novel drying and finishing technologies that require less energy [58, 60]. The integration of renewable energy sources, such as solar and wind power, has been advanced to reduce the textile industry's reliance on fossil fuels and lower its carbon footprint. By addressing the barriers to energy efficiency in textile production, the industry can potentially reduce its environmental impact and move towards more sustainable practices. While the textile industry faces significant challenges in improving its energy efficiency and environmental sustainability, researchers have been exploring various strategies and technologies to overcome these barriers [61].

Although prior studies have contributed greatly to highlighting barriers for energy efficiency in textile production and recommendations to go further, they leave some critical gaps which need to be explored further. One significant area is the exploration of the efficacy and viability of such energy-efficient/sustainable technologies emerging in the context of the textile sector. Though innovations like natural dyes [57], membrane bioreactors [58] and advanced energy management systems [60] may present effective solutions, but they are often burdened by high upfront costs and require major infrastructure changes for widespread adoption. So, without a compelling financial incentive or policy support, many textile manufacturers, especially the small and medium-sized enterprises (SMEs) may find it hard to finance these transitions [59]. Further economic analysis is required to determine how much of a beneficial impact these technologies can offer, and whether it justifies the costs in terms of reduced site costs and environmental footprint, or enhanced efficiency [62]. Note, these high-level firm-level characteristics are also characterised by the second process of exploring global supply chain dynamics, which is underexplored in terms of how it may impact the adoption of energy-efficient practices. The textile industry is highly interconnected with the processes of raw material sourcing, production, and distribution occurring across many countries. Trade rules, labour costs, demand for green products and renewable energy supply all govern how far manufacturers in particular areas can go in introducing sustainable solutions. Moreover, greater emphasis needs to be paid to policymakers' role in promoting energy efficiency through regulatory measures, incentives, and public–private partnerships. Policies that incentivise energy audits, subsidise sustainable technologies or enforce stricter efficiency standards could help accelerate the industry's sustainability transition [63].

The important subject of regional and contextual differences in energy efficiency challenges and solutions in the textile sector needs to be further studied. Implementation of energy-efficient technologies is geographically dependent, regulatory-driven, development-driven, and resource-dependent (clean energy, water, etc.) [64]. Developing countries have challenges such as a lack of infrastructure, a lack of technical expertise, and financial constraints in applying green technologies on a large scale, further impeding green practices in the textile hubs. On the other hand, jurisdictions with more rigorous regulation and access to R&D capabilities may be able to absorb innovations more easily. These differences are critical to the creation of customised responses that meet the individual challenges presented to the different branches of the textile industry [65]. Addressing barriers to energy efficiency will require a multidisciplinary approach that integrates economic analysis, policy evaluation, and technological assessment. In the long run, and at a larger scale, these efforts will enable the textile industry to move toward a more sustainable future with its environmental impact reduced without sacrificing economic viability in an increasingly competitive global market.

11.6 Challenges and Regulatory Issues

Lack of Standardised Energy Efficiency Regulations

The textile industry is a rapidly growing energy-consuming industry, and the energy efficiency regulations vary for each region and country, which creates a major challenge for the industry to standardise and strengthen energy efficiency regulations that are deemed comparable. Because different countries have different policies, which lead to disparities in the standards for energy usage, this could halt the global textile manufacturers from complying with different regulations when investing in other countries [61]. For example, developed nations can have highly-enforced, and sometimes stringent, policies that strongly compel energy efficiency, whereas developing countries are often much less equipped to secure compliance [58]. This lack of consistency results in inefficiencies in implementation that make it challenging for businesses to plan long-term energy-saving investments [66]. Moreover, the lack of a global framework for energy efficiency benchmarks means that the industry does not have the opportunity to adopt best practices on a wider scale. Governments and international organisations must work together to elaborate on countries' flexible standards. A global energy efficiency certification for textile manufacturing could enforce the requirement for a sustainable production process irrespective of its location.

High Compliance Costs and Bureaucratic Hurdles

Meeting energy efficiency regulations in the textile industry is often associated with high costs and complex bureaucratic processes. Manufacturers must obtain multiple certifications, comply with environmental regulations, energy audits and so on, all of which require heavy investments [61]. The compliance costs constitute a significant burden upon small and medium enterprises (SMEs) that comprise much of the textile industry. As many SMEs work with tight profit margins, the cost of implementing energy efficiency technologies or complying with regulations can be especially significant [58]. Moreover, bureaucratic biases like lengthy approval processes, vague regulatory laws, and frequent changes in policies make it complex [66]. From energy-efficient plumbing to the use of new materials in construction, the regulatory landscape for energy efficiency can be fragmented, forcing businesses to traverse a maze of agencies for permits and approvals. Clear guidance and streamlined regulatory processes can go a long way toward making manufacturers feel less burdened by compliance and encouraging better compliance with energy efficiency regulations across the board [65].

Limited Government Incentives and Support

Limited government incentives for businesses are one of the biggest reasons why only a few businesses are going towards energy-efficient practices in the textile sector. Some countries offer financial support for sustainability projects, but these incentives tend to be either insufficient or not universally available to all manufacturers [67]. For instance, tax benefits and subsidies for upgrading machines to more energy-efficient models are not universally available, making it hard for companies to justify the high upfront cost [52]. Again, energy support programs are often aimed at large enterprises, leaving SMEs with little or no support for the transition to energy-efficient operability. To promote widespread uptake of energy-efficient behaviours, governments should broaden and deepen financial incentives. The provision of low-interest loans, grants, and reduced taxes on businesses adopting sustainable technologies may encourage the textile industry to improve energy efficiency across the board.

Resistance to Regulatory Changes

Another obstacle is the resistance to regulatory changes in the textile industry when it comes to efficiency in energy consumption. The reason that many manufacturers are resistant to new policies is that they fear production will stop, costs will rise, and known workflows established for years will have to change [68]. Textiles are challenging to produce; changing energy consumption practices in factories is no easy feat. Manufacturers are afraid of the compliance process as it could mean downtime, which can hit production schedules and, in turn, profitability. There is scepticism in some quarters about the efficacy of certain energy efficiency standards, if they are perceived to be more policy-driven system than an industry need-driven. Third, industry associations and policymakers must collaborate to resolve these issues through educational programs, training workshops, and financial incentives. Providing clarity on the long-term advantages of regulatory compliance, including cost savings and environmental sustainability, can aid in overcoming resistance to adopting energy-efficient practices [69].

Environmental and Sustainability Challenges

The textile industry is one of the most polluting industries in the world, consuming huge amounts of energy, water, as well as raw materials. While implementing energy efficiency measures is critical to reducing the sector's carbon footprint, the progress on this front is limited by several challenges. One of the principal issues is the shift to energy renewal sources. Fossil fuels are still used to power so many textile factories that sustainability commitments are hard to hit. Investing in solar or wind generators can alleviate some of this, but in many regions, the need for existing infrastructure and local weather or soil conditions can act as a hurdle [70]. Waste management is also still a major concern. Textile manufacturing produces significant amounts of discarded chemicals, dyes, and waste. It is vital to adopt sustainable practices to enhance waste management, like the reuse and recycling of materials. However, these practices require considerable modification to production processes, complicating the transition to an energy-efficient enterprise [71]. To encourage the shift towards more sustainable production methods, the government and industry stakeholders should work together and promote cleaner production methods, invest in renewable energy infrastructure, and provide incentives for sustainable initiatives. Public awareness campaigns can use it to encourage people to buy eco-friendly textile products [72].

Supply Chain and Market Pressures

The competitive nature of the globalised textile industry has demanded lower prices, with energy efficiency becoming a secondary concern. Many manufacturers bear harsh pressure to create textiles for as little cash as possible, and so it is arduous to prioritise sustainability initiatives [52]. Most textile suppliers work in tightly controlled supply chains which demand cost-effectiveness. Fashion brands and other retailers that are buyers often focus on lower prices in a supplier base that has limited ability to invest in energy-efficient solutions, especially under budget constraints [58]. Furthermore, it can be confusing to ensure that energy efficiency regulations are adhered to all over the global supply chains. Global textile companies struggle to make all their factories equally energy efficient because different countries have different regulatory requirements. From the suppliers' side, they might be less willing to leverage new technologies owing to their own financial or logistical constraints. In response to these issues, brands and retailers need to play an active role in driving energy-efficient practices across their supply chains. Contractual agreements, as well as the financial resourcing and collaboration that make such agreements possible, can encourage suppliers to undertake sustainability initiatives that result in widespread improvement in energy efficiency. Moreover, consumer knowledge and demand for environmentally friendly products can provide market-driven incentives for manufacturers to focus on energy-efficient production [73].

11.7 Conclusions and Future Research Opportunities

The textile industry is responsible for contributing to GHG (greenhouse gas) emissions, making this industry a major consumer of global energy and an environmental polluter. Yarn spinning and textile wet processing are energy-intensive processes that have aroused criticism of the sustainability of the industry. Apart from energy use, the textile sector is also the largest consumer of water resources and a major contributor to polluted effluents that add to the environmental burden. By viewing the textile industry through an industrial ecology perspective, we can start to make sense of the energy use and subsequent environmental impacts of the sector, which examines the movements of materials and energy through and between industrial sectors [69]. As such, ensuring energy efficiency in the various production processes plays a critical role in the textile industry to ensure social responsibility, economic and environmental sustainability. This is important because the textile industry consumes high energy to power various machinery in various production processes. This leads to a high cost of production and contributes to the carbon footprint globally. Furthermore, by implementing energy efficiency strategies, the industry can ensure savings, reduction in greenhouse gas emissions, and operational costs, hence improving profitability. These key benefits are important for every manufacturer and country. In addition to these benefits, there are multiple barriers, challenges, and regulatory issues associated with energy efficiency in the textile industry. These challenges include the high costs of energy-efficient technologies, the requirement for skilled workers, issues related to scalability and standardisation, the need for innovation advancements, and insufficient government support and collaboration. These factors impact the comprehensive adoption and execution of energy-efficient strategies throughout the textile production processes.

The future of energy efficiency in the textile production process can be advanced in key areas like optimisation of the process, integrating renewable energy and innovative technologies like AI and IoT. Subsequently, future studies can report on the effects of regulations and policies by the governments to identify relevant barriers that limit the full implementation and utilisation of energy efficiency practices. By grasping these issues, policymakers, stakeholders, and manufacturers can promote essential energy-efficient strategies and practices, paving the way for a greener and more sustainable future in the textile sector.

References

1. Ozturk, E., N. C. Cinperi, and M. Kitis. 2020. Improving energy efficiency using the most appropriate techniques in an integrated woolen textile facility. *Journal of Cleaner Production* 254: 120145.
2. Aranda-Usón, A., et al. 2012. Energy consumption analysis of Spanish food and drink, textile, chemical and non-metallic mineral products sectors. *Energy* 42 (1): 477–485.
3. Leal Filho, W., et al. 2022. An overview of the contribution of the textiles sector to climate change. *Frontiers in Environmental Science* 10.
4. UNFCCC. 2018. UN helps fashion industry shift to low carbon. https://unfccc.int/news/un-helps-fashion-industry-shift-to-low-carbon
5. Muneer, T., S. Maubleu, and M. Asif. 2006. Prospects of solar water heating for textile industry in Pakistan. *Renewable and Sustainable Energy Reviews* 10 (1): 1–23.
6. Abdel-Dayem, A. M., and M. A. Mohamad. 2001. Potential of solar energy utilization in the textile industry—a case study. *Renewable Energy* 23 (3): 685–694.
7. Muhib, M. I., and M. M. Rahman. 2024. Water and energy efficiency in textile production. In *Dye pollution from textile industry: challenges and opportunities for sustainable development*, ed. P. Singh, 279–305. Singapore: Springer Nature Singapore.
8. Bannister, K. E., and J. Alexander. 2009. *Industrial energy efficiency handbook*. McGraw-Hill Professional.
9. Rasul, M. G. 1994. A review on rational use of energy and its conservation measures in process industry: a case of textile industry. In *Mech 94: international mechanical engineering congress and exhibition: preprints of papers, resource engineering, volume 1 conference 1, construction, maintenance and operation*. Institution of Engineers, Australia Barton, ACT.

10. Palanichamy, C., et al. 2001. Budget constrained energy conservation-an experience with a textile industry. *IEEE Transactions on Energy Conversion* 16 (4): 340–345.
11. Tang, Z.-X., X. Wang, and B. Fraser. 2004. Minimizing energy consumption of yarn winding in ring spinning. *Textile Research Journal* 74 (12): 1097–1103.
12. Pardo Martínez, C. I. 2010. Energy use and energy efficiency development in the German and Colombian textile industries. *Energy for Sustainable Development* 14 (2): 94–103.
13. Jiang, Q., et al. 2019. A sustainable low temperature yarn reinforcing process to reduce water and energy consumptions and pollution in the textile industry. *Journal of Cleaner Production* 210: 646–652.
14. Hasanbeigi, A., and L. Price. 2012. A review of energy use and energy efficiency technologies for the textile industry. *Renewable and Sustainable Energy Reviews* 16 (6): 3648–3665.
15. Ozturk, H. K. 2005. Energy usage and cost in textile industry: a case study for Turkey. *Energy* 30 (13): 2424–2446.
16. Ozturk, E., et al. 2016. Sustainable textile production: cleaner production assessment/eco-efficiency analysis study in a textile mill. *Journal of Cleaner Production* 138: 248–263.
17. Kirchain, R., et al. 2015. Sustainable apparel materials. Materials Systems Laboratory, Massachusetts Institute of Technology, Cambridge, 1–34. https://search.issuelab.org/resources/29099/29099.pdf.
18. Sudhakara Reddy, B., and B. Kumar Ray. 2011. Understanding industrial energy use: physical energy intensity changes in Indian manufacturing sector. *Energy Policy* 39(11):7234–7243.
19. Farhana, K., et al. 2022. Energy consumption, environmental impact, and implementation of renewable energy resources in global textile industries: an overview towards circularity and sustainability. *Materials Circular Economy* 4 (1): 15.
20. Palamutcu, S. 2010. Electric energy consumption in the cotton textile processing stages. *Energy* 35 (7): 2945–2952.
21. Angelova, R. A., et al. 2021. Consumption of electric energy in the production of cotton textiles and garments. *IOP Conference Series: Materials Science and Engineering* 1031 (1): 012030.
22. Koç, E., E. Kaplan. 2007. An investigation on energy consumption in yarn production with special reference to ring spinning. *Fibres & Textiles in Eastern Europe*, 15(4):63.
23. Wang, L., Y. Li, and W. He. 2017. The energy footprint of china's textile industry: perspectives from decoupling and decomposition analysis. *Energies* 10 (10): 1461.
24. Li, K., and B. Lin. 2018. How to promote energy efficiency through technological progress in China? *Energy* 143: 812–821.
25. Dragović, N., S. Urošević, M. Vuković. 2021. Increasing energy efficiency of textile industry. In *5th international scientific conference contemporary trends and innovations in textile industry.* CT&ITI 2022.
26. Lin, B., L. Zhang, and Y. Wu. 2012. Evaluation of electricity saving potential in China's chemical industry based on cointegration. *Energy Policy* 44: 320–330.
27. Lin, B., Y. Chen, and G. Zhang. 2018. Impact of technological progress on China's textile industry and future energy saving potential forecast. *Energy* 161: 859–869.
28. Zhao, H., and B. Lin. 2020. Impact of foreign trade on energy efficiency in China's textile industry. *Journal of Cleaner Production* 245: 118878.
29. Hasanbeigi, A., and L. Price. 2015. A technical review of emerging technologies for energy and water efficiency and pollution reduction in the textile industry. *Journal of Cleaner Production* 95: 30–44.
30. Yoon, H.-S., et al. 2015. Towards greener machine tools–a review on energy saving strategies and technologies. *Renewable and Sustainable Energy Reviews* 48: 870–891.
31. Khattab, T. A., M. S. Abdelrahman, and M. Rehan. 2020. Textile dyeing industry: environmental impacts and remediation. *Environmental Science and Pollution Research* 27 (4): 3803–3818.
32. Sundaramoorthy, S., et al. 2023. Energy efficiency as a foundational technology pillar for industrial decarbonization. *Sustainability* 15 (12): 9487.
33. Yılmaz, K., et al. 2024. Sustainable textile manufacturing with revolutionizing textile dyeing: deep learning-based, for energy efficiency and environmental-impact reduction, pioneering green practices for a sustainable future. *Sustainability* 16 (18): 8152.

34. Afonso, T. L., A. C. Marques, and J. A. Fuinhas. 2017. Strategies to make renewable energy sources compatible with economic growth. *Energy Strategy Reviews* 18: 121–126.
35. Shrouf, F., and G. Miragliotta. 2015. Energy management based on internet of things: practices and framework for adoption in production management. *Journal of Cleaner Production* 100: 235–246.
36. Gavrila, H., et al. 2017. New trends in energy efficient electrical machines. *Procedia Engineering* 181: 568–574.
37. Dhayaneswaran, Y., Ashokkumar, L. 2013. A study on energy conservation in textile industry. *Journal of The Institution of Engineers (India): Series B* 94(1), 53–60.
38. Nunes, L. J. R., R. Godina, and J. C. D. O. Matias, Technological innovation in biomass energy for the sustainable growth of textile industry. *Sustainability* 11 (2): 528.
39. Kousar, S., et al. 2022. Sustainable energy consumption model for textile industry using fully intuitionistic fuzzy optimization approach. *Computational Intelligence and Neuroscience* 2022 (1): 5724825.
40. Kumar, P., et al. 2022. Economics of implementing solar thermal heating systems in the textile industry. *Energies* 15 (12): 4277.
41. Petrillo, A., M. Rehman, and I. Baffo. 2024. Digital and sustainable transition in textile industry through internet of things technologies: a Pakistani case study. *Applied Sciences* 14 (13): 5380.
42. Hossein Motlagh, N., et al. 2020. Internet of things (IoT) and the energy sector. *Energies* 13 (2): 494.
43. Aydin, M., et al. 2023. The role of green innovation in achieving environmental sustainability in European union countries: testing the environmental Kuznets curve hypothesis. *Gondwana Research* 118: 105–116.
44. Teklie, D. K., and M. H. Yağmur. 2024. The role of green innovation, renewable energy, and institutional quality in promoting green growth: evidence from African countries. *Sustainability* 16 (14): 6166.
45. Saqib, Z. A., G. Xu, and Q. Luo. 2024. Green manufacturing for a green environment from manufacturing sector in Guangdong province: mediating role of sustainable operations and operational transparency. *Applied Sciences* 14 (22): 10637.
46. Sarfraz, M., et al. 2022. Exploring a pathway to sustainable performance in manufacturing firms: the interplay between innovation capabilities, green process, product innovations and digital leadership. *Sustainability* 14 (10): 5945.
47. Hasanbeigi, A., C. Menke, and P. du Pont. 2010. Barriers to energy efficiency improvement and decision-making behavior in Thai industry. *Energy Efficiency* 3 (1): 33–52.
48. Peng, L., et al. 2015. Energy efficiency and influencing factor analysis in the overall Chinese textile industry. *Energy* 93: 1222–1229.
49. Sardianou, E. 2008. Barriers to industrial energy efficiency investments in Greece. *Journal of Cleaner Production* 16 (13): 1416–1423.
50. Alam, M. F. B., et al. 2023. Assessing the barriers of integrating technological innovations in textiles sector: implications towards sustainable production. *Green Technologies and Sustainability* 1 (3): 100039.
51. Pathirana, S., and M. Yarime. 2018. Introducing energy efficient technologies in small-and medium-sized enterprises in the apparel industry: a case study of Sri Lanka. *Journal of Cleaner Production* 178: 247–257.
52. Hasan, A. S. M. M., et al. 2019. Drivers and barriers to industrial energy efficiency in textile industries of Bangladesh. *Energies* 12 (9): 1775.
53. Kazancoglu, I., et al. 2020. A conceptual framework for barriers of circular supply chains for sustainability in the textile industry. *Sustainable Development* 28 (5): 1477–1492.
54. Kazancoglu, I., et al. 2022. Investigating barriers to circular supply chain in the textile industry from Stakeholders' perspective. *International Journal of Logistics Research and Applications* 25 (4–5): 521–548.
55. Vishwakarma, A., et al. 2024. Breaking barriers: paving the path to sustainable fashion–insights for a greener apparel and textile sector. *Journal of Advances in Management Research* 21 (3): 376–397.

56. Lin, B., and H. Zhao. 2016. Technology gap and regional energy efficiency in China's textile industry: a non-parametric meta-frontier approach. *Journal of Cleaner Production* 137: 21–28.
57. Saxena, S., A. S. M. Raja, and A. Arputharaj. 2017. Challenges in sustainable wet processing of textiles. In *Textiles and clothing sustainability: sustainable textile chemical processes*, ed. S. S. Muthu, 43–79. Singapore: Springer Singapore.
58. Zhang, Y., B. R. Bakshi, and E. S. Demessie. 2008. Life cycle assessment of an ionic liquid versus molecular solvents and their applications. *Environmental Science & Technology* 42 (5): 1724–1730.
59. Gulzar, T., et al. 2019. 1−green chemistry in the wet processing of textiles. In *The impact and prospects of green chemistry for textile technology*. ed. Shahid ul, I., B. S. Butola, 1–20. Woodhead Publishing.
60. Khatri, Z., et al. 2011. Cold pad-batch dyeing method for cotton fabric dyeing with reactive dyes using ultrasonic energy. *Ultrasonics Sonochemistry* 18 (6): 1301–1307.
61. Shah, I. A., et al. 2018. Experimental study on NOx emission characteristics of oxy-biomass combustion. *Journal of Cleaner Production* 199: 400–410.
62. Hussain, T., and A. Wahab. 2018. A critical review of the current water conservation practices in textile wet processing. *Journal of Cleaner Production* 198: 806–819.
63. Gopalakrishnan, M., V. Punitha, and D. Saravanan. 2019. 8-Water conservation in textile wet processing. In *Water in textiles and fashion*, ed. S. S. Muthu, 135–153, Woodhead Publishing.
64. Schramm, W., and J. Jantschgi. 1999. Comparative assessment of textile dyeing technologies from a preventive environmental protection point of view. *Coloration Technology* 115 (4): 130–135.
65. Tong, O., et al. 2012. An AHP-based water-conservation and waste-reduction indicator system for cleaner production of textile-printing industry in China and technique integration. *Clean Technologies and Environmental Policy* 14 (5): 857–868.
66. Khatri, A., et al. 2015. A review on developments in dyeing cotton fabrics with reactive dyes for reducing effluent pollution. *Journal of Cleaner Production* 87: 50–57.
67. Worrell, E., et al. 2009. Industrial energy efficiency and climate change mitigation. *Energy Efficiency* 2 (2): 109–123.
68. Hochman, G., and G. R. Timilsina. 2017. Energy efficiency barriers in commercial and industrial firms in Ukraine: an empirical analysis. *Energy Economics* 63: 22–30.
69. Trianni, A., and E. Cagno. 2012. Dealing with barriers to energy efficiency and SMEs: some empirical evidences. *Energy* 37 (1): 494–504.
70. Sharma, M., et al. 2023. Advancement in solar energy-based technologies for sustainable treatment of textile wastewater: reuse, recovery and current perspectives. *Journal of Water Process Engineering* 56: 104241.
71. Biyada, S., and J. Urbonavičius. 2025. Circularity in textile waste: challenges and pathways to sustainability. *Cleaner Engineering and Technology* 24: 100905.
72. Nikam, J. 2023. Gaps, challenges and drivers for environmentally sustainable textile and garment manufacturing in India. In *SEI Working Paper*.
73. Karim, M. R., et al. 2024. Analyzing the factors influencing sustainable supply chain management in the textile sector. *Cleaner Logistics and Supply Chain* 13: 100183.

Chapter 12
Lifecycle Assessment: Measuring Sustainability in Textiles and Fashion Production

Benjamin Tawiah, Emmanuel Abankwah Ofori, Raphael Kanyire Seidu, and Bin Fei

Abstract As the textile and fashion industry faces increasing scrutiny over its environmental and social impacts, LCA emerges as an essential tool for measuring and mitigating these effects across the entire product lifecycle. This chapter provides a holistic overview of LCA in the textiles and fashion sustainability discourse, outlining the key stages of textile and fashion production, from raw material extraction to manufacturing, distribution, use, and end-of-life management. The unique challenges and opportunities of each phase in the sustainability discourse are examined in detail with particular emphasis on the ecological footprint of textiles and fashion products, concerning the environmental impact categories, including resource depletion, water use, greenhouse gas emissions, and chemical toxicity. Additionally, the chapter addresses the social and economic implications of production practices, such as labour rights and community impacts, underscoring the interconnectedness of textiles and fashion sustainability. Also, the methodologies for conducting LCA, including ISO standards and software tools, along with the relevant case scenarios that illustrate successful applications in the fashion industry, have been discussed. The challenges and limitations of LCA, such as data quality and variability in product types, are critically assessed to highlight areas for improvement. Finally, the chapter concludes by presenting a succinct discussion on the future trends of LCA in the textiles and fashion industry, focusing on innovations in sustainable practices and

B. Tawiah (✉)
Department of Industrial Art (Textiles and Fashion), College of Art and Built Environment, Kwame Nkrumah University of Science and Technology, PMB, Kumasi, Ghana
e-mail: benjamin.tawiah@knust.edu.gh

B. Tawiah · R. K. Seidu · B. Fei (✉)
School of Fashion and Textiles (SFT), The Hong Kong Polytechnic University, Hong Kong SAR, China
e-mail: bin.fei@polyu.edu.hk

E. A. Ofori
School of Automotive Engineering, Hubei University of Automotive Technology, Shiyan, China

Hubei Key Laboratory of Energy Storage and Power Battery, Shiyan, China

Shiyan Key Laboratory of Quantum Information and Precision Optics (QIPO), Shiyan, China

S.-X. Jiang et al. (eds.), *Advances in Textile Materials and Processing Techniques for Sustainability*, SDGs and Textiles, https://doi.org/10.1007/978-981-95-0469-5_12

circular economy models. The chapter will serve as a valuable resource for textile and fashion sustainability education.

Keywords Lifecycle assessment (LCA) · Sustainability · Textiles · Fashion production · Environmental impact · Circular economy

12.1 Introduction

The textiles and fashion industry stands at a critical juncture in its evolution, facing unprecedented scrutiny regarding its environmental and social impacts. With global concerns about climate change, resource depletion, and social equity becoming increasingly urgent, stakeholders across the value chain, from manufacturers to consumers, have been seeking effective ways to assess and improve sustainability practices in recent years [1, 2]. Within this framework, Lifecycle Assessment (LCA) has become an essential instrument for monitoring and controlling the sustainability of apparel and textile goods.

Lifecycle Assessment is a systematic approach that evaluates the environmental, social, and economic impacts of a product from its inception to its disposal [3]. This methodology makes it possible to fully comprehend the environmental impact of a product, considering all phases of its lifecycle, including the extraction of raw materials, production, distribution, usage, and end-of-life management [4, 5]. Stakeholders can target measures that improve sustainability by using LCA to identify resource usage and emissions hotspots. The difficulties with sustainability are best exemplified by the textile and apparel industries [6]. It is distinguished by quick fashion cycles, disposable culture, and rapid production cycles, all of which seriously harm the environment. A single cotton t-shirt, for example, can consume more than 2,700 L of water during production-enough for one person to drink continuously for 2.5 years [7, 8]. Furthermore, the industry plays a significant role in climate change, accounting for 10% of greenhouse gas emissions worldwide [9, 10]. Because of this, there is a greater need than ever for reliable evaluation tools like LCA.

LCA provides a framework for quantitatively understanding these complex issues. By analysing the impacts associated with each phase of a product's lifecycle, LCA enables designers, manufacturers, and policymakers to make informed decisions that can lead to more sustainable practices [11, 12]. For example, LCA can reveal that while organic cotton may be perceived as a more sustainable option, its water-intensive cultivation may still pose significant ecological challenges. This kind of insight is crucial for developing truly sustainable solutions. Moreover, LCA extends beyond environmental considerations; it also encompasses social and economic dimensions. Hence, the fashion industry is not only a significant economic driver but also a sector that heavily influences labour practices and community well-being. By integrating social impact assessments into the LCA framework, stakeholders can evaluate labour conditions, community engagement, and the economic sustainability of production practices. This holistic approach fosters a more equitable and sustainable industry [13].

In recent years, the adoption of LCA in the textiles and fashion sector has gained momentum, driven by increasing consumer awareness and regulatory pressures. Brands are now more frequently employing LCA to substantiate their sustainability claims, engage consumers, and differentiate themselves in a crowded marketplace [1]. However, challenges remain in standardising LCA methodologies, ensuring data quality, and addressing variability among products and processes.

This chapter, therefore, provides a comprehensive overview of LCA in textiles and fashion production. It will explore the theoretical foundations of LCA, the various stages of a textile product's lifecycle, and the environmental and social impact categories that are critical to understanding sustainability. Furthermore, the chapter discusses the methodologies for conducting LCA, including best practices and tools available to industry stakeholders. It further illustrates the potential of LCA to drive meaningful change in the textiles and fashion industry and address the challenges and limitations of the methodology, offering insights into how these can be overcome to enhance the effectiveness of sustainability assessments. Finally, the chapter discusses the future trends and innovations in LCA, highlighting the role of technology and consumer engagement in fostering a more sustainable fashion ecosystem in navigating the complexities of environmental issues associated with textiles and fashion production.

12.2 Understanding Lifecycle Assessment (LCA)

The life cycle assessment (LCA) serves as a structured method for assessing the environmental effects of a product, process, or activity, spanning from acquiring raw materials to their ultimate disposal. This approach aids in comprehending ecological repercussions and facilitates informed strategic planning for future endeavors [14, 15] (see Fig. 12.1). LCA is acknowledged worldwide and standardised by the International Organisation for Standardisation (ISO) 14,040 series, which comprises four primary stages: defining the goal and scope, conducting life inventory analysis, performing impact assessment, and interpreting the results [16, 17]. An in-depth understanding of the fundamentals of LCA is advantageous. With its extensive history, the LCA protocols offer a reliable and consistent framework for assessments across diverse industries and regions. This empowers designers, manufacturers, and policymakers to advance sustainability and the tenets of a circular economy.

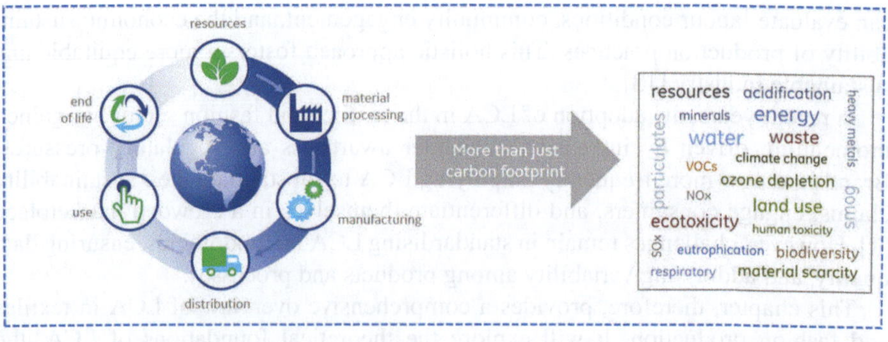

Fig. 12.1 The "cradle-to-grave" accounting concept of LCA. Available from https://www.cytivalifesciences.com/en/us/Solutions/Bioprocessing/Knowledge-center/Single-use-and-sustainability

12.2.1 Historical Context of LCA

LCA has been used since its origins in the 1960s, when concerns about resource scarcity and environmental damage were prevalent [18, 19]. Originating in industrial packaging research, life cycle assessment (LCA) first concentrated on energy consumption and a few emissions, which led to a mostly disorganised technique development in the US and Northern Europe [20]. LCA has gone through several phases of development and adoption, as shown in Fig. 12.2. These include (i) the early years (1960–1970), (ii) the conception period (1970–1990), (iii) the standardisation period (1990–2000), (iv) the elaborations period (2000–2010), and (v) the concept broadening period (2010–until now) [19, 21].

The concept of LCA was first proposed in the 1960s as Eco balance in Europe or Resource and Environmental Profile Analysis (REPA) in the US [22]. Due to its private nature, one of the first thorough LCA studies was carried out for Coca-Cola's beverage bottle selection in 1969, but it was never published [23]. Solid waste management emerged as a key concern in the early 1970s. Influential committees and publications from that era, such as the "Committee on the Survey of Materials Science and Engineering", "A Blueprint for Survival", and "The Limits to Growth", reflect the growing body of knowledge on environmental issues. Later in the decade, the energy crisis prompted businesses to adopt energy management strategies rooted in life cycle thinking. This period saw the development of foundational methodologies, setting the stage for more extensive assessments in the future. Towards the end of the 1980s and into the 90 s, global environmental issues heightened the agenda [24], expanding from local and regional issues to encompass more international ones, including those garnering worldwide attention, such as ozone depletion and climate change. Around this time, 150 countries gathered in Rio for the UN Conference on Environment and Development to establish global warming standards and assess the effects of environmental pollution and hazards. The UN's 1987 Brundtland report sparked a surge in interest in sustainability and sustainable development [25, 26],

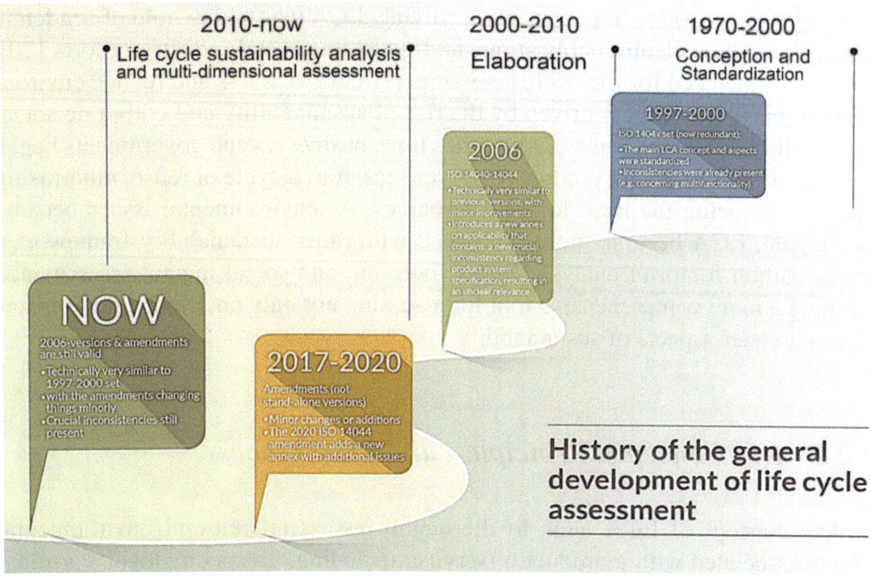

Fig. 12.2 The overall development history of LCA: development stages of LCA. Reproduced from [5], 2024. An open access article by Elsevier

and as LCA was viewed as a useful tool for calculating impact across a range of issues, scholarly interest in it started to grow as well.

The Society of Environmental Toxicology and Chemistry (SETAC) standards served as a direct inspiration for the development and publication of the life cycle assessment (LCA) concept in the early 1990s [27]. These were based on the establishment of a retrospective tool to quantify the impacts of a particular product [28]. Additionally, it was linked to EU integrated product policy-based (IPP) rules, packaging laws, marketing laws, and eco-labelling laws. The ISO Technical Committee (TC 207) Subcommittee SC 5 in Paris adopted and modified the SETAC standards to become International Organisation for Standardisation (ISO) standards in November 1993 [29]. This resulted in the ISO 14040 series (ISO 14040–49 in 1997–2000), which was subsequently revised in 2006 [28, 30]. Important initiatives in the late twentieth century included the introduction of the International Journal of Life Cycle Assessment, the publication of the LCA sourcebook A European Guide to Life Cycle Assessment, the introduction of the International Organisation for Standardization's (ISO) standards 14,040 and 14,041, the development of the Dutch Eco-Indicator 99 method, and the introduction of normalised assessment scores for environmental impact categories (CML-92 method) [31, 32].

The scope and volume of LCA advancements have increased over time, making it a more comprehensive and extensive environmental instrument that can even be used to determine policy [33, 34]. These policy-triggered jumps are connected to energy and bioenergy rules for greenhouse gas (GHG) accounting. However, while this regulatory-to-policy trajectory was evolving, LCA was taking hold of academic interest through academic publications and government and industry reports [20]. Importantly, the need for life cycle assessment (LCA) to assess and reduce environmental impacts was largely driven by the rise of sustainability and corporate social responsibility (CSR) initiatives. During this time, businesses and governments began to look into ways to design products that were easier to recycle or reuse, minimising waste and reducing the need for new resources. As environmental issues became more urgent, LCA became more integrated with other sustainability frameworks, such as carbon footprint analysis, water footprint, and social impact assessments, becoming a more comprehensive tool for assessing not only environmental impacts but also broader aspects of sustainability.

12.2.2 *Key Concepts, Principles, and LCA Process*

The key concept of LCA aims to thoroughly assess all relevant environmental impacts associated with a product or service, including effects on global warming, ecosystems, human health, and resource depletion. This "cradle-to-grave" approach (see Fig. 12.3a) ensures a complete understanding of environmental interactions, preventing the unintended displacement of environmental burdens from one life cycle stage to another [35–37].

Transparency is a fundamental principle in life cycle assessment (LCA), highlighting the significance of adequately documenting and presenting all decisions made during the evaluation by clearly defining the boundaries of the system [38]. Depending on the purpose and extent, the system boundaries can be established at various degrees of detail; for instance, the analysis can be restricted to a product's production phase (cradle-to-gate) or encompass the product's full life cycle (cradle-to-grave) [20]. Based on the principles of a standardised methodology described by the International Organisation for Standardisation (ISO) in the ISO 14040 series, the

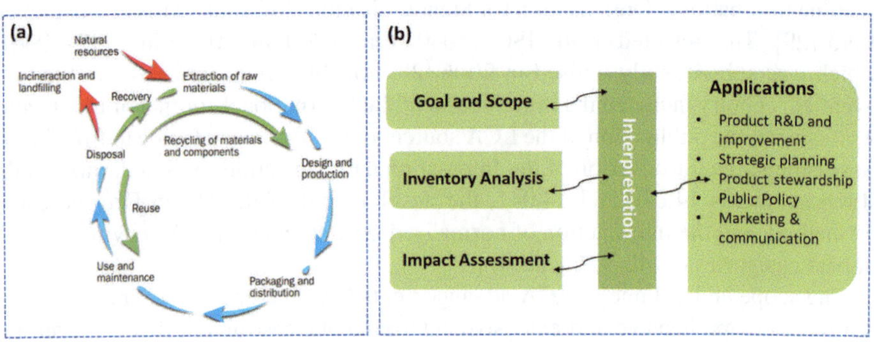

Fig. 12.3 **a** Life Cycle Assessment iterative process in a cradle-to-grave concept **b** Structured framework of LCA carried out in four basic, interdependent stages [37]

clear and defined boundaries make up a scientific approach in guaranteeing consistency and preventing bias in the results [39]. This standardisation, which is grouped into a four-phase organised framework shown in Fig. 12.3b [37], allows for consistent application and comparison of outcomes. These are: goal and scope definition, life inventory analysis, impact assessment, and interpretation [16, 17].

LCA's purpose and scope, according to ISO 14040, comprises the target audience, the context of the work conducted, and the causative elements driving the LCA, together with the expected utility of the results for decision making [40, 41]. Inventory analysis serves as the basis for a subsequent effect assessment by obtaining data from sources, generating flowcharts, and using computation techniques [42]. These procedures provide precise estimates of pertinent inputs and outputs, such as raw materials, energy use, emissions, and waste production, related to a product and its system. Crucially, several life cycle inventory databases covering various industrial sectors and run by multiple institutions and organisations were established in the early 1990s. The resource usage and emissions of a single industrial process may, however, vary significantly across databases due to variations in data standards and quality; however, at this stage of development, the emphasis was on broadening the coverage, and for many processes, there was no data at all [43]. These issues were improved with the launch of the first ecoinvent database (v1.01) in 2003, which covered industrial sectors and sought to achieve standardised data standards and quality [18]. While the impact assessment considers the possible environmental effects of the data gathered, the interpretation examines the findings in light of the stated purpose and scope, pointing out important problems and offering suggestions for improvement [44]. Not surprisingly, these four stages informed the texts of each of the ensuing documents: 14,040 General Principles in 1997, 14,041 Life Cycle Inventory Analysis (LCI) in 1998, as well as both 14,042 Life Cycle Impact Assessment (LCIA) and 14,043 Interpretation in 2000 [29].

Furthermore, the functional unit is a key idea in LCA that offers a benchmark for contrasting various systems. By standardising the inputs and outputs of various products, this standardised measure enables meaningful comparisons [45]. Additionally, LCA uses an iterative approach, acknowledging that as new information becomes available, the assessment method may need to be modified and improved. Because of this flexibility, practitioners can go back and change parts of the research in light of new information discovered during the evaluation [46]. LCA's iterative process guarantees that the final findings appropriately represent the most up-to-date and pertinent data available. The LCA principle is used to evaluate the environmental footprints of products in a variety of industries, including manufacturing, building, food production, the automotive, textile, and fashion sectors, due to its importance in fostering sustainability, developing policies, and decision-making [47].

12.3 Lifecycle Stages of Textiles and Fashion Products

The textile value chain, which supplies fabrics and products needed in daily life, from apparel to household goods and other textile products, is essential to modern society [48, 49]. Since the year 2000, the world's textile production has nearly doubled, with total fibre production rising from 5.9 kg per capita in 1975 to 13 kg per capita in 2018 [50]. Hence, it has been predicted that the volume of apparel manufacturing will increase by 81% between 2018 and 2030 [51]. The linear model, which is currently dominated by rapid fashion, is characterised by low rates of repair, reuse, and fibre-to-fibre recycling, which encourages consumers to buy cheap, short-lived clothing and exacerbates the industry's environmental effects [52]. Furthermore, wastewater pollution from textile dyeing and finishing operations is difficult to manage, detrimental to aquatic habitats, and dangerous to human health [53]. The fashion sector has a significant impact on the environment, producing about 92 million tonnes of garbage annually and using 79 trillion litres of water [54]. Furthermore, coal remains an important component of the operations of major textile-producing nations like Bangladesh, India, and China [55]. These environmental issues are exacerbated by energy-intensive clothing maintenance procedures like washing, drying, pressing, and dry cleaning [56, 57]. Additionally, fewer than 1% of clothing is recycled worldwide due to inadequate technology, and while used clothing can be transported outside the European Union (EU), the great majority (87%) is burned or dumped in landfills. Thus, harmful chemicals and greenhouse gases (GHG) are released into the environment when textile waste is improperly disposed of at landfills or incineration facilities. It is estimated that each year, Europeans utilise approximately 26 kg of textiles and trash approximately 11 kg. It is clear from the aforementioned submissions that a complex web of processes is involved in the various phases of the lifetime of textiles and fashion goods [48]. These include several key stages which start from raw material extraction and proceed to fabric manufacturing, clothing manufacturing, retailing, use and end of life stages [58]. Each phase, from design to end-of-life, presents opportunities for improvement in sustainability practices. Figure 12.4 depicts a generic garment life cycle, with the different stages being explained in the ensuing subsections.

12.3.1 *Raw Material Acquisition*

The initial phase of textile production for fashion items and textiles is material sourcing. The extraction, harvesting, or purchase of materials needed to make textiles is known as raw material sourcing. The final goods' sustainability, quality, and environmental impact are all greatly impacted by this step. One or a combination of multiple fibre types can be used as the raw material for a fashion garment [51]. As shown in Fig. 12.5, textile fibres can be categorised as natural fibres (animal and vegetable) and manufactured fibres, which include but are not limited to synthetics,

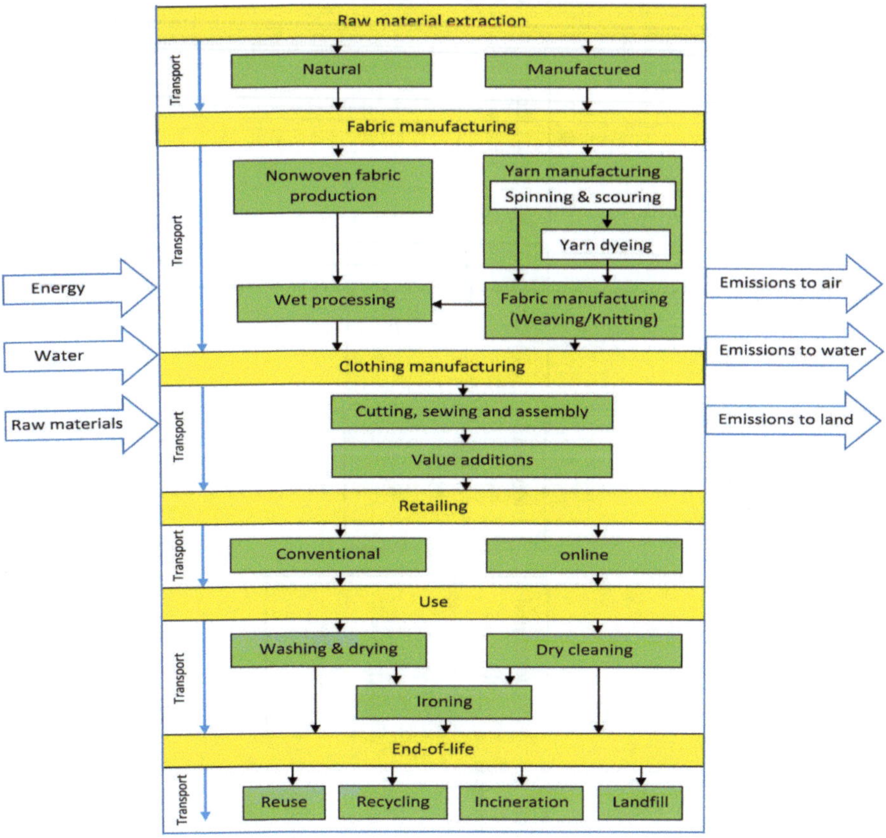

Fig. 12.4 A generic life cycle diagram for clothing. Reproduced from [51], 2021. An open access article by Elsevier

regenerated cellulosic, inorganic, and recycled fibres, and biobased, including organic cotton and bamboo [59, 60].

12.3.1.1 Natural versus Synthetic Fibres

Natural fibres have been an important component of human civilisation as sustainable and environmentally acceptable sources of raw materials for making eco-friendly products [61]. Due to their lower density, lighter weight, reduced cost, biodegradability, abundance, low CO_2 emissions, energy requirements, and minimal health risks during processing, natural fibres are preferred bio-sourced materials [62–64]. Natural fibre materials can be categorised by their source, including minerals, cellulose (found in plants), and protein (found in animals). Plant fibres consist of the following: linen, hemp, and jute are bast fibres, which are made up of overlapping

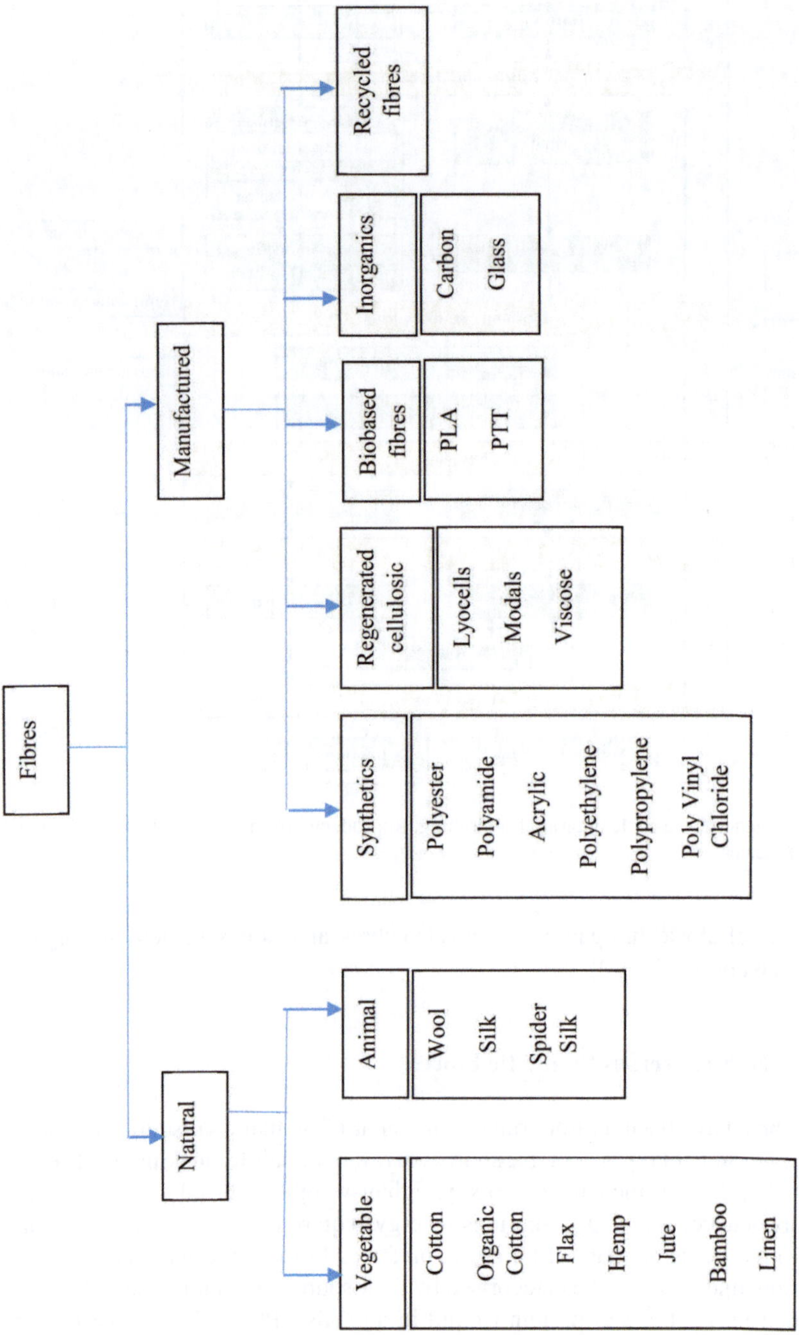

Fig. 12.5 Types of fibres for material sourcing. Reproduced from [51], 2021. An open access article by Elsevier

cells in the inner bast tissue of plant stems; sisal, abaca, and henequen are leaf fibres that occur as part of the fibrovascular systems of leaves; husk fibres such as coir from coconuts; and seeds, such as cotton and kapok, each of which is made up of a single, long, and narrow cells [60]. Chemically, all vegetable fibres consist mainly of cellulose (refer to Fig. 12.6a), although they also contain varying amounts of substances such as hemicellulose (see Fig. 12.6b), pectin (see Fig. 12.6c), wax and fat (refer to Fig. 12.6d), and lignin (refer to Fig. 12.6e) [61, 65–67].

Hemicellulose, which has a far lower molecular weight than cellulose, is firmly bound to cellulose fibrils, most likely via hydrogen bonds [69]. Hemicellulose is partially soluble in water and hygroscopic because its open structure contains numerous hydroxyl and acetyl groups. There is an aromatic polymer of phenylpropane units in lignin, which is amorphous and extremely complex [70]. Due to its ability to retain water, protect against biological threats, and fortify the stem against wind and gravity, lignin is a crucial component of plant fibre. Animal fibres such as wool, hair, fur, and secretions like silk are made of proteins and have a very high strength-to-weight ratio. These fibres are produced by worms, spiders, and other species [71]. The most widely used natural fibres are wool and cotton. Although asbestos is the only significant mineral fibre, its economic effects are currently minimal due to the related health issues. As shown in Fig. 12.6f are some natural fibre resources [63]; whereas Table 12.1 below shows the major natural fibre producers in the world, their potential applications, and associated by-products [72].

However, since the middle of the 1990s, synthetic fibres, which are manufactured from petrochemical-based basic ingredients and polymerised into long, linear fibres, have dominated the market for fibre, surpassing cotton [73]. Currently, synthetic fibres account for around 64% of the world market for textile fibre, with 111 million metric tons of fibre produced each year [74]. This demand stems from their exceptional feel, softness, and attractiveness as well as specially designed qualities like stain resistance, waterproofing, and frequently stretchability, which make them very consumer-friendly. Nylon, polyester, spandex, acrylic, and olefin are the "big five" petroleum-based synthetic fibres; other types include glass, metallic fibres, acrylics, polyurethanes, and polypropylene. Of them, polyester alone accounts for around 51% of the world's fibre production, with over 53 million metric tons manufactured each year [72]. Crucially, the numerous types of fibres derived from synthetic polymers have been assessed as potentially profitable commercial goods; the three most popular ones are nylon, polyester, and acrylic [75]. Despite their dominance, substantial carbon dioxide (CO_2) loads are produced by the high temperatures during polycondensation [76]. Hence, different organisations have expressed a variety of worries about the detrimental effects of the clothing and textile sector, regarding the natural and synthetic classes of fibres. This is because, whereas polyester and acrylic fibres are made using energy-intensive petrochemicals [77–79], cotton and cellulose fibre production begins with agricultural cultivation, which requires a lot of water and chemicals [79, 80]. As a result, the concepts of sustainability, eco-friendly materials, circular business models, and ethical sourcing have emerged [81].

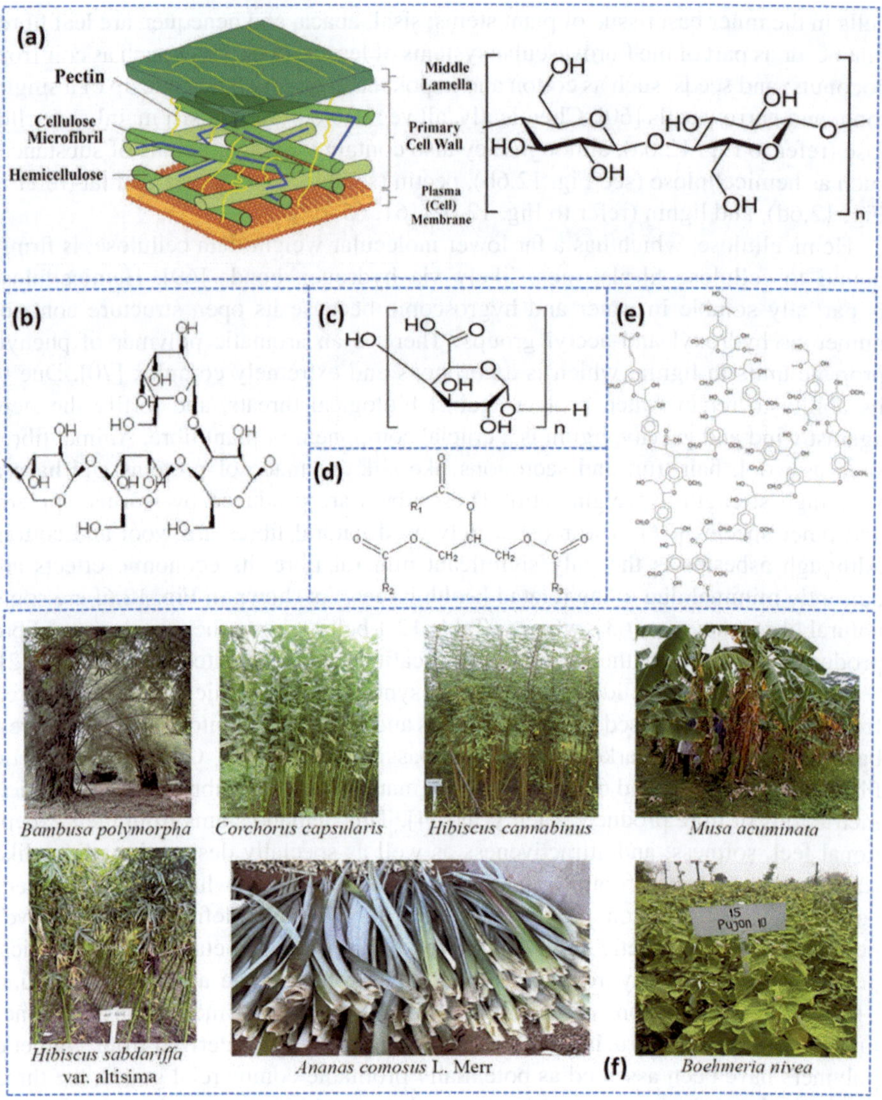

Fig. 12.6 **a** Plant cell walls with cellulose as the main component. Reproduced from [68], 2024. An open access article by Taylor and Francis, **b** hemicellulose **c** pectin **d** lignin **e** waxes and fats. Reproduced from [67] 2021, an open-access article by MDPI **f** Some natural fibre resources. Reproduced from [63], 2020. An open-access article by MDPI

Table 12.1 Natural fibre types, producers, and markets

Natural fibre	Main producers	Fibre market	By-product
Cotton	China, USA, India, Pakistan	Textile fabric: apparel, home furnishing, upholstery, non-woven, cellulose, medical and hygienic supplies	Linter, cottonseed, stalks
Kapok	Indonesia	Pillow, mattress	Seeds, wood
Jute	India, Bangladesh	Hessian, sacks, and carpet backing	Stalks (sticks)
Kenaf	China, India, Thailand		
Flax	China, France, Belgium, Belarus, Ukraine	Textile fabric, composites of non-woven, insulation mats	Seeds, shives
Hemp	China		
Ramie	China	Textile fabric	Leaves, stem
Abaca	Philippines, Ecuador	Specialty paper, tea bags	Leaves, juice
Coir	India, Sri Lanka	Twine, ropes, carpets, brushes, mattresses, geotextile, and horticultural products	Copra, water, shell, pith, wood and leaves
Sisal	Brazil, China, Tanzania, Kenya	Twine and ropes	Short fibre, juice, poles, and stem
Henequen	Mexico		
Wool	Australia, China, New Zealand	Knitted wear	Lamb meat, cheese
Silk	China, India	Fine garments, veils, and handkerchiefs	Worms, cocoons, fruits, and wood

12.3.1.2 Sustainable Sourcing Practices

The fashion industry, which ranks second in the world for pollution after the oil industry, is searching for sustainable materials derived from plants, animals, and microbes to replace the current materials [82]. In response to the industry's problems, sustainable sourcing has surfaced, emphasising the reduction of the sector's environmental impact while advancing moral labour standards and economic justice [83]. According to Balchandani et al. [84], it is emphasised that in 2025, consumers' decisions to purchase mass-market clothing will be heavily influenced by sustainability. Therefore, global fashion retailers such as Nike, Zara, Uniqlo, Doodlage, Bodice Raw Mango, Rent the Runway, H&M, Patagonia, Stella McCartney, and others are steadfastly dedicated to implementing long-term initiatives to change their material sourcing and downstream supply chains to a more responsible approach [85–87].

In particular, converting to sustainable, eco-friendly fabrics like hemp, lyocell, organic cotton, linen, and recycled polyester/nylon decreases emissions from the extraction and cultivation of raw materials upstream [88]. Also, organic bamboo, banana, and merino wool fibres serve as alternative materials for the sustainable textile industry [89]. The underground root structure of mushrooms, known as mycelium, was found during excavations and is used as a raw material by the fashion industry as a fast-growing alternative to synthetic leather [90]. Conversely, newer materials are sourced by bacterial fermentation, crop/food waste upcycling, and carbon-sequestering algae cultures, as well as alternative materials like Piñatex (pineapple leaf leather). Similarly, bacterial cellulose for remodelling and mending is sourced from agro-industrial waste materials, which lowers prices and increases product accessibility [91]. Adjusting the organism's growth process makes it possible to create naturally biodegradable materials that are flexible, durable, and breathable without requiring the soil disturbance or irrigation that comes with animal husbandry. Importantly, organic cotton has emerged as the most representative and fastest-growing market at 39.5% from 2013 to 2014, an estimated US$15.7 billion in 2014 [92]. Presently, a compound annual growth rate of 6.8% is anticipated for organic cotton between 2022 and 2029 [76]. In 2004, Nudie Jeans produced their first organic cotton t-shirt. By 2017, they had switched to using only organic cotton in all of their cotton items [76, 93]. In addition, Ellis, McCracken and Skuza [94] discovered that consumers would pay 25% more for organic cotton t-shirts. Table 12.2 presents some sustainable materials for textile production.

Table 12.2 Summary of sustainable materials, their sources, properties, benefits and features

Material	Source	Properties	Environmental benefits	Distinct features
Organic cotton	Natural fibre from cotton plants grown without synthetic chemicals	Soft, breathable, hypoallergenic, biodegradable	Reduces water pollution, no toxic pesticides or fertilisers	Better for sensitive skin, requires significant water use
Recycled polyester	Post-consumer plastics	Durable, resistant to stretching and shrinking, moisture-wicking	Reduces landfill waste, lowers carbon footprint	Reuses non-biodegradable plastics, suitable for activewear
Hemp	Stalks of the hemp plant	Strong, durable, breathable, naturally anti-microbial	Low water use and improves soil health	UV-resistant, grows quickly without pesticides
Lyocell	Wood pulp	Soft, breathable, moisture-wicking, biodegradable	Closed-loop process reduces chemical waste	Smooth surface ideal for sensitive skin and luxury textiles
Bamboo	Bamboo plant	Soft, silky texture, moisture-wicking, antibacterial	Grows rapidly, no pesticides, renewable	Can be processed sustainably or chemically

(continued)

Table 12.2 (continued)

Material	Source	Properties	Environmental benefits	Distinct features
Recycled wool	Pre and post-consumer wool waste	Warm, durable, and breathable	Diverts waste from landfills, reduces demand for virgin wool	Preserves natural wool properties, suitable for cold climates
Pinatex	Pineapple leaf	Leather-like, lightweight, water-resistant	Repurposes agricultural waste, no additional land or water use	Vegan leather alternative with a unique texture
Econyl	Recycled nylon	Strong, elastic, lightweight, resistant to abrasion and chemicals	Reduces ocean plastics, lowers demand for virgin nylon	Circular production ideal for swimming outfits

12.3.2 Manufacturing and Production

12.3.2.1 Processes and Technologies

In the textile business, manufacturing and production involve several intricate procedures that transform raw fibres into finished textiles and products (see Fig. 12.7) [53]. Natural fibres like cotton, wool, silk, and linen undergo initial dry and wet processing techniques to clean and prepare them for spinning. For example, cotton is ginned to separate the fibres from the seeds [95, 96], whereas wool must be scourged to remove lanolin and impurities using a continuous five to six-bowl line with 1% surfactant to remove the dirt, sand, oil, and suint [97, 98]. Conversely, synthetic fibres such as acrylic, polyester, and nylon are extruded and polymerised to create filaments [75]. Following preparation, the fibres go through a process called spinning, where they are extracted and twisted into skeins. Many technologies, such as ring spinning, open-end spinning, and air-jet spinning, are used in modern spinning mills to produce yarns with a variety of properties [99, 100]. The spinning process establishes crucial yarn characteristics, including thickness, strength, and homogeneity, before fabric construction, which mostly entails knitting, weaving, and non-woven techniques [101, 102]. Weaving uses looms to interlace two sets of threads (warp and weft) at right angles. To create intricate patterns and structures, modern weaving uses advanced equipment such as rapier, projectile, or air-jet looms. Knitting creates fabric by forming interconnected loops of yarn, either using weft knitting (common in t-shirts) or warp knitting (used in lingerie and technical textiles) [103, 104]. In non-woven manufacture, fibres are bound together by mechanical, thermal, or chemical methods, completely avoiding the development of yarn [105]. In contrast, there are several ways that sustainable textile processing varies from wet processing, such as the use of enzymes, environmentally friendly dyeing, plasma treatment, supercritical fluid technology, digital inkjet printing, using

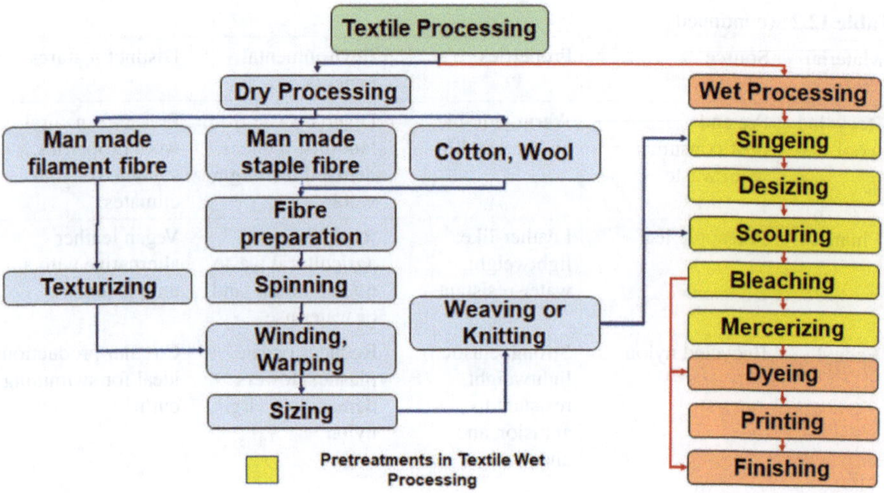

Fig. 12.7 Process flow chart for textile processing Reprinted with permission from [53] 2021, Elsevier

ultrasonic waves instead of thermal energy, recycling process inputs, electrochemical dyeing, foam finishing, and advancements in dyeing and printing equipment [106–109]. Sustainable advancements in textile dyeing and printing have been widespread in addressing environmental issues and improving the economy, quality, and energy conservation.

12.3.2.2 Energy Consumption and Emissions

Energy is one of the main cost factors in the textile industry due to the long manufacturing flow chain and the intensive use of machines, all of which require significant energy to create fabrics [76, 110]. The share of total manufacturing energy consumed by the textile industry in a particular country depends on the structure of the manufacturing sector in that country [107]. For instance, in 2010, the textile sector consumed less than 2% of the total energy used in manufacturing in the United States, compared to roughly 4% in China [107, 111]. Fuel energy accounted for 53% of the total energy consumed in the U.S. textile industry in 2010, while electricity accounted for 47%, according to manufacturing census data [112]. However, the U.S. textile industry is classified as the 6th highest steam user amongst 15 major industrial sectors investigated in the U.S. The same study found that roughly 48% of the energy input to the textile industry is lost onsite (e.g. in boilers, motor systems, distribution, etc.) [107].

Approximately 70% of the energy utilised by the textile industry is generated by electricity, and over 50% from thermal energy. For example, spinning and weaving are completely powered by electricity, but sizing, dyeing, and rewinding use thermal energy to heat the water or dry the fabric [113, 114]. Furthermore, spinning cotton can consume up to 3,984 kWh of energy, knitting uses 926 MJ, and weaving consumes roughly 10,430 MJ [115]. However, Moazzem et al. [116] report that spinning is most burdensome to climate change potential, abiotic depletion, and acidification categories. According to a Geneva Environmental Network report, the textile and apparel sector has been noteworthy for its substantial yearly contribution of 1.7 billion tons of CO_2, or roughly 6 to 10% of world carbon emissions [117–119].

Importantly, in the raw material extraction stage, conventional cotton cultivation generates a large number of ammonia emissions and requires extensive energy use [120, 121]. Furthermore, finishing and dying have a big influence on emissions and the environment. Due to the large quantities of coal, steam, water, dyes, and auxiliaries needed, the dyeing stage of a cotton t-shirt is where the majority of abiotic depletion damage occurs [122]. While yarns negatively impact the ability of life to live in the environment, batch dyeing uses 150 L of water every kilogram of cloth dyed, or 20% of the world's water pollution [10, 123].

A major issue is the growing amounts of textile waste, which are disposed of annually at a rate of about 92 million tons and 79 trillion litres of water [54]. A typical cotton t-shirt, based on 50 washes, provides 35% of carbon dioxide (CO_2) emissions during the production phase and 52% during the use phase, according to studies of life cycle emissions. Furthermore, the production of polyester, one of the most popular synthetic materials in the world, necessitates the use of non-renewable resources like petroleum and energy-intensive manufacturing techniques. Similarly, the expansion of fibre crops, especially cotton, has been accompanied by excessive pesticide production and the depletion of natural resources and energy [124]. Hence, the manufacturing of textiles has a significant environmental impact, with greenhouse gas emissions generated at every stage of a garment's lifecycle.

12.3.3 Distribution and Retail

The distribution and retailing phases are pivotal in the textile and fashion industry, bridging manufacturers and consumers [125]. These phases dictate the delivery of fashion items, final products, and raw resources to consumers. As the demands of consumers throughout the world increase and change, the fashion and textile industries are adjusting to satisfy customer demands for sustainability and convenience [126]. This is compounded by the growing role of e-commerce and digital platforms in shaping modern retailing practices [127, 128]. Crucially, distribution networks and worldwide markets have been impacted by globalisation and textile production. With an export value of over USD 176 billion as of 2021, China was the largest exporter of all textile types. The European Union came in second with an export rate of USD 151 billion, and the top 5 countries were Bangladesh, Vietnam, and Turkey [129]. Thus, it is projected that the global fashion market would increase from $1.5 trillion in 2020 to approximately $2.25 trillion in 2025 [130–133].

Therefore, to ensure seamless distribution, manufacturers, wholesalers, and retailers must overcome obstacles including tariffs, cultural differences, and supply chain interruptions. Moving completed goods from producers to different retail locations is known as distribution in the textile and fashion industries. Offering a vast array of products and services to customers, retailing is a vibrant and diverse industry. To stay competitive, international merchants frequently use a just-in-time inventory model, which lowers stock holdings and streamlines operations [134]. Thus, retailing of fashion clothing occurs in two ways: conventional retailing and online retailing [135]. Retailing primarily affects the environment through the energy required in stores, and traditional retailing has been shown to have a greater environmental impact than internet retailing [127, 136]. For brand positioning, customer experience, and eventually sales, the retail stage is therefore essential. Fashion retailers are urging customers to visit their stores more often by promoting the concept of "Here Today, Gone Tomorrow" in conjunction with the rise of small merchandise collections [137]. However, fashion stores have been competing with others in recent years by guaranteeing speed to market through their capacity to deliver the current trends that are showcased at runways and fashion shows quickly [138]. Retailers' operations have a variety of ways to impact environmental and social sustainability [139]. For instance, merchants can enhance the sustainability of their offerings in terms of packaging, delivery, and assortment. As a result, understanding the industry's supply chain results is important.

12.3.3.1 Transportation Impacts

Transport, as a sector of the national economy, has an irreplaceable role, as it connects people, cities, and countries. Transport consumes resources, but it is essential for the functioning of human society. It is one of the main pillars of the economy, giving people access to goods and services as well as work. From a global perspective, transport is crucial for international trade, which would otherwise not be possible [140].

Given that transportation accounts for 15% of anthropogenic GHG emissions, 23% of global energy-related emissions, and 8.7 Gt CO_2-eq emissions, it has emerged as a key theme in the fight against GHG emissions and their effects on human health and the environment [141]. Although the precise amount of fuel used is impossible to measure, container ships deliver almost 90% of clothing each year. This demonstrates that transportation plays a significant role in the production of textiles, involving every step of the garment life cycle [142], from the supply chain to the final consumer, and requiring a significant amount of petroleum [143, 144]. Moreover, the production and burning of fossil fuels are the main causes of transportation's climatic impact. Large volumes of greenhouse gases, including carbon dioxide (CO_2), methane (CH_4), nitrous oxide (N_2O), and hydrofluorocarbons (HFCS), are released during this process, which contributes to environmental deterioration and global warming [145]. The clothing supply chain is global, and air freight, when used, adds to the significant contributions of greenhouse gas (GHG) emissions [136]. Businesses in the fast fashion sector select rapid and efficient modes of transportation since they regularly switch up their collections and strive for prompt and effective deliveries.

Because of its enormous carbon emissions, air freight has a much greater environmental impact than road transport. The European Environment Agency (EEA) claims that per ton km, air freight emits up to ten times as many emissions as road transport. Approximately 23% of global energy-related CO_2 emissions in 2019 came from direct greenhouse gas (GHG) emissions from the transportation sector. Road vehicles accounted for 70% of these emissions, with rail, shipping, and aviation contributing 1%, 11%, and 12%, respectively. In contrast, a much slower ocean vessel has a much higher cargo capacity. Despite its high CO_2 emissions, it transports thousands of tons of freight, which results in comparatively low emissions per ton of cargo. Air travel contributes the most to greenhouse gas emissions, while maritime transport is the most ecologically benign of the three. The industry supply chains are highly fragmented, which makes it easy to decide which mode of transportation should be used primarily for long distances from an environmental standpoint. On the one hand, this allows them to maintain fast, cost-effective production, but on the other hand, it creates many challenges related to transparency, monitoring, securing proper labour conditions, consistency, and a high shock sensitivity level [146, 147]. For example, a quantitative study by Zamani et al. [148] demonstrated the potential risk of problem shifting from increased customer transportation, which can completely offset the benefits gained from reduced production and increased emissions. In another study, [149] establishes that the production and cross-broad transportation of jeans contribute to 91% of the carbon footprint of fast fashion consumption. Furthermore, the transportation process uses a lot of packaging materials, which increases its environmental impact [127].

12.3.3.2 Packaging and Waste

In the textile industry, packaging uses plastics like polypropylene (PP), high-density polyethylene (HDPE), low-density polyethylene (LDPE), wrapping films, and paper materials to present and protect clothing and fabric during manufacturing, shipping, and retail sales [150, 151]. However, when discarded, these packing materials frequently become non-biodegradable waste, which greatly adds to the waste generated by the textile industry. According to Lüttin [152], the amount of packaging trash produced annually in the European Union increased by 22.5% per person between 2010 and 2021, to an alarming 84.3 million tons.

12.3.4 Use Phase

Textile consumption is projected to rise by around 63% by 2030, leading to a significant increase in textile waste, an issue that is drawing global concern [153]. The use stage is a crucial phase in the lifecycle of clothing, during which time water and energy are expended on dry cleaning, ironing, drying, and washing [154–156]. The types of appliances used (or the absence of their use) can make a significant difference to the environmental impacts of this phase [157, 158]. During the user phase, the quantity of washing cycles is crucial since it impacts durability and raises energy use. A garment's colour should typically hold up for at least 20 washings. Yet, there are guidelines, like the Durawash test method, which is popular among mass-market fashion firms and takes into account 45 domestic washes when evaluating the durability of clothing [51].

12.3.4.1 Consumer Behaviour and Product Lifespan

The average service lifetime of textile products has been estimated to be about five years [159]. However, consumers now have the power to demand and obtain new goods and services that satisfy social and environmental standards in addition to their long-standing preferences for costs, styles, and quality [160]. Numerous elements impact purchasing patterns, such as consumer demographics, geographic location, economic stability, and the launch of new products. Nevertheless, the quick introduction of fashion and textile products has fuelled their intense desire to get rid of their outdated, out-of-style items to make room for new, trendy ones. Zara's 200 in-house designers, for example, select 12,000 styles annually from a total of 40,000 varieties. Zara introduces 24 new clothing lines, whereas H&M launches 12 to 16 and changes them regularly throughout the year [161]. Because consumers increasingly perceive inexpensive apparel as "almost disposable" and perishable, they are more likely to make impulsive purchases and produce more trash [162, 163]. Not only are these tendencies harming the environment, but they are also decreasing the industry's prospects of long-term prosperity. This means that now is the finest time to act ever. To achieve "climate justice," people are demonstrating in the streets, saying that "beautiful fashion should not cost the Earth" and urging businesses to be "a force of cultural change."

12.3.4.2 Care and Maintenance

Textile cleaning, which includes both wet and dry methods, makes up the majority of "care and maintenance" [164]. The cleaning procedure generates effluent and utilises electricity, detergents, washing machines, and water or organic solvents [165]. In contrast to wet cleaning, which requires water and additional materials, dry cleaning has been used for clothing that needs extra attention to prevent harm to the fabric while being cleaned. Furthermore, how consumers behave while buying clothes during the "use phase" is crucial to the upkeep and care of textiles. Burton et al. [166] explore how consumer perceptions and practices regarding the quality and care of clothing can influence sustainability outcomes. Their findings suggest that increased awareness of care practices, such as creative reuse or resale, can significantly reduce the carbon footprint associated with clothing consumption.

12.3.5 End-of-Life Management

The effects of textile production and disposal on the environment are becoming a major concern as the demand for textiles rises worldwide. For example, according to Bukhari et al. [167], who analysed population and per capita disposal, the United States, China, Hong Kong, the EU-27, France, and the United Kingdom disposed of a total of 34.6 million tons of landfilled textile waste in 2016. Nationwide, 11 million tons of textile waste were dumped in US landfills in 2018, according to the U.S. Environmental Protection Agency [74, 168]. However, Jönsson et al. [169] estimate that the amount of textile waste will have accumulated to almost 4.5 Gt by 2040. To solve this significant waste problem, lessen environmental effects, and enhance textile waste handling, end-of-life (EOL) management solutions are required. Additionally, the industry must transition from a linear to a circular economy (see Fig. 12.8).

12.3.5.1 Disposal Methods

Globally, around 92 million tons of textile waste are accumulated at different phases of their life cycles; amongst those, 50% of the trash is generated at a post-consumption level [170]. This stage has a major negative influence on the environment since post-consumer textiles and apparel are disposed of in an unsustainable manner [171]. For instance, landfill is generally considered to have the worst environmental impact. This is because landfilling of natural and bio-based fibres can emit greenhouse gases (GHGs) while synthetics can remain in the soil for centuries without decomposing

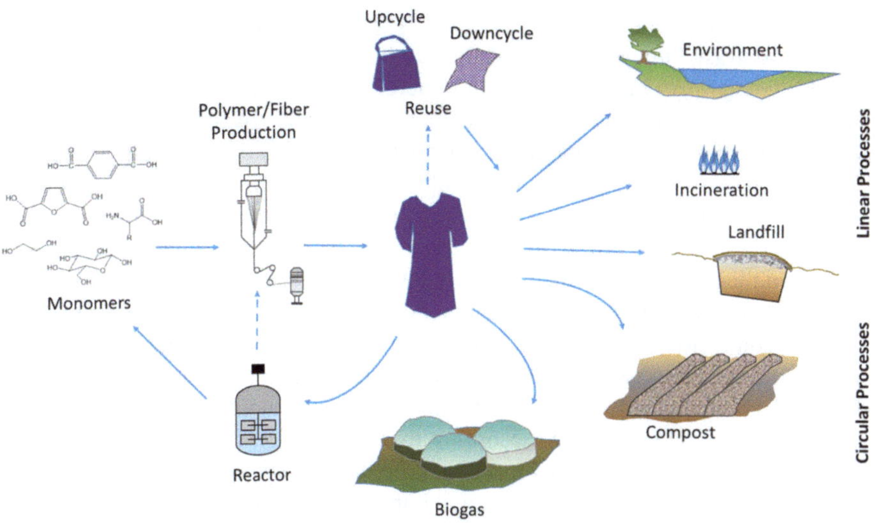

Fig. 12.8 Textile disposal by linear and circular processes. Reproduced from [74] 2022, an open access article by Springer Nature

[172, 173]. Disposed textiles that are sent to landfills represent lost value and persist there because landfills are primarily designed to contain rather than to degrade [174]. Microorganisms mediate the anaerobic breakdown of organic materials in landfills, producing carbon dioxide and methane [175]. Furthermore, when the necessary facilities are available, some textiles are burned to produce electricity [176]. The incineration method reduces the volume of textile waste by burning it. Nevertheless, it releases harmful pollutants, including dioxins and greenhouse gases. Therefore, burning synthetic materials releases chemicals generated from plastic, increasing air pollution and endangering the health of those nearby. This necessitates the adoption of circular economy principles to help curb waste.

12.3.5.2 Recycling and Circular Economy Practices

The "take-make-dispose" strategy, which has historically defined the fashion and textiles sector, is changing dramatically in the direction of sustainability. Recycling and circular economy (CE) principles, which seek to decrease waste, prolong product lifecycles, and lessen environmental effects, are at the core of this shift. According to De Felice [177], a linear economy works by converting natural resources into waste via production, whereas CE avoids the deterioration of the environment and reduces the pollution produced through the removal of natural capital from waste. The three main categories of textile recycling methods are mechanical, chemical, and, less commonly, thermal. The fact that recycling pathways frequently involve a combination of mechanical, chemical, and thermal processes makes this a simplification of reality in many circumstances [178]. Recycling can include down-cycling to wipe and polish cloths that are thrown away after use, up-cycling via monomer, oligomer, and polymer to value-added products [179], or converting to fibre by shredding [178]. In contrast, "upcycling" is a viable method of moving toward a more restorative and regenerative circular economy. Therefore, CE practices have the potential to greatly increase sustainability. These practices include novel dyeing methods that use enzymes and other bio-based chemicals to reduce water and energy consumption, as well as technology that allows closed-loop recycling of textile fibres into new fabrics. Additional strategies that result in transformative improvements include fashion rentals, resale, zero-waste methods, and take-back and collection initiatives. Because of this, Patagonia, which is well-known for its sustainability initiatives, has a *Worn Wear* scheme (Fig. 12.9a) that enables customers to exchange their old or undesired Patagonia clothing for credit. These used clothing items are recycled, repaired, and repurposed into new items that are then sold in their stores. Another company, RE/DONE, collaborated with Levi's to create new styles out of their antique jeans (see Fig. 12.9b). This practice, according to a publication by Weavabel [180] has "diverted over 145,000 garments from landfills" and "saved over 2.5 million gallons of water" to help reduce the environmental impacts. Additionally, designers like Zero discarded Daniel (New York), Christopher Raeburn (British designer), and Reet Aus (Estonia) [181] have reused discarded materials for new clothing to help lessen the environmental impact. In a similar vein, European researchers found that repairing

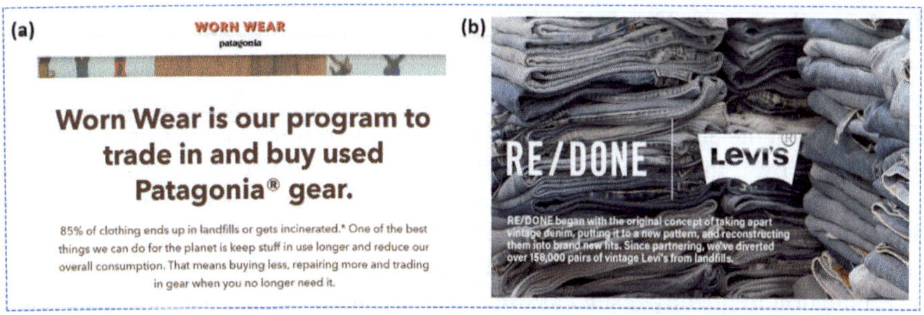

Fig. 12.9 **a** The worn wear program by Patagonia. Image credit: https://wornwear.patagonia.com **b** A collaboration between RE/DONE and Levi's to upcycle denim jeans. Image credit: https://shopredone.com/en-hk/collections/re-done-levis

clothing is crucial to sustainable clothing usage [182]. In a different study, Chan [183] disclosed that the fashion brand is developing a service that would allow customers to register their jeans online and receive a complimentary denim repair. This strategy aims to inspire consumers to develop the practice of upcycling or exchanging their used clothing for new items. They are compelled by this circumstance to participate in the effort to lessen environmental trash.

12.4 Environmental Impact Categories of LCA in Textiles and Fashion

A reduction in environmental impacts from the textiles sector requires a change toward circularity. The European Commission, in the framework of the circular economy action plan, presented the EU Strategy for Sustainable and Circular Textiles, addressing practical actions to make textiles more durable, repairable, reusable, and recyclable, to tackle the problem of fast fashion and stimulate innovation within the sector [184]. Decisions on how best to reduce environmental impacts require comprehensive data concerning inputs (such as water, energy, chemicals, and raw materials) and outputs (such as releases to air, land, and water) that arise at all stages of the life cycle of clothing [51]. Therefore, throughout the lifecycle of textile and fashion products, the LCA approaches are used to detect and quantify important environmental consequences. These impacts are categorised to provide a comprehensive understanding of the industry's footprint, treated as free-standing below.

12.4.1 Resource Depletion

The overproduction issue in the fashion and textile industries has a significant negative impact on the environment, leading to pollution, resource depletion, and textile waste [185, 186]. Interestingly, the industry's resource depletion happens at several levels of the supply chain, from the extraction of raw materials to manufacture and disposal, which leads to increasing greenhouse gas emissions, excessive water use, and deforestation. For example, cotton is one of the most water-intensive crops, and its production frequently results in water scarcity in the areas where it is cultivated. As reported in Galloway [187], the United Nations Environment Programme (UNEP) estimates that cotton agriculture covers 2.5% of the world's arable land yet requires 200,000 tons of insecticides and 8 million tons of fertilisers yearly, accounting for 16 and 4% of worldwide use, correspondingly. Furthermore, according to the Synthetics Anonymous 2.0 report on fashion sustainability, the annual production of synthetic fibre uses more oil than Spain consumes, which is a startling statistic that highlights the industry's dependence on fossil fuels. This is because fast fashion frequently utilises synthetic fibres like polyester, which are made from petroleum-based goods, increasing the carbon footprint of the sector. Concurrently, more than 50% of thermal energy and around 70% of electricity are used in various processes of the textile industry [188]. The energy-intensive processes of spinning, weaving, and dyeing are among the main causes of greenhouse gas emissions [189]. Nonetheless, transporting goods from supply chains to consumers requires a substantial quantity of gasoline, whereas a pound of cloth requires 132.5 L of water to dye [143, 144], which harms water bodies due to improper disposal.

12.4.2 Water Use and Pollution

The textile & clothing industry is the second-highest consumer of water globally [11]. Furthermore, it contributes significantly to the world's water consumption and pollution, impacting aquatic ecosystems and freshwater supplies as a result of contaminants discharged throughout a textile product's longer lifecycle [190]. According to current estimates, the textile sector utilises between 79[2] and 93[1] billion cubic meters of water annually. However, the share of cotton consumption is 2.6% in global water usage, as it consumes almost 10,000 L of water to produce one kilogram of cotton fabric [191, 192], which means it takes about 8000 L of water for a pair of cotton jeans. The WWF estimates that it takes over 2,700 L (713 gallons) of water to produce one cotton t-shirt, which is enough to nourish a person for two and a half years. Environmental stress is greatly exacerbated by this high water use [187]. As a result, irrigation-intensive farming causes significant water stress in key cotton-producing nations like China, India, and Pakistan.

Nonetheless, the production of waste contributes to a myriad of negative environmental impacts, including decreasing air and water quality [193]. According to Han et al. [194], the UK fashion and textile sector produces 3.1 million tonnes of carbon dioxide (CO_2) and 20 million tonnes of wastewater yearly. In Bangladesh, where the textile industry contributes approximately 82% of GDP, rivers near manufacturing centres show extreme pollution levels [195]. Okafor et al. [196] report that annually, washing clothes discharges 500,000 tons of plastic micro-fibres into the ocean, an equivalent of 50 billion plastic bottles. Furthermore, 92 million tons of textile waste are discarded by society each year [54], amounting to more than $400 billion in clothing, and this amount is expected to rise by 60% between 2015 and 2024 [197]. A World Bank estimate predicts that by 2025, the amount of municipal solid trash produced worldwide will have increased by 70%, from the present 1.3 billion tonnes to 2.2 billion tonnes annually [163]. According to reports, Bangladesh's textile and garment industry alone generates more than 577,000 tonnes of waste per year, of which 250,000 tonnes, or half of the total, are made entirely of recyclable cotton valued at approximately US$100 million. Of the 16 million tonnes of waste generated in the USA in 2015, only 15.28% of that amount was recycled, 19.03% was burned, and the remainder was disposed of in landfills [198], contributing to carbon emissions.

12.4.3 Greenhouse Gas Emissions

The textile and garment industry accounts for a significant proportion of greenhouse gas (GHG) emissions, with estimates ranging from 6 to 8% of global carbon emissions [54, 117]. This corresponded to 1.7 billion tons of CO_2 emissions in 2015, more than all foreign travel and cargo combined [199]. The energy-intensive operations, dependency on fossil fuels, and intricate supply chains of this business are the main causes of its carbon footprint. Especially in nations with laxer environmental laws, factories frequently use coal-fired electricity, which releases a lot of CO_2. Additionally, the fast fashion model exacerbates the problem by encouraging overproduction and overconsumption, leading to shorter product lifecycles and increased waste [200]. Following discard after only a few uses and ending up in landfills, these textiles decompose and release methane, another potent greenhouse gas. As reported by Schröcker et al. [201], the European Environment Agency stated that textile purchases in the EU during 2020 resulted in approximately 270 kg of CO_2 emissions per person, amounting to a total of 121 million tonnes. A typical cotton t-shirt, based on 50 washes, generates 35% of CO_2 emissions during the production phase and 52% during the use phase, according to studies on life cycle emissions of clothing [202]. Again, non-biodegradable synthetic fibres further compound this issue. To produce blocks for PET and other industrial uses, crude oil must be extracted and refined [123]. This requires high temperature, resulting in significant CO_2 loads [76].

The fashion and textile industries are making more and more efforts to cut their GHG emissions. The goal of the Paris Agreement is to keep global warming to less than 2 degrees over pre-industrial levels, with 1.5 degrees being the ideal limit. The reduction of emissions required to meet this target is substantial; to achieve net zero emissions by 2050, global emissions must decrease by almost 45% by 2030 [203]. The carbon footprint of the sector can be reduced by supply chain optimisation and the transition to renewable energy in industrial facilities. Furthermore, employing sustainable practices—like using recycled polyester, organic cotton, and hemp or bamboo can drastically reduce emissions by 40% CO_2, 60% energy use, and 80% toxin release [204, 205]. As a result, life cycle assessment (LCA) standards like ISO 14067 concentrate on the carbon footprint of products by offering precise guidelines and procedures for the measurement and sharing of the carbon footprint of products throughout their existence. According to Wiedmann and Minx [206], clothing and textiles account for approximately 4% of an individual's secondary carbon footprint in the developed world. LCA was used to establish criteria for an environmental product declaration for textile products, and five European textile industries provided unique, first-hand industrial data. Regarding woollen products, Barber and Pellow [207] detailly report on the inventory of resource inputs for New Zealand Merino wool and assessed its total energy use profile relative to other textiles. Results of this study show that New Zealand Merino fibre production and early-stage processing uses significantly less energy than synthetic fibres.

12.4.4 Chemical Use and Toxicity

The manufacture of textiles necessitates using hazardous chemicals in several manufacturing and production phases, such as dyeing, bleaching, and finishing (refer to Table 11.3) [208]. The global textile chemicals market was estimated to be worth USD 21.23 billion as of 2020, yet of the 8000 chemicals used, 10% are harmful to the environment and human health because of effluent discharge [208, 209]. Ahsan et al. [210] report that up to 80% of untreated textile effluent is used to irrigate crops in countries such as Pakistan, where residents have reported health issues. Similarly, the subjection of Jew fishes to triphenylmethane dye causes the loss of their lives within 24 h [211]. Furthermore, the World Health Organisation reported in 2019 that chemical exposure resulted in the loss of 53 million disability-adjusted life years and two million lives [212]. Tounsadi et al. [209] claim that these chemicals have a detrimental effect on the environment when they are released because they produce colourful wastewater that ranges from 2000 to 3000 m^3/j and contains persistent dyes including Blue 124, Blue 106, and Yellow 3. However, the careless release of dyes into water bodies limits photosynthesis, increases the need for chemical and biological oxygen, decreases the amount of sunlight that reaches the water's surface, and limits plant growth [211]. In response to these challenges, Life Cycle Assessment (LCA) has emerged as a valuable tool for evaluating and mitigating the environmental impact of textile production. Hence, in a recent study, Roos et al. [213] investigated the value of LCA in assessing the chemical performance of textile products and the impact of toxicity issues on environmental performance rankings (Table 12.3).

Table 12.3 Chemicals used in different stages of textile manufacturing

Process step	Chemicals or chemical groups used	Function/product specifics
Fibre production	Pesticides, detergents	Growing of vegetable plants and removal of lanolin
	Heavy metals	Viscose, polyester
Knitting	Mineral oils, including polyaromatic hydrocarbons	Lubricating/emulsifying
Washing	Organic solvents, nonylphenols/nonylphenol ethoxylates (NPEOs)	Detergents in washing
Desizing	Pentachlorophenol	Remove starch sizes
Dyeing or printing	Azo dyes	Used for dyeing
	Heavy metals	Attach dyes to the fibre
	Organochlorines (chlorinated solvents, chlorinated benzenes)	Carriers
	Solvents, formaldehydes, NPEOs	Auxiliary substances
Stabilizing	Formaldehyde, triazones, carbamates	Stabilising of cellulose fibre
Biocide treatment	Metals, SCCPs, triclosan, organotins	Anti-mould or anti-microbial
Anti-pilling, waterproofing	Phthalates, heavy metals, organotins, perfluorinated compounds (including PFOS, PFOA)	Cotton, polyester
Protective printing inks	Phthalates, heavy metals, organotins	Depending on the fabric and use
Transport and storage	Chlorinated phenols	Added as biocides

12.5 Social and Economic Impacts of Textile and Fashion Production

The global textile and fashion industry, as a multi-trillion-dollar sector, plays a crucial role in economic development, employment, and cultural expression. The World Bank estimates that the global fashion business is worth over $2.5 trillion and supports over 75 million jobs globally [132]. The manufacturing of textiles and clothing is a gateway to international trade for many low- and middle-income nations, fostering industrial growth, producing jobs, and generating export money. Bangladesh, Vietnam, India, and Ethiopia have all experienced significant economic growth, partly as a result of the rise of their textile industries. However, there are also serious social and economic issues brought on by its quick expansion, such as economic inequality, environmental damage, and labour exploitation [132,

214]. Thus, the industry often operates under conditions that prioritise profit over people and sustainability. Hence, it is important to assess the industry through all three dimensions of sustainability: economic, environmental, and social [215–217]. Thus, companies alter their conduct and take responsibility for their actions through policing for many reasons, such as improving and protecting their brand, reducing stakeholder pressure, expanding into new markets, and gaining a competitive advantage in the market [218] while consuming fewer natural resources [219]. Together, the observance and practice of sustainability provide social and economic benefits by reducing environmental transportation costs; facilitating the regulation and oversight of employment conditions in supply chains, as well as the environmental performance of manufacturers. These related social, environmental, and economic issues are treated as free-standing below.

12.5.1 Labour Rights and Working Conditions

The textile industry faces many important issues, particularly in developing countries where labour-intensive garment manufacturing operations are common. These issues include migrant worker abuse, product safety, labour rights, safe housing, product service life, and community involvement [220]. Some areas of the textile industry, particularly those that manufacture ready-made clothing, have hazardous and unpleasant working conditions, as reported by Chattopadhyay and Gangopadhyay [221]. For instance, in India, in some factories with large orders, including child labour, workers are forced to work continuously for 36 h [222]. Moreover, in Bangladesh, female workers in fast-fashion factories frequently face discrimination, including violations of human rights, gender inequality, exploitation, and low wages. Many of these women are unaware of their rights and often experience harassment and physical abuse, making it difficult for them to stand up for themselves in such challenging conditions [222–224]. Similarly, the production of fibres affects the environment, and the textile industry currently favours high-yield cotton due to its profitability. The exposure to cotton dust or other fine particles and harmful chemicals used during dyeing leads to respiratory disorders and other health problems [225, 226]. The safety of working environments is also subpar in some low-income nations; for instance, inadequate ventilation can cause respiratory illnesses in employees as well as lung disease or cancer [227]. Hence, a study by Paul-Majumder [228], cited in Ahmed and Raihan [229], illustrated various diseases and illnesses among female garment workers. Notable injuries include musculoskeletal, eye, and other physical injuries due to problems such as poor lighting, prolonged repetition of an action, and vision fatigue. Due to these, the UNGPs urge states to help companies recognise the increased risks of sexual and gender-based violence and to provide guidance on how to evaluate "gender, vulnerability, and marginalisation" issues [223]. In light of this, the International Labour Organisation (ILO) established the historic Convention on the Elimination of Violence and Harassment in the World of Work in 2019 [230]. In addition to enforcing stringent anti-harassment regulations, proactive welfare officer appointments, the establishment of women's helplines, and high-level gender sensitisation initiatives aid in addressing complaint resolution bottlenecks.

12.5.2 Community Impact and Stakeholder Engagement

The fashion and textile industry has reached a pivotal moment where regulatory measures and policy actions are urgently required to curb its high levels of resource extraction and emissions, driven by rapidly increasing waste. Active participation from all stakeholders, brands, manufacturers, multi-stakeholder groups, non-profits, policymakers, investors, and consumers is essential to drive meaningful change [231, 232]. As a result, non-governmental organisations were founded to promote remedies for the adverse consequences. Governments, NGOs, and the private sector have joined forces in countries including the UK, Germany, and the Netherlands to create non-binding agreements to address issues like dangerous working conditions and supply chain human trafficking [233]. Additionally, implementing regulatory tools lowers the upfront costs associated with implementing circular business models, thus encouraging suppliers to meet environmental procurement standards and encouraging businesses to invest in and adopt circular economy methods [234]. Thus, ecolabels have been divided into three primary categories by the International Organization for Standardization (ISO), which are as follows: Type II (ISO 14021) ecolabels are self-declared environmental claims; Type III (ISO 14025) ecolabels are environmental product declarations containing quantitative environmental information; and Type I (ISO 14024) ecolabels are broad-based third-party verified ecolabels such as the EU Ecolabel and the Swan [235]. In comparison to the existing state of affairs, the effective implementation of the CE idea could lead to a global reduction in emissions of 48% by 2030 and 83% by 2050, promoting sustainable economic and environmental development [236]. To further establish a strategic vision and significantly lessen the effects of human activity on the environment, long-term programs are being released by the governments of Europe and other countries [237].

12.5.3 Economic Viability and Market Trends

The textile industry in the European Union employs 1.5 million people, with the fashion sector making up about half of this workforce. The industry generates about EUR 160 billion in revenue a year, with small enterprises accounting for 90% of the amount [238]. Importantly, the economic viability and market trends of the textile and fashion industry are influenced by a multitude of factors, including consumer behaviour, supply chain dynamics, sustainability practices, and regional characteristics. A key economic benefit is the potential for cost savings. In terms of digitisation, it is seen as a significant force for economic growth and is thus encouraged by policies and investments [239, 240]. Thus, radio frequency identification (RFID) tags and blockchain technology are facilitating improved supply chain and material flow tracking, promoting circular business models [241] while minimising costs and maximising efficiency [242]. Additionally, the conversion of waste into high-value products presents an innovation in production. The economic potential of this aligns with consumer demand for environmentally responsible products [243]. Similarly, new revenue opportunities further boost viability where rental allows monetising the use phase after design for durability and modular components permit upgrades to extend product lifecycles rather than accelerating disposal. Via these appropriate strategies and market conditions, sustainability becomes economically viable and thus contributes to growth and development.

12.6 Methodologies for Conducting LCA in Fashion

LCA, as a systematic approach, is used to evaluate the environmental impacts of a product throughout its entire life cycle. Given the industry's significant environmental footprint, LCA helps brands identify hotspots, improve sustainability, and make data-driven decisions. Numerous LCA models have been developed over time, each with different technical assumptions and methodologies [244]. Streamlining LCA through probabilistic under-specification has been explored to enhance the credibility of results, especially in the context of materials production in consumer products [245]. Nonetheless, bridging the gap between LCA, Life Cycle Costing (LCC), and Cost–Benefit Analysis (CBA) as sustainability assessment tools is essential for interpreting and integrating assessment results effectively [246]. The continuous improvement in data accuracy and circular design will further refine LCA applications in fashion. Significantly, by following standardised methodologies (ISO 14040/44), brands can enhance sustainability, comply with regulations, and meet consumer demand for eco-friendly products.

12.6.1 ISO Standards and Best Practices

The International Organisation for Standardisation (ISO) has played a significant role in establishing benchmarks for assessing management techniques in a range of sectors. This published regulatory framework outlines four stages of a life cycle assessment [76]. Importantly, ISO 14044 outlines specifications and criteria for Life Cycle Assessment (LCA), whereas ISO 14040 essentially covers the ideas and framework for LCA [247]. Although ISO 9001 focuses mainly on improving internal processes and collaboration within supply chain links, it also has some influence on environmental processes. The ISO standards are divided into six major groups covering various management and safety areas, and they apply to organisations looking to show that they can consistently deliver goods and services that satisfy customers and adhere to applicable laws and regulations [248].

Additionally, ISO 45001:2018 provides guidelines for an occupational health and safety management system by preventing work-related illnesses and injuries and by improving health and safety. Similarly, two ISO standards, ISO 14000 and ISO 14001, are closely aligned with SDG 13, a climate action that aims to combat climate change and its negative effects. In particular, ISO 14000 equips organisations with useful tools to manage their environmental footprint, while ISO 14001 provides guidelines and requirements for environmental management systems (EMSs), making it a universally recognised standard. The inclusion of traceability within the scope of ISO/TC 38 indicates that the committee recognises the significance of tracking and verifying the origins and processing of fibres in the textile supply chain, thereby promoting accountability, transparency, and ethical sourcing practices.

Hence, employed approaches such as adopting non-toxic green chemistry in finishing reduce toxicity. In contrast, improved water management systems, zero waste pattern cutting [249, 250], ethical treatment of workers, and facilities compliance audits all contribute to cleaner production processes. Comparably, biotechnology is employed to produce low-carbon, biodegradable, and safe textiles, reducing the environmental damage caused by textiles made from petroleum and enhancing the sustainability potential of the industry. Furthermore, utilising an efficient system with cutting-edge energy absorption and storage technologies, such as wind or hydropower, is a novel way to produce electricity while reducing reliance on fossil fuels [251]. On the other hand, alternative materials for the sustainable textile and apparel sector include bacterial cellulose, organic bamboo, mycelium, kombucha, and banana fibres [89, 90]. However, in the supply chain sector, export markets have implemented radio frequency identification (RFID) tags to support circular business models and standardised sustainability labelling verifications through ecolabels to safeguard producers' integrity [241, 252].

12.6.2 Software Tools and Applications

As the fashion industry grapples with its vast environmental and social impacts, standards and certifications, along with important software to drive and validate responsible practices across the complex global supply chains, are increasingly adopted and are presented in Table 12.4 below [48, 253–256].

12.7 Challenges and Limitations of LCA

Life Cycle Assessment (LCA) has become an invaluable tool for evaluating the environmental impact of projects. However, its application in the complex landscape of the textiles and fashion industry is not without challenges and limitations. In this section, we unravel the various challenges, limitations, and variability that can shape the outcomes of LCA.

Table 12.4 LCA software tools and their specific application

Software tools	Application
SimaPro	Best known for its LCI and LCA capabilities integrated into its commonly used package
GaBI	Provide sectoral LCA solutions, including textiles, and consider supply chain databases
OpenLCA	Free software used for conducting comprehensive LCA in various sectors
Carbon trail	An advanced LCA tool focusing on carbon footprint tracking and sustainability insights for the textile industry
Higgs index	Evaluates social and environmental sustainability at every stage of the production process, including factories, packaging, materials, dyeing and finishing, and branding

12.7.1 Data Gaps and Quality Issues

According to studies, the textile sector has a major negative influence on the environment, with ramifications for many different environmental impact categories [178]. Notably, the textile and fashion industry faces significant challenges in terms of comprehensive and reliable data, which creates gaps and quality issues in life cycle assessments (LCAs) [257, 258]. These challenges include missing data, data accessibility, data management, reliability, and relevance [257], contributing to uncertainty in LCA results, particularly in inventory data and the impact assessment indicators due to errors and normal fluctuations in the data [259]. Many LCAs are produced without reporting the uncertainty of the data, which affects transparency. Furthermore, data quality can be influenced by a lack of data, wrong and ambiguous data, inaccurate measurements, and model assumptions [260–262]. However, there is a lack of consensus regarding the systematic methodology of how data quality can be assessed. Additionally, the limitation on standardised databases for environmental impact data and outdated data further amplify the challenge. This creates a significant hurdle, hindering the acquisition of accurate information for various materials and processes, potentially leading to skewed LCA outcomes and hampering the ability of stakeholders to benchmark environmental performance. This, in turn, makes it harder to identify best practices and impedes progress toward more sustainable practices.

12.7.2 Variability in Textiles and Fashion Products

Textile raw materials, obtained from diverse sources, impact the environment differently, due to variations in cultivation, energy consumption, water use, chemical processing, and complexity in the supply chain [263, 264]. Hence, conducting a standardised LCA across the broad spectrum of materials is challenging to identify and quantify the environmental impacts of these products. Furthermore, life cycle assessment (LCA) studies have highlighted the environmental issues that non-polyester/cotton textiles have, which highlights the difficulties in determining the sustainability of these products [264]. Inconsistencies in data and effect assessment approaches can make it difficult to compare various textile products using life cycle assessment (LCA), according to the International Wool Textile Organisation (IWTO). Bio-based textiles, recycled fibres, and biodegradable polymers are examples of new and sustainable materials that are being developed more frequently. However, their environmental impact assessments are still developing, and there is a lack of long-term data to assess their true sustainability compared to traditional textiles. This highlights the need for standardised approaches and guidelines in conducting LCA for textiles and fashion products [265]. Despite the advantages of LCA in providing valuable environmental data for products and processes, these limitations and challenges must be addressed to effectively integrate sustainability practices in the fashion industry.

12.7.3 LCA Integration into Textiles and Fashion Business Models

The integration of LCA into textiles and fashion business models is crucial for addressing and mitigating the environmental impact of the industry [266, 267]. However, there are limitations to the current LCA datasets when applied to fibres and raw materials used in the fashion industry. The short life span of textile end products poses a barrier to implementing sustainable approaches and LCA in the industry [268]. In a similar vein, using specialised LCA software necessitates substantial human and financial resources. Small and medium-sized fashion businesses (SMEs), however, frequently lack the funding and technological know-how necessary to conduct thorough evaluations. Furthermore, the textile disposal phase, which includes a variety of processes like landfilling, incineration, recycling, and biodegradation, makes it more difficult for LCA models to take into consideration customer behaviour, product reuse, and the differing environmental effects of various disposal techniques. Despite growing interest in sustainable fashion, many consumers still prioritise cost and aesthetics over environmental impact. Without strong consumer demand for LCA-backed sustainable products, businesses may lack incentives to integrate LCA into their decision-making.

12.7.4 Policy Development and Regulatory Compliance

In the textile and apparel sector, policy creation and regulatory compliance are essential elements that affect economic viability, environmental sustainability, and overall industry growth. Through the United Nations Framework Convention on Climate

Change (UNFCCC) Fashion Industry Charter for Climate Action, stakeholders in the apparel industry united in 2018 to commit to climate action. By 2050, this seeks to attain carbon neutrality by reducing emissions by 30%. This is a major obstacle because, to achieve this 30% reduction in emissions, the sector would need to reduce more than half a billion tonnes of carbon dioxide annually by 2030. Governments are being compelled by the growing necessity of addressing environmental problems and climate change to reduce greenhouse gas emissions by enacting more stringent environmental laws. For example, one of the primary causes of air pollution and greenhouse gas emissions is road transportation. Electric vehicles (EVs), which are now powered by batteries, are replacing internal combustion engines as a result of supply chain methods intended to lessen the effects of distribution [269]. However, because the industry operates internationally, it can be challenging to comply with laws in different jurisdictions, resulting in inconsistent labour laws, environmental policies, and safety standards; implementing sustainable materials, ethical labour practices, and safety standards requires a substantial investment that may be unaffordable for smaller industry players; and implementing new policies frequently encounters opposition from industry stakeholders due to operational disruptions and financial concerns.

12.8 Future Directions in LCA for Textiles and Fashion

The textile and fashion industry is witnessing a significant evolution in LCA and its methodologies in assessing the industry's footprint. There are several avenues for future studies on LCA for sustainability in the clothing and textile industry. One area of research could focus on addressing data gaps and quality issues in LCA, which can help identify opportunities for improving water management practices and reducing the industry's overall environmental footprint. Additionally, proposed business model life cycle assessment (BM-LCA) as a method in analysing the environmental impact of different business models can overcome the scale-related limitations of mainstream LCA. To enhance the effectiveness of LCA in shaping policy and practice, there is a growing need for transparent, traceable, and up-to-date datasets. This would allow for better benchmarking and help integrate LCA findings into regulatory frameworks. Likewise, research efforts are needed to establish recommended practices for uncertainty analysis and to elaborate guidance for practitioners and method developers on estimating, communicating, interpreting, and managing uncertainty in life cycle inventory (LCI) and life cycle impact assessment (LCIA). Equally important for the future of LCA is consumer awareness. While LCA provides valuable insights into environmental impacts, these insights must be benchmarked and communicated effectively to consumers to drive behavioural change in consumption [270, 271]. Educating consumers on the environmental impact of textile products across their life cycle can foster more sustainable consumption habits and enhance demand for responsibly produced fashion. Future directions should emphasise translating complex LCA data into accessible information through mobile applications, QR codes, and eco-labels.

12.8.1 Innovations in Sustainable Practices

The textile and fashion industry is profoundly transforming through numerous sustainability innovations across its value chain. These innovations address the industry's substantial environmental footprint while creating new business opportunities. For example, enzymatic recycling techniques and green chemistry-based techniques have become viable eco-innovative solutions with reduced environmental impact and high recovery rates. Furthermore, biotechnology is now able to produce safe, low-carbon, and biodegradable textiles, which increases the sustainability potential of the fashion and textile industries and lessens the environmental damage caused by textiles derived from petroleum. In dyeing, the use of enzymes, plasma, and liquid CO_2 is reducing water and chemical use. Nonetheless, computer-aided design (CAD) platforms, digital prototyping, virtual fittings using body scanning, and 3d design tools are making design faster, more flexible, and less wasteful by enabling designers to digitally develop and test clothing on virtual avatars before generating actual prototypes. Similarly, laser-cut garments using 3d virtual prototyping have become popular as a production activity following sustainability. The use of laser technology has a very beneficial effect on the environment because it does away with inks, chemicals, and solvents. As a controlled process, it also operates precisely, generates less waste, has superior treatment, processes quickly, and consumes less energy [272]. Furthermore, the sustainable textile sector relies on organic bamboo and banana fibres as substitute materials and practices sustainable sourcing. This has been accomplished by greatly increasing the efficiency and rate of sorting processes through the integration of NIR spectroscopy with AI. These advancements enable the more accurate identification and separation of fibre types, laying the groundwork for more effective recycling methods.

12.8.2 Advances in Technology and Digital Fashion Innovations and Data Analytics

The textile and fashion industry is been significantly transformed by artificial intelligence (AI), which has revolutionised several fields within the sector. The extensive application of intelligent design assistants, fashion style analysis, creative generation, and size customisation has resulted in more inventive and productive methods for clothing design [273]. Additionally, AI algorithms help the sector make better use of its resources, which lowers costs and increases production speed, efficiency, and effectiveness. Thus, enhancing sustainability performance in the sector requires the digitisation of manufacturing and supply chain procedures as well as effective monitoring made possible by Industry 4.0 tools [266]. Therefore, blockchain technology with radio frequency identification (RFID) tags enables better supply chain and material transfer tracking, supporting circular business models. Similar to this, blockchain, a digital ledger composed of blocks of interconnected data, guarantees traceability systems where each action and transaction is visible and can be monitored and recovered when goods are transferred between parties [274, 275]. This reduces expenses and guards against errors and tampering. Furthermore, BCT assists suppliers and customers in creating smart contracts that measure supplier performance [276]. A more transparent transaction is achieved by using data analytics technology to collect logistics and supplier data and identify high-risk areas for ethical or environmental violations. The fourth industrial revolution is greatly impacted by the wide range of capabilities that advanced technologies provide when combined with different solutions like the Internet of Things (IoT) and artificial intelligence (AI). These capabilities include virtual reality, device-to-device communications, and cyber-physical systems [277].

12.9 Conclusion

The textile and fashion industry has long been associated with significant environmental and social impacts, from raw material extraction to manufacturing, distribution, usage, and eventual disposal. As global consciousness shifts toward sustainable development, Lifecycle Assessment (LCA) has emerged as a critical tool for evaluating and improving the sustainability performance of textile and fashion production systems. This enables stakeholders to quantify environmental impacts at every stage of a product's life, identify areas for improvement, and make informed decisions that support eco-conscious innovation and responsible consumption. Unlike traditional evaluation methods that often focus solely on production processes, LCA encapsulates data from cradle to grave, including raw material sourcing, water and energy use, chemical treatments, transportation, consumer use, and end-of-life disposal. Shedding light on environmental costs, LCA empowers manufacturers and designers to implement targeted changes such as substituting harmful chemicals, adopting renewable energy sources, or choosing biodegradable materials. Moreover, LCA supports the transition to a circular economy in the textile sector by promoting resource efficiency, waste reduction, and product longevity. In addition to environmental metrics, LCA approaches are increasingly integrating social and economic dimensions, giving rise to more comprehensive sustainability assessments. These expanded frameworks can assess labour conditions, community impacts, and economic viability, providing a multidimensional view of sustainability performance. As the fashion industry grapples with mounting pressure from consumers, regulators, and investors to be more transparent and accountable, LCA offers a standardised and scientifically grounded means of communicating sustainability credentials. However, challenges remain in fully realising the potential of LCA in textiles and fashion. Data availability, quality, and comparability continue to be limiting factors, especially for small and medium enterprises. Additionally, the complexity and variability of global supply chains can make it difficult to obtain accurate data and ensure consistency across assessments. Despite these limitations, collaborative efforts internationally and the ongoing advancements in digital tools, databases, and standardised methodologies are helping to streamline LCA practices and broaden their accessibility. This enhances the reliability of LCA to promote sustainable practices and propel the textile and fashion industry toward a more ethical, resource-efficient, and environmentally responsible future.

References

1. Rahaman, M. T., and M. S. H. Khan. 2025. Green merchandising of textiles and apparel in a circular economy: recent trends, framework, challenges and future prospects towards sustainability. *Journal of Open Innovation: Technology, Market, and Complexity* 11 (1): 100457.
2. Leal Filho, W., et al. 2024. Reducing the carbon footprint of the textile sector: an overview of impacts and solutions. *Textile Research Journal* 94 (15–16): 1798–1814.
3. Herrera Almanza, A. M., and B. Corona. 2020. Using social life cycle assessment to analyze the contribution of products to the sustainable development goals: a case study in the textile sector. *The International Journal of Life Cycle Assessment* 25(9): 1833–1845.
4. Curran, M. A. 2016. Life-cycle assessment☆. In *Encyclopedia of ecology (second edition)*, ed. B. Fath, 359–366. Oxford: Elsevier.
5. Liu, M., G. Zhu, and Y. Tian. 2024. The historical evolution and research trends of life cycle assessment. *Green Carbon* 2 (4): 425–437.
6. Luo, Y., et al. 2021. Environmental sustainability of textiles and apparel: a review of evaluation methods. *Environmental Impact Assessment Review* 86: 106497.
7. Ahmed, R. 2024. *Sustainable clothing systems: a comparative life cycle assessment of t-shirts made from primary and recycled materials.* NTNU.
8. Gugenishvili, I. 2025. *Woven for change: nearly new cashmere's sustainable fashion legacy.*
9. Samant, L., et al. 2024. Impact of the textile industry on global climate change. In *Climate action through eco-friendly textiles, Sadhna,* ed. R. Kumar and S. Greeshma, 11–26. Singapore: Springer Nature Singapore.
10. Coscieme, L., et al. 2022. A framework of circular business models for fashion and textiles: the role of business-model, technical, and social innovation. *Sustainability: Science, Practice and Policy* 18 (1): 451–462.
11. Sakib, M. N., G. Kabir, and S. M. Ali. 2024. A life cycle analysis approach to evaluate sustainable strategies in the furniture manufacturing industry. *Science of The Total Environment* 907: 167611.
12. Pryshlakivsky, J., and C. Searcy. 2021. Life Cycle Assessment as a decision-making tool: Practitioner and managerial considerations. *Journal of Cleaner Production* 309: 127344.
13. Mármol, C., et al. 2023. Social life cycle assessment for industrial product development: A comprehensive review and analysis. *Heliyon* 9(12).
14. Brusseau, M. L. 2019. Chapter 32-sustainable development and other solutions to pollution and global change, 3rd edn. In *Environmental and pollution science*, ed. M. L. Brusseau, I. L. Pepper, and C. P. Gerba, 585–603, Academic Pres.
15. Radovanović, M. 2023. Chapter 3-strategies for sustainable energy management. In *Sustainable Energy Management (Second Edition)*, ed. M. Radovanović, 35–94. Boston: Academic Press.

16. Finkbeiner, M. 2014. *The international standards as the constitution of life cycle assessment: The ISO 14040 series and its offspring*, 85–106.
17. Bergerson, J., and L. Lave. 2004. Life cycle analysis of power generation systems. In *Encyclopedia of Energy*, ed. C. J. Cleveland, 635–645. New York: Elsevier.
18. Bjørn, A., et al. 2018. LCA History. In *Life cycle assessment: theory and practice*, ed. M. Z. Hauschild, R. K. Rosenbaum, and S. I. Olsen, 17–30. Cham: Springer International Publishing.
19. Amahmoud, A., M. M. El Attar, and A. Meleishy. 2022. The evolution of life cycle assessment approach: a review of past and future prospects. *IOP Conference Series: Earth and Environmental Science* 992 (1): 012002.
20. Villagrán, E., et al. 2024. Life cycle assessment in protected agriculture: where are we now, and where should we go next? *Horticulturae* 10. https://doi.org/10.3390/horticulturae10010015.
21. McManus, M. C., and C. M. Taylor. 2015. The changing nature of life cycle assessment. *Biomass and Bioenergy* 82: 13–26.
22. Hunt, R. G., J. D. Sellers, and W. E. Franklin. 1992. Resource and environmental profile analysis: a life cycle environmental assessment for products and procedures. *Environmental Impact Assessment Review* 12 (3): 245–269.
23. Jensen, A. A. 1998. *Life cycle assessment (LCA): a guide to approaches, experiences and information sources*. European Environment Agency.
24. Mazur, A., and J. Lee. 1993. Sounding the global alarm: environmental issues in the US national news. *Social Studies of Science* 23 (4): 681–720.
25. Redclift, M. 2005. Sustainable development (1987–2005): an oxymoron comes of age. *Sustainable Development* 13 (4): 212–227.
26. Valdivia, S., et al. 2021. Principles for the application of life cycle sustainability assessment. *The International Journal of Life Cycle Assessment* 26 (9): 1900–1905.
27. Klöpffer, W. 2006. The role of SETAC in the development of LCA. *The International Journal of Life Cycle Assessment* 11 (1): 116–122.
28. Liu, M., G. Zhu, and Y. Tian. 2024. *The historical evolution and research trends of life cycle assessment.* Green Carbon.
29. Pryshlakivsky, J., and C. Searcy. 2013. Fifteen years of ISO 14040: A review. *Journal of Cleaner Production* 57: 115–123.
30. Standard, I. 2006. *Environmental management-life cycle assessment-requirements and guidelines*, vol. 14044. ISO.
31. Da Silva, F. M., et al. 2017. *Challenges for technology innovation: an agenda for the future: proceedings of the international conference on sustainable smart manufacturing (S2M 2016), October 20–22, 2016, Lisbon, Portugal.* CRC Press.
32. Guinée, J. B., et al. 2011. Life cycle assessment: past, present, and future. *Environmental Science & Technology* 45 (1): 90–96.
33. Witcover, J., S. Yeh, and D. Sperling. 2013. Policy options to address global land use change from biofuels. *Energy Policy* 56: 63–74.
34. Yeh, S., and D. Sperling. 2013. Low carbon fuel policy and analysis. *Energy Policy* 56: 1–4.
35. Pennington, D. W., and T. Rydberg. 2005. Life cycle assessment. In *Encyclopedia of toxicology (Second Edition)*, ed. P. Wexler, 715–719. New York: Elsevier.
36. Finnveden, G., and J. Potting. 2014. Life cycle assessment. In *Encyclopedia of toxicology (Third Edition)*, ed. P. Wexler, 74–77. Oxford: Academic Press.
37. Guine, J., et al. 2002. *Life cycle assessment*, 1–3. An operational guide to ISO standards.
38. Mudersbach, M., et al. 2023. Life cycle assessment in a nutshell—best practices and status quo for the plastic sector. *Macromolecular Rapid Communications* 2300466.
39. Pålsson, A.-C., and E. Riise. 2011. Principles for life cycle assessment. In *Tosca Sustainability framework*.
40. Torrellas, M., A. Antón, and J. Montero. 2013. An environmental impact calculator for greenhouse production systems. *Journal of environmental management* 118C: 186–195.
41. Khasreen, M. M., P. F. G. Banfill, and G. F. Menzies. 2009. Life-cycle assessment and the environmental impact of buildings: a review. *Sustainability* 1, 674–701 https://doi.org/10.3390/su1030674.

42. de Bruijn, H., et al. 2002. Inventory analysis. In *Handbook on life cycle assessment: operational guide to the ISO standards,* ed. H. de Bruijn, 41−62. Springer Netherlands: Dordrecht.
43. Heijungs, R., et al. 1992. *Environmental life cycle assessment of products: guide and backgrounds (part 1).* Centre of Environmental Science, Leiden, The Netherlands.
44. Muralikrishna, I. V., and V. Manickam. 2017. Chapter five-life cycle assessment. In *Environmental management,* ed. I.V. Muralikrishna, and V. Manickam, 57–75. Butterworth-Heinemann.
45. Hauschild, M. Z., R. K. Rosenbaum, and S. I. Olsen. 2018. *Life cycle assessment.* Springer.
46. Ferdous, J., et al. 2024. Use of process simulation to obtain life cycle inventory data for LCA: a systematic review. *Cleaner Environmental Systems* 14: 100215.
47. Yi, Y., et al. 2023. Integration of life cycle assessment and system dynamics modeling for environmental scenario analysis: A systematic review. *Science of The Total Environment* 903: 166545.
48. Fonseca, A., et al. 2023. Systematic insights into a textile industry: reviewing life cycle assessment and eco-design. *Sustainability* 15. https://doi.org/10.3390/su152115267.
49. Costa, C., et al. 2020. Textile industry in a changing world: Challenges of sustainable development. *U. Porto Journal of Engineering* 6 (2): 86–97.
50. Peters, G. M., G. Sandin, and B. Spak. 2019. Environmental prospects for mixed textile recycling in Sweden. *ACS Sustainable Chemistry & Engineering* 7 (13): 11682–11690.
51. Munasinghe, P., A. Druckman, and D. G. K. Dissanayake. 2021. A systematic review of the life cycle inventory of clothing. *Journal of Cleaner Production* 320: 128852.
52. Smith, P., J. Baille, and L.-S. McHattie. 2017. Sustainable design futures: an open design vision for the circular economy in fashion and textiles. *The Design Journal* 20 (sup1): S1938–S1947.
53. Panda, S. K. B. C., K. Sen, and S. Mukhopadhyay. 2021. Sustainable pretreatments in textile wet processing. *Journal of Cleaner Production* 329: 129725.
54. Niinimäki, K., et al. 2020. The environmental price of fast fashion. *Nature Reviews Earth & Environment* 1 (4): 189–200.
55. Konwar, M., and R. R. Boruah. 2020. Textile industry and its environmental impacts: a review. *Indian Journal of Pure & Applied Biosciences* 8: 134–139.
56. Broega, A. C., C. Jordão, and S. B. Martins. 2017. Textile sustainability: Reuse of clean waste from the textile and apparel industry. *IOP Conference Series: Materials Science and Engineering* 254 (19): 192006.
57. Yasin, S., et al. 2016. Statistical analysis of use-phase energy consumption of textile products. *The International Journal of Life Cycle Assessment* 21 (12): 1776–1788.
58. O'Rourke, D. 2014. The science of sustainable supply chains. *Science* 344 (6188): 1124–1127.
59. Rahaman, M. T., et al. 2024. Green production and consumption of textiles and apparel: importance, fabrication, challenges and future prospects. *Journal of Open Innovation: Technology, Market, and Complexity* 10 (2): 100280.
60. Bullon, J., et al. 2017. Manufacturing processes in the textile industry. Expert Systems for fabrics production. *ADCAIJ: Advances In Distributed Computing And Artificial Intelligence Journal* 6: 15.
61. Karimah, A., et al. 2021. A review on natural fibers for development of eco-friendly bio-composite: Characteristics, and utilizations. *Journal of Materials Research and Technology* 13: 2442–2458.
62. Nagaraja, S., et al. 2024. Synergistic advances in natural fibre composites: A comprehensive review of the eco-friendly bio-composite development, its characterization and diverse applications. *RSC Advances* 14 (25): 17594–17611.
63. Karimah, A., et al. 2021. A comprehensive review on natural fibers: technological and socio-economical aspects. *Polymers* 13. https://doi.org/10.3390/polym13244280.
64. Maiti, S., et al. 2022. Sustainable fiber-reinforced composites: a review. *Advanced Sustainable Systems* 6 (11): 2200258.
65. Thyavihalli Girijappa, Y. G., et al. 2019. Natural fibers as sustainable and renewable resource for development of eco-friendly composites: a comprehensive review. *Frontiers in Materials* 6.

66. da Costa, C., B. Del Curto, and A. Ritti. 2014. *Mechanical and chemical characteristics of vegetable fibers, and some applications*.
67. Romanowska, B., W. Różańska, and M. Zimniewska. 2024. The influence of the chemical composition of flax and hemp fibers on the value of surface free energy. *Materials* 17. https://doi.org/10.3390/ma17051104.
68. Mohammadi, M., M. R. Ishak, and M. T. H. Sultan. 2024. Exploring chemical and physical advancements in surface modification techniques of natural fiber reinforced composite: a comprehensive review. *Journal of Natural Fibers* 21 (1): 2408633.
69. Ben Sghaier, A. E. O., et al. 2012. Morphological and crystalline characterization of NaOH and NaOCl treated Agave americana L. fiber. *Industrial Crops and Products*, 36(1): 257–266.
70. Ahmad, U. M., et al. 2021. Can lignin be transformed into agrochemicals? recent advances in the agricultural applications of lignin. *Industrial Crops and Products* 170: 113646.
71. Han, D., and A. J. Steckl. 2019. Coaxial electrospinning formation of complex polymer fibers and their applications. *ChemPlusChem* 84 (10): 1453–1497.
72. Jatinder Singh, D. 2019. *Natural fibers: applications, in generation, development and modifications of natural fibers*, ed. Mudassar, A., and J. Han-Yong, IntechOpen: Rijeka. Ch. 2.
73. Suaria, G., et al. 2020. Microfibers in oceanic surface waters: a global characterization. *Science Advances* 6(23): eaay8493.
74. Egan, J., and S. Salmon. 2021. Strategies and progress in synthetic textile fiber biodegradability. *SN Applied Sciences* 4 (1): 22.
75. Jamir, M. R. M., M. S. A. Majid, and A. Khasri. 2018. 8-natural lightweight hybrid composites for aircraft structural applications. In *Sustainable composites for aerospace applications*, ed. M. Jawaid, and M. Thariq, 155–170, Woodhead Publishing.
76. Gonzalez, V., X. Lou, and T. Chi. 2023. Evaluating environmental impact of natural and synthetic fibers: a life cycle assessment approach. *Sustainability* 15, https://doi.org/10.3390/su15097670.
77. Petrescu, L., M. Fermeglia, and C.-C. Cormos. 2016. Life cycle analysis applied to acrylic acid production process with different fuels for steam generation. *Journal of Cleaner Production* 133: 294–303.
78. Yacout, D. M. M., M. A. Abd El-Kawi, and M. S. Hassouna. 2016. Cradle to gate environmental impact assessment of acrylic fiber manufacturing. *The International Journal of Life Cycle Assessment* 21(3): 326–336.
79. Muthu, S. S., et al. 2012. Quantification of environmental impact and ecological sustainability for textile fibres. *Ecological Indicators* 13 (1): 66–74.
80. Esteve-Turrillas, F. A., and M. de la Guardia. 2017. Environmental impact of recover cotton in textile industry. *Resources, Conservation and Recycling* 116: 107–115.
81. Lawless, E., and K. Medvedev. 2016. Assessment of sustainable design practices in the fashion industry: Experiences of eight small sustainable design companies in the Northeastern and Southeastern United States. *International Journal of Fashion Design, Technology and Education* 9 (1): 41–50.
82. Rathinamoorthy, R., and T. Kiruba. 2022. Bacterial cellulose-a potential material for sustainable eco-friendly fashion products. *Journal of Natural Fibers* 19 (9): 3275–3287.
83. Abuzawida, S. S., A. B. Alzubi, and K. Iyiola. 2023. Sustainable supply chain practices: an empirical investigation from the manufacturing industry. *Sustainability* 15, https://doi.org/10.3390/su151914395.
84. Balchandani, A., et al. 2020. *The state of fashion 2021: In search of promise in perilous times*, 8–118. McKinsey: Atlanta GA, USA.
85. Bhandari, N., et al. 2022. Barriers to sustainable sourcing in the apparel and fashion luxury industry. *Sustainable Production and Consumption* 31: 220–235.
86. Kim, Y., and K. W. Oh. 2020. Which consumer associations can build a sustainable fashion brand image? evidence from fast fashion brands. *Sustainability* 12. https://doi.org/10.3390/su12051703.

87. Karim, M. R., et al. 2024. Analyzing the factors influencing sustainable supply chain management in the textile sector. *Cleaner Logistics and Supply Chain* 13: 100183.
88. Vadicherla, T., and D. Saravanan. 2014. Textiles and apparel development using recycled and reclaimed fibers. In *Roadmap to sustainable textiles and clothing: eco-friendly raw materials, technologies, and processing methods*, ed. S. S. Muthu, 139–160. Singapore: Springer Singapore.
89. Thakker, A. M., and D. Sun. 2022. Ecological application of natural biomaterial on natural fibres. *Cleaner Materials* 3: 100038.
90. Amobonye, A., et al. 2023. Fungal mycelium as leather alternative: a sustainable biogenic material for the fashion industry. *Sustainable Materials and Technologies* 38: e00724.
91. da Silva Junior, C.J., et al. 2022. Design of a naturally dyed and waterproof biotechnological leather from reconstituted cellulose. *Journal of Functional Biomaterials* 13. https://doi.org/10.3390/jfb13020049.
92. Hong, H., and J. H. Kang. 2019. The impact of moral philosophy and moral intensity on purchase behavior toward sustainable textile and apparel products. *Fashion and Textiles* 6 (1): 16.
93. Saric, M., and M. Nellström. 2019. *A comparative life cycle assessment of nudie jeans' repair and reuse concept.*
94. Ellis, J. L., V. A. McCracken, and N. Skuza. 2012. Insights into willingness to pay for organic cotton apparel. *Journal of Fashion Marketing and Management: An International Journal* 16 (3): 290–305.
95. Heinze, T., and T. Liebert. 2012. 10.05-celluloses and polyoses/hemicelluloses. In *Polymer science: a comprehensive reference*, ed. Matyjaszewski, K., and M. Möller, 83–152. Elsevier: Amsterdam.
96. Abuhay, A., et al. 2021. Opportunities for new biorefinery products from Ethiopian ginning industry by-products: current status and prospects. *Journal of Bioresources and Bioproducts* 6 (3): 195–214.
97. Möller, M., and C. Popescu. 2012. 10.16-natural fibers, in polymer science: a comprehensive reference, ed. Matyjaszewski, K., and M. Möller, 267–280. Elsevier: Amsterdam.
98. Halliday, L. 2002. *Woolscouring, carbonising and effluent treatment*, 21–59.
99. Alagirusamy, R., and A. Das. 2011. *2-yarns: production, processability and properties.* In *Fibrous and composite materials for civil engineering applications*, ed. R. Fangueiro, 29–61. Woodhead Publishing.
100. Kiron, M. I. 2022. *Unconventional and new spinning systems and technologies in textile learner*.
101. Tyagi, G. K. 2010. *Yarn structure and properties from different spinning techniques*, 119–154. Advances in Yarn Spinning Technology.
102. Patti, A., and D. Acierno. 2023. Materials, weaving parameters, and tensile responses of woven textiles. *Macromol* 3: 665–680. https://doi.org/10.3390/macromol3030037.
103. Khanna, S., and A. Kaur. 2022. 4-innovation and technology of knitted apparels. In *Advanced knitting technology*, ed. S. Maity, et al., 95–140. Woodhead Publishing.
104. Pavko Čuden, A. 2023. Knitting towards sustainability, circular economy and Industry 4.0. *Applied Research* 2(6): e202200087.
105. Gaminian, H., et al. 2024. Revolutionizing sustainable nonwoven fabrics: the potential use of agricultural waste and natural fibres for nonwoven fabric. *Biomass* 4: 363–401. https://doi.org/10.3390/biomass4020018.
106. Kumar, P., and E. Gunasundari. 2018. *Sustainable wet processing—an alternative source for detoxifying supply chain in textiles.*
107. Hasanbeigi, A., and L. Price. 2015. A technical review of emerging technologies for energy and water efficiency and pollution reduction in the textile industry. *Journal of Cleaner Production* 95: 30–44.
108. Senthil Kumar, P., and E. Gunasundari. 2018. *Sustainable wet processing—an alternative source for detoxifying supply chain in textiles*, in *detox fashion: sustainable chemistry and wet processing*, ed. S. S. Muthu, 37–60. Springer Singapore: Singapore.

109. Tausif, M., et al. 2018. Cotton in the new millennium: advances, economics, perceptions and problems. *Textile Progress* 50 (1): 1–66.
110. Hasanbeigi, A., and L. Price. 2012. A review of energy use and energy efficiency technologies for the textile industry. *Renewable and Sustainable Energy Reviews* 16 (6): 3648–3665.
111. Bravo, M. V. C., and W. M. P. Iturralde. 2022. Energy efficiency in the competitiveness of the textile industry. *Open Journal of Business and Management* 10 (4): 1755–1767.
112. Hasanbeigi, A. 2013. *Alternative and emerging technologies for an energy-efficient, water-efficient, and low-pollution textile industry.*
113. Reddy, B. S., and B. K. Ray. 2011. Understanding industrial energy use: physical energy intensity changes in Indian manufacturing sector. *Energy Policy* 39 (11): 7234–7243.
114. Marquesone, R. D. and T. C. Carvalho. 2022. Examining the nexus between the versus of big data and the sustainable challenges in the textile industry. *Sustainability* 14. https://doi.org/10.3390/su14084638
115. Nayak, R., T. Panwar, and L. V. T. Nguyen. 2020. 1-sustainability in fashion and textiles: a survey from developing country. In *Sustainable technologies for fashion and textiles*, ed. R. Nayak, 3–30. Woodhead Publishing.
116. Moazzem, S., et al. 2021. Assessing environmental impact reduction opportunities through life cycle assessment of apparel products. *Sustainable Production and Consumption* 28: 663–674.
117. Imran, S., et al. 2023. Assessing the potential of GHG emissions for the textile sector: A baseline study. *Heliyon* 9 (11): e22404.
118. Roy Choudhury, A. K. 2014. Environmental impacts of the textile industry and its assessment through life cycle assessment. In *Roadmap to sustainable textiles and clothing: environmental and social aspects of textiles and clothing supply chain*, ed. S. S. Muthu, 1–39. Singapore: Springer Singapore.
119. Papachristou, E., D. Kalaitzi, and V. Pissas. 2023. A methodological framework for the integration of 3D virtual prototyping into the design development of laser-cut garments. *Journal of Engineered Fibers and Fabrics* 18: 15589250231194620.
120. La Rosa, A. D. and S. A. Grammatikos. 2019. Comparative life cycle assessment of cotton and other natural fibers for textile applications. Fibers 7. https://doi.org/10.3390/fib7120101.
121. Baydar, G., N. Ciliz, and A. Mammadov. 2015. Life cycle assessment of cotton textile products in Turkey. *Resources, Conservation and Recycling* 104.
122. Zhang, Y., et al. 2015. Life cycle assessment of cotton T-shirts in China. *The International Journal of Life Cycle Assessment* 20 (7): 994–1004.
123. Palacios-Mateo, C., Y. van der Meer, and G. Seide. 2021. Analysis of the polyester clothing value chain to identify key intervention points for sustainability. *Environmental Sciences Europe* 33 (1): 2.
124. Algahni, H. and M. A. Al-Dabbagh. 202. Assessing the impact of social media in the consumer trend towards sustainable clothing. *Fibres and Textiles* 3.
125. Abdelmeguid, A., M. Afy-Shararah, and K. Salonitis. 2024. Towards circular fashion: management strategies promoting circular behaviour along the value chain. *Sustainable Production and Consumption* 48: 143–156.
126. Vassalo, A., et al. 2024. Sustainability in the fashion industry in relation to consumption in a digital age. *Sustainability* 16: 5303.
127. Bertram, R., and T. Chi. 2018. A study of companies' business responses to fashion e-commerce's environmental impact. *International Journal of Fashion Design, Technology and Education* 11: 254–264.
128. Burt, S., and L. Sparks. 2003. E-commerce and the retail process: a review. *Journal of Retailing and Consumer Services* 10: 275–286.
129. Smith, P. 2022. Leading exporting countries of textile worldwide by value in 2021. 2022 August 12, 2023]. https://www.statista.com/statistics/1054452/textile-apparel-goods-leading-exporters-worldwide/.
130. Centobelli, P., et al. 2022. *Slowing the fast fashion industry: An all-round perspective*, 100684. Current Opinion in Green and Sustainable Chemistry.

131. Papamichael, I., et al. 2022. Building a new mind set in tomorrow fashion development through circular strategy models in the framework of waste management. *Current Opinion in Green and Sustainable Chemistry* 36: 100638.
132. Centobelli, P., et al. 2022. Slowing the fast fashion industry: an all-round perspective. *Current Opinion in Green and Sustainable Chemistry* 38: 100684.
133. Ikram, M. Transition toward green economy: technological innovation's role in the fashion industry. *Current Opinion in Green and Sustainable Chemistry* 37.
134. Shukaili, S., Z. Jamaluddin, and N. Zulkifli. 2023. The impact of strategic inventory management on logistics organization's performance. *International Journal of Business and Technology Management* 5: 288–298.
135. Wrigley, N., et al. 2015. *British high streets: from crisis to recovery? a comprehensive review of the evidence.*
136. Eryuruk, S. 2012. Greening of the textile and clothing industry. *Fibres and Textiles in Eastern Europe* 95: 22–27.
137. Yang, S., Y. Song, and S. Tong. 2017. Sustainable retailing in the fashion industry: a systematic literature review. *Sustainability* 9: 1266.
138. Bhardwaj, V. 2010. Fast fashion: Response to changes in the fashion industry. The International Review of Retail. *Distribution and Consumer Research.* 165–173.
139. MacCarthy, B. L., and P. Jayarathne. 2013. Supply network structures in the international clothing industry: differences across retailer types. *International Journal of Operations & Production Management* 33 (7): 858–886.
140. Chovancová, J., M. Popovičová, and E. Huttmanová. 2023. Decoupling transport-related greenhouse gas emissions and economic growth in the European Union countries. *Journal of Sustainable Development of Energy, Water and Environment Systems* 11 (1): 1–18.
141. Ferrer, A. L., and A. M. Thomé, 2023. *Carbon emissions in transportation: a synthesis framework. Sustainability* 15. https://doi.org/10.3390/su15118475.
142. Tseng, S.-C., and S.-W. Hung. 2014. A strategic decision-making model considering the social costs of carbon dioxide emissions for sustainable supply chain management. *Journal of Environmental Management* 133: 315–322.
143. Hiller Connell, K. Y., and J. M. Kozar. 2012. Sustainability knowledge and behaviors of apparel and textile undergraduates. *International Journal of Sustainability in Higher Education* 13(4): 394–407.
144. Tebaldi, L., A. Brun, and E. Bottani. 2022. Evidences on sustainability issues in the fashion supply chain: an empirical study in Italy. *Sustainable Production and Consumption* 33: 651–663.
145. Roy Choudhury, A. 2014. Environmental impacts of the textile industry and its assessment through life cycle assessment. In *Roadmap to sustainable textiles and clothing: environmental and social aspects of textiles and clothing supply chain*, 1–39.
146. Turker, D., and C. Altuntas. 2014. Sustainable supply chain management in the fast fashion industry: an analysis of corporate reports. *European Management Journal* 32 (5): 837–849.
147. Lee, M. S., M. Seifert, and H. Cherrier. 2017. Anti-consumption and governance in the global fashion industry: transparency is key. In *Governing corporate social responsibility in the apparel industry after Rana Plaza*, 147–174.
148. Zamani, B., G. Sandin, and G. M. Peters. 2017. Life cycle assessment of clothing libraries: Can collaborative consumption reduce the environmental impact of fast fashion? *Journal of cleaner production* 162: 1368–1375.
149. Li, Z., et al. 2024. The carbon footprint of fast fashion consumption and mitigation strategies-a case study of jeans. *Science of The Total Environment* 924: 171508.
150. Sulochani, R. M. N., et al. 2024. Waste-based composites using post-industrial textile waste and packaging waste from the textile manufacturing industry for non-structural applications. *Sustainable Chemistry for the Environment* 8: 100163.
151. Ćirović, N., and L. Trumbulović. 2023. Waste management in the textile industry. *Reciklaza i odrzivi razvoj* 16: 99–107.

152. Lüttin, L. 2025. [Textile Industry] The EU's packaging and packaging waste regulation (PPWR) explained.
153. Biyada, S., and J. Urbonavičius. 2025. Circularity in textile waste: challenges and pathways to sustainability. *Cleaner Engineering and Technology* 24: 100905.
154. Druckman, A., and T. Jackson. 2010. An exploration into the carbon footprint of UK households. *Social Research* (New York) 02–10.
155. Yun, C., et al. 2016. Assessment of environmental and economic impacts made by the reduced laundering of self-cleaning fabrics. *Fibers and Polymers* 17 (8): 1296–1304.
156. Yasin, S., et al. 2016. Statistical analysis of use-phase energy consumption of textile products. *The International Journal of Life Cycle Assessment* 21.
157. Hicks, A. L., et al. 2015. Life cycle payback estimates of nanosilver enabled textiles under different silver loading, release, and laundering scenarios informed by literature review. *Environmental Science and Technology* 49 (13): 7529–7542.
158. Kim, J., et al. 2015. Post-consumer energy consumption of textile products during 'use' phase of the lifecycle. *Fibers and Polymers* 16 (4): 926–933.
159. Geyer, R., J. R. Jambeck, and K. L. Law. 2021. Production, use, and fate of all plastics ever made. *Science Advances* 3(7): e1700782.
160. Koszewska, M. 2013. A typology of Polish consumers and their behaviours in the market for sustainable textiles and clothing. *International Journal of Consumer Studies* 37 (5): 507–521.
161. Ofori, E. A., et al. 2025. A review on upcycling as a strategic marketing tool for sustainability in the fashion industry. *Journal of Responsible Production and Consumption* 2 (1): 1–24.
162. Nathalie, R., S. Eveline, and S. Steven. 2023. Style that's sustainable: a new fast-fashion formula. *Sustainability 2016*. https://www.mckinsey.com/capabilities/sustainability/our-insights/style-thats-sustainable-a-new-fast-fashion-formula.
163. Yalcin-Enis, I., M. Kucukali-Ozturk, and H. Sezgin. 2019. Risks and management of textile waste. In *Nanoscience and biotechnology for environmental applications*, ed. K. M. Gothandam, et al., 29–53. Cham: Springer International Publishing.
164. Samui, A. B., P. E. Hande, and M. I. H. Mondal. 2022. 25-care, maintenance and disposability of medical and protective textile products. In *Protective textiles from natural resources*, ed. M. I. H. Mondal, 793–837. Woodhead Publishing.
165. Classen, E. 2024. 16-sustainable textile care and maintenance. In *Sustainable innovations in the textile industry*, ed. Paul, R., and T. Gries, 427–456. Woodhead Publishing.
166. Burton, M. D., et al. 2020. The consumer's perspective: Care and maintenance in 'use phase' of clothing apparel life cycle. *Sustainability in Fashion* 1(1).
167. Bukhari, M. A., R. Carrasco-Gallego, and E. Ponce-Cueto. 2018. Developing a national programme for textiles and clothing recovery. *Waste Management & Research* 36 (4): 321–331.
168. Powell, J. T., and M. R. Chertow. 2019. Quantity, components, and value of waste materials landfilled in the United States. *Journal of Industrial Ecology* 23 (2): 466–479.
169. Jönsson, C., et al. 2021. Biocatalysis in the recycling landscape for synthetic polymers and plastics towards circular textiles. *Chemsuschem* 14 (19): 4028–4040.
170. Shamsuzzaman, M., et al. 2025. Fashion and textile waste management in the circular economy: a systematic review. *Cleaner Waste Systems* 11: 100268.
171. Hole, G., and A. S. Hole. 2019. Recycling as the way to greener production: a mini review. *Journal of Cleaner Production* 212: 910–915.
172. Muthu, S. S. 2020. *Assessing the environmental impact of textiles and the clothing supply chain*. Woodhead publishing.
173. Weber, S., J. Lynes, and S. B. Young. 2017. Fashion interest as a driver for consumer textile waste management: reuse, recycle or disposal. *International Journal of Consumer Studies* 41 (2): 207–215.
174. Vaverková, M. D. 2019. Landfill impacts on the environment—review. *Geosciences* 9. https://doi.org/10.3390/geosciences9100431.
175. Ransom-Jones, E., et al. 2017. Lignocellulose-degrading microbial communities in landfill sites represent a repository of unexplored biomass-degrading diversity. *mSphere* 2: e00300–17.

176. Degenstein, L., R. McQueen, and N. Krogman. 2021. 'What goes where'? characterizing Edmonton's municipal clothing waste stream and consumer clothing disposal. *Journal of Cleaner Production* 296: 126516.
177. De Felice, F., et al. 2025. Circular economy practices in the textile industry for sustainable future: a systematic literature review. *Journal of Cleaner Production* 486: 144547.
178. Sandin, G., and G. M. Peters. 2018. Environmental impact of textile reuse and recycling–a review. *Journal of Cleaner Production* 184: 353–365.
179. Singh, J., et al. 2019. Challenges and opportunities for scaling up upcycling businesses–the case of textile and wood upcycling businesses in the UK. *Resources, Conservation and Recycling* 150: 104439.
180. Weavabel. 2023. 9 Brands that are embracing fashion upcycling. *Sustainability 2023*. https://www.weavabel.com/blog/brands-that-are-embracing-fashion-upcycling.
181. Giordano, D. 2019. Weaving the circle: exploring the potential and challenges to upcycle post-industrial textile waste of an outdoor apparel company. In *The international institute for industrial environmental economics (IIIEE)*, 88. Lund University: Sweden
182. Norum, P. S. 2013. Examination of apparel maintenance skills and practices: implications for sustainable clothing consumption. *Family and Consumer Sciences Research Journal* 42 (2): 124–137.
183. Chan, E. 2021. How re/done became the sustainable jeans brand fashion insiders can't live without. *Sustainability* 2023. https://www.vogue.co.uk/fashion/article/redone-sustainable-jeans-brand.
184. Bianco, I., et al. 2023. Environmental impacts in the textile sector: a life cycle assessment case study of a woolen undershirt. *Sustainability* 15. https://doi.org/10.3390/su151511666.
185. Chakravarty, D., I. Satpathy, and B. Patnaik. 2024. Embellishing fashion with sustainable goals: challenges in fashion supply chain. In *Illustrating digital innovations towards intelligent fashion: leveraging information system engineering and digital twins for efficient design of next-generation fashion*, 179–202. Springer.
186. Rana, N. 2024. *The environmental impact of fashion*, in *threaded harmony: a sustainable approach to fashion*, 5–16. Emerald Publishing Limited.
187. Galloway, K. 2024. The environmental impact of cotton production. In *International Science Council*.
188. Farhana, K., et al. 2022. Energy consumption, environmental impact, and implementation of renewable energy resources in global textile industries: an overview towards circularity and sustainability. *Materials Circular Economy*, 4.
189. Singh, J., and S. Bansal. 2024. The impact of the fashion industry on the climate and ecology. *World Journal of Advanced Research and Reviews* 21 (1): 210–215.
190. Stone, C., et al. 2020. Natural or synthetic–how global trends in textile usage threaten freshwater environments. *Science of The Total Environment* 718: 134689.
191. Chapagain, A., R. Mathews, and G. Zhang. *A guide to reducing the water footprint of cotton cultivation in India*. Water Footprint Network.
192. Singh, N., and V. Shinde. 2024. Sustainable cotton cultivation unveiled: rainfed farming and the NICE standard. In *Control union global*.
193. Bailey, K., A. Basu, and S. Sharma. 2022. The environmental impacts of fast fashion on water quality: a systematic review. *Water* 14: 1073.
194. Han, S., D. Tyler, and P. Apeagyei. 2015. *Upcycling as a design strategy for product lifetime optimisation and societal change.*
195. Sakamoto, M., et al. 2019. Water pollution and the textile industry in bangladesh: flawed corporate practices or restrictive opportunities?. *Sustainability* 11. https://doi.org/10.3390/su11071951.
196. Okafor, C. C., et al. 2021. Sustainable management of textile and clothing. *Clean Technologies & Recycling* 1: 70–87.
197. Shirvanimoghaddam, K., et al. 2020. Death by waste: fashion and textile circular economy case. *Science of The Total Environment* 718: 137317.

198. Stanescu, M. D. 2021. State of the art of post-consumer textile waste upcycling to reach the zero waste milestone. *Environmental Science and Pollution Research International* 28 (12): 14253–14270.
199. Sharpe, S., M. Retamal, and M. C. Martinez-Fernandez. 2022. *Assessing the impact: Environmental impact assessment in the textile and garment sector in Bangladesh, Cambodia, Indonesia and Viet Nam.* ILO Working Paper.
200. Bailey, K., A. Basu, and S. Sharma. 2022. The environmental impacts of fast fashion on water quality: a systematic review. *Water* 14. https://doi.org/10.3390/w14071073.
201. Schröcker, D., et al. 2023. *ERA industrial technology roadmap for circular technologies and business models in the textile, construction and energy-intensive industries.*
202. Boykoff, M., et al. 2021. Examining climate change and sustainable/fast fashion in the 21st century: 'Trash the Runway'. *Oxford Open Climate Change* 1(1): kgab003.
203. Dominish, E. 2022. *Taking climate action: measuring carbon emissions in the garment sector in Asia.* International Labour Organization. https://webapps.ilo.org/static/english
204. Jonášová, S., T. Zoumpalová, and B. Moldan. 2022. Analysis of the amount of textile waste in mixed municipal waste in the Czech Republic between 2016 and 2021. In *Waste Forum.*
205. Zoumpalova, T., S. K. Jonášová, and B. Moldan. Barriers to the circular economy in textile industry: a case study of the Czech Republic. *WASTE.*
206. Wiedmann, T., and J. Minx. 2008. A definition of 'carbon footprint.' *Ecological economics research trends* 2008 (1): 1–11.
207. Barber, A., and G. Pellow. 2006. LCA: New Zealand merino wool total energy use. In *5th Australian life cycle assessment society (ALCAS) conference.* Melbourne.
208. Gaonkar, O. 2021. *An overview of toxic chemicals in textiles.*
209. Tounsadi, H., et al. 2020. Impact of chemical substances used in textile industry on the employee's health: Epidemiological study. *Ecotoxicology and Environmental Safety* 197: 110594.
210. Ahsan, A., et al. 2023. Wastewater from the textile industry: Review of the technologies for wastewater treatment and reuse. *Korean Journal of Chemical Engineering.*
211. Ardila-Leal, L. D., et al. 2021. A brief history of colour, the environmental impact of synthetic dyes and removal by using laccases. *Molecules* 26 (13): 3813.
212. Zajac, G. 2023. *Toxic fashion: evaluation of chemicals in clothing and recommendations for the amended TSCA.*.
213. Roos, S., et al. 2015. Is unbleached cotton better than bleached? exploring the limits of life-cycle assessment in the textile sector. *Clothing and Textiles Research Journal* 33.
214. Fang, B. 2023. Environmental and human impacts of fast fashion. *Communications in Humanities Research* 15: 166–171.
215. Chen, L., et al. 2017. Supply chain collaboration for sustainability: a literature review and future research agenda. *International journal of production economics* 194: 73–87.
216. Allen, L. 2022. What are the three pillars of sustainability?, ed. Truehugger, E. MacLennan. Truehugger.
217. de Brito, M. P., V. Carbone, and C. M. Blanquart. 2008. Towards a sustainable fashion retail supply chain in Europe: organisation and performance. *International Journal of Production Economics* 114 (2): 534–553.
218. Thorisdottir, T. S. and L. Johannsdottir. 2020. Corporate social responsibility influencing sustainability within the fashion industry. A systematic review. *Sustainability* 12. https://doi.org/10.3390/su12219167.
219. Nassivera, F., et al. 2017. Willingness to pay for organic cotton: Consumer responsiveness to a corporate social responsibility initiative. *British Food Journal* 119 (8): 1815–1825.
220. Unilorin, N. 2022. Assessing social sustainability performance in textile industry using integrated fuzzy best-worst method and fuzzy inference system 19: 287−297.
221. Chattopadhyay, A., and S. Gangopadhyay. 2020. *Trends of alterations in pulmonary function and symptoms in garment workers of Garden- Reach-Metiabruz area of Kolkata.*
222. Mukherjee, S. 2015. *Environmental and social impact of fashion: Towards an eco-friendly, ethical fashion.* 2: 22–35.

223. Liu, M., and R. Vijeyarasa. 2022. Fast fashion for 2030: using the pattern of the sustainable development goals (SDGs) to cut a more gender-just fashion sector. *Business and Human Rights Journal* 7 (1): 45–66.
224. Prasad, A. 2018. De-feminization and (Dis) empowerment of women workers in garment factories. *ANTYAJAA: Indian Journal of Women and Social Change* 3(1): 12–23.
225. Seidu, R. K., et al. 2024. A systematic review of work-related health problems of factory workers in the textile and fashion industry. *Journal of Occupational Health* 66(1): uiae007.
226. Gam, H. J., U.-J. Yu, and S. Yang. 2020. The effects of health consciousness on environmentally sustainable textile furnishing product purchase. *Family and Consumer Sciences Research Journal* 49 (1): 84–100.
227. Bick, R., E. Halsey, and C. C. Ekenga. 2018. The global environmental injustice of fast fashion. *Environmental Health* 17 (1): 92.
228. Paul-Majumder, P. 2003. *Health status of the garment workers in Bangladesh*, vol. 1. Bangladesh Institute of Development Studies.
229. Ahmed, S., and M. Z. Raihan. 2014. Health status of the female workers in the garment sector of Bangladesh. *Journal of Economics Finance and Administrative Science* 4 (1): 43–58.
230. Sahan, M. 2020. The first international standard on violence and harassment in the world of work. *Business and Human Rights Journal* 5 (2): 289–295.
231. Martínez-Peláez, R., et al. 2023. Role of digital transformation for achieving sustainability: mediated role of stakeholders, key capabilities, and technology. *Sustainability* 15 (14): 11221.
232. Sani, D., et al. 2021. An overview of the transition to a circular economy in emilia-romagna region, Italy considering technological, legal–regulatory and financial points of view: a case study. *Sustainability* 13 (2): 596.
233. Wu, J., and L. Li. 2019. *Sustainability initiatives in the fashion industry.*
234. Manshoven, S., and W. Van Opstal. 2022. The carrot or the stick? stakeholder support for mandatory regulations towards a circular fashion system. *Sustainability* 14. https://doi.org/10.3390/su142214671.
235. Clancy, G., M. Fröling, and G. Peters. 2015. Ecolabels as drivers of clothing design. *Journal of Cleaner Production* 99: 345–353.
236. Tleuken, A., et al. 2022. Legislative, institutional, industrial and governmental involvement in circular economy in central Asia: a systematic review. *Sustainability* 14 (13): 8064.
237. Rossi, M., et al. 2021. Life cycle assessment of a leather shoe supply chain. *International Journal of Sustainable Engineering* 14 (4): 686–703.
238. Szegedi, K., T. Németh, and D. Körtvési. 2023. Employer branding in the fashion industry: CSR actions by fashion SMEs. *Sustainability* 15. https://doi.org/10.3390/su15031827.
239. Beier, G., et al. 2022. Impact of industry 4.0 on corporate environmental sustainability: comparing practitioners' perceptions from China, Brazil and Germany. *Sustainable Production and Consumption* 31.
240. Villa Todeschini, B., et al. 2017. *Innovative and sustainable business models in the fashion industry: Entrepreneurial drivers, opportunities, and challenges.* Business Horizons.
241. Islam, M. D., H. Shen, and S. Badsha. 2022. Integrating blockchain into supply chain safeguarded by PUF-enabled RFID. *Internet of Things* 18: 100505.
242. Carter, C. R., D. S. Rogers, and T. Y. Choi. 2015. Toward the theory of the supply chain. *Journal of Supply Chain Management* 51 (2): 89–97.
243. Pensupa, N., et al. 2017. Recent trends in sustainable textile waste recycling methods: current situation and future prospects. *Topics in Current Chemistry (Cham)* 375 (5): 76.
244. Gentil, E. C., et al. 2010. Models for waste life cycle assessment: Review of technical assumptions. *Waste Management* 30 (12): 2636–2648.
245. Olivetti, E., S. Patanavanich, and R. Kirchain. 2013. Exploring the viability of probabilistic under-specification to streamline life cycle assessment. *Environmental science & technology* 47.
246. Hoogmartens, R., et al. 2014. Bridging the gap between LCA, LCC and CBA as sustainability assessment tools. *Environmental Impact Assessment Review* 48: 27–33.

247. Finkbeiner, M., et al. 2006. The new international standards for life cycle assessment: ISO 14040 and ISO 14044. *The International Journal of Life Cycle Assessment* 11 (2): 80–85.
248. Salim, S., and S. Ismail. 2024. ISO certifications and sustainable fashion practices in Malaysia: a legal perspective. *IIUM Law Journal* 32: 193–230.
249. Golob, U., et al. 2022. Brand management and sustainability: exploring potential for the transformative power of brands. *Journal of Brand Management* 29 (6): 513–519.
250. Chan, C. K., J. Shin, and S. X. K. Jiang. 2017. Development of tailor-shaped bacterial cellulose textile cultivation techniques for zero-waste design. *Clothing and Textiles Research Journal* 36 (1): 33–44.
251. Kabeyi, M. J. B., and O. A. Olanrewaju. 2022. Sustainable energy transition for renewable and low carbon grid electricity generation and supply. *Frontiers in Energy Research* 9.
252. Hayat, N., A. Hussain, and H. D. Lohano. 2020. Eco-labeling and sustainability: a case of textile industry in Pakistan. *Journal of Cleaner Production* 252: 119807.
253. Rai, M. 2025. Life cycle assessment (LCA) for textiles and clothing. In *Carbon trail*.
254. Zamani, B. 2023. *Life cycle assessments for textile and fashion product*.
255. Chun, E., et al. 2021. Business transparency and willingness to act environmentally conscious behavior: applying the sustainable fashion evaluation system "Higg Index." *Journal of Global Scholars of Marketing Science* 31 (3): 437–452.
256. Watson, K. J., and S. G. Wiedemann. 2019. Review of methodological choices in LCA-based textile and apparel rating tools: key issues and recommendations relating to assessment of fabrics made from natural fibre types. *Sustainability* 11. https://doi.org/10.3390/su11143846.
257. Niinimäki, K., et al. 2024. *Identifying data gaps in the textile industry and assessing current initiatives to address them*.
258. Mutambo, N., et al. 2024. Mapping the environmental impact assessment landscape in the fashion and textile industries: critical gaps and challenges. *Sustainability* 16. https://doi.org/10.3390/su16198377.
259. Finkbeiner, M., et al. 2014. Challenges in life cycle assessment: an overview of current gaps and research needs. In *Background and future prospects in life cycle assessment*, ed. W. Klöpffer, 207–258. Springer Netherlands: Dordrecht.
260. Baker, J., and M. Lepech. 2009. Treatment of uncertainties in life cycle assessment. In *10th international conference on structural safety and reliability (ICOSSAR09)*. Osaka, Japan.
261. Barahmand, Z., and M. S. Eikeland. 2022. Life cycle assessment under uncertainty: a scoping review. *World* 3: 692–717. https://doi.org/10.3390/world3030039.
262. Ciroth, A., G. Fleischer, and J. Steinbach. 2004. Uncertainty calculation in life cycle assessments. *The International Journal of Life Cycle Assessment* 9: 216–226.
263. Kazdin, T. 2024. *ASTM releases textile circularity standards report*. recyclingtoday.com.
264. Arun, C. M., et al. 2025. Eco-innovative approaches for recycling non-polyester/cotton blended textiles. *Waste Management Bulletin* 3 (1): 255–270.
265. Benoît, C., et al. 2010. The guidelines for social life cycle assessment of products: just in time! *The International Journal of Life Cycle Assessment* 15 (2): 156–163.
266. Liscio, M., and P. Sospiro. 2023. *Life cycle assessment on fashion industry: four case studies*.
267. Olatubosun, P., E. Charles, and T. Omoyele. 2021. Rethinking luxury brands and sustainable fashion business models in a risk society. *Journal of Design, Business & Society* 7: 49–81.
268. Jhanji, Y. 2023. 18-life cycle analysis of textiles and associated carbon emissions. In *Sustainable fibres for fashion and textile manufacturing*, ed. R. Nayak, 403–431. Woodhead Publishing.
269. Bonsu, N. O. 2020. Towards a circular and low-carbon economy: Insights from the transitioning to electric vehicles and net zero economy. *Journal of Cleaner Production* 256: 120659.
270. Nissinen, A., et al. 2007. Developing benchmarks for consumer-oriented life cycle assessment-based environmental information on products, services and consumption patterns. *Journal of Cleaner Production* 15 (6): 538–549.
271. Ryding, S.-O. 1999. ISO 14042 Environmental management life cycle assessment life cycle impact assessment. *The International Journal of life cycle assessment* 4 (6): 307–307.

272. Yuan, G. X., et al. 2013. Application of laser engraving for sustainable fashion design. *Research Journal of Textile and Apparel* 17 (2): 21–27.
273. Deng, M., Y. Liu, and L. Chen. 2023. AI-driven innovation in ethnic clothing design: an intersection of machine learning and cultural heritage. *Electronic Research Archive.*
274. Guo, S., X. Sun, and H. K. S. Lam. 2023. Applications of blockchain technology in sustainable fashion supply chains: operational transparency and environmental efforts. *IEEE Transactions on Engineering Management* 70 (4): 1312–1328.
275. Wang, B., et al. 2020. Blockchain-enabled circular supply chain management: a system architecture for fast fashion. *Computers in Industry* 123: 103324.
276. Kouhizadeh, M., and J. Sarkis. 2018. Blockchain practices, potentials, and perspectives in greening supply chains. *Sustainability* 10. https://doi.org/10.3390/su10103652.
277. Fernández-Caramés, T. M., and P. Fraga-Lamas. 2018. Towards the internet of smart clothing: a review on IoT wearables and garments for creating intelligent connected e-textiles. *Electronics* 7. https://doi.org/10.3390/electronics7120405.

Dr. Benjamin Tawiah holds a B.A. in Textiles and a Master of Engineering from Kwame Nkrumah University of Science and Technology, Ghana, and Jiangnan University, China, respectively. He also has a Postgraduate Diploma in Management Information Systems from GIMPA and a PhD in Textile Technology (Chemistry) from Hong Kong Polytechnic University. Previously, he was a postdoctoral fellow at Hong Kong Polytechnic University and City University of Hong Kong. He lectures at Kwame Nkrumah University of Science and Technology, focusing on polymer nanomaterial composites, flame retardant polymer composites, and second-generation batteries for advanced grid energy storage.

Emmanuel Abankwah Ofori is a postgraduate student studying material science and engineering at the Hubei University of Automotive Technology, Shiyan-Hubei, China. He is currently under the supervision of Dr. Hou Yongdan. Emmanuel's research interest focuses on batteries and supercapacitors for energy storage, green transportation, textile science, sustainable environmental practices, and flame-retardant polymer composites. Recently, he co-authored a paper that was adjudged the "Second-Best Paper" presented at the 12th Sustainable Education and Development Research Conference in Somanya, Ghana. He has actively participated in several research projects, working with renowned authors to explore various facets of knowledge. Emmanuel worked as a Teaching and Research Assistant under Dr. Benjamin Tawiah, who helped him develop his writing abilities. His research contributions are noted in prestigious scientific journals like the Journal of Energy Storage, MDPI Sustainability and Engineering Proceedings, Journal of Occupational Health, and Research Journal of Textile and Apparel.

Raphael Kanyire Seidu holds a Ph.D. in Fashion and Textile Design at the School of Fashion and Textiles at The Hong Kong Polytechnic University, Hong Kong. Before this, he gained both his MFA degree in Textile Design and BA. Degree in Industrial Art (Textile option) from Kwame Nkrumah University of Science and Technology, Ghana. He is also a Visiting Research Student at Nottingham Trent University (NTU) in the UK. He has several publications in leading journals in textile, design, technology, and material science. His recent paper was adjudged the "Second-Best Paper" presented at the 12th Sustainable Education and Development Research Conference in Somanya, Ghana. His research interest focuses on retro-reflective materials, smart textiles, cloth culture, weaving, and print technologies. He has several papers and has filed for a patent from his PhD project which aims at developing smart interactive retro-reflective clothing for pedestrians' safety at night-time.

Bin Fei is a Professor in the School of Fashion and Textiles at The Hong Kong Polytechnic University. Prof. Fei got a B.S. from the Department of Applied Chemistry, University of Science and Technology of China in 1996 and a Ph.D. from Changchun Institute of Applied Chemistry, Chinese Academy of Sciences in 2003. With more than two decades of experience in polymer and composites research, Prof. Fei has been engaged in fundamental and applied research on sustainable manufacturing of bio-composites, flame retardant materials, and energy devices.

Chapter 13
Upcycling and Recycling Revolution: Techniques for Textile Waste Management Towards Sustainability

Benjamin Eghan and Kyungeun Sung

Abstract The escalating challenges of environmental concerns regarding textile waste on a global scale have intensified the need for sustainable waste management strategies. The textile industry contributes approximately 92 million tonnes of textile waste annually, contributing 17–20% of the global solid waste. This is characterised by a linear system that encourages frequent order and disposal of clothing produced to meet changing consumer demand and trends. This leads to landfills or incineration every second. To address the textile industry's waste problem, sustainable solutions through the circular framework are ushering in a revolution towards "Upcycling and Recycling". Taken together, these approaches offer new strategies for reducing the environmental footprint of the textile and fashion industry, which is a major contributor to global waste and pollution. This book chapter addresses various techniques of upcycling and recycling, underlining their potential for effective waste reduction, resource conservation, and promotion of circular economy principles. This chapter delves into the various innovative techniques for managing textile waste via upcycling— creative reuse of textile waste into more valuable products, thereby reducing waste and extending the lifespan of materials—and recycling, which focuses on the processing of deconstructing waste into reusable raw materials. Both ways help to prolong the life of textiles, reduce waste and the need for virgin materials in production, thereby improving the environment. Additionally, technologies advancing these techniques are explored, while highlighting the importance of consumer awareness. The adoption of these sustainable practices helps reduce environmental footprint and conserve resources, which is crucial in moving towards a greener and more sustainable future.

Keywords Upcycling · Recycling · Textile waste · Sustainability · Waste management

---

B. Eghan (✉) · K. Sung
Faculty of Arts, Design and Humanities, De Montfort University, Leicester, UK
e-mail: bennelson2224@gmail.com

K. Sung
e-mail: Kyungeun.sung@dmu.ac.uk

S.-X. Jiang et al. (eds.), *Advances in Textile Materials and Processing Techniques for Sustainability*, SDGs and Textiles, https://doi.org/10.1007/978-981-95-0469-5_13

13.1 Introduction

The global textile and fashion sector is confronted with a grave paradox: it drives economic growth and consumerism on the one hand, but it so happens to be one of the largest environmental polluters [1]. As textile fibre production and consumption of apparel increase, textile waste has become escalating across the globe, placing unsustainable pressure on nature resources and ecosystems [2]. Fashion waste is no longer a secondary problem but a central environmental and economic concern today. The dominant fashion model—fast fashion and linear consumption culture—has increased resource extraction, waste production, pollution and made waste management a major concern for sustainability in the industry.

The ecological footprint of textile waste is staggering. Up to 75% of global textile waste is incinerated or sent to landfills, releasing hazardous chemicals, greenhouse gases, and microplastics that undermine human health and ecosystems [2]. Already at capacity, landfills are being plugged with non-biodegradable textile waste that will take decades to decompose, meanwhile spewing toxins and releasing methane—a toxic greenhouse gas [1]. This is driven yet more by growing global population and economic development, which both continue to increase demand for inexpensive, disposable apparel [3].

One of the primary drivers of the crisis is the rise of fast fashion. With fashion becoming increasingly affordable and accessible, particularly in the UK and Europe [4], consumer attitudes have shifted to frequent buying and quick disposal [5]. Fashion trends now cycle so quickly that clothing is discarded before the end of its functional lifespan. This overconsumption has resulted in a high rate of used textiles being exported to developing countries, incinerated, or landfilled—rarely recovered or recycled back into the production cycle [6]. Furthermore, the composite material composition and complex design of modern clothing items pose technical challenges for recycling, and they are generally destined for the waste stream regardless of their intrinsic value [2].

Added to this is the linear model of resource usage—take, make, dispose—that characterizes present textile production and consumption patterns. This model is inherently unsustainable, as it repeatedly extracts finite natural resources and disposes of enormous amounts of waste [5], leading to raw material depletion, climate change, and loss of biodiversity [7]. To resolve this, the sector needs to shift towards a circular economy that emphasizes resource efficiency, and regenerative design.

This chapter, therefore, explores approaches to managing textile waste through recycling—which transforms waste into reusable raw materials; and upcycling, which creatively repurposes used textiles into higher-value products. Both methods help reduce waste generation, extend the lifespan of materials, and lessen the demand for virgin resources [8]. In today's rapidly evolving world, upcycled or refashioned clothing is becoming increasingly important [9]. Both consumers and stakeholders in the textile and fashion industries share the responsibility of promoting sustainability and safeguarding the environment.

13.2 The Environmental Impact of Textile Waste and Fast Fashion

Before the 1800s, possessions were cherished, carefully maintained, and used to the fullest extent of their utility [5]. Quality and durability were central to both production and consumption practices [3]. Today, the fashion and textile industry (FTI) operates within a long and complex supply chain that starts with the extraction of natural resources for fibre production and extends through manufacturing, logistics, and retail. With rapid production, consumption, and disposal embedded in the industry, FTI is confronted with significant environmental and resource concerns. Global textile fibre production, consumption, and waste continue to rise, with fibre output projected to reach 130 Mt by 2025, from about 85.5 Mt in 2013 [4]. This surge is driving a proportional increase in the industry's environmental footprint, doubling its production, consumption, and disposal rates [6]. Although the development of textile composites may offer temporary relief, without sustainable manufacturing and effective recycling practices, these composites are ultimately destined for landfills at the end of their useful lives [1].

Studies estimate that only 20–23% of the globally discarded fashion and textile (FT) items are collected for potential recycling, yet the fibre-to-fibre recovery rate remains as low as 1%. As a result, over 70% of these items are landfilled or incinerated [10]. Figure 13.1 shows the material flow for FT production through to end use. Despite its economic significance and capacity to provide employment, the FTI faces ongoing criticism for being one of the most polluting sectors globally [9].

In a world where change happens quickly, the fast fashion industry has adapted to consumers' need for vast quantities of cheap clothing. Nevertheless, this has obscured the growing environmental costs associated with the industry's waste management

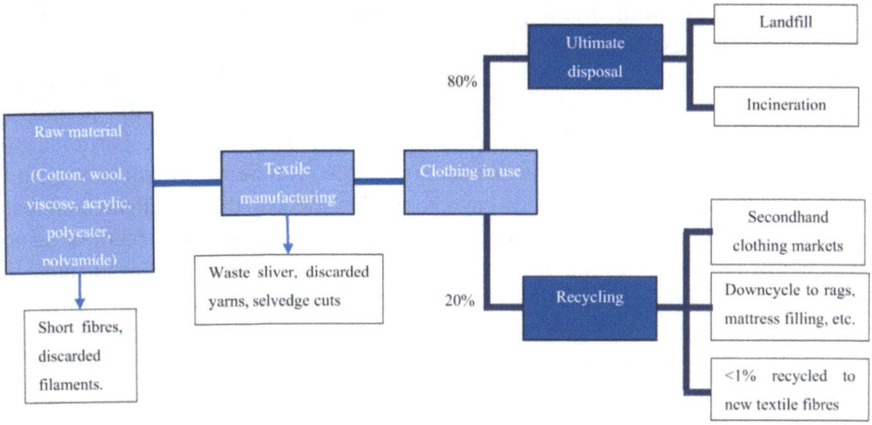

Fig. 13.1 A schematic diagram of textile material flow to the end of life [11] (Open-access image distributed under the Creative Commons Attribution 4.0 License (https://creativecommons.org/licenses/by/4.0/) (© The Author(s) 2023)

[2]. In 2012, clothing expenditure accounted for 2.5% of household expenditure in Finland compared to 4.2% in the EU. There is compelling evidence that Europeans consume much more clothing today than they did a few decades ago [4], which has a significant impact on the environment.

In terms of manufacturing, fast fashion is not a favourable concept, although it is profit-driven. It utilises low-grade materials and cheap labour to deliver inferior goods to thirsty consumers at unexpectedly low prices. These elements form the backbone of the fast fashion business today, prioritising profit over ethical and sustainable production. The General Accounting Office (GAO) of the United States defines fast fashion manufacturing hubs as workplaces that violate multiple federal or state labour laws governing workers' compensation, occupational safety and health, child labour, industrial homework, minimum wage and overtime, or industry regulation [8]. Global warming is also accelerated by the fast-fashion culture, which ends up producing enormous quantities of clothing. The low-quality materials quickly end up in landfills, where they pose a threat to both environment and human health [12].

Consumer behaviour plays a significant role in both the purchase and disposal of clothing. Evidence show that garments worn for longer periods are typically better maintained [4]. In contrast, the fast fashion model encourages excessive consumption, leading to the overuse of energy, raw materials, and water resources, resulting in the production of low-quality products [2]. The environmental impact of textiles varies significantly across the value chain, depending on the types and combinations of fibres used, as well as the associated processes of manufacturing, use, and disposal [4]. The industry's linear model has exacerbated its negative environmental effects [2]. The sheer volume of waste generated further contributes to the rising levels of textile waste [2]. Yarn and fabric, in addition to the basic materials (cotton, polyester, nylon, acrylic, and elastane) used, determine the environmental impact. For example, cotton is linked to heavy fertilisers and pesticides, while synthetic fibres often rely on toxic chemicals and non-renewable resources [4].

Most of the environmental impact categories have significantly greater effects at later stages of the textile production process [3]. Petrochemicals, from which 63% of textile fibres are made, result in significant carbon dioxide (CO_2) emissions during their production and disposal. Cotton, which accounts for 24% of the remaining 37%, is a thirsty crop linked to hazardous pollution from extensive pesticide use and water depletion, of which the desiccation of the Aral Sea is the most notorious example [3]. Over 1,900 chemicals have been identified by the Swedish Chemicals Agency as being utilised in the garment manufacturing process, and 165 of these are classified as hazardous to the environment or health in the EU [4]. Fossil fuels are typically used to spin yarns and weave/knit fabrics, resulting in emissions such as CO_2 and particulates. Wet processing operations (dyeing, finishing, printing, etc.) are also significant sources of harmful emissions. The main environmental issues in the textile industry are waste, hazardous chemicals, greenhouse gas emissions, and water use [3]. An assessment of the life cycle impacts of textiles shows that the production phase is responsible for the majority of harmful emissions and impacts [2, 4].

The fashion industry is facing increasing international scrutiny for its environmentally damaging supply chain practices [6]. All stakeholders, especially designers and manufacturers, have a responsibility to recognise the scale of textile waste and pursue solutions that introduce innovative alternatives [9]. The growing trend of repurposing textile waste not only addresses environmental issues but also fosters a more resource-efficient and sustainable industrial ecosystem [2].

13.3 Circular Economy and the Role of Design in Minimising Textile Waste

13.3.1 The Circular Economy: A Solution to the Waste Problem

The notion of the Circular Economy (CE) has grown in popularity, stemming from the perspectives of ecological and environmental economists. The CE currently embodies a paradigmatic shift in the discourse surrounding the pillars of sustainability, including People, Profit, and Planet (the 3Ps) [13] as shown in Fig. 13.2. This responds to the implications of a linear model of production and consumption [14], proposing that products are uniquely made to enable either a biological or a technical cycle [13]. At its core, CE reconfigures growth by decoupling economic progress from the depletion of finite natural resources, thereby addressing the inherent unsustainability of linear economic frameworks. The traditional 'take-make-dispose' model that has long underpinned industrial activity is based on an extractive logic that not only depletes raw materials but also exacerbates environmental degradation and waste accumulation [2]. In contrast, CE advocates a regenerative system in which resource flows are meticulously optimised to ensure continuous reutilisation, extending material lifecycles and reducing environmental impact. Figure 13.3 illustrates the continuous use of resources, both renewable and non-renewable, which are kept in the system for as long as possible, ensuring that the 3Ps are achieved in the long term through remanufacturing, reuse, and resale of used or discarded goods [15].

Crucially, the CE goes beyond simplistic waste reduction strategies by fostering an integrated design philosophy that prioritises durability, repairability, and material reintegration at the end of life. This approach necessitates a fundamental restructuring of industrial processes, where product design shifts from disposability to longevity, embedding sustainability at the start of production rather than as an afterthought [1]. By embedding principles of circularity into the textile industry, for instance, stakeholders can cultivate a closed-loop system that not only reduces raw material dependency but also minimises the carbon footprint associated with textile waste. Such a systemic reorientation underscores the potential of CE as an indispensable mechanism for sustainable development, requiring cross-sectoral collaboration and policy innovation to achieve tangible environmental benefits [2].

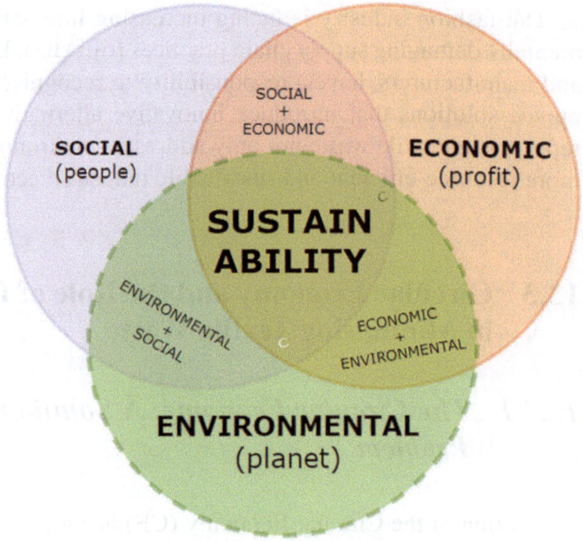

Fig. 13.2 The 3Ps of circularity. Adapted from [16]. The Creative Commons Attribution-ShareAlike 4.0 International License governs this work. © 2021 Muhammad Rijal, Muhammad Rafi Ihsan, and Muhd. Arief Al Husaini

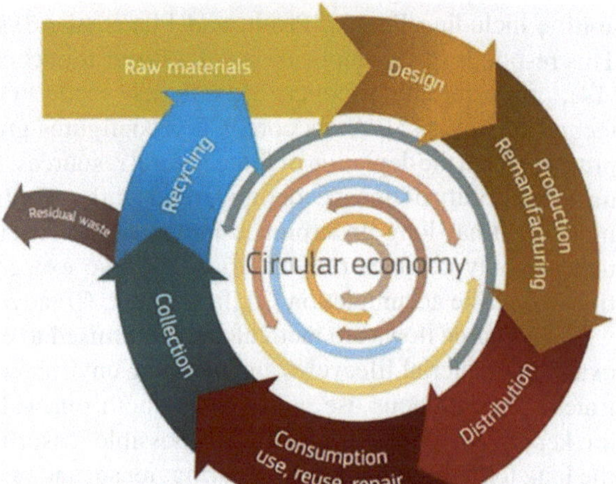

Fig. 13.3 The circularity concept. Reprinted from [17]. Authorised with acknowledgement. © European Union, 2015

Moreover, the success of CE hinges on a broader cultural and behavioural shift that redefines consumption patterns. While technological advances in recycling and upcycling are instrumental, they must be complemented by consumer engagement and regulatory frameworks that incentivise circular practices. Without a change in societal attitudes towards resource use, even the most sophisticated CE models risk stagnation. This underscores the need for interdisciplinary approaches that integrate economic incentives, legislative measures, and educational campaigns to cultivate a collective commitment to sustainability [1]. In this light, CE is not merely a theoretical construct, but a pragmatic necessity to meet the environmental and economic challenges of the twenty-first century.

13.3.2 Upcycling and Recycling as Pillars of the Circular Economy

Textile upcycling and recycling are pivotal pillars of the circular economy (CE), providing tangible solutions to the mounting crisis of textile waste and its detrimental impact on the environment. These strategies represent a necessary departure from the linear take-make-dispose model, advocating instead a system in which materials retain their utility and value for as long as possible.

Upcycling and recycling of textiles are fundamental strategies within the CE framework, addressing the urgent need to reduce the environmental degradation caused by excessive textile waste. Upcycling, as a method of waste valorisation, not only extends the lifecycle of materials but also enhances their functional and aesthetic value, often surpassing that of the original products [3]. By turning discarded textiles into new, high-value products, upcycling promotes sustainable innovation while reducing pressure on raw material extraction. Conversely, recycling plays a crucial role in resource efficiency by converting textile waste via mechanical or chemical means into raw materials for the manufacture of new products, thereby reducing dependence on virgin resources [4].

Recycling is integral to resource efficiency, converting textile waste into raw materials suitable for producing new items. By minimising reliance on virgin resources, recycling alleviates pressure on ecosystems, conserves finite materials, and substantially reduces pollution [4]. Whether through mechanical or chemical processes, textile recycling diverts waste from landfills, curbs greenhouse gas emissions, and conserves critical resources such as water and energy. However, while the theoretical benefits of recycling are well-documented, practical challenges persist. The efficacy of recycling efforts is often hindered by variations in textile quality, complex material compositions, and limitations in existing processing technologies. These factors underscore the necessity for continued advances in materials science and infrastructure investment to optimise recycling systems. It is worth noting that recycling is often referred to as 'downcycling' because the derived product is of a lower value than the original, compared to upcycling. For instance, according to Schmidt et al. [18], discarded apparel and home textiles are sometimes downcycled into rags and low-grade insulation materials. However, while both upcycling and recycling contribute to sustainability, their effectiveness is influenced by the quality of the textile waste, the material composition, and technological advances in processing techniques.

13.3.3 *The Role of Sustainable Design in Waste Prevention*

Sustainable design is a fundamental strategy for reducing textile waste by addressing inefficiencies at the source. By integrating sustainability principles into design methodologies, fashion designers can create products that are not only aesthetically pleasing, but also durable, repairable, and recyclable. Sustainable design challenges the prevailing culture of disposability and aligns the fashion industry with circular economy principles, promoting long-term environmental responsibility and resource efficiency.

13.3.3.1 Designing for Longevity

The longevity of textile products is crucial in minimising waste, as extended product lifespans reduce consumption rates and overall material disposal. Achieving this objective necessitates a comprehensive approach that incorporates high-quality materials, structural integrity, and user engagement. The selection of superior materials plays a defining role in product durability, yet the concept of quality extends beyond fibre composition to include resilience, wear resistance, and consumer perception [5, 6]. Understanding the psychological and behavioural dimensions of material selection is vital, as consumer preferences often dictate whether a product remains in use or is prematurely discarded.

In addition to material considerations, reparability is a key determinant of product lifespan. Garments engineered for ease of maintenance and repairability discourage premature disposal, thereby reducing textile waste [7]. This aspect is particularly important in the case of technologically integrated textiles, where modularity ensures that individual components can be replaced or upgraded without having to discard the entire garment. Through deliberate design choices that prioritise long-term usability, the fashion industry can disrupt the cycle of short-lived clothing consumption.

13.3.3.2 Modular Design and Design for Disassembly

The principles of modular design and Design for Disassembly (DfD) are increasingly recognised as transformative solutions for textile waste management. By facilitating the easy separation of components, these strategies increase material recovery rates and contribute to a regenerative textile system [9, 19]. The implementation of DfD is particularly relevant in the fast fashion sector, where garments are often made from blended fibres that complicate recycling processes.

In addition to recyclability, modularity extends the life of garments by enabling the replacement of worn or obsolete components rather than discarding entire products [20]. This innovation promotes resource efficiency by ensuring that textile products remain functional and adaptable over time. By embedding modular and disassembly-oriented designs into mainstream production, the fashion industry can transition from a linear consumption model to a more circular and sustainable framework.

13.3.3.3 Sustainable Material Selection

Material innovation is central to sustainable design, with increasing emphasis on biodegradable and recyclable fibres that minimise environmental impact. Biodegradable polymers derived from renewable sources, such as starch and cellulose, present a promising alternative to synthetic textiles. These materials decompose naturally, mitigating long-term pollution and reducing accumulation in landfills [21–23]. Moreover, their use contributes to sustainable industrial practices by reducing the environmental impact of non-degradable waste streams [11, 24].

In parallel with biodegradability, recyclable materials play a pivotal role in establishing both closed-loop and open-loop textile economies based on the expected end product [25]. In an open-loop system, recyclable polymers can be reprocessed into new products (see Fig. 13.4), other than their original appearance, thus reducing dependence on virgin resources and fossil fuel-based materials [11, 26, 27]. In this way, waste material is converted into new material forms with different properties, which become the raw material for a unique product manufacturing process. Closed-loop recycling, on the other hand, presents recovered materials as substitutes for their virgin counterparts. This means that textile waste is returned to its original product, e.g. fibre-to-fibre or fabric-to-fabric, or garment-to-garment. Figure 13.4 illustrates the closed-loop and open-loop textile waste management systems, where discarded textile garments are recycled and further processed to make a new garment (closed-loop) and insulating padding for a mattress (open-loop) [25, 28].

13.4 Textile Upcycling

The growing environmental crisis caused by textile waste requires a shift towards more sustainable waste management strategies. Among these, upcycling and recycling have emerged as fundamental and promising solutions, offering a means of transforming discarded materials into valuable products while conserving natural resources [29]. Unlike conventional recycling, which often degrades material integrity, upcycling preserves and even enhances the value of waste textiles. This approach is particularly relevant in the context of fast fashion, where overproduction and rapid consumption contribute to excessive waste [30]. By transforming discarded materials into high-quality items, upcycling presents an opportunity to promote a more ethical and resource-efficient fashion industry.

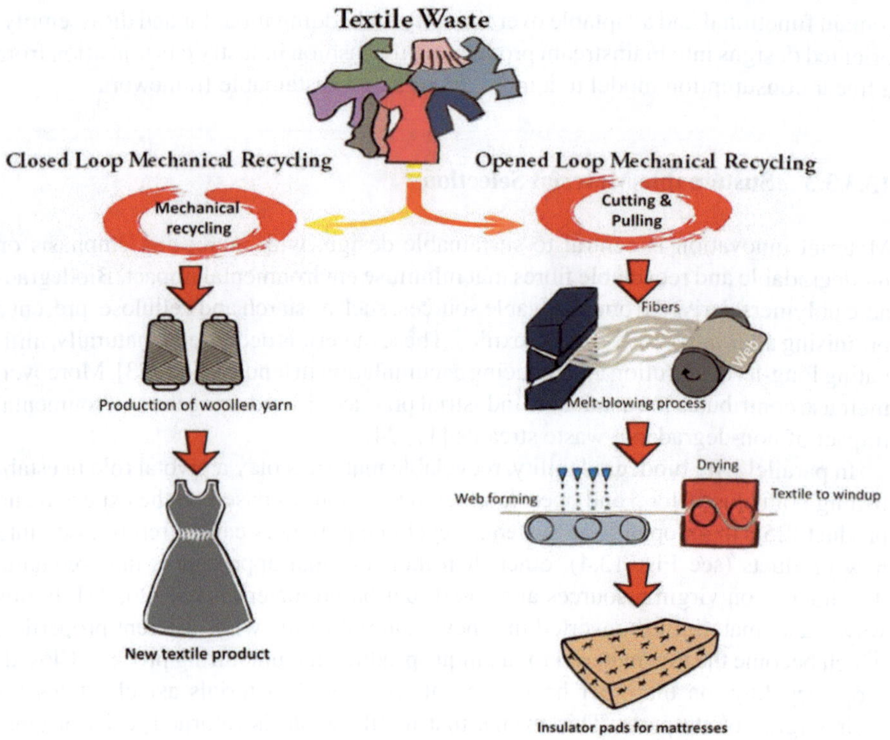

Fig. 13.4 The loop systems of textile waste, reprinted from [25]. (Open-access image distributed under the Creative Commons Attribution 4.0 License (http://creativecommons.org/licenses/by/4.0/). © 2021 by the authors

13.4.1 Approaches to Upcycling Textile Waste

A variety of innovative techniques have been developed to maximise the potential of discarded textiles. Traditional methods such as patchwork and quilting, as highlighted by Rohan [29], not only prevent the accumulation of waste, but they also preserve craftsmanship. These methods, deeply rooted in cultural heritage, offer a bridge between sustainability and tradition [31]. In a globalised fashion industry that often overlooks local craftsmanship, upcycling through patchwork and quilting revives historic textile practices while reducing environmental impact.

Beyond traditional methods, contemporary approaches such as garment reconstruction have gained traction. Cassar [32] underscores how this technique fosters creativity and sustainability by repurposing material remnants into unique, customised designs. Han et al. [33] expand on this concept, illustrating how dismantling and reassembling old garments allows for personalised fashion that meets both aesthetic and ethical concerns. This approach not only supports sustainability but also challenges the fast-fashion model by advocating slow, deliberate consumption.

Moreover, upcycling extends beyond apparel into broader applications such as textile art and interior design. Bridgens et al. [5] explored how discarded fabrics can be transformed into artistic and decorative pieces, reimagining textile waste as a resource rather than a burden. By integrating sustainability with artistic expression, upcycling fosters a creative culture where discarded materials are given new life, reinforcing the principles of the circular economy.

13.4.2 Challenges and Future Directions in Textile Upcycling

Despite its many advantages, upcycling faces significant challenges that hinder its widespread adoption. One of the main barriers is the complexity of processing mixed fibre textiles, as identified by Rohan [29]. Many textiles contain blends of synthetic and natural fibres, making it difficult to effectively separate and reuse them [34]. To overcome this challenge, advances in fibre separation technologies are needed to enable the efficient reuse of complex textile compositions.

Economic viability is also a barrier to mainstreaming upcycling practices. As Savas et al. [35] noted, establishing profitable business models for upcycled products remains a challenge, particularly in the competitive, low-cost fashion industry. While luxury and niche markets have demonstrated demand for high-quality upcycled goods, scalability remains an issue [36]. However, emerging consumer trends indicate a growing preference for sustainable fashion, suggesting that targeted investment and strategic branding could facilitate the expansion of upcycling practices into the broader market [37].

The integration of upcycling into high-end fashion offers an example of its commercial potential. Roungpaisan et al. [38] examined how upcycled tweed fabric, inspired by Chanel's iconic designs, was successfully incorporated into luxury fashion. This case highlights the ability of upcycling to combine sustainability with prestige, proving that eco-conscious fashion does not have to compromise on quality or aesthetics. Such innovations demonstrate the feasibility of upcycling as a mainstream practice, provided there is sufficient industry commitment to research and development.

13.4.3 Environmental and Social Impacts of Textile Upcycling

Beyond its material benefits, upcycling plays a crucial role in environmental protection and social sustainability. Milam-Guerrero et al. [39] emphasise that upcycling significantly reduces the accumulation of textile waste in landfills, thereby mitigating the environmental impact of disposal. This process aligns with the principles of the circular economy by diverting waste from landfills and reducing reliance on virgin materials, ultimately promoting a more sustainable production cycle.

Socially, upcycling promotes community engagement and cultural reconnection. Bridgens et al. [40] explored how creative upcycling initiatives reconnect individuals with materials and places, thereby fostering local sustainability efforts. At a time when consumerism often disconnects individuals from the production process, upcycling offers an opportunity for greater appreciation of craftsmanship and responsible consumption. This approach is particularly relevant in Western societies, where overconsumption exacerbates waste problems. By encouraging mindful engagement with textiles, upcycling cultivates a culture of sustainability that extends beyond fashion into broader consumer habits.

13.5 Case Studies of Upcycling in Sustainable Textiles

13.5.1 Re/Done: Upcycling Vintage Denim for Modern Consumption

The fashion industry is undergoing a paradigm shift, with sustainability and circularity at the heart of emerging innovations. One such example is Re/Done, a brand that seamlessly integrates upcycling into modern fashion by transforming vintage denim into contemporary apparel. This approach significantly reduces reliance on virgin materials and addresses the environmental impact of traditional denim production, which is notorious for excessive water consumption—approximately 7,500 L per pair of jeans—as well as heavy use of pesticides and synthetic dyes [33, 41, 42].

Re/Done's collaboration with Levi's exemplifies the potential of extending textile lifecycles while maintaining high aesthetic and functional value. By repurposing pre-existing denim, the brand not only minimises textile waste but also enhances discarded materials, transforming them into high-end, unique pieces that appeal to contemporary consumers [43]. This practice directly supports circular economy principles by reducing the carbon footprint of virgin cotton production, which is a highly resource-intensive process involving water, fertiliser, and chemical treatments [33, 44]. In synthesis, Re/Done's innovative method of upcycling jeans blends sustainability and exclusivity, demonstrating the commercial potential of eco-friendly design. Despite its limitations, such as labour-intensive hand deconstruction and a lack of quality vintage denim sourcing, the brand's success highlights the promise of high-value upcycling in the fashion sector.

13.5.2 Tonlé: A Zero-Waste Fashion Upcycling Model

While Re/Done focuses on post-consumer textile waste, Tonlé, a Cambodia-based brand, takes a different approach by repurposing pre-consumer fabric scraps discarded by mass producers. This not only diverts a significant amount of textile waste from landfills, but also follows the principles of zero-waste fashion by utilising even the smallest remnants to create hand-woven textiles [45].

In addition to waste reduction, Tonlé's sustainability model incorporates ethical labour practices that ensures fair wages and safe working conditions for artisans. This emphasis on social responsibility complements its environmental commitment and illustrates how sustainable fashion can drive both environmental and economic change [44, 46, 47]. Unlike Re/Done, which operates in the high-end market, Tonlé prioritises affordability and accessibility, demonstrating that upcycled fashion can be inclusive while maintaining ethical and environmental integrity.

This review of Tonlé's sustainable fashion model challenges the assumption of higher costs by embracing low-cost upcycling, demonstrating accessible and financially sustainable practices while challenging the fast fashion industry's dominance of high-volume production.

13.5.3 Comparative Analysis: The Viability and Limitations of Textile Upcycling

Both Re/Done and Tonlé present upcycling as a viable path to sustainability, yet their models reveal contrasting strengths and challenges. Re/Done's emphasis on exclusivity and premium branding attracts consumers seeking uniqueness and traditional craftsmanship, whereas Tonlé focuses on ethical labour and waste reduction. This divergence reflects wider industry tensions between sustainability and mass accessibility, between high-quality craftsmanship and scalable production.

Despite their success, both brands highlight the fundamental limitations of upcycling as a large-scale solution. A key challenge is the inconsistency of available materials. Unlike industrial recycling, which processes raw textiles into standardised fibres, upcycling relies on pre-existing garments and fabric scraps, making it difficult to create consistent products at scale. Moreover, the high labour costs associated with deconstruction, redesign, and manual reassembly hinder wider industrial adoption.

For upcycling to become more mainstream, structural changes in the industry are needed. Advances in automated textile processing, greater regulatory incentives for circular fashion and a shift in consumer attitudes towards valuing sustainable practices are critical [48]. While upcycling alone cannot replace large-scale industrial recycling, it remains an essential strategy within a diversified approach to textile sustainability.

13.6 Textile Recycling

While upcycling has gained traction as an innovative solution for sustainable fashion, its reliance on handcrafting and design adaptation has inherent limitations, particularly in terms of scalability [48]. As a result, the textile industry is increasingly turning to recycling technologies that convert both post-consumer and pre-consumer textile waste into reusable raw materials. These recycling methods primarily fall into two categories—mechanical and chemical recycling, each with their own advantages and challenges [15, 49]. A closer look at these techniques below reveals not only their technical feasibility, but also the environmental and economic factors influencing their adoption.

13.6.1 Mechanical Recycling

Mechanical recycling is currently the most widely adopted method of textile waste management and is particularly effective for natural fibres such as cotton, linen, and other bast fibres [15, 49]. This process involves shredding discarded textiles (e.g., using a rotor shredder), hard water separation, blending, tearing, baling, or re-spinning the fibres into yarn, which can then be woven or knitted into new fabrics [50] (see Figs. 13.5 and 13.6).

One of the most significant advantages of mechanical recycling is that it provides a low-energy alternative to producing virgin fibres, making it an appealing option from an environmental perspective [51]. However, despite its benefits, mechanical recycling is not without challenges. A major limitation of this method is fibre degradation. Each recycling cycle reduces the length and strength of the fibres, diminishing their

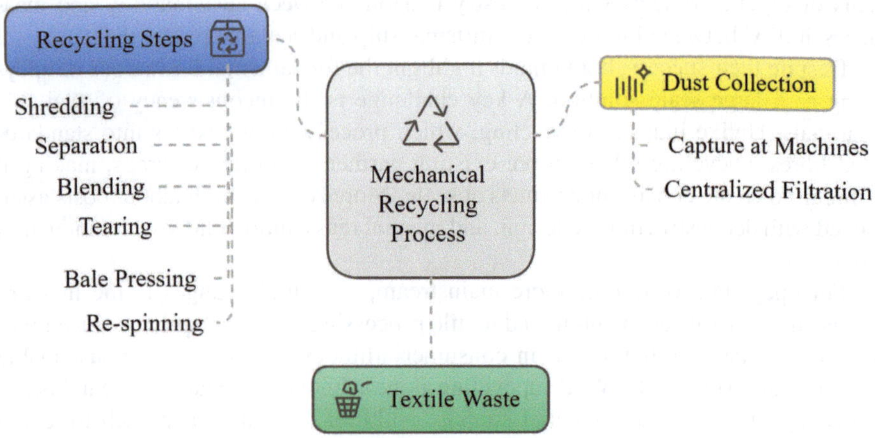

Fig. 13.5 Typical mechanical recycling of textile waste [50]

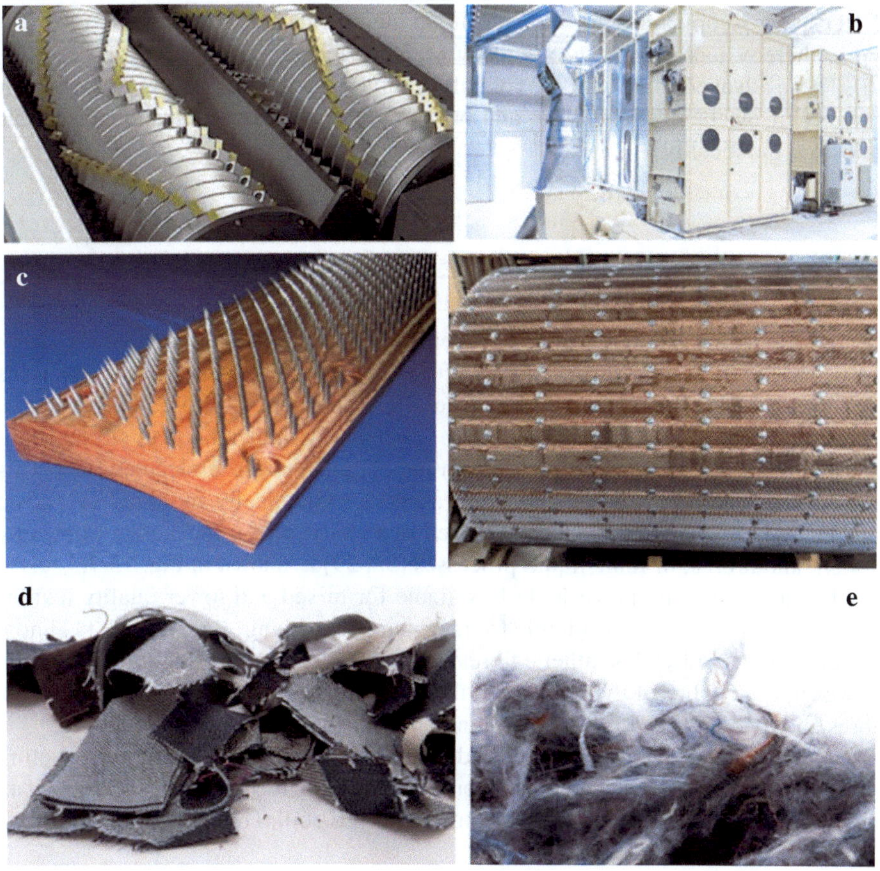

Fig. 13.6 **a** A rotor shredder, **b** blender for mixing shredded textile pieces, **c** Tearing devices, **d** shredded fabric pieces, **e** textile torn into fibres [50]. Open-access image distributed under the Creative Commons Attribution 4.0 License (https://creativecommons.org/licenses/by/4.0/). © 2024 Authors. The Union of Textile Engineers and Technicians of Serbia is the publisher

quality and relegating them to lower-value applications such as insulation, padding, or industrial cleaning cloths rather than high-end textiles [49, 52, 53]. In addition, mechanically recycled fibres struggle with uniformity, making them less suitable for the production of garments that require high durability and longevity. Another significant barrier is the difficulty of processing blended fabrics, particularly those made up of synthetic and natural fibres. For example, polyester-cotton blends are among the most common textiles found in fast fashion, but separating the synthetic components from the natural fibres remains a complex and inefficient process [54]. Without effective separation, mechanical recycling of blended fabrics results in mixed fibre outputs of lower quality, limiting their usability.

Nonetheless, recent advances in sorting and fibre identification technologies offer promising solutions to these challenges. Innovative techniques such as near-infrared

(NIR) spectroscopy allow for more accurate identification and separation of textile types, thereby increasing the efficiency of the mechanical recycling process [55, 56]. These advances highlight the role of technological innovation in improving the viability of mechanical recycling and bridging the gap between sustainability and industrial practicality.

13.6.2 Chemical Recycling

In contrast to mechanical recycling, chemical recycling involves breaking down textiles at the molecular level, allowing for the recovery of valuable raw materials such as cellulose from plant fibres or keratin from wool [46, 57]. For example, cotton, a difficult plant fibre to dissolve, is chemically induced due to strong intra- and inter-molecular hydrogen bonding in the cellulose macromolecules [58]. This is either dissolved in solvent (to obtain a pure or chemically modified cellulosic fibre) or depolymerised, which can be used as a feedstock for regenerated fibres by lyocell, alkali/urea, or ionic liquid processes [58, 59] as shown in Fig. 13.7.

Chemical recycling is particularly suitable for mixed and lower quality textiles that would otherwise be unsuitable for mechanical recycling. Baloyi et al. [11] note that they work well with synthetic fibres, allowing the fibres to retain their integrity. However, the quality of the recovered fibres is usually lower than that of the virgin fibres and may require the incorporation of virgin fibres for rejuvenation [41].

Unlike mechanical recycling, which degrades fibres over time, chemical recycling allows materials to be regenerated with properties similar to virgin fibres, making it an attractive option for high quality textile production [10, 60]. This capability represents

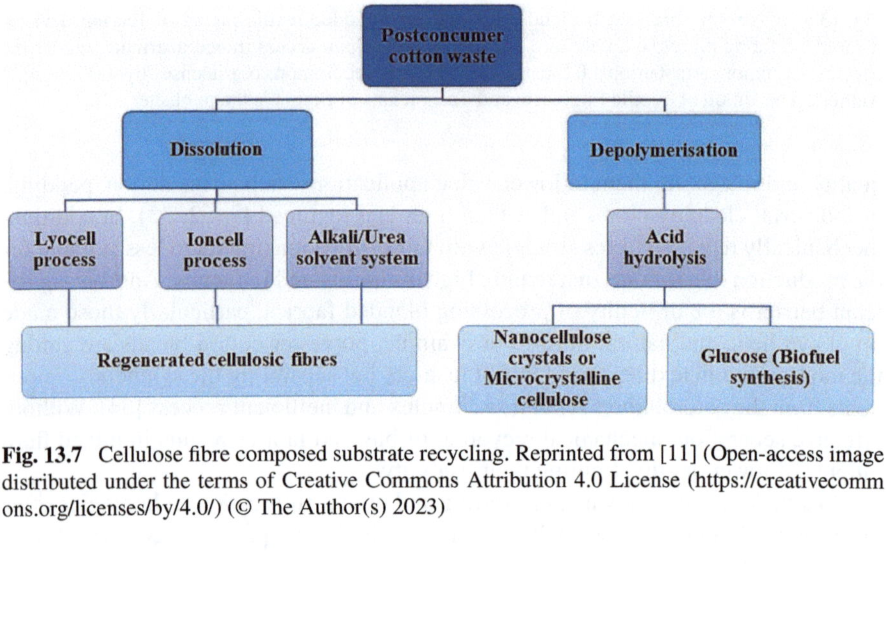

Fig. 13.7 Cellulose fibre composed substrate recycling. Reprinted from [11] (Open-access image distributed under the terms of Creative Commons Attribution 4.0 License (https://creativecommons.org/licenses/by/4.0/) (© The Author(s) 2023)

a breakthrough in achieving a circular economy for textiles, enabling a closed-loop system where discarded garments can be transformed into new materials without compromising quality. According to Hann and Connock [61], the chemical recycling processes fall into three broad categories: (i) solvent purification (employing the solubility principle to selectively separate plastic polymer mixed with other fibres by dissolving the plastic component from the contaminants); (ii) chemical depolymerisation (using chemical to break down the polymer chain into monomers, recover and purify; and (iii) thermal depolymerisation (breaking down polymer chains by heat treatment) [61].

However, the chemical recycling approach encounters challenges. The process is often energy intensive and costly, requiring sophisticated infrastructure and high operational inputs [3, 46, 60, 62]. Additionally, concerns about the environmental impact of chemical solvents and waste disposal raise questions about the long-term sustainability of this method. Certain chemical recycling processes rely on hazardous substances that, if not properly managed, can offset their environmental benefits. To address these concerns, researchers and industry leaders are exploring closed-loop systems and green chemistry approaches that aim to minimise environmental harm while maintaining economic feasibility [63, 64]. These strategies include developing biodegradable solvents and optimising chemical reactions to reduce energy consumption, making chemical recycling a more sustainable long-term solution.

Recent innovations further highlight the potential of chemical recycling. For example, advances in acid hydrolysis have made it possible to convert cotton fibres into glucose, which can be used in the production of bio-based chemicals or for energy generation [65]. Similarly, microwave-assisted glycolysis has shown promising results in breaking down polyester and spandex, effectively separating mixed fibres for reuse [66]. Such developments illustrate how chemical recycling, despite its current limitations, has the potential to revolutionise textile waste management when integrated with sustainable technologies.

Meanwhile, Fig. 13.8 illustrates a biochemical process, a greener approach that relies on enzymes or microbes to decompose natural textiles, including cotton and wool, into their simplest components, such as glucose and amino acids, from which valuable products, such as bioethanol, can be produced. This technology requires mild conditions, low energy consumption, renewable or natural resources and replaces harsh chemicals to improve the comfort and quality of fabrics [67]. However, it is limited to natural fibres that require pre-treatments to remove contaminants or finishes, and is capital intensive [11, 15, 68].

Although enzyme-based methods are being investigated for the degradation of synthetic fibres such as polyester and polyamides, they currently face challenges due to limited penetration and surface-only action. Studies show that cellulase and keratinase enzymes can effectively reduce cellulose and wool respectively, including blends [11, 67, 69]. Hu et al. [69] claim that enzymes derived from *Aspergillus niger* can potentially break down cotton, with a glucose yield of approximately 70%, while proteases can separate wool/polyester blends.

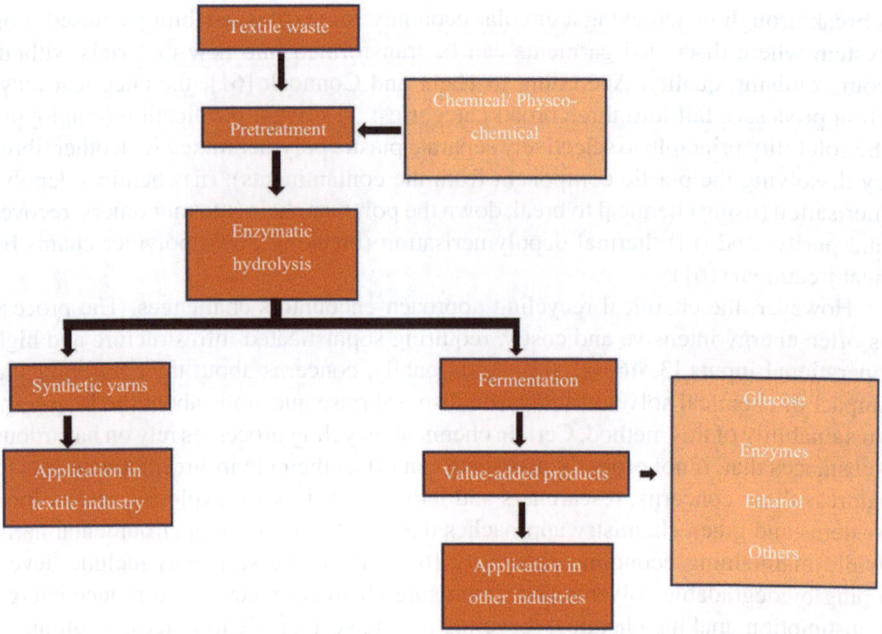

Fig. 13.8 Biochemical recycling process. Reprinted from [11] (Open-access image distributed under the Creative Commons Attribution 4.0 License (https://creativecommons.org/licenses/by/4.0/) (© The Author(s) 2023)

13.6.3 Integrating Mechanical and Chemical Recycling into a Circular Textile Economy

Both reviews show that mechanical and chemical recycling are essential for textile waste management, but neither is a stand-alone solution. Mechanical recycling is practical but limited by fibre quality and blended fabrics. Chemical recycling offers high-quality fibre regeneration but requires energy input. It is therefore worth noting that the future of textile recycling depends on three key developments:

1. **Advances in Sorting and Separation Technologies**: Improvements in fibre detection and sorting, such as automated NIR spectroscopy and enzyme-based separation methods, can improve the efficiency of both mechanical and chemical recycling processes [55, 56].
2. **Investment in Green Chemistry and Closed-Loop Systems**: The adoption of environmentally friendly chemical processes, such as solvent-free recycling methods, can mitigate the drawbacks of chemical recycling while maintaining its advantages [57, 64]. According to Baloyi et al. [11], unlike biochemical recycling, pure chemical recycling is ideal for closed-loop recycling, where cellulosic textile waste is processed from fibre to fibre while maintaining fibre properties.

3. **Policy and Consumer Awareness Initiatives**: Government regulations and consumer education programmes that encourage responsible textile disposal and recycling can help scale up these technologies and create demand for recycled textiles in mainstream fashion markets [15, 70].

At the end, a circular economy requires a balance between technological advancement, environmental responsibility, and economic feasibility. Integrating advanced recycling systems into textile waste management can ensure a sustainable future by rethinking textile production, consumption, and reuse.

13.7 The Role of Technology and Consumer Awareness in Reducing Textile Waste

13.7.1 Technological Advances in Textile Recycling and Upcycling

Textile recycling is experiencing a renaissance, demonstrating cutting-edge advances, such as chemical, biological and other processes that can break down complex textile materials, and automated sorting systems that improve efficiency and precision. These developments set the stage for a sustainable future in the FTI by addressing the long-standing waste management challenges [26]. Building on this, technological advances in recycling and upcycling have shown ways in which innovation can turn waste into valuable resources [71], with huge potential to revolutionise industry approaches to sustainability and stimulate the creation of a circular economy [29].

Recent studies have shown advances in technologies for textile recycling [71]. These technologies include, but are not limited to, mechanical [30] and chemical [31]. The collection of discarded textiles through sorting, grinding, carding and blending into new fibres and re-spinning are typical steps in mechanical recycling, without altering the chemical structure of the material. This process can regenerate textile waste as a feedstock for future production. Shredded material is sometimes needle-punched (to form webs) or chemically bonded to make insulation pads or disposable masks, respectively [11] (see Fig. 13.9). This was followed by the thermo-mechanical technology phase [33], suitable for both synthetic textile waste [11] and natural fibres [40]. For instance, Savas et al. [35] ground textile (plastic) waste into powder and mixed it with potassium acetate, which was pyrolysed at 600 °C to obtain a porous carbon sorbent for CO_2 capture, which can contribute to the reduction of toxic atmospheric gases. Simiarly, Roungpaisan et al. [38] melted polyethene terephthalate (PET) waste and cooled it by hot pressing at 250 °C to form a compact sheet, which was further processed into r-PET fibres by melt spinning. This resulted in improved crystallinity and tensile strength. Chemically, synthetic second-hand clothing (SHC)

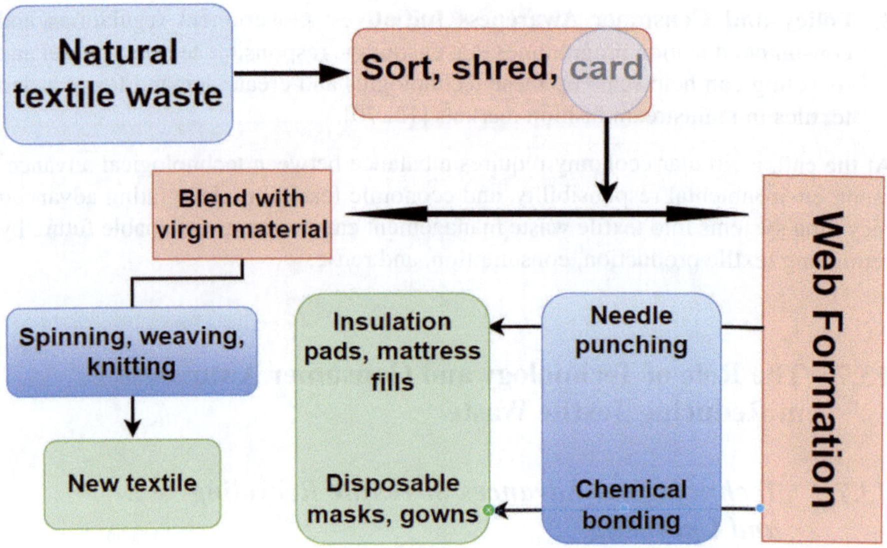

Fig. 13.9 Mechanical recycling and repurposing of natural textile waste in a schematic diagram. Taken from [11] (Open-access image distributed under the Creative Commons Attribution 4.0 License (https://creativecommons.org/licenses/by/4.0/)) (© The Author(s) 2023)

of acrylic and nylon origin was dissolved in formic acid and homogenised by stirring at room temperature. The solution was characterised for viscosity and electrical conductivity, and electro-spun into a filament. The outcome has great potential for smart fashion innovation [39]. Research, however, shows that mechanical and chemical recycling degrades the quality of the derived material; hence, additional virgin material is added to achieve the desired quality [41].

Premise to achieving the desired quality of recycled materials, Darwesh et al. [42] present a biological means of disabling the complexities that characterise fibre blends and maintaining the quality of the recovered fibre. For the selective dissolution of textile fibre blends, Mihalyi and colleagues [43] utilised the enzymatic approach to select and decompose viscose and recover it from other blends by simultaneous saccharification and fermentation. Cellulose can be successfully degraded to glucose, which is then converted to lactic acid by the Weizmannia coagulant microorganism [43]. Newer developments such as thermo-chemical [46] are beginning to gain traction in textile recycling, involving processes such as pyrolysis and carbonisation to produce activated carbon for dye removal in wastewater treatment [45] to adsorb and remove methyl orange and blue dyes in wastewater, and conductive materials suitable for e-textile fabrication [47].

13.7.1.1 Textile Waste Selection Systems

Upcycling and recycling seem to be a compelling option for textile waste management, offering a new life for old material [44]. However, a common challenge [48] regarding textile recycling is that manual material identification and sorting are required due to the complexity of the material structure [49], including the presence of foreign elements such as garment accessories (e.g., buttons, zips, etc.) [15]. These have a direct impact on material processing and the possibility of complete technological advancement for automation, according to Kozlowski et al. [52]. Microscopy is an essential tool for identification of textile materials, but the need for human intervention is a discouraging setback that frustrates recyclers and upcycles when dealing with larger volumes of waste and similar materials such as cotton from viscose or variants of synthetic fibres that cannot be accurately distinguished by appearance [52].

For textile identification, Near-Infrared (NIR) technology has shown over 90% accuracy in cotton fibre prediction and 100% accuracy in acrylic, polyester, polyamide, silk, and wool fibre prediction. Makela et al. [53] further confirm that although NIR is unable to detect waste fabrics with surface coatings, decorative and core yarns, its imaging spectroscopy is reliable for perfect fibre classification [54, 55]. Peets et al. [49] conclude that the introduction of Fourier Transfer Infrared (FTIR) spectroscopy for identification is proposed to offset the challenges of NIR, which is not reliable for synthetic counterparts. They further experimented with Reflectance Fourier Transform Infrared (r-FTIR) spectroscopy and Attenuated Total Reflectance Fourier Transform Infrared (ATR-FTIR) spectroscopy to identify wool, silk, and polyamide 6 6, concluding r-FTIR is capable of distinguishing polyamide fibres. A previous study by Riba et al. [55] successfully demonstrated excellent fibre identification using several algorithms of the ATR-FTIR instrument. However, these approaches are time consuming and destructive to the thermal behaviour of textile fibres during identification [56]. This makes them equally challenging for efficient fibre sorting prior to recycling.

13.7.1.2 Sorting of Textile Waste

Textile waste sorting plays an essential role, critical for textile upcycling and/or recycling for added value [44]. Successful fibre identification and sorting determine the quantity and quality of the recycled textile products, as well as the added value, given the current challenges of different fibre types, embellishments and contaminants [44, 57]. Technologies have been developed to facilitate the sorting of textile waste. For example, Du et al. [65] demonstrated the synergy of hyperspectral imaging and artificial intelligence (AI) technology that utilises wavelengths to capture images of textile waste, which are further analysed and classified based on fibre content. They claim that the technology is fast and provides high accuracy in fibre sorting [65]. Cura et al. [44] also combines NIR spectroscopy, which emits radiation, and optical sorting, which uses cameras, sensors, and lighting systems to classify and separate materials based on visual and compositional characteristics. The spectral signatures enable the detection of fibres—cotton, polyester, wool and silk—including the presence of dyes, surface coatings and finishes, which are precisely sorted by a dedicated device called PICVISA[1] [64] and WASTEX[2] [72].

Although these technologies may be capital-intensive, they could provide accurate and automated garment sorting based on high-quality image capture [70] of fibre composites and colour, facilitating the efficient segregation of large quantities of textile waste suitable for recycling [57]. Even with the amazing capabilities described above, integrating AI algorithms to analyse massive amounts of spectral data requires sensitive detectors, large data storage, and fast processors for automation [70]. NIR and hyperspectral imaging are susceptible to interference, which hinders the functionality of AI algorithms and causes misclassification [3, 57]. This integration has, however, revolutionised textile sorting processes, significantly improving the efficiency of recycling operations while reducing labour costs and human error in the sorting process.

13.7.1.3 Digital Manufacturing and Upcycling

Digital manufacturing in upcycling is transforming industries and promoting sustainability. A good example is the virtual 3D prototype that is being adopted for apparel redesign. This technology allows ideas, fabrics, and patterns to be explored in a virtual environment before physical items are produced. This significantly reduces designers' labour time and costs, while minimising waste in production [63, 66]. A recent study synergised the modular design method and 3D digital technology in upcycled apparel, which enhanced the effective redesign and reuse of used clothing for fashion upcyclers. Digital inkjet printing is also a key resource-efficient production technique. It requires less operational input (energy, water, etc.) and achieves high material efficiency compared to traditional methods. Dulal et al. [73] pointed out that this method of producing conductive e-textiles is best suited for synthetic upcycling when combined with melt spinning. Such a development not only reduces environmental impact, but also expands the functional potential of recycled materials, placing digital manufacturing at the forefront of circular fashion systems. It is worth noting that the thermoplastic nature of PET makes it prominent in 3D printing (additive manufacturing) applications, to some extent controlling and reusing polyester waste [38, 74]. Interestingly, for the first time, single-screw extrusion without fibre segregation [75] transformed discarded cotton-polyester blend into a filament with good mechanical properties suitable for engineering applications.

---

[1] The PICVISA brand introduces ECOCLIP optical separator, a cutting-edge technology that enables circularity in the textile and fashion industry by automating the sorting process of textile offcuts in recycling centres worldwide. It performs full-surface analysis to sort textiles based on fibre composition and colour and automatically detects and separates hard points and metals.

[2] WASTEX, a joint venture between Barcelona-based companies Coleo and Picvisa, combines their expertise in textile manufacturing, recycled fibres, and cutting-edge sorting machines.

13.7.2 Digital Platforms and Consumer Engagement in Circular Fashion

For the implementation of circularity in diverse industries and challenges, digital technologies, especially digital platforms, facilitate interactions and value exchange between actors in a network within industries—this is common with industry brands and consumers to build trust and credibility, as well as to educate for awareness [76, 77]. Several scholars have presented digital technologies and their role in circularity around textile waste, such as recovery, since the rampant disposal of second-hand clothing (SHC) is a concern. For instance, an online marketing platform such as Consumer-to-Consumer (C2C) appeals to eco-conscious consumers with a strong belief in reducing the production of new products that may only be discarded in a short time, as some selected SHC are still functional and cost-effective [77]. Also, the increasing pace and trendy fashion industry makes it challenging for consumers to choose and purchase products. They are likely to purchase based on affordability, regardless of the consequences [78]. This brings to light the Artificial Intelligence-driven Recommender System (AI-RS) that analyses preferences and purchase history to suggest products that customers need [79]. This is an experience enhancement [80] feature of the existing Complimentary Item Recommendation (CIR)—which suggests products to enhance the value of the specific item that potential consumers will purchase [81].

Environmental knowledge is vital to waste generation and prevention by end users. A behavioural study by Okur et al. [81] found that consumers' knowledge of environmental issues is instrumental on their attitude towards socially and environmentally friendly products. As a way of creating environmental awareness, gaming elements become essential that engage consumers, pricking their consciousness for a sustainable shift. Gamification offers a fun experience and a more engaging strategy, encompassing pressing subjects [82]. A report from Technical Textiles Today [72, 83] conclude that "*gamification can be a useful tool in raising awareness through positive reinforcement about the consequences of fashion over-consumption*". Social media platforms offer a similar context. It is an avenue for the exchange of knowledge that educates consumers on behavioural patterns that impact the environment [84], eventually leading to a shift in consumer attitudes and behaviour towards circularity. Meanwhile, Okur et al. [81] reiterate that branded products based on responsible practices are a motivating factor influencing consumers' sustainable purchasing decisions. However, a Portuguese study by Ribeiro et al. [85] points out that although customers may be attracted to greener brands, they are mostly not responsible for the well-being of the environment. They practically refrain from sharing and supporting second-hand clothes, but accept the concept of extending product lifespan for their possessions. Ribeiro et al. [85] also found that budget, quality, comfort, and functionality take precedence over sustainability. It may be that potential consumers have little information about their behaviour and its relational impact on the environment in the midst of industry practices. It is assumed that if end users are well informed, they are more likely to be environmentally aware when purchasing products.

Amidst the rapid growth of the FTI, consumer behaviour increasingly reflects a gradual gravitation towards technological advancement that is somehow pivotal in influencing their purchase choices, highlighting the significant role of technological advancement in shaping consumer choices. This is in line with Lou et al. [86] who investigated the trade-offs consumers make when purchasing sustainable denim jeans, examining important attributes such as price, brand name, material types, and eco-labels. They reiterate that despite the growing awareness of sustainability issues, eco-labels have a relatively low impact (15.10% total impact) on purchase decisions. This may suggest that the dynamics of cost and brand perception still play a dominant role. However, they found that consumers prefer blockchain eco-labelled garments, recording a path-worth utility value (0.142) that is more appealing, especially to the younger, digital generation (Gen Z), against Fair-Trade Certification (0.102) and Certified B Corporation (−0.165). This preference is in line with the findings of Priporas et al. [87], who characterise Gen Z as a digitally native cohort with a strong inclination towards new technology, which enhances the credibility of the eco-label by presenting transparency and traceability data, thus eliminating uncertainty about greenwashing [88]. Guo et al. [89] added that the integration of blockchain technology is, to an extent, closely related to a customer's willingness to patronise goods, as the technology displays the promotional activities on the environmental quality of a fashion product.

13.8 Policy, Regulation, and Initiatives Promoting Textile Sustainability

Promoting textile sustainability requires the management of potential and actual waste. Nadeera et al. [90] explain that the waste management hierarchy provides the basic understanding of waste handling principles (Fig. 13.10), upheld by textile industries to set up systems for appropriate waste management. The inverted pyramid outlines the strategies in order of importance, with waste prevention being the highest priority. Although following this scheme sets the stage for safe waste reduction to promote environmental health, prevention seems impossible because the multiple processing stages and the supply chain of the textile industry result in a certain amount of waste at the end. Hence, a more feasible approach is to reduce—minimising the quantity of waste generated at each production line. However, to achieve waste prevention, Curran and Williams propose the adoption of Zero-waste, "*to eliminate rather than manage waste*" [91], diversifying waste resources for a closed-loop circular system to eliminate environmental blueprints [92]. This encourages industrial transition, away from the traditional linear system [14, 93], and governmental policies have been ignited in a circular direction.

Governmental policies reflect national and international commitments that drive the development of technologies, and the operationalisation of textile waste management has become increasingly important. For example, the UK government initiated

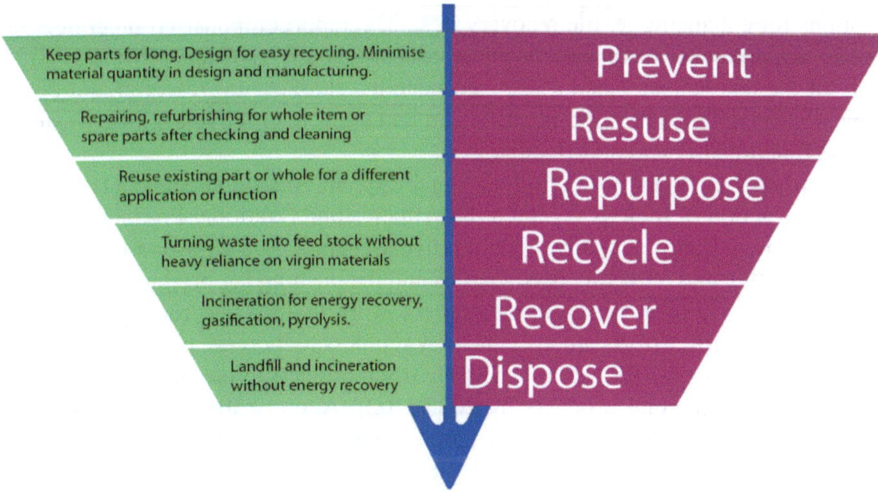

Fig. 13.10 Waste management hierarchy. Adapted from [94] © 2020 The Author(s). Published by Elsevier Ltd. This is an open-access article under the CC BY license (http://creativecommons.org/licenses/by/4.0/)

the Waste Prevention Program in 2021 to mitigate the environmental and social effects of textile waste. Textiles via repair, reuse, consumer information and upscaling the circular business model [95]. Dawson et al. [93] assert that the UK government has instigated the "Polluter Pays" Principle (PPP) (which has been in force since 1972) from the Extended Producer Responsibility (EPR)—a regulatory tool to compel manufacturers to be continuously responsible for their products through their useful life and at the end of life [93, 96]. Dawson et al. [93] reiterates that the EPR ensures that the product manufacturer is liable for the cost of product end-of-life management, public information, and is the primary culprit for any environmental damage caused by the product throughout its life. This is, however, not the global case, underscoring the need for a global perspective as developing countries are at the mercy of the global North production industries. In support of this, an investigation by the Australian Broadcasting Corporation (ABC) concluded that clothing waste diverted from landfills in the Global South may still end up overseas [97]. This is evident in the linear approach championed by the fast fashion [98].

The European Union (EU) unveiled a new Circular Economy Action Plan in 2020, outlining a thorough plan to improve the sustainability and circularity of the EU textile industry, building on the need for thorough consumer education. This calls for the introduction of new regulations on product labelling and traceability to give customers accurate and trustworthy information about the sustainability of textile products [57]. This focus on sustainability is echoed in Tang's [57] review in 2018, revealing that the US alone generated 5.3% of textile waste to the global municipal solid waste, of which 15.2% was recycled and the remaining landfilled. This prompted the release of a report on the *Textile Recovery in US* that proposed solutions for enhancing textile recovery [99]. The report emphasised on collection, sorting, recycling of waste and consumer education on the sustainability and circularity of the textile sector and associated benefits, sparking up research and development, driving innovations to identify, sort and upcycle or recycle textile waste for diverse functionality and added value [57].

Regarding initiatives promoting sustainability, an Australian study [100] highlighted the action of their government investment fund (National Product Stewardship Investment Fund), assisting collaborative associations with over $1,500,000 to aid in collecting and recycling used textiles and the pushing for product stewardship [101] just like the take-back programmes for worn textiles and rewards customer loyalty upon return. Efforts are also made by retailers to increase recycled materials of commercial goods—for example, H&M Group projects to use 30% of recyclable materials of commercial products by 2025 [102]. Additionally, for the first time, waste cotton fabrics are considered a good resource for landfilling, which could impact the environment positively. An experiment (a one-year trial) conducted on a cotton farm diverted two tonnes of shredded waste cotton fabrics that were spread on the soil to decompose. This successfully demonstrated improved soil health, inspiring microbial activity and carbon capturing of 2.25 metric tons (Co2e) [102].

13.9 Barriers to Scaling up Textile Upcycling and Recycling

To become more circular, many fast fashion firms have attempted to include upcycling into their textile value chains [103]. However, some companies are still unsure about how to take advantage of upcycling to improve waste management [104]. Theoretically, recycling is a good idea, but more work needs to be done to improve the efficiency of upcycling fashion products [105]. As consumer demand for sustainability increases, brands must prioritise sustainable options to remain competitive in the marketplace [104].

One of the biggest challenges in upscaling upcycling is access to sufficient quantities of good quality used materials. Unlike the stable and predictable supply chains of linear production, based on factory waste, by-products, and unsold or defective products, upcycling is constrained by contamination from mixed materials and chemical residues. Maintaining the quality of collected waste is crucial to ensuring successful upcycling processes [106]. In addition, sorting textiles is a major bottleneck. Clothing labels often do not provide accurate information on fibre content, making it difficult to sort the fabrics appropriately. Despite technological advances in sorting, most plants still use manual sorting, which leads to inefficiencies and misclassification. Sorting needs to be automated to produce high quality material fractions and increase the scalability of upcycling [106].

Beyond material quality, other critical issues for an upcycling business are access to waste resources, the energy used in the transformation processes, and the quality of finished upcycled products [104]. Even with new recycling technologies, the fashion industry remains plagued by inefficiencies in waste collection and sorting due to lack of coordination of waste management and recycling infrastructures contributes to the complexity of managing post-consumer fashion waste [36, 106].

Consumer attitudes also stand in the way of widespread uptake of upcycled products. While demand for sustainable fashion is growing, most consumers remain averse to upcycled products compared to mass-produced products. Scepticism about reused materials, uncertainty about product durability, and availability of sizes limit the commercial scope of upcycling. Without wider consumer acceptance, upcycling cannot make the transition from niche activity to mass business [36]. This shift towards circularity in textile recycling is a milestone for the global fashion industry [106]. However, for upcycling to be a sustainable and scalable approach, systemic changes are needed. Effective waste sorting, mechanisation, consumer attitudes, and a supportive business environment are essential to make upcycling an accepted and competitive business strategy.

In addition, there are significant barriers to scaling up textile upcycling and recycling in terms of economic viability, market demand, regulatory support, and technological limitations. Consumers in general have a genuine perception of quality amidst the awareness of some consumers about sustainability and the control of textile waste. Existing studies claim that products made from upcycled or recycled discarded textiles have environmental benefits as well as are being perceived to be inferior to new conventional products [107, 108]. Earlier studies are of the view that perceived risks, including functional, financial, and safety, are related to products that promote these strategies [109–111]. These perceptions, coupled with fast fashion, impact the scalability of upcycled and recycled goods by reducing consumer acceptance and preference for virgin materials over used goods.

Additionally, mass textile upcycling and recycling is faced with some significant challenges due to the complexity of the material as well as technical limitations. The most significant among these is the incompatibility or blending of textiles, which makes processing extremely difficult. For example, Wojciechowska and Kowaluk [112] refer to the difficulties of recycling upholstery materials that contain adhesives, foams, fabrics, and flame retardant finishes that require complicated separation procedures. Moreover, cotton/polyester blends pose a further hurdle, as their highly coherent fibre structures make them difficult to recycle on a large scale, making separation virtually impossible [113, 114]. Wagaw et al. [115] further emphasise that mixed hues and chemical residues in discarded clothing impede purity and make it difficult to produce recycled goods in the same colours. Though promising, existing recycling methods—mechanical, chemical, and biochemical—are not easily scalable due to technical drawbacks—such as mechanical degradation of fibre length that impacts fibre quality [11, 62]; chemical release of harmful substances [11, 116]; and the biochemical method, which is costly, slow and limited to natural fibres [11]. Furthermore, the lack of infrastructure in most places [117, 118] and the non-recyclable nature of textiles make widespread adoption of circular practices even more difficult. These are major obstacles to the industrial-scale advancement in sustainable textile waste management.

On the policy front, regulatory gaps and weak waste control are among the major setbacks to scaling up recycling and upcycling in the textile industry, particularly in less developed areas. For instance, waste generation in Asia is expected to increase to 1.8 billion tonnes per year in 2025 duc to poor implementation of waste management legislation [119]. In sub-Saharan Africa, hesitance by leadership to embrace circular economy (CE) approaches [120], and a lack of monetary and non-monetary incentives to promote waste reduction in manufacturing are hampering progress [121]. In addition, the lack of clear government action plans continues to impede effective textile waste management, contributing to the continent's failure to meet environmental targets such as MDG (Millennium Development Goals) 7 on sustainability [118, 122, 123]. The declining availability of sanitation infrastructure, especially in urban areas of sub-Saharan Africa, from 74% in 2000–2005 to 53% in 2010–2015, further illustrates the wider systemic problems [124]. These cumulatively form infrastructural barriers and disincentives to facilitating mass upcycling and recycling.

13.10 Implications of Findings

The implications of the findings stress the need for systemic change across the textile supply chain that seeks to address inefficiencies in upcycling and recycling. Manual sorting of materials presents one of the biggest bottlenecks in recycling, making the process inefficient and prone to error. The integration of AI technologies based on Near-Infrared (NIR) spectroscopy and hyperspectral imaging for textile identification could be a great application in this area, as it would allow for improved purity in the resulting quality of recovered materials and less reliance on laborious methods. Automation of sorting could save more than a thousand hours at individual recycling plants, in addition to scaling up recycling efforts.

Regulatory measures and policies are critical to ensuring sustainability in the apparel industry. The adoption of EPR and PPP would hold fashion brands accountable for the waste they generate, leading to more sustainable production and waste management practices. Improved labelling of fibre composition on clothing would also enable recyclers to implement more effective recycling and sorting practices, making it easier for recyclers to process the material. Consumer behaviour, on the other hand, remains a key challenge for the success of upcycling projects. While environmental awareness of fast fashion is growing, economic benefits and convenience remain the main drivers for consumers, confining upcycled products to a niche market. These issues can be tackled by tailored awareness campaigns, improved transparency in eco-labelling, and the use of digital engagement strategies that can build consumer trust and encourage sustainable purchasing behaviour.

Another key challenge in scaling up upcycling is the inconsistent supply of high quality textile waste. Unlike linear production, which benefits from a steady stream of standardised materials, upcycling depends on the availability of used fabrics that meet specific quality criteria. The developing of standardised waste collection and pre-processing systems could help ensure a more reliable supply of materials. Collaboration between manufacturers, retailers, and waste management companies is also essential to create an efficient supply chain for upcycled textiles.

Finally, the integration of upcycling and recycling into a circular economy model is crucial for the long-term sustainability of the textile sector. Businesses need to invest in innovative ways of reusing textile waste, focusing on scalable and cost-effective solutions. Fostering partnerships between industry stakeholders, policy makers, and research institutions can drive innovation and accelerate the transition toward circularity. Overall, while upcycling and recycling offer promising solutions for textile waste management, overcoming technological, regulatory, and consumer barriers remains essential for their widespread adoption.

13.11 Conclusion

The textile industry has reached a critical juncture, where technological innovation meets consumer habits and political frameworks to address two very important issues: waste management and sustainability. Advanced recycling technologies, including mechanical, chemical, and biological methods, have indeed proved their promising potential to create value from textile waste. Amidst the cost, access, and technical challenges associated with textile waste management, automated, advanced imaging techniques and AI-driven material sorting offer improvements in this area. Technological advancement and environmental awareness are gradually shifting consumer behaviour toward sustainability. This is particularly evident in the younger generation, which is more open to technological solutions, although price still dominates purchasing decisions. The role of digital platforms, gamification, and social media has become central to raising awareness, however their ability to bring about considerable behavioural change varies. Additionally, regulatory frameworks such as EPR and PPP are designed to encourage a shift from traditional linear models to circular systems, yet this is happening in a very uneven way at a global level, particularly between developed and developing countries.

References

1. Rashid, Md. E, Md. R. Khan, R. U. Haque, Md. Hasanuzzaman. 2023. Challenges of textile waste composite products and its prospects of recycling. *Journal of Materials Cycles and Waste Management* 25: 1267–1287. https://doi.org/10.1007/s10163-023-01614-x.
2. Stella, F., S. Fraterrigo Garofalo, N. Cavallini, D. Fino, and F. A. Deorsola. 2024. Closing the loop: Analysis of biotechnological processes for sustainable valorisation of textile waste from the fast fashion industry. *Sustainable Chemistry and Pharmacy* 38: 101481. https://doi.org/10.1016/j.scp.2024.101481.
3. Sandin, G., and G. M. Peters. 2018. Environmental impact of textile reuse and recycling—A review. *Journal of Cleaner Production* 184: 353–365. https://doi.org/10.1016/j.jclepro.2018.02.266.
4. Dahlbo, H., K. Aalto, H. Eskelinen, and H. Salmenperä. 2017. Increasing textile circulation—consequences and requirements. *Sustain Prod Consum* 9: 44–57. https://doi.org/10.1016/j.spc.2016.06.005.

5. Bridgens, B., M. Powell, G. Farmer, C. Walsh, E. Reed, M. Royapoor, et al. 2018. Creative upcycling: Reconnecting people, materials and place through making. *Journal of Cleaner Production* 189: 145–154. https://doi.org/10.1016/j.jclepro.2018.03.317.
6. Niinimäki, K., G. Peters, H. Dahlbo, P. Perry, T. Rissanen, and A. Gwilt. 2020. The environmental price of fast fashion. *Nat Rev Earth Environ* 1: 189–200. https://doi.org/10.1038/s43017-020-0039-9.
7. Wegener C. 2016. 22 Upcycling.
8. Williams, E. 2022. Appalling or advantageous? Exploring the impacts of fast fashion from environmental, social, and economic perspectives. *Journal for Global Business and Community* 13. https://doi.org/10.56020/001c.36873.
9. Pandit, P., K. Singha, L. Kumar, S. Shrivastava, and V. Yashraj, *Business paradigm shifting. Recycling from waste in fashion and textiles*, 151–176. Wiley. https://doi.org/10.1002/9781119620532.ch7.
10. Juanga-Labayen, J. P., I. V. Labayen, and Q. Yuan. 2022. A review on textile recycling practices and challenges. *Textiles* 2: 174–188. https://doi.org/10.3390/textiles2010010.
11. Baloyi, R. B., O. J. Gbadeyan, B. Sithole, and V. Chunilall. 2024. Recent advances in recycling technologies for waste textile fabrics: A review. *Textile Research Journal* 94: 508–529. https://doi.org/10.1177/00405175231210239.
12. Aishwariya, S., and M. J. Jaisri. 2020. Harmful effects of textile waste Article published in fiber2fashion. *Harmful Effects of Textile Wastes.*
13. Wautelet, T. 2018. The concept of circular economy: Its origins and its evolution. https://doi.org/10.13140/RG.2.2.17021.87523.
14. D'Itria, E., and R. Aus. 2023. Circular fashion: evolving practices in a changing industry. *Sustainability: Science, Practice and Policy* 19. https://doi.org/10.1080/15487733.2023.2220592.
15. Damayanti, D., L. A. Wulandari, A. Bagaskoro, A. Rianjanu, and H.-S. Wu. 2021. Possibility routes for textile recycling technology. *Polymers (Basel)* 13: 3834. https://doi.org/10.3390/polym13213834.
16. Ihsan, M. R. 2021. Penerapan Aspek Environmental Dalam Arsitektur Sustainable Pada Perancangan Stasiun Kereta Api Pekanbaru. MARKA (Media Arsitektur Dan Kota). *Jurnal Ilmiah Penelitian* 4: 100–113. https://doi.org/10.33510/marka.2021.4.2.100-113.
17. Hekkert, M. P., and E. Worrell. 2017. Circular economy: Measuring innovation in the product chain.
18. Schmidt, A., D. Watson, S. Roos, C. Askham, and P. B. Poulsen. 2016. Gaining benefits from discarded textiles. *Nordic Council of Ministers*. https://doi.org/10.6027/TN2016-537.
19. Patti, A., and D. Acierno. 2022. Towards the sustainability of the plastic industry through biopolymers: Properties and potential applications to the textiles world. *Polymers (Basel)* 14: 692. https://doi.org/10.3390/polym14040692.
20. Gonçalves, A., and C. Silva. 2021. Looking for sustainability scoring in apparel: A review on environmental footprint, social impacts and transparency. Energies (Basel) 14: 3032. https://doi.org/10.3390/en14113032.
21. Fråne, A., C. Askham, S. Gíslason, N. Kiørboe, H. Ljungkvist, D. McKinnon, et al. 2017. *The Nordic textile reuse and recycling commitment—a certification system for used textiles and textile waste.* Nordic Council of Ministers. https://doi.org/10.6027/TN2017-545.
22. Zamani, B. 2011. Carbon footprint and energy use of textile recycling techniques. *Case study.*
23. Shen, M., W. Xiong, B. Song, C. Zhou, E. Almatrafi, G. Zeng, et al. 2022. Microplastics in landfill and leachate: Occurrence, environmental behavior and removal strategies. *Chemosphere* 305: 135325. https://doi.org/10.1016/j.chemosphere.2022.135325.
24. Hole, G., and A. S. Hole. 2019. Recycling as the way to greener production: A mini review. *Journal of Cleaner Production* 212: 910–915. https://doi.org/10.1016/j.jclepro.2018.12.080.
25. Dissanayake, D. G. K., and D. U. Weerasinghe. 2021. Fabric waste recycling: A systematic review of methods, applications, and challenges. *Materials Circular Economy* 3: 24. https://doi.org/10.1007/s42824-021-00042-2.
26. Supriya, S., and A. Ashok. 2024. Recycling and upcycling of technical textiles.

27. Geissdoerfer, M., P. Savaget, N. M. P. Bocken, and E. J. Hultink. 2017. The circular economy—a new sustainability paradigm? *Journal of Cleaner Production* 143: 757–768. https://doi.org/10.1016/j.jclepro.2016.12.048.
28. Larrain, M., S. Van Passel, G. Thomassen, U. Kresovic, N. Alderweireldt, E. Moerman, et al. 2020. Economic performance of pyrolysis of mixed plastic waste: Open-loop versus closed-loop recycling. *Journal of Cleaner Production* 270: 122442. https://doi.org/10.1016/j.jclepro.2020.122442.
29. Rohan. 2024. Innovations in reducing textile waste—knowing fabric.
30. Kirstein, M., C. Lücking, L. Biermann, E. Brepohl, V. Salikov, C. Eichert, et al. 2023. Monomer recycling and repolymerization of post-consumer polyester textiles. *Chemie Ingenieur Technik* 95: 1290–1296. https://doi.org/10.1002/cite.202200197.
31. Negrier, M., E. El Ahmar, R. Sescousse, M. Sauceau, G. Bouet, D. Eglin, et al. 2024. Upcycling cellulose waste textile into aerogel beads via prilling technique. *Cellulose* 31: 823–839. https://doi.org/10.1007/s10570-023-05659-x.
32. Cassar, R. 2024. Upcycling with material debris: Nurturing the creative process through responsible handling of waste materials. *Fashion Theory* 28: 175–207. https://doi.org/10.1080/1362704X.2024.2325805.
33. Wang, Y., C. Liu, X. Zhang, and S. Zeng. 2023. Research on sustainable furniture design based on waste textiles recycling. *Sustainability* 15: 3601. https://doi.org/10.3390/su15043601.
34. Mazzitelli, M., F. Papile, and B. Del Curto. 2024. Materials selection and fashion design: Strengthening reflections on fibre's nature in fibres and textiles selection. *Discover Sustainability* 5(1): 180. https://doi.org/10.1007/s43621-024-00294-3.
35. Savas, P. E., W. A. Algozeeb, Z. Wang, C. Kittrell, X. Wu, and J. N. Hall, et al. 2023. Carbon dioxide sorbent from construction and textile plastic waste. *Advanced Sustainable Systems* 7. https://doi.org/10.1002/adsu.202200436.
36. Singh, J., K. Sung, T. Cooper, K. West, and O. Mont. 2019. Challenges and opportunities for scaling up upcycling businesses—The case of textile and wood upcycling businesses in the UK. *Resources, Conservation and Recycling* 150: 104439. https://doi.org/10.1016/j.resconrec.2019.104439.
37. Ofori, E. A., B. D. Akweley, B. Eghan, R. K. Seidu, and R. Acquaye. 2025. A review on upcycling as a strategic marketing tool for sustainability in the fashion industry. *Journal of Responsible Production and Consumption* 2(1): 1–24. https://doi.org/10.1108/JRPC-02-2024-0007.
38. Roungpaisan, N., N. Srisawat, N. Rungruangkitkrai, N. Chartvivatpornchai, J. Boonyarit, T. Kittikorn, et al. 2023. Effect of recycling PET fabric and bottle grade on r-PET fiber structure. *Polymers (Basel)* 15: 2330. https://doi.org/10.3390/polym15102330.
39. Milam-Guerrero, J., D.-J. Kwon, Y. Y. Choi, F. Al-badani, J. Jiang, J. Schaefer, et al. 2023. From waste to wearable: An alternative waste stream for unusable textiles turned into piezoelectric textiles. *RSC Sustainability* 1: 326–334. https://doi.org/10.1039/D2SU00068G.
40. Liu, T., P. Miao, Y. Shi, K. H. D. Tang, and P.-S. Yap. 2022. Recent advances, current issues and future prospects of bioenergy production: A review. *Science of The Total Environment* 810: 152181. https://doi.org/10.1016/j.scitotenv.2021.152181.
41. Akram, H. M. A., F. E. Elawad, A. M. Salah, E. H. Hago, A. E. Isam, and H. Elsir. 2021. Textile recycling—A review. *Global Journal of Engineering and Technology Advances* 6: 69–74. https://doi.org/10.30574/gjeta.2021.6.3.0015.
42. Darwesh, O. M., N. G. Al-Balakocy, A. Ghanem, and I. A. Matter. 2023. Application of microalgal-ZnO-NPs for reusing polyester/cotton blended fabric wastes after modification by cellulases enzymes. *Waste Dispos Sustain Energy* 5: 471–482. https://doi.org/10.1007/s42768-023-00170-2.
43. Mihalyi, S., M. Tagliavento, E. Boschmeier, V.-M. Archodoulaki, A. Bartl, F. Quartinello, et al. 2023. Simultaneous saccharification and fermentation with Weizmannia coagulans for recovery of synthetic fibers and production of lactic acid from blended textile waste. *Resources, Conservation and Recycling* 196: 107060. https://doi.org/10.1016/j.resconrec.2023.107060.

44. Cura, K., N. Rintala, T. Kamppuri, E. Saarimäki, and P. Heikkilä. 2021. Textile recognition and sorting for recycling at an automated line using near infrared spectroscopy. *Recycling* 6: 11. https://doi.org/10.3390/recycling6010011.
45. Parmakoğlu, E. Ü., A. Çay, and J. Yanık. 2023. Valorization of solid wastes from textile industry as an adsorbent through activated carbon production. *AATCC Journal of Research* 10: 133–143. https://doi.org/10.1177/24723444221147983.
46. Payel, S., Md. A. Hasan, F. Pahlevani, D. Prasher, A. Ghose, and V. Sahajwalla. 2024. From closet to contaminant to control: Unveiling microplastic sources in household textiles and potential for environmental application. *Journal of Water Process Engineering* 68: 106400. https://doi.org/10.1016/j.jwpe.2024.106400.
47. Wicaksono, I., A. Maheshwari, D. D. Haddad, J. Paradiso, and A. Danielescu. 2024. Design and fabrication of multifunctional E-textiles by upcycling waste cotton fabrics through carbonization. *Proceedings of the ACM Interactive, Mobile, Wearable and Ubiquitous Technologies* 8: 1–31. https://doi.org/10.1145/3659588.
48. Peets, P., I. Leito, J. Pelt, and S. Vahur. 2017. Identification and classification of textile fibres using ATR-FT-IR spectroscopy with chemometric methods. *Spectrochimica Acta Part A: Molecular and Biomolecular Spectroscopy* 173: 175–181. https://doi.org/10.1016/j.saa.2016.09.007.
49. Peets, P., K. Kaupmees, S. Vahur, and I. Leito. 2019. Reflectance FT-IR spectroscopy as a viable option for textile fiber identification. *Heritage Science* 7: 93. https://doi.org/10.1186/s40494-019-0337-z.
50. Nemeša, I., M. Pešić, and V. Bozoki. 2024. Mechanical recycling of textile waste. *Tekstilna Industrija* 72: 24–28. https://doi.org/10.5937/tekstind2404024N.
51. Enking, J., A. Becker, G. Schu, M. Gausmann, S. Cucurachi, A., Tukker, and T. Gries .2025. Recycling processes of polyester-containing textile waste–A review. *Resources, Conservation and Recycling* 219: 108256. https://doi.org/10.1016/j.resconrec.2025.108256.
52. Kozłowski, R. M., and M. Mackiewicz-Talarczyk. 2020. Introduction to natural textile fibres. In *Handbook of Natural Fibres* 1–13. Elsevier. https://doi.org/10.1016/B978-0-12-818398-4.00001-3.
53. Mäkelä, M., M. Rissanen, and H. Sixta. 2020. Machine vision estimates the polyester content in recyclable waste textiles. *Resources, Conservation and Recycling* 161: 105007. https://doi.org/10.1016/j.resconrec.2020.105007.
54. Liu, Z., W. Li, and Z. Wei. 2020. Qualitative classification of waste textiles based on near infrared spectroscopy and the convolutional network. *Textile Research Journal* 90: 1057–1066. https://doi.org/10.1177/0040517519886032.
55. Riba, J.-R., R. Cantero, T. Canals, and R. Puig. 2020. Circular economy of post-consumer textile waste: Classification through infrared spectroscopy. *Journal of Cleaner Production* 272: 123011. https://doi.org/10.1016/j.jclepro.2020.123011.
56. Nedumpillil, N. N., A. Sankaran, S. Jose, S. George, and S. Thomas. 2022. Destructive analytical techniques for the analysis of historic textiles. In *Handbook of museum textiles*, 215–234. Wiley. https://doi.org/10.1002/9781119983903.ch30.
57. Tang, K. H. D. 2023. State of the art in textile waste management: A review. *Textiles* 3: 454–467. https://doi.org/10.3390/textiles3040027.
58. Liu, W., S. Liu, T. Liu, T. Liu, J. Zhang, and H. Liu. 2019. Eco-friendly post-consumer cotton waste recycling for regenerated cellulose fibers. *Carbohydrate Polymers* 206: 141–148. https://doi.org/10.1016/j.carbpol.2018.10.046.
59. Haule, L. V., C. M. Carr, and M. Rigout. 2014. Investigation into the removal of an easy-care crosslinking agent from cotton and the subsequent regeneration of lyocell-type fibres. *Cellulose* 21: 2147–2156. https://doi.org/10.1007/s10570-014-0225-3.
60. Ribul, M., A. Lanot, C. Tommencioni Pisapia, P. Purnell, S. J. McQueen-Mason, and S. Baurley. 2021. Mechanical, chemical, biological: Moving towards closed-loop bio-based recycling in a circular economy of sustainable textiles. *Journal of Cleaner Production* 326: 129325. https://doi.org/10.1016/j.jclepro.2021.129325.
61. Hann, S., and T. Connock. 2020. Chemical recycling: State of play report for CHEM trust.

62. Celep, G., D. G. Tetik, and F. Yilmaz. 2023. Limitations of textile recycling: The reason behind the development of alternative sustainable fibers. In *Next-generation textiles*. IntechOpen. https://doi.org/10.5772/intechopen.105118.
63. Papahristou, E. 2016. Can 3D virtual prototype conquer the apparel industry? *Journal of Fashion Technology & Textile Engineering* 4. https://doi.org/10.4172/2329-9568.1000134.
64. Veltkamp, T. 2023. Recycling post-industrial waste in the textile industry: A circular future can advanced sorting contribute to plastic circula.
65. Du, W., J. Zheng, W. Li, Z. Liu, H. Wang, and X. Han. 2022. Efficient recognition and automatic sorting technology of waste textiles based on online near infrared spectroscopy and convolutional neural network. *Resources, Conservation and Recycling* 180: 106157. https://doi.org/10.1016/j.resconrec.2022.106157.
66. Choi, K.-H. 2024. 3D digital technology in upcycling apparel design: the creation of a modular redesign system and designer perspectives. *International Journal of Fashion Design, Technology and Education* 1–14. https://doi.org/10.1080/17543266.2024.2413568.
67. de Souza Moreira, L. R., Lo Sciuto, D., and Filho, E. X. F. 2016. An overview of cellulose-degrading enzymes and their applications in textile industry. In *New and future developments in microbial biotechnology and bioengineering*. 165–175. Elsevier. https://doi.org/10.1016/B978-0-444-63507-5.00014-9.
68. Piribauer, B., and A. Bartl. 2019. Textile recycling processes, state of the art and current developments: A mini review. *Waste Management & Research: The Journal for a Sustainable Circular Economy* 37: 112–119. https://doi.org/10.1177/0734242X18819277.
69. Hu, Y., C. Du, S.-Y. Leu, H. Jing, X. Li, and C. S. K. Lin. 2018. Valorisation of textile waste by fungal solid state fermentation: An example of circular waste-based biorefinery. *Resources, Conservation and Recycling* 129: 27–35. https://doi.org/10.1016/j.resconrec.2017.09.024.
70. Kim, D., and Lee, D. 2024. *International journal of intelligent systems and applications in engineering fabric defect detection based on artificial intelligence*, vol. 2024.
71. Cobb, D. 2024. *Advances in textile recycling—textile technology source*.
72. Technical Textiles Today. 2023. Wastex revolutionizes textile recycling with cutting-edge sorting technology. *Technical Textiles Today*.
73. Dulal, M., S. Afroj, J. Ahn, Y. Cho, C. Carr, I.-D. Kim, et al. 2022. Toward sustainable wearable electronic textiles. *ACS Nano* 16: 19755–19788. https://doi.org/10.1021/acsnano.2c07723.
74. Vaucher, J., A. Demongeot, V. Michaud, and Y. Leterrier. 2022. Recycling of bottle grade PET: Influence of HDPE contamination on the microstructure and mechanical performance of 3D printed parts. *Polymers (Basel)* 14: 5507. https://doi.org/10.3390/polym14245507.
75. Apostolopoulou-Kalkavoura, V., N. Fijoł, S. Lombardo, M. Ruiz-Caldas, and A. P. Mathew. 2024. In situ functionalisation and upcycling of post-consumer textile blends into 3D printable nanocomposite filaments. *Adv Sustain Syst*. https://doi.org/10.1002/adsu.202400132.
76. Khan, S. A. R., A. Z. Piprani, and Z. Yu. 2022. Digital technology and circular economy practices: Future of supply chains. *Operations Management Research* 15: 676–688. https://doi.org/10.1007/s12063-021-00247-3.
77. Moriuchi, E., and I. Takahashi. 2022. The role of perceived value, trust and engagement in the C2C online secondary marketplace. *Journal of Business Research* 148: 76–88. https://doi.org/10.1016/j.jbusres.2022.04.029.
78. Boström, M., and M. Micheletti. 2016. Introducing the sustainability challenge of textiles and clothing. *Journal of Consumer Policy (Dordr)* 39: 367–375. https://doi.org/10.1007/s10603-016-9336-6.
79. Shirkhani, S., H. Mokayed, R. Saini, and H. Y. Chai. 2023. Study of AI-driven fashion recommender systems. *SN Computer Science* 4: 514. https://doi.org/10.1007/s42979-023-01932-9.
80. Cheng, W.-H., S. Song, C.-Y. Chen, S. C. Hidayati, and J. Liu. 2022. Fashion meets computer vision. *ACM Computing Surveys* 54: 1–41. https://doi.org/10.1145/3447239.
81. Okur, N., and C. Saricam. 2019. *The impact of knowledge on consumer behaviour towards sustainable apparel consumption*, 69–96. https://doi.org/10.1007/978-981-13-1265-6_3.

82. Shahzad, M. F., S. Xu, O. ul. Rehman, and I. Javed. 2023. Impact of gamification on green consumption behavior integrating technological awareness, motivation, enjoyment and virtual CSR. *Scientific Reports* 13(1): 21751. https://doi.org/10.1038/s41598-023-48835-6.
83. Waydel-Bendyk, M. 2020. *Evaluating potential of gamification to facilitate sustainable fashion consumption*, 205–215. https://doi.org/10.1007/978-3-030-50341-3_16.
84. Du, S., G. Yalcinkaya, and L. Bstieler. 2016. Sustainability, social media driven open innovation, and new product development performance. *Journal of Product Innovation Management* 33: 55–71. https://doi.org/10.1111/jpim.12334.
85. Ribeiro, P. R., P. Batista, F. Mendes-Palma, M. Pintado, and P. Oliveira-Silva. 2023. Consumers' engagement and perspectives on sustainable textile consumption. *Sustainability* 15: 15812. https://doi.org/10.3390/su152215812.
86. Lou, X., and Y. Xu. 2024. Consumption of sustainable denim products: the contribution of blockchain certified eco-labels. *Journal of Theoretical and Applied Electronic Commerce Research* 19: 396–411. https://doi.org/10.3390/jtaer19010021.
87. Priporas, C.-V., N. Stylos, and A. K. Fotiadis. 2017. Generation Z consumers' expectations of interactions in smart retailing: A future agenda. *Comput Human Behav* 77: 374–381. https://doi.org/10.1016/j.chb.2017.01.058.
88. Navas, R., H. J. Chang (Julie), S. Khan, J. W. Chong. 2021. Sustainability transparency and trustworthiness of traditional and blockchain ecolabels: A comparison of generations X and Y consumers. *Sustainability* 13: 8469. https://doi.org/10.3390/su13158469.
89. Guo, S., X. Sun, and H. K. S. Lam. 2023. Applications of blockchain technology in sustainable fashion supply chains: operational transparency and environmental efforts. *IEEE Transactions on Engineering Management* 70: 1312–1328. https://doi.org/10.1109/TEM.2020.3034216.
90. Nadeera, W. A. A. P., D. D. A. Karunarathna, D. M. H. Madushan, G. D. V. Perera, and W. L. K. Abeyunawardhana. 2022. Solid waste management in textile industry.
91. Curran, T., and I. D. Williams. 2012. A zero waste vision for industrial networks in Europe. *Journal of Hazardous Materials* 207–208: 3–7. https://doi.org/10.1016/j.jhazmat.2011.07.122.
92. Ma, S., W. Ding, Y. Liu, Y. Zhang, S. Ren, X. Kong, et al. 2024. Industry 4.0 and cleaner production: A comprehensive review of sustainable and intelligent manufacturing for energy-intensive manufacturing industries. Journal of Cleaner Production 467: 142879. https://doi.org/10.1016/j.jclepro.2024.142879.
93. Dawson, L. 2019. 'Our waste, our resources; A strategy for England'—Switching to a circular economy through the use of extended producer responsibility. *Environmental Law Review* 21: 210–218. https://doi.org/10.1177/1461452919851943.
94. Nagle, A. J., E. L. Delaney, L. C. Bank, and P. G. Leahy. 2020. A comparative life cycle assessment between landfilling and co-processing of waste from decommissioned Irish wind turbine blades. *Journal of Cleaner Production* 277: 123321. https://doi.org/10.1016/j.jclepro.2020.123321.
95. Department of Environment Food and Rural Affairs. 2021. *Consultation on the waste prevention programme for England: Towards a resource-efficient economy.*
96. Nilisha, M., P. Desai, M. Uday, and D. Gaikwad. 2013. Polluter pay principle (PPP): Tool for environment pollution control and management "A Case Study on Pollution of Bandi River by Textile industries in Pali, Rajasthan".
97. Times News Group. 2021. Textile waste growing threat to ocean's health—Bellarine Times.
98. Dissanayake, K., and R. Pal. 2023. Sustainability dichotomies of used clothes supply chains: A critical review of key concerns and strategic resources. *The International Journal of Logistics Management* 34: 75–97. https://doi.org/10.1108/IJLM-10-2022-0410.
99. Adler, M. 2020. Textile recovery in the US.
100. Khan, M. I., L. Wang, and R. Padhye. 2023. Textile waste management in Australia: A review. *Resources, Conservation & Recycling Advances* 18: 200154. https://doi.org/10.1016/j.rcradv.2023.200154.
101. Department of Climate Change Energy the Environment and Water. 2022. Investing in Australia's waste and recycling infrastructure—DCCEEW.

102. Cotton Research and Development Corporation. 2022. Australian Cotton _ Ground breaking trial returning cotton textile…
103. Runstrand, A., and I. A. Johansson. 2020. Obstacles in the textile upcycling chain—a case study of the communication between small-scaled upcycling actors and their processes.
104. Caldera, S., R. Jayasinghe, C. Desha, L. Dawes, and S. Ferguson. 2022. Evaluating barriers, enablers and opportunities for closing the loop through 'Waste Upcycling': A systematic literature review. *Journal of Sustainable Development of Energy, Water and Environment Systems* 10: 1–20. https://doi.org/10.13044/j.sdewes.d8.0367.
105. Runstrand, A., and I. A. Johansson. n.d. Obstacles in the textile upcycling chain-a case study of the communication between small-scaled upcycling actors and their processes.
106. Saif, M., R. Blay-Roger, M. Zeeshan, L. F. Bobadilla, T. R. Reina, M. A. Nawaz, et al. 2024. Navigating the legislative interventions, challenges, and opportunities in revolutionizing textile upcycling/recycling processes for a circular economy. *ACS Sustainable Resource Management* 1: 2338–2349. https://doi.org/10.1021/acssusresmgt.4c00242.
107. Hamzaoui-Essoussi, L., and J. D. Linton. 2014. Offering branded remanufactured/recycled products: At what price? *Journal of Remanufacturing* 4: 9. https://doi.org/10.1186/s13243-014-0009-9.
108. Polyportis, A., R. Mugge, and L. Magnier. 2022. Consumer acceptance of products made from recycled materials: A scoping review. *Resources, Conservation and Recycling* 186: 106533. https://doi.org/10.1016/j.resconrec.2022.106533.
109. Calvo-Porral, C., and J.-P. Lévy-Mangin. 2020. The circular economy business model: Examining consumers' acceptance of recycled goods. *Administrative Sciences* 10: 28. https://doi.org/10.3390/admsci10020028.
110. Park, H. J., and L. M. Lin. 2020. Exploring attitude–behavior gap in sustainable consumption: Comparison of recycled and upcycled fashion products. *Journal of Business Research* 117: 623–628. https://doi.org/10.1016/j.jbusres.2018.08.025.
111. Kuah, A. T. H., and P. Wang. 2020. Circular economy and consumer acceptance: An exploratory study in East and Southeast Asia. *Journal of Cleaner Production* 247: 119097. https://doi.org/10.1016/j.jclepro.2019.119097.
112. Wojciechowska, M., and G. Kowaluk. 2024. Challenges and opportunities in recycling upholstery textiles: Enhancing high-density fiberboards with recycled fibers. *Fibers* 12: 105. https://doi.org/10.3390/fib12120105.
113. Kahoush, M., and N. Kadi. 2022. Towards sustainable textile sector: Fractionation and separation of cotton/ polyester fibers from blended textile waste. *Sustainable Materials and Technologies* 34: e00513. https://doi.org/10.1016/j.susmat.2022.e00513.
114. Hossain, M. T., M. A. Shahid, M. G. M. Limon, I. Hossain, and N. Mahmud. 2024. Techniques, applications, and challenges in textiles for a sustainable future. *Journal of Open Innovation: Technology, Market, and Complexity* 10: 100230. https://doi.org/10.1016/j.joitmc.2024.100230.
115. Wagaw, T., and K. M. Babu. 2023. Textile waste recycling: A need for a stringent paradigm shift. *AATCC Journal of Research* 10: 376–385. https://doi.org/10.1177/24723444231188342.
116. Lahl, U., and B. Zeschmar-Lahl. 2024. Material recycling of plastics—a challenge for sustainability. *Sustainability* 16: 6630. https://doi.org/10.3390/su16156630.
117. Hajam, Y. A., R. Kumar, and A. Kumar. 2023. Environmental waste management strategies and vermi transformation for sustainable development. *Environmental Challenges* 13: 100747. https://doi.org/10.1016/j.envc.2023.100747.
118. Asante, K. A., Y. Amoyaw-Osei, and T. Agusa. 2019. E-waste recycling in Africa: Risks and opportunities. *Current Opinion in Green and Sustainable Chemistry* 18: 109–117. https://doi.org/10.1016/j.cogsc.2019.04.001.
119. Agamuthu, P., and S. Babel. 2023. Waste management developments in the last five decades: Asian perspective. *Waste Management & Research: The Journal for a Sustainable Circular Economy* 41: 1699–1716. https://doi.org/10.1177/0734242X231199938.

120. Debrah, J. K., G. K. Teye, and M. A. P. Dinis. 2022. Barriers and challenges to waste management hindering the circular economy in sub-Saharan Africa. *Urban Science* 6: 57. https://doi.org/10.3390/urbansci6030057.
121. Roy Choudhury, A. K. 2014. *Environmental impacts of the textile industry and its assessment through life cycle assessment*, 1–39. https://doi.org/10.1007/978-981-287-110-7_1.
122. United Nations. 2012. *The millennium development goals report 2012*. UN. https://doi.org/10.18356/32f1e244-en.
123. Glazebrook, T., and E. Opoku. 2020. Gender and sustainability: Learning from women's farming in Africa. *Sustainability* 12: 10483. https://doi.org/10.3390/su122410483.
124. Armah, F. A., B. Ekumah, D. O. Yawson, J. O. Odoi, A.-R. Afitiri, and F. E. Nyieku. 2018. Access to improved water and sanitation in sub-Saharan Africa in a quarter century. *Heliyon* 4: e00931. https://doi.org/10.1016/j.heliyon.2018.e00931.

Benjamin Eghan is a Ph.D. candidate at DMU, UK, where he is based in the Faculty of Arts, Design and Humanities (ADH). His research focuses on sustainability in the textiles and fashion industry. Specifically, his doctoral work explores innovative techniques and technologies for transforming textile and fashion waste into new products, with the aim of generating social and environmental impact. Benjamin currently holds a position as a Research Assistant in Art and Design within the ADH Faculty at DMU. He also serves as a peer reviewer for the Fashion and Textile Review journal, contributing his expertise to the academic community.

Kyungeun Sung is a Senior Lecturer in Product Design at De Montfort University, UK. She has research interests and expertise in sustainable design, craft, production, business, supply chains, behaviour, and consumption, focusing on upcycling, circular economy and net zero. Her research uses design as a tool for new product development, service innovation, behaviour change, scaling up niche sustainable practices/behaviours, and transition to the circular economy and net zero. Her recent books are Research Journeys to Net Zero: Current and Future Leaders (2024) and State-of-the-Art Upcycling Research and Practice (2021). She has published over thirty peer-reviewed academic papers and book chapters.

Chapter 14
Regulatory Standards and Certifications for Sustainable Textiles

Chipambwa Walter and Dzingai Pethile

Abstract Globally, the textile industry is encountering increased scrutiny as it has a significant environmental impact, as it contributes to water pollution, excessive waste production, widespread use of hazardous chemicals, and elevated energy consumption during manufacturing. In recent years, sustainability has become crucial in the textile industry, motivated by eco-conscious customers, environmental concerns, and evolving legal demands. To support this shift, a variety of standards and certifications have been introduced by regulatory bodies and organizations worldwide. This chapter takes a closer look at key global sustainability standards and the specific certifications designed to help the textile sector operate more responsibly. This chapter also looks at how textile certifications are being put into practice, using a range of case studies to highlight key takeaways and real-world insights. It explores how different levels of regulation, regional, national, and international, interact to shape sustainability standards and ensure industry-wide compliance. The chapter also focuses on the role of technology in the certification process, including a review of current tools and emerging trends. The chapter concludes by addressing some of the major challenges, such as aligning various standards, compliance verification, and cost implications for manufacturers. At the same time, it also highlights how these standards and regulations create opportunities for innovation and boost competitiveness in the market.

Keywords Sustainability · Textiles · Certification · Standards · Technology

---

C. Walter (✉)
Department of Clothing and Textile Technology, School of Art and Design, Chinhoyi University of Technology, Chinhoyi, Zimbabwe
e-mail: wchipambwa@cut.ac.zw

D. Pethile
Department of Fibre and Polymer Materials Engineering, Faculty of Engineering, National University of Science and Technology, Bulawayo, Zimbabwe
e-mail: pethile.dzingai@nust.ac.zw

S.-X. Jiang et al. (eds.), *Advances in Textile Materials and Processing Techniques for Sustainability*, SDGs and Textiles, https://doi.org/10.1007/978-981-95-0469-5_14

14.1 Introduction

The textile sector significantly contributes to economic growth and employment in various nations, making it one of the most influential sectors globally. However, despite its significance globally, it consumes a lot of resources, which makes it a highly polluting industry, raising critical concerns about its environmental and social impact [1, 2]. Sustainability in textiles is a broad topic, covering aspects of the environment, social and economic. The textile sector faces considerable challenges related to the fast fashion model. However, various strategies and innovations are emerging to manage these challenges and encourage sustainable practices in this industry. Nonetheless, the sector has emerging innovative methods to tackle these challenges and promote the best sustainability practices. With increasing critical attention given to the environmental footprint of the textile sector, international responses have been made with regulatory standards and certifications to help achieve sustainability. This chapter examines the evolution of these standards and how they have been applied in the textile industry so that various players along the supply chain participate in promoting sustainable business practices. The chapter begins by discussing sustainable regulatory issues in textiles and then expands into regional and specific certifications and standards. Concerns over pollution, resource depletion, and labour exploitation in the sector have led to the development of numerous legislations over time. Figure 14.1 illustrates a typical textile product life cycle. All the stages in the cycle have varying degrees of sustainability hotspots that include water usage, use of pesticides, chemical pollution, energy usage, waste generation, and labour concerns. To mitigate the effects of these various stages, textile standards and regulations have been crafted to promote sustainable practices in textile manufacturing.

There is a growing expectation for textile products to meet sustainability standards as set by regulatory bodies. Textile standards guide manufacturers, fashion brands, and consumers toward more environmentally and socially responsible choices. By clearly defining what sustainability looks like, they also make it easier for third parties to hold companies accountable to ethical and environmental standards. This chapter explores some of the leading regulatory standards and certification schemes, examining how they contribute to the push for sustainability in the textile industry. It also explores how these systems promote both environmental care and social responsibility throughout the supply chain.

14.2 Regulatory Standards

The demand for eco-friendly textiles has been necessitated by growing consumer awareness of environmental and sustainability issues. Despite this positive trend, the textile industry remains hindered by the absence of a cohesive regulatory framework and standardised certification procedures, which complicates efforts to address the

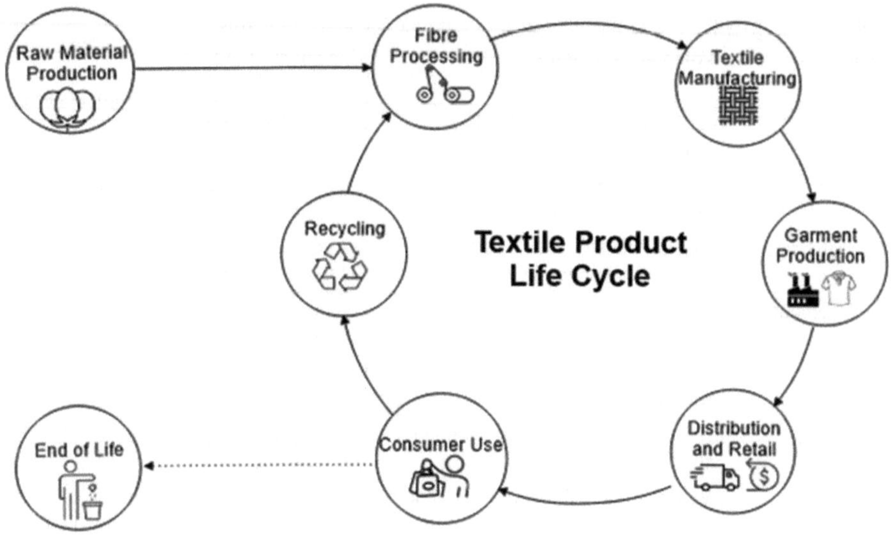

Fig. 14.1 Textile product life cycle

multifaceted challenges of textile sustainability. Although numerous labels purport to signal environmental responsibility, as [3] notes, there is a lack of universally accepted criteria that are meaningful and practical for both consumers and producers. These inconsistencies contribute to widespread confusion among stakeholders regarding the definition and scope of genuine sustainability. While many voluntary sustainability standards (VSS) have emerged in response to these challenges, their effectiveness remains contested due to variable implementation practices and inconsistent evaluation mechanisms. Consequently, there is a critical need for more nuanced and comprehensive policy guidance to support informed and responsible consumption within the textile sector. Existing voluntary certification schemes like Oekotex and Global Organic Textile Standard (GOTS) demonstrate potential benefits in promoting corporate social responsibility but often lack uniformity in their implementation and effectiveness [4]. Table 14.1 summarises the major differences between voluntary and mandatory standards in textiles citing the examples for each aspect.

14.2.1 Environmental Issues in Textiles

The extensive and complex supply chains characteristic of the textile sector contribute to its substantial ecological footprint, positioning the industry as a major driver of environmental degradation. According to Lundy [5], the United States generates over 100 pounds of textile waste per person annually, with the majority either ending up in landfills or being exported to developing countries. The trade in second-hand

Table 14.1 Voluntary versus mandatory sustainability standards in textiles

Aspect	Voluntary standards	Mandatory standards
Definition	Adopted by choice (e.g., certifications like GOTS, Fair Trade)	Legally enforced by governments (e.g., EU REACH, CPSIA)
Purpose	Market differentiation, brand reputation, consumer trust	Legal compliance, public health/ environmental protection
Governance	NGOs, industry consortia (e.g., Textile Exchange, OEKO-TEX)	Governments, regulatory bodies (e.g., EU Commission, EPA)
Enforcement	Third-party audits; non-compliance leads to loss of certification	Penalties, fines, or bans for non-compliance
Scope	Often sector-specific (e.g., organic cotton, recycling)	Broad (e.g., chemical restrictions, labour laws)
Flexibility	High (companies choose standards aligning with goals)	Low (uniform requirements for all market players)
Transparency	Varies; some publish reports, others lack supply chain visibility	Publicly accessible regulations; enforcement data may be published
Examples	– GOTS (organic textiles) – Bluesign® (chemical management) – Cradle to cradle (circularity)	– EU REACH (chemical restrictions) – CPSIA (U.S. product safety) – France AGEC Law (anti-waste)

clothing (SHC) represents a rapidly expanding sector in many developing economies, presenting both opportunities and challenges from a textile industry perspective. Globally, the SHC market exceeds $1 billion in value, contributing to greater affordability of clothing and generating employment opportunities within these countries [6, 7]. However, the absence of formal recycling facilities in many of these regions [8, 9] poses significant threats to domestic textile industries and employment prospects, thereby constraining economic development and undermining the livelihoods of local populations.

14.2.1.1 Water Consumption

The textile industry ranks among the most water-intensive sectors globally, contributing significantly to environmental degradation through excessive water consumption and the discharge of untreated or inadequately treated wastewater. Processes such as dyeing, bleaching, and finishing are particularly water-intensive stages in textile production, often accounting for a substantial portion of the industry's overall water footprint. Regulations for water consumption in textiles aim to reduce freshwater consumption and promote water recycling. For instance, the use of waterless dyeing methods through supercritical fluid, plasma technology, and carbon dioxide reduces water, energy, and chemical consumption [10]. Water footprint assessment has also been introduced to measure and lower freshwater consumption in textile production [11, 12]. Textile wastewater contains harmful pollutants, including dyes, pigments heavy metals, and other organic compounds [13]. The quality of textile wastewater is usually measured through its biochemical oxygen demand (BOD), and chemical oxygen demand (COD) [14]. Despite statutes in various countries to control wastewater, pollution of water from textiles is still evident [13, 14]. In Turkey permissible limits for pollutants like sulphates, COD, and BOD are enforced as guided by the Istanbul Water and Sewerage Administration (ISKI) discharge standards [15]. Regulations in some regions mandate the adoption of the Zero Liquid Discharge (ZLD) method to eliminate wastewater release [16]. ZLD systems can recover water and valuable by-products through membrane filtration, evaporation, and crystallisation. European Union's Water Framework Directive (EU-WFD) controls chemical emissions into water bodies the region seeks to control textile wastewater treatment.

14.2.1.2 Chemical Usage in Textiles

The regulation of chemical usage in textile manufacturing serves as a critical mechanism for promoting environmental sustainability and safeguarding consumer health. A range of international and national legislative frameworks has been established to control the use of hazardous substances in textile production, aiming to mitigate both environmental degradation and human health risks associated with chemical exposure. These regulations adopt a comprehensive approach, incorporating tools such as Restricted Substances Lists (RSLs) and Manufacturing Restricted Substances Lists (MRSLs), which play a central role in governing the permissible use of chemicals throughout the textile supply chain. In Europe, REACH regulates chemical manufacturing and use and their health and environmental effects [17]. REACH requires companies to identify chemicals they produce or import into Europe. RSL and MRSL lists identify chemicals that are restricted or banned in textile production due to their potential health and environmental hazards [18, 19]. They include substances like azo dyes, formaldehyde, and phthalates, which have adverse effects on the environment. According to Ivy [19], there is a need for effective chemical management systems that can ensure compliance with regulations and protect workers and the environment. These systems involve monitoring chemical use, training workers, and adopting cleaner technologies for chemicals used in the textile sector.

14.2.1.3 Energy Consumption in Textiles

Textile production consumes a large amount of energy and depletes natural resources. One of the most important policies is The EU Ecodesign Directive which controls the energy-using efficiency of Ecodesign textiles products [20]. This directive attempts to mitigate the adverse environmental effects of a product's life cycle by regulating the design and production processes. The EU's product policy mandates the supplier to provide customers with information on the environmental performance of the specific item, which includes energy labelling and other transparency measures. The US-Environmental Protection Agency (EPA) also provides a voluntary labelling program Energy Star which encourages the conservation of energy in textile industry products and garments. The Energy Star label is granted to products that meet energy efficiency standards, providing consumers with information to make more sustainable choices [21].

14.2.1.4 Carbon Footprint of Textile Industry

According to [22] the textile industry contributes about 8–10% of the global carbon emissions due to its manufacturing and transportation processes. The Paris Agreement's global objective is to keep global warming below 2 degrees Celsius. Several steps are being implemented to control carbon emissions, and companies are now adopting processes such as carbon offsetting and renewable energy use to decrease their footprint. Table 14.2 summarizes some of the documents in the form of standards that have been tailored for the textile industry about carbon emission control in greenhouse gas (GHG) emissions management.

14.2.1.5 Microplastic Pollution

Microplastic pollution from textiles, particularly synthetic fibres, has prompted the need for specific regulatory requirements and standards. The washing of synthetic fabrics is the source of about 35% of the microplastics found in the world's oceans

Table 14.2 Carbon footprint certifications and standards

Standard/framework	Carbon footprint component	Key focus
Science based targets initiative (SBTi)	Brands commit to GHG reduction targets aligned with Paris agreement	Supply chain emissions reduction strategies
Carbon trust footprint label	Independently verified product carbon footprint (kg CO_2e per item)	Transparency to consumers
ISO 14067:2018	Guidelines for calculating product carbon footprint (PCF)	Life cycle GHG assessment methodology
Higg index and brand modules	Carbon footprint assessment at material and facility level	Benchmarking environmental impacts of fibres, processes
Cradle to cradle certified	Includes climate protection category	Carbon neutrality and offsetting
Climate neutral certification	Covers all emissions, requires offsetting and reduction plans	Full organizational carbon accounting

[22]. Synthetic fibres such as polyester and nylon release microplastics during laundering, and these enter waterways and harm marine life. There is a growing call for regulations to reduce microplastic emissions from textiles, though specific standards are still evolving. The EU Microplastics Restriction Proposal under REACH came into force in the year 2023 with a phased implementation for different product categories. In France, the Microfibre Filters Mandate requires that all washing machines be equipped with microfiber filters under the Anti-Waste Law (AGEC). The OEKO-TEX standard also encourages synthetic fibre management and wastewater treatment for particle capture. To curb the issue of microplastics, there is a need for the development of innovative textiles that minimise fibre shedding [23].

14.2.2 Need for Regulations and Certifications

The demand for textiles continues to grow globally due to the increasing population and industrial applications of various textile-related products. This growth has made the industry face significant challenges in ensuring sustainable practices and development [24]. This has necessitated the need for strict laws and certifications that promote social and environmental responsibility in textile manufacturing. The shift in the textile business landscape requires the development of robust regulation and certification mechanisms to guarantee adherence to safety standards throughout the production process. The development of policies and regulatory frameworks that seek to enforce compliance in the textile industry through established standards can help businesses to reduce the sustainability-related challenges they might face. Existing regulations in the textile industry such as those outlined by [25] demonstrate that while some measures are in place to ensure product safety through labelling and flammability standards, inconsistencies remain. These inconsistencies can lead to consumer confusion and undermine trust in textile products. An effective regulatory system must not only enforce compliance but also foster transparency throughout supply chains to achieve ethical production standards [3, 26].

14.2.3 Consumer Influence

Consumer behaviour is one of the main reasons for implementing regulations in the textile business. Consumers are turning towards fast fashion, defined as cheap and quickly produced clothing with a significant environmental impact [27]. Fast fashion increases waste accumulation as customers make quick purchases and quickly discard these clothing items, thus contributing to pollution and unethical labour practices. Therefore, consumer influence is critical in promoting necessary regulatory frameworks that can inspire sustainable practices in the textile sector. Fast fashion's appeal lies in its affordability and the variety it gives to fashion consumers; however, this convenience disguises significant environmental costs. Fashion consumers from developed world economies often dispose of garments after only a few uses [28], contributing to massive textile waste as two-thirds of this waste goes to landfills [29]. The current purchasing practices show how consumer actions fail to connect with the environmental consequences of their purchasing decisions. An increase in public understanding of fast fashion damage to the environment has led to a strong need for regulatory measures that direct customers toward sustainable choices. Regulations would benefit the environment while educating consumers about buying actions that preserve the planet.

14.3 Global Regulatory Standards for Sustainable Textiles

Throughout the world, the textile industry holds an essential position in commercial activity yet it receives rising scrutiny because of its social and environmental impact concerns. Global regulatory standards are becoming more key to industry transformation because sustainable practices have risen to a crucial level of necessity. These regulations in the textile industry seek to alleviate adverse effects such as excessive water consumption, greenhouse gas emissions, and labour exploitation [30, 31]. Despite the importance of sustainability in textiles, the enforcement of various frameworks and standards across borders faces challenges. National policies and voluntary standards presently exist as a disorganised combination in the current settings because of their insufficient harmonisation capability, alongside weak enforcement systems [32]. The inconsistencies between different jurisdictions create challenges for larger companies to comply and for small companies to adopt sustainable practices because of the varied regulatory demands in diverse locations. The emergence of ecolabeling programs has become a tool for sustainability transparency [3], yet they exist outside conventional regulatory systems. The assessment of sustainability progress in textile manufacturing needs to examine regulatory effects as it determines how well these standards achieve industry-wide transformation.

The absence of uniform international regulatory requirements for sustainable textiles is one of the main obstacles in this respect. The sector operates across multiple jurisdictions, each with its own regulations and requirements, making it difficult to establish a unified framework for sustainable production. Nonetheless, there are some initiatives underway to address this issue. International Organisation for Standardization (ISO) and the United Nations Framework Convention on Climate Change, have developed guidelines and standards for sustainable textile production that seek to reduce water usage, save energy or use renewable energy, and reduce the use of hazardous chemicals [3, 33]. Generally, to attain a certification in the textile industry, the organisation has to follow the processes shown in Fig. 14.2. If the decision is a fail the process has to be started all over again though this can also be dependent on the certification body's guidelines. The majority of certifications in textiles usually cover a period of between one year to three years (1–3 years). Follow-up audits for compliance are carried out during this period to ensure compliance with the standard.

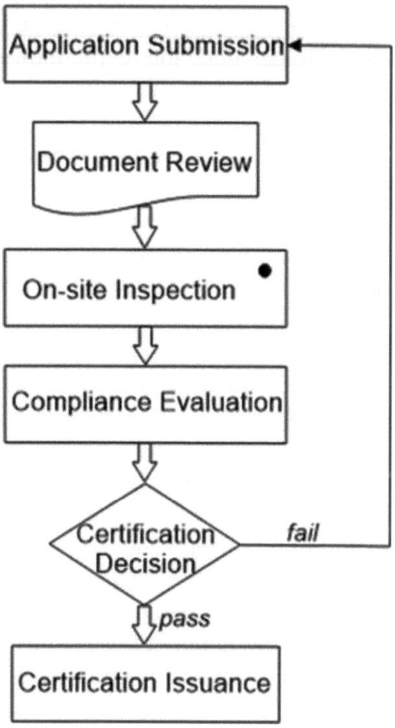

Fig. 14.2 Flowchart of the textile certification process

14.3.1 International Organisation for Standardisation (ISO)

ISO is essential in promoting sustainability in the textile sector by creating guidelines for eco-friendly certifications and practices. ISO has formulated several procedures to help the textile sector adopt environmentally friendly practices to address sustainability issues related to the sector. Among the crucial ISO standards for the textile industry is ISO 14001, which seeks to establish and maintain environmental management systems. This standard addresses every facet of a textile company's activities, from the procurement of materials, production, and distribution, up to end-of-life care. By implementing ISO 14001, textile companies can identify and mitigate their environmental impact, to improve their sustainability performance. Based on their literature search Zimon et al. [34] lament a lack of recent studies or literature that discusses the effects of ISO 14001 on the textile sector. Scholars and researchers should therefore treat this gap in literature as a possible area of study so that more literature can be made available for reference depending on context. In addition to ISO 14001, the ISO has developed more specific standards to address sustainability issues facing the textile sector such as ISO 14064, which offers standards for measuring and disclosing greenhouse gas emissions. ISO 20121 also specifies requirements for sustainable event management, which can be applied to fashion shows and other industry events [35]. Global fashion weeks in major city centres such as Paris, London, Milan, and New York significantly increase carbon emissions. To calculate the textile industry's carbon footprint, [36] proposes using three methods namely the Tier approach, Life Cycle Assessment (LCA), and the DEFRA method.

The ISO 21021 standard seeks to promote sustainable event management through the three pillars of sustainability. Gucci, an Italian luxury brand is an example of a fashion brand that has chosen to get certified under the ISO 21021 certification for all its fashion shows beginning in the year 2019 [37]. ISO 9001, a standard for quality management systems (QMS), encourages sustainability in the textile sector. As global awareness of environmental issues and social responsibility rise, industries are compelled to adopt practices that not only enhance product quality but also minimise ecological impact. Incorporation of ISO 9001 into textile manufacturing processes can foster a culture of continuous improvement and accountability, ultimately leading to more sustainable practices throughout the supply chain [34]. The ISO 9001 framework includes the assessment of suppliers based on their sustainability practices, promoting more responsible choices in both materials and manufacturing methods. From a social responsibility standpoint, ISO 26000 serves as a crucial guideline for organisations seeking to integrate corporate social responsibility (CSR) into their operational strategies. Although ISO 26000 is not a certifiable standard and does not impose specific requirements, it offers comprehensive guidance on embedding social responsibility within business practices. The framework emphasises the significance of ethical conduct, stakeholder engagement, and transparency, thereby fostering more socially responsible organisational behaviour [38].

The textile industry provides job opportunities around the world, so it is known to be a labour-intensive sector. The adoption of ISO 26000 can potentially change textile production by encouraging companies to adopt more sustainable methods that consider economic viability alongside environmental stewardship and social equity. Working conditions such as those prevalent in wet processing systems in textiles pose a serious threat to the well-being of the employees. According to [39], the introduction of ISO 45001, Occupational Health and Safety Management has been one of those progressive steps in textiles aimed at reducing workplace risks. ISO 45001 seeks to ascertain that there are no hazards and risks within the various stages of textile production that will be encountered by the workers, in other words, the employers must provide safe working conditions. Adhering to the ISO 45001 standards, textile manufacturers can establish safer work environments that prioritise employee welfare while improving productivity and morale. ISO 45001 emphasises continuous development using the Plan-Do-Check-Act (PDCA) cycle. Through this approach, the organisation can systematically evaluate its health and safety practices regularly so that they create a conducive and safe working environment for its employees.

14.3.2 EU Textile Regulations

Through several laws and initiatives designed to promote a circular economy, the European Union (EU) is actively promoting sustainability in the textile business. The EU's Strategy for Sustainable and Circular Textiles (EU-SSCT) emphasises the need for eco-design standards [40], mandating that textile products be more durable, repairable, and recyclable. Studies by [3, 41] underscore the significance of this EU strategy in mitigating the environmental footprint of textiles. The EU fosters circularity within the textile sector by incentivising manufacturers to invest in technologies that transform used textiles into new fibres, which can then be spun and woven into new textile products [42]. The establishment of sustainability standards and ecolabels has empowered consumers to pinpoint environmentally friendly items, thereby boosting the demand for sustainable textile products. To oversee and regulate these hazardous chemical releases, the EU enacted the REACH regulation back in 2007. REACH's restrictions on specific harmful chemicals in textiles have reduced the industry's harmful impact on both the environment and human health. The EU-SSCT employs a variety of regulatory measures and campaigns to encourage sustainability in the EU textile sector, including the Eco-design for Sustainable Products Regulation, the Textile Labelling Regulation, and the "Reset the Trend" campaign [40]. These policies have far-reaching consequences, going beyond mere adherence and compelling EU member states to fundamentally alter the methods of textile production, marketing, and disposal. The EU aims to foster cooperation between producers, policymakers, and consumers, creating a landscape where sustainable practices are not just encouraged, but mandated.

14.3.3 United States Textile Regulations

The United States utilizes a mix of federal and state regulations and initiatives to oversee the textile industry, aiming to promote sustainability. While past methods have tended to favour immediate economic benefits over lasting environmental consequences, more recent rules hint that the industry is beginning to grasp the importance of eco-friendly practices. The Environmental Protection Agency (EPA) broadly watches over the environmental footprint of various industries across the country. EPA formulated several federal regulations specifically for the US textile industry such as Clean Water Act (CWA), Toxic Substances Control Act (TSCA), Resource Conservation and Recovery Act (RCRA), and Clean Air Act (CAA). Textile processing uses a lot of water therefore textile companies that are involved in wet processing companies in the US are required to abide by CWA rules in discharging their wastewater [43]. Textile manufacturers are required by law to set up wastewater treatment facilities to remove harmful substances like heavy metals before releasing the water. The TSCA also governs the chemicals used in textiles such as dyes, finishing chemicals, and flame retardants.

In addition to requiring testing and risk assessments, the EPA can prohibit or limit the use of textile chemicals if they endanger human health or the environment. The TSCA laws have impacted the textile sector as evidenced by the development of safer substitutes for several chemicals used in manufacturing [44]. However, issues persist in detecting and regulating all potentially harmful substances. Textile waste, which includes fabric scraps, dyeing sludge, and other by-products, may be subject to RCRA laws. The act demands the treatment of hazardous textile waste to avoid environmental damage. While many textile wastes are non-hazardous, other procedures can produce dangerous wastes that require specific handling and disposal. The EPA has also formulated standards for managing textile waste, promoting recycling and reuse whenever possible [43]. The EPA in the US administers several programs and initiatives that support sustainability in the textile industry mainly focusing on pollution prevention, waste management, and technological advancements. The agency has been key in the management of grants given to companies in the US to develop cleaner production technologies, technical assistance to textile manufacturing companies, and outreach programs to educate stakeholders about best management practices. Other regulations such as the Fabric Act regulate and promote sustainable textiles in the US, as it seeks to improve working conditions and environmental standards through the implementation of sustainable practices [31, 45]. The US-Textile Fibre Products Identification Act also mandates that all textile items have unambiguous labels that reveal the manufacturer's name, country of origin, and fibre content [31]. This gives customers the ability to make wise choices every time they buy something. Through the prohibition of child labour and the regulation of working conditions for workers, the Fair Labour Standards Act (FLSA) guarantees adherence to labour standards [46]. State-level regulations across the US are also tackling specific regional environmental worries and industry-specific hurdles. For instance, states with a large textile manufacturing presence might enforce tighter wastewater discharge standards or have particular rules about the use of certain chemicals [5, 46]. California has been a leading state in textile regulation, focusing on enhancing labour rights and promoting ethical supply chain practices. The Transparency in Supply Chains Act and the Garment Worker Protection Act are two important pieces of legislation that demonstrate the state's commitment to guaranteeing equitable treatment of employees and accountability from businesses operating inside its borders [5, 31, 45, 46]. State-level initiatives in the US are essential in promoting innovation in sustainable textile processes and modifying national policies to fit local settings [45, 46]. Sustainability is a priority for the US textile industry, which, in turn, has taken steps at self-regulation for its promotion. These include industry standards on environmental performance, certifications of sustainable products, or codes of conduct on responsible sourcing and manufacturing. The effectiveness of these initiatives varies as they depend upon the approach of every company to sustainability.

14.3.4 Sustainability in Asian Textiles

Textile manufacturing shifted from Europe and the US to Asia, as countries like China, India, and Bangladesh have become global powerhouses in textiles. China and Bangladesh collectively represent over 60% of global textile output as such, they carry a heavy burden to adopt eco-friendly measures that curb harmful practices while improving the lives of a mostly female workforce. China, indisputably the world's largest textile exporter, has established a complex and multifaceted regulatory framework to address the environmental challenges posed by its huge textile sector. Research work by Xu et al. [47] reveals a notable trend in the development of China's textile environmental policy system. This policy predominantly focuses on mitigating water pollution, a critical environmental concern in the textile manufacturing process. However, the study also points to a significant deficiency: the lack of effective regulation of environmental issues at the consumer stage of the textile lifecycle. Except for China, many of these Asian nations are ranked as developing nations and the study by Hashim et al. [48] on supply chain risks in Pakistan's textile sector offers insightful information about the larger obstacles that developing countries must overcome to adopt sustainable manufacturing techniques. These challenges, such as managing endogenous environmental risks and navigating complex supply chain vulnerabilities, are highly relevant to understanding the complexities of China's regulatory journey. With up to 70% of export earnings depending on the nation, the textile industry is a vital economic sector in many Asian nations. The FABRIC initiative supported by Germany seeks to help the Asian textile sector transition to environmentally and socially responsible production [49]. China has the largest textile industry in the world, making it a key participant in the global textile chain. This also translates to being the highest contributor to textile manufacturing waste in the form of greenhouse gases, wastewater pollution, and solid textile waste. China's textile industry responded by being the first to support the industry-level zero-carbon target [50]. Bangladesh is the second-largest exporter of apparel and textiles, following China, hence the sector serves as a crucial source of revenue for the nation.

The major destination of textile products from Bangladesh is the EU, which has made several changes to legislation governing the sustainability of imported textiles. This has forced textile companies in Bangladesh to invest in eco-friendly manufacturing procedures and also introduce fair working conditions for the employees. In particular, it is essential to create laws that support a circular economy, which reduces, reuses, and repurposes materials. Such policies will help reduce waste and consumption of resources. Asian textile giants are trying to conform to changing requirements as sustainability becomes a higher priority in important garment export markets including the US and the EU [49]. However, while some of these Asian key textile players have begun working towards sustainability, there is still a gap between policy making and policy practicing. The existing legislation in textile manufacturing in many Asian countries lacks enforcement mechanisms, making it difficult to implement sustainable practices.

14.4 Certifications for Sustainable Textiles

Certifications are essential in the expanding sector of sustainable textiles because they guarantee that goods fulfill the set ethical, social, and environmental requirements. These certifications act as trusted benchmarks for consumers, manufacturers, and retailers, providing transparency and accountability in the complex textile value chain. Certification aids in directing the textile sector toward more ethical and sustainable practices as awareness of the effects on the environment and ethical behaviour grows. The development of organic and natural fibres can have some processes that can be harmful to the environment, therefore certification of such textile products aims to control the use of pesticides, water consumption, carbon footprint, ethical labour practices, and the ecological impact along the whole process.

14.4.1 Organic and Natural Fibre Certifications

In the textile industry certifications play a key role as they ensure the product is being made following certain standardised practices. Certifications can differ depending on a specific area of application for example there are quality certifications, management certifications, and environmental certifications. Textile fibres can be grouped into two broad areas that are, natural and synthetic fibres. Synthetic fibres are considered as unsustainable as they are not biodegradable but they are cheaper than natural fibres which makes them a lucrative option for fast fashion brands. As environmental concerns gain more attention, natural fibres are becoming more and more popular in the textile business. Global Organic Textile Standard (GOTS) accreditation is one of the most reputable and well-known for organic fibres. The GOTS certification covers the entire process of cultivating organic fibres and processing them into textiles [51]. According to the GOTS certification, organic fibres used in textile items must be produced without employing genetically modified organisms, synthetic fertilizers, or pesticides. The GOTS certification also assesses how manufacturing operations affect the environment more specifically on water use, energy use, and waste management. The social criteria are also included in the standard, guaranteeing that workers in the textile supply chain receive equitable treatment while maintaining safe conditions at work. GOTS-certified products contain at least 70% organic fibres with higher percentages leading to higher certification levels for example organic versus made with organic [52]. Similarly, the OEKO-TEX Standard 100, though not specifically for organic or natural fibres plays a key part in certifying that textiles are free of harmful substances and chemicals. The OEKO-TEX certification checks for dangerous material residues, such as formaldehyde, heavy metals, and pesticides, in raw materials and the final goods [53]. Since the lack of synthetic chemicals is a key selling point of organic textiles, this certification becomes very crucial for organic and natural fibres. Ibrahim, Fawzy Ibrahim [54] reports that textile companies that earn OEKO-TEX certification, show their dedication to lessening their impact on the environment, giving customers a trustworthy way to select eco-friendly products, and setting them apart from traditional textile manufacturers. OEKO-TEX has also introduced the new Sustainable Textile Production (STeP) certification system targeting fashion brands, retailers, and textile manufacturers who seek to share their sustainability progress with the public [55]. STeP certification aims to ensure the long-term application of environmentally sustainable industrial methods and, an ideal working environment in terms of health and safety as well as social acceptance. Unlike other certification systems that primarily consider specific facets of sustainability, STeP facilitates a thorough investigation and assessment of sustainable production conditions. The objectives and criteria for the STeP accreditation are specifically tailored to the textile and garment industry. The STeP requirements are standardised globally to ensure comparability. They are continuously reviewed in case of market changes, government regulations, and any other relevant technical changes that impact the textile regulations.

In the United States, the USDA Organic certification certifies agricultural products such as organic fibres. Certification means these products, whether cotton, wool, or any other fibres, have met very strict and high standards to attain such as being grown without having any synthetic chemicals used, including GMOs, along with the prohibition of mostly all fertilizers [52]. To achieve such certification from the USDA, traceability shall be complete at each step along the supply chain. Mainly, this focuses on the agri-aspect of fibre, but also entails that the material is grown compatible with the natural environment in favour of soil quality, biodiversity, and saving on water. In the EU a similar certification the EU-Organic Logo is another significant certification for organic textiles in the European market [56]. The logo indicates that a product has been produced in line with the EU's stringent organic farming regulations, that emphasise biodiversity, soil health, and the avoidance of synthetic pesticides and fertilizers. The EU-Organic logo on a product is a mark that guarantees that the raw materials used in textiles have been grown in a sustainable and environmentally conscious manner, similar to the USDA Organic certification. Although certifications for natural and organic fibres are crucial for advancing sustainability, their application comes with difficulties and objections. For small businesses in the textile industry, who might be keen to become certified issues such as costs involved for the certification become a hurdle. Costs associated with certification training, regularising non-compliant issues, and audits for certification limit the small-scale players who might otherwise be committed to sustainable practices. Furthermore, certification procedures frequently entail several verification steps, which can be laborious and cause delays in the release of the product into the market. Certifications in textile products are useful instruments, but they are not always enforced or accepted in all markets due to differences in the measurable parameters. These inconsistencies in can act as a barrier to international trade, particularly for textile manufacturers aiming to enter regions with different standards and regulations.

14.4.2 Recycled Textiles Certification

As the textile sector adopts a circular economy approach, new products being made through techniques such as recycling and upcycling must also be certified. To improve accountability and transparency in textiles, several initiatives have been developed to standardise and authenticate the usage of recycled textile materials [21, 57]. Schumacher et al. [48] cite that a lack of a universal standard, however, remains a significant hurdle for recycled textile products, leading to a fragmented landscape of certifications with varying levels of rigour and scope. Frane et al. [49] report that the Nordic Textile Reuse and Recycling Commitment exemplifies how a regional approach to textile certification can improve the competitiveness of the textile industry. This voluntary certification promotes recycling and reusing all textile-related trash produced by the Nordic region's numerous textile firms. Third parties certify recycled textile products following the guidelines of the Global Recycled Standard (GRS) and the Recycled Claim Standard (RCS). Textile Exchange, one of the most significant non-profit groups encourages ethical and sustainability issues in the textile sector, through the promotion the Global Recycled Standard (GRS). Through the implementation of this standard, Textile Exchange recognises that recycling is essential to creating a model of sustainable production and consumption [58]. GRS's primary goals are to improve the quality of recycled textile goods and encourage less resource usage. GRS certification is available for any textile product that contains at least 20% pre- and post-consumer textile recycled materials. Braun et al. [59] report that Wear2Wear, a collaborative initiative focused on closing the loop for polyester textiles intends to create a closed-loop system for collecting, processing, and recycling worn polyester clothing into new clothing. The Wear2Wear model, while promising, requires further evaluation to assess its scalability, applicability to other textile materials, and overall environmental impact. Several governments worldwide are actively implementing policies to support the transition of the textile industry to adopt circularity [21, 59]. These policies vary significantly in their scope and approach, ranging from extended producer responsibility (EPR) schemes [60] to regulations on textile waste management, eco-design requirements, and incentives for recycling and reuse. Despite significant progress in developing certifications and initiatives for recycled textiles and circular economy practices, several challenges and limitations hinder their effectiveness in driving widespread adoption of sustainable practices. As mentioned by Schumacher et al. [48], the absence of a universally accepted standard for recycled textile content and production processes is a major impediment to the effectiveness of certification schemes.

14.4.3 Ethical and Social Compliance

Ethical issues such as low pay and long working hours in the textile industry, continue to be a serious issue that needs urgent address. Several studies, [3, 61, 62] indicate the widespread occurrence of exploitative labour practices, including cases of forced labour and hazardous working conditions in the textile value chain. According to [63], textile workers lack adequate representation and have limited bargaining power thus leaving workers vulnerable to exploitation. Initiatives aimed at improving wages and working conditions, such as the SA8000 standard [63], while laudable in their intent, face substantial challenges in their implementation and effective enforcement, highlighting the need for stronger oversight and accountability mechanisms. Tragic incidents like building collapses and factory fires serve as harsh reminders of the hazardous working conditions in many textile industries. The Rana Plaza building, a garment manufacturing facility in Bangladesh collapsed in 2013 claiming the lives of over a thousand workers [64]. This serves as an illustration of the dangers that the industry faces when certain standards are not adopted in manufacturing. The incident underscores the need for improved workplace safety regulations and their effective enforcement. The absence of adequate safety measures, combined with insufficient regulatory oversight, contributes to a high rate of workplace accidents and injuries, resulting in significant human suffering and economic losses [64]. The Fair Trade certification has been getting more attention within the textile industry, as it aims to ensure that farmers and workers are accorded due respect. Fair Trade standards guarantee sustainable environmental practices, safe working conditions, fair salaries, and social and economic fairness. Many textile brands manufacturing products from textile fibres also seek Fair Trade certification to position their products with ethical sourcing practices. The certification defends the rights not only of the workers themselves but also contributes to community development and improving livelihoods for those involved in natural fibre production. As part of their investigation into the equity of textile industry regulation Arenas et al. [65] argues that while textile companies in developing economies claim to improve working conditions for workers, there is a need for more transparency and accountability in meeting these criteria for fair trade. The Worldwide Responsible Accredited Production (WRAP) is a social compliance accreditation for apparel manufacturers. Textile companies that want to export to the US market must have this certification and the certification is growing in acceptance in the EU region [66]. WRAP incorporates the spirit or language of relevant International Labour Organization (ILO) conventions. It is based on 12 principles derived from well-recognised global workplace standards, local laws, and workplace regulations.

14.4.4 Other Certifications

The Sustainable Apparel Coalition (SAC), an association of worldwide fashion brands, developed the Higg Index. The index gives fashion and textile enterprises a comprehensive method to assess and enhance their economic, social, and environmental effects. The Higg Index contains three distinct modules: the facility module, the product module, and the brand and retail module. Each of these modules offers a score that represents the sustainability performance of the business or product. These scores serve as a baseline for companies to benchmark their practices, identify areas for improvement, and track their progress as they implement sustainability initiatives. Despite measuring sustainable practices in textiles, Tabuchi [67] argues that the Higg Index favours synthetic materials more, instead of promoting natural fibres. The Higg Index is voluntary and only identifies areas or opportunities for improvement as it gives a rating of the various stages of production in textiles. Much like the Higg Index, the Bluesign system represents a proactive strategy in textile manufacturing, aimed at eliminating hazardous chemicals and materials from production processes before they even enter the supply chain. Bluesign is founded upon five core principles: consumer safety, control of water and air emissions, resource efficiency, worker protection, and the removal of harmful substances. By adhering to these principles, textile manufacturers can substantially reduce their environmental impact while ensuring that their products are produced in ways that safeguard both human health and the environment. Textile standards and certifications differ in terms of their scope. Table 14.3 gives a summary of some of the certifications popular in the textile sector and their scope.

Initiatives like Cotton Made in Africa (CmiA) offer an alternative method of certifying sustainable textiles in Africa. Unlike GOTS, which prioritises organic production, CmiA focuses on improving the livelihoods of cotton farmers in Africa while promoting sustainable agricultural practices. CmiA prioritises the training of farmers on techniques that can improve productivity and reduce the environmental impact of cotton farming in Africa [68]. Africa is a continent with a rich historical background, deeply embedded in the continent's cultural heritage and economic development. Despite being one of the major sources of natural fibres that are utilised to make various textiles worldwide, the continent lags in the promotion of sustainability initiatives in textiles. The African Organisation for Standardisation (ARSO) has been at the forefront of developing textile standards that seek to harmonise the sector at continent level after realising that the African textile industry lacks standardisation.

While certification is crucial for encouraging eco-friendly practices in the textile sector, there is a problem with "greenwashing," where companies may use certification labels without fully complying with the standards. Some of the certifications discussed depend on data or information from the manufacturers themselves thus providing an opportunity to misrepresent information by the fashion or textile companies. Certifications such as GOTS, Fair Trade, and Higg index are key in promoting sustainability in the textile industry. In cases where fashion brands or textile companies misrepresent their data, such certifications become prone to misuse. To mitigate against such malpractices the certifying bodies must maintain an objective and rigorous audit process so that consumers are always informed correctly about what each specification represents.

Table 14.3 Textile certifications and their scope

Certification	Sign	Scope
FAIR TRADE	FAIR TRADE CERTIFIED	– Better prices for all players in value chain – Decent working conditions for employees – Fair terms of trade for farmers and workers
GOTS	GLOBAL ORGANIC TEXTILE STANDARD · GOTS ·	– Ensure organic status of textiles – Environment and social impact – Correct labelling of products
GRS	Global Recycled Standard	– Verifies the recycled content of textile products – Asses the production process for any harmful chemicals – Increase recycled materials in production – Reduce harmful chemicals
HIGG INDEX	Higg Index	– Identify areas of improvement – Benchmark processes against competitors – Report the score of each business process
BLUESIGN	bluesign®	– Compliance to environmental and health – Consumer safety, waste control and occupational health – Hazardous chemicals in production
OEKO-TEX	OEKO TEX® STANDARD 100	– Product safety standard – Product type and use versus harmful substances
STeP	SUSTAINABLE TEXTILE PRODUCTION STeP CERTIFIED BY OEKO-TEX	– Social and environmental aspects in textile production – Responsibility towards employees and the environment
Cradle to Cradle (C2C)	CERTIFIED cradletocradle	– Focuses on the circularity of products – Safe materials for humans and environment – Reutilisation of material through design – Efficient use of water and renewable energy – Social responsibility of the company
CMiA	COTTON MADE IN AFRICA	– Ensure that cotton is grown under rain-fed conditions – Responsible and effective cultivation – Harvested by hand

(continued)

Table 14.3 (continued)

Certification	Sign	Scope
EU organic label		– Sustainable fibre production – Minimum pollution during production – Control of hazardous substances during manufacturing – Durable products
Fair wear		– Code of labour practices – Targets the garment supply chain – Decent pay for employees – Respect of human rights
EU eco label		– Less harmful to the environment than similar products – Allow customers to recognise ecologically responsible products

14.5 Standards and Sustainable Practices in Textiles

The nature of the textile industry's production methods has a significant impact on the environment [41, 44, 45, 62]. Various processes are carried out from the cultivation of the natural fibres to the processing of the synthetic fibres. In addition to vast volumes of water and energy, numerous dyes and chemicals are used in all activities. This is further worsened by the fast fashion model adopted by the big fashion brands which supports high-volume production of inexpensive garments with very short life spans. The sector has had a detrimental effect on the environment as a result, as shown by the enormous volumes of waste in landfills and greenhouse gas emissions. These challenges have led to more scrutiny from consumers, regulators, and climate activists as issues of climate change are affecting everyone prompting many companies to implement sustainability criteria to enhance their environmental and social performance [69]. Certifications and standards such as GOTS, OEKO-TEX Standard 100, Fair Trade Certification, Bluesign, and Cradle to Cradle [52, 53, 56] have been adopted by some textile manufacturing companies. In addition to adhering to standards, the textile industry is also adopting various sustainable practices that include eco-friendly materials, circular economy principles, transparency and traceability, and ethical labour practices [54, 70] to minimise its environmental and social impact. Textile production is considerably more concentrated in developing nations, where issues such as poor working conditions, low salaries, and child labour are common in the industry, necessitating an urgent call to action for sustainable practices that address environmental and social concerns.

14.5.1 Textile Certification and Product Quality

Certifications in the textile industry, such as ISO 9000 and OEKO-TEX Standard 100, not only improve operational processes but also align products with consumer expectations for sustainability and safety. Customers are more likely to buy a product that is safe for their use. This is very important for textile products such as those for infants, newly born babies and other individuals who might have skin problems. The adoption of the ISO 9000 certification in the textile industry has resulted in improved processing methods and product characteristics [71]. The OEKO-TEX Standard 100 promotes consumer safety on textile products they purchase as it ensures that they are free from harmful substances [53]. This certification of textile products appeals to environmentally conscious consumers, thus enhancing brand reputation and marketability. Compliance with such eco-labels supports sustainable practices within the supply chain, contributing to overall product quality. The new developments in e-textiles also call for standardised procedures to ensure the safety of these new textiles. According to Iftekhar Shuvo [72], there are several standards in e-textiles depending on the area of application such as those developed by ASTM, and AATCC. Through textile certification, consumers are assured that the products have been independently assessed and meet the predetermined benchmarks for various attributes. This promotes trust and transparency in the textile business as it helps the consumer to make a better-informed decision about the products they purchase.

14.5.2 Textile Standards and Sustainable Development Goals (SDGs)

The link between textile standards and the Sustainable Development Goals (SDGs) is important for the development of sustainable practices. The textile industry significantly relates to various SDGs, particularly in areas such as responsible consumption, environmental protection, and social equity. Sustainability standards like GOTS, RCS, and GRS provide frameworks for aligning textile production with SDGs [73]. By addressing poverty alleviation, water conservation, clean energy adoption, ethical labour practices, and climate action, the textile industry can contribute to achieving global sustainability goals by 2030. Table 14.4 summarises the scope of SDGs relevant to the textile industry and their corresponding certification or standard.

14.6 Case Studies

With over 5000 outlets across 74 countries, H&M is one of the largest fashion stores globally [74]. H&M promotes sustainability in the textile value chain through various standards and certifications they have adopted. As one of its pledges in the year 2025,

Table 14.4 Textile Standards and SDGs

UN-SDG	Scope	Textile standard/certification
SDG 1	No poverty	Africa organic cotton, better initiative cotton (BCI)
SDG 6	Clean water and sanitation	Zero discharge of hazardous chemicals (ZHDC)
SDG 7	Affordable and clean energy	Higg index, Bluesign, global organic textile standard (GOTS)
SDG 8	Decent work and economic growth	Global Organic Textile Standard (GOTS), Fair Wear Foundation (FWF), Fair Trade Certification (FTC)
SDG 12	Responsible consumption and production	Recycled Claim Standard (RCS), Global Recycled Standard (GRS)
SDG 13	Climate action	Global Organic Textile Standard (GOTS), OEKO-TEX STeP, EU Ecolabel, Cradle to Cradle (C2C)

H&M has committed to sourcing regenerated cellulosic fibres with a Forest Stewardship Council (FSC) certification. H&M hopes to address deforestation and forest depletion with this pledge. In 2022, H&M's newborn fashion collection was certified for the Cradle to Cradle standard. There are five quality categories included in the standard: material health, product circularity, renewable energy, water consumption, and social justice. Sofia Löfstedt, Head of Kidswear Design at H&M, claims they earned the gold certification by ensuring the non-presence of harmful chemicals throughout the supply [75]. All the pieces for the newborn collection were 100% biodegradable and used 100% recycled water and renewable energy. This collaboration not only ensured the safety of the products but also promoted eco-friendly practices within the fashion industry. Despite the positive drive of H&M on sustainability in the textile value chain, the company faces inherent tensions in its quest to seek financial growth versus its commitment to sustainability. As pointed out by [76], while H&M has made noteworthy attempts to improve its sustainable performance, including initiatives related to recycled materials and carbon neutrality, controversies surrounding labour practices and environmental impacts in its supply chain persist. According to Ponte [77] despite the positive sustainability drive of H&M, the Quartz investigation on the Higg Index score of H&M raises more concerns as they claim H&M scores were inaccurate or inflated. Given the size of H&M's business and its global reach, challenges are common and present in the management of its supply chains. The intricacy of its supply chain makes it challenging to monitor and implement sustainability standards across all tiers of its manufacturing process. Cases like those highlighted by Quartz about H&M's sustainability certification method put the brand in the limelight. The question to the company becomes, "What can the brand do to improve the certification and enforce standards to its various suppliers in the chain." This is a major challenge facing most fashion brands that operate in similar environments. Despite reported unsafe working conditions and low wages given to textile employees in Bangladesh, brands such as H&M heavily depend on the country for apparel manufacturing. This highlights some of the challenges facing fashion brands in managing ethical and environmental standards in a global supply chain.

A study by Zimon et al. [34], reports on the application of ISO 9001 and ISO 14001 in textile firms in three European countries: Poland, Slovakia, and the Czech Republic. The study revealed that there were huge improvements in textile supply chains after adopting ISO 9001 and ISO 14001. Integrating quality and environmental management systems is essential to achieving holistic improvements in sustainability performance as shown by these ISO standards. Various measures were used to assess how ISO standards affected different supply chain processes in the textile industry. The case study demonstrates the importance of integrating quality and environmental management systems (EMS) into sustainability management strategies. The simultaneous implementation of ISO 9001 and ISO 14001 highlighted the interconnectedness of various business operations and the benefits of a comprehensive approach to sustainability. Findings from the three case studies as reported by Zimon et al. [34] have implications for companies seeking to make supply chains more sustainable, suggesting that the combined implementation of these standards can result in notable gains in sustainability performance and efficiency.

As reported by Narayanasamy [78], Armstrong Knitting Mills in India is a vertically integrated textile company that has embraced sustainability from the cultivation of the fibres to the actual production of the fabric. Among the certifications achieved by the company are the GOTS, FAIRTRADE, OEKO-TEX, GRS, and EU flower certification. Armstrong invested in renewable energy with solar and wind being the key sources of power for its manufacturing and they produce more power than they consume [78]. After realising the environmental damage caused by plastic waste the company has embraced using recycled PET polyester in its production thus attaining GRS certification for its products. The company has been one of the first textile entities to be awarded the Carbon Neutral Footprint mark in India. Armstrong's GOTS certification demonstrates their commitment to environmentally sustainable production methods. In keeping with Armstrong's mission to protect the environment, it guarantees customers that the textiles are made with natural resources in a responsible and considerate manner. Such deliberate efforts by textile manufacturing companies to attain sustainable certifications increase customer confidence in purchasing textile products as they view this as key in managing environmental impact.

According to the GOTS report for 2023/2024, there was an increase of 21% in textile companies certified under GOTS in Sri Lanka [79]. GOTS claims their certification reduces risk, enhances wastewater management, reduces expenses, and provides value while ensuring regulatory compliance, efficiency, investor trust, quality assurance, and third-party credibility [79]. Fashion brands such as POMP, known for its organic cotton and recycled products have achieved certification under the OEKO-TEX Standard 100 and SA 8000 standard. POMP has also taken a step further by establishing a recycling program that enables customers to return their used, unwanted clothing by simply following the directions on the care label, which include a Quick Response (QR) code. Danish brand Organic Basics is known for its sustainable underwear products. This fashion brand is GRS and GOTS certified as they use organic and recycled materials such as nylon and tencel which provide good comfort for undergarments. It is common for fashion brands to proudly display their certification mark on the sewn-in garment label and the detachable sales tag. It is crucial to highlight that certification and standards in the textile industry should be content-specific as companies have unique challenges and processes.

The cases presented illustrate that successful textile sustainability standards demand a multifaceted approach. While the adoption of specific standards, such as ISO 9001 and ISO 14001, can contribute to improvements in textile supply chain management, a holistic strategy that goes beyond compliance is crucial. This may entail effective stakeholder involvement, management's dedication to ongoing improvement, and incorporating sustainability into the fundamental business model. The fast fashion model's inherent problems underscore the necessity of a fundamental change in patterns of production and consumption toward more sustainable ones. For this shift to be successful, the textile industry needs to collaborate in crafting standards and certifications that can have a positive outcome in the sector's sustainability drive.

14.7 Role of Technology in Compliance

Several changes in the textile industry have been due to technological innovations. With consumer interest in high-quality textiles that are eco-friendly and ethically produced rising, the need for compliance and stringent oversight comes into focus. As regulations and consumer expectations mount, standards and certifications in textiles seek to mitigate concerns regarding product quality, occupational health and safety, environmental impacts, and labour practices so that textile businesses achieve compliance with regulations and meet consumer demands. Due to the intricate web of certifications and standards within the textile industry, effective enforcement and monitoring systems are necessary. Also, compliance has been quite challenging for manufacturers, brands, and other stakeholders from the global textile community because of the intricacy of the global supply chains and the constant changes to rules that need to be met. Funding technological investments has become a key attribute for companies that seek to improve their standards and compliance procedures. Several digital tools such as AI predictive analytics and blockchain for supply chain have become crucial in textile certification. Technology integration in the textile industry increases accuracy and efficiency and also promotes better sustainable behaviour and accountability.

14.7.1 Blockchain Technology in Textiles

Agrawal et al. [69] define blockchain technology in textiles as "*a decentralised ledger that ensures transparency and material traceability throughout the textile industry's supply chain*". This technology aids in tracking standard compliance and confirming the legitimacy of certifications. The inability to change data once stored on the blockchain increases confidence in certification procedures. Blockchain generates a verifiable and impenetrable record by documenting every transaction and movement of textiles, from raw material sourcing to the final product's sale [80, 81]. This increased transparency is instrumental in ensuring compliance with ethical sourcing standards, environmental regulations, and product quality certifications [82]. A fashion consumer can scan a QR code on a garment and trace its journey, verifying the authenticity of organic cotton claims or confirming fair labour practices. One of the world's leading fibre producers Lenzig has adopted blockchain technology to facilitate greater traceability of its fibres throughout the supply chain [83]. The inherent immutability of blockchain data significantly reduces the risk of data manipulation, fostering trust among all stakeholders, including consumers, brands, manufacturers, and suppliers. This enhanced trust is critical in addressing counterfeit products and unethical labour practices, common in opaque supply chains. The potential applications of blockchain extend beyond simple product tracking; it can also be used to manage contracts, track certifications, and verify payments, thereby streamlining compliance processes across the entire supply chain. However, investment in blockchain technology has its challenges such as the scalability of blockchain technology for the massive volumes of data generated by global textile supply chains [81], the need for widespread adoption across the industry, and the potential for technical complexities and high implementation costs especially in developing economies where textile production thrives. In Fig. 14.3 a typical blockchain system in the textile business summarising the role of each key player in the chain and the data they have to input to the system. There are real time updates in all the stages and the information can be accessed from the cloud storage.

14.7.2 Real-Time Tracking in Textiles

Internet of Things (IoT) gadgets and radio-frequency identification (RFID) tags provide an additional means of tracking and monitoring textile products in real time during their lifetime [84, 85]. RFID tags are tiny electronic tags which may be sewn onto individual garments or bales of fabric and they broadcast identifiers readable by RFID scanners. Attaching RFID tags to the textile item makes it possible to track each item meticulously through the supply chain, eliciting any real-time data about its whereabouts, movements, and status [86]. Sensors deployed all over production lines can collect information on temperature, humidity, and material characteristics so that manufacturers can timely detect and ensure corrective actions for any deviation

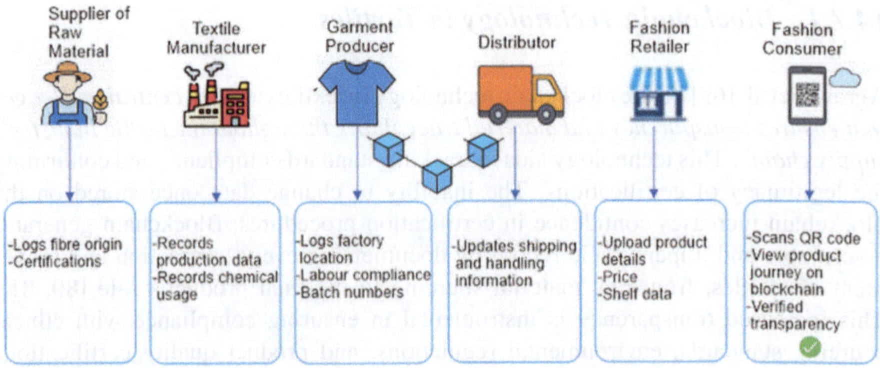

Fig. 14.3 Digital traceability in the textile supply chain using blockchain

from the set standard. The IoT networked devices installed inside the production or logistics system have access to an extremely large volume of data ranging from environmental parameters to machine performance and energy consumption. This huge volume of data can ensure environmental standard compliance, production process optimisation, and overall efficiency improvement. Artificial Intelligence (AI) has become one of the key technological advances impacting various industrial sectors worldwide. Analysis carried out on data from textile industrial processes by AI systems can identify any deviations from the established quality standard or parameter hence allowing immediate adherence and corrections. Historical data can also be utilized by AI to predict any possible future challenges for the textile industry regarding compliance with specific standards so that manufacturers take anticipative actions in managing such issues. The integration of RFID, AI, GPS, and other IoT technologies provides a very robust tool for improving visibility throughout the supply chain which in turn would help companies in textiles to be more proactive on potential issues of compliance waste and inefficiency deliver on time. Investing in these technologies improves visibility and transparency in the textile industry. Beyond ensuring certification of textile products it also helps management in making better decisions that impact the quality of their products. Investment in technology has a financial cost implication for the company, thus it is the duty of top management to look at its long-term objectives and how it views standardisation and certification expectations.

14.7.3 Automation and Quality Control

Automation and robotics are transforming multiple facets of textile manufacturing, resulting in substantial enhancements in efficiency, quality, and compliance. Traditionally quality control methods employed in the textile industry heavily rely on manual inspection which is time-consuming, subjective, and prone to mistakes.

Automated quality control systems that leverage machine learning, computer vision, and AI [80, 86] offer possible improvements to the quality process for analysing textile products for defects, inconsistencies, and deviations from specified standards. According to Sikha et al. [75], machine learning algorithms can be trained to identify specific types of defects, improving the accuracy and consistency of inspections. In textile manufacturing, integrating automation with artificial intelligence (AI) enhances accuracy, expedites the inspection process, and lowers the possibility of non-compliant products coming off the production lines. Automated quality control in the textile industry is crucial in guaranteeing adherence to safety requirements, as even minor product faults can result in significant repercussions to the company. The textile industry is also synonymous with repetitive tasks that require high attention to detail along the value chain. According to Oliveira Neto [87], such repetitive tasks can be assigned to robots as they can improve the consistency and efficiency of the production system. This increased consistency in textile production contributes significantly to compliance with standards related to product specifications and manufacturing processes. In textile wet processing, especially the dyeing process, colour matching is key for the reproducibility of results. The established colour standards have to be matched with a slight tolerance deviation allowed. Using robots or automated systems increases the accuracy of matching these colours thus maintaining the standard. Robots can also be used in other processes in garment construction pattern cutting and sewing especially garments with 3D shapes which might be difficult to handle. For fashion brands that have a global presence which requires quick and accurate logistics automation and robotics improve the warehousing operations thus reducing the risk of damage or loss of materials. These enhancements to the textile value chain directly affect the company's quality rating, but they are also expensive because they call for a large investment and training.

14.7.4 Digital Platforms for Compliance in Textiles

Digital platforms and software solutions are becoming increasingly vital in optimising compliance management operations in textile business. Textile Exchange is a non-profit organisation that strives to assist the fashion, textile, and associated industries in lowering their carbon footprint. They designed a digital platform, Digital Trackit (dTrackit) to enable users to authenticate the scope of any standard and transaction certifications within the textile value chain. Fashion brands certified with Textile Exchange can easily access their certifications and traceability data about their suppliers for all their products in one place. Specialised software solutions are being developed to assist the textile industry in better managing its compliance expectations. The Eco-design for Sustainable Products Regulation (ESPR), by the EU, has established strict guidelines for the fashion and textile sector. Despite the commendable efforts by the EU [88] argues that the EU does not provide the textile industry with the necessary resources to help them comply. Consequently, textile companies and brands collaborate with third-party solution providers to establish the mandatory processes and infrastructure for compliance. Solutions such as Libryo's regulatory tracking software help textile and clothing companies easily understand their compliance status and relevant legal requirements throughout the supply chain. Retraced is also another compliance management platform for the textile and fashion industry which allows them to digitise and trace their supply chains. Brands such as VAUDE, Desigual, Calzedonia, Victoria Secrets, and Marc O'Polo have used the Retraced platform in their operations as it empowers them to control their compliance information and obtain complete transparency regarding the raw materials utilised in each of their product lines. Quality management systems such as QUONDA, iAuditor, TrackWise, and C2P do provide options for managing quality in the whole textile system to ensure it adheres to the set standards and can alert the user of any deviations in real-time. These systems can generate notifications of potential compliance problems, enabling the textile company to handle these issues proactively and avert non-compliance. These digital compliance management solutions can streamline compliance programs for all textile processes which helps to cut down costs and improve process efficiency.

Implementation of technology for compliance in textile business depends on several factors such as regulatory environment and technological infrastructure. To promote fair competition and trade in the global textile sector, there is a need for harmonisation of standards and regulations. Technological platforms can be used to support harmonisation as they can create an opportunity for the exchange of knowledge and best practices between diverse textile companies across different countries and regions. To keep textile businesses worldwide informed of the most recent needs, digital platforms can be utilised to spread information on best practices, international standards, and regulatory advancements. Achieving this harmonisation of textile standards and regulations calls for all players from government, industry and regulatory bodies to work together towards a common solution.

14.8 Challenges in Adopting Sustainable Certifications in Textiles

The pressure facing the textile industry to adopt sustainable practices stems from a confluence of factors such as growing environmental consciousness among consumers and social costs associated with manufacturing textiles [89], increased regulatory scrutiny of industry practices [28, 90], and the inherent unsustainability of current linear production models [89, 91]. Eco-friendly products are gaining popularity due to consumers' increasing demands for ethical sourcing and transparency in textile products. But to meet these expectations the textile manufacturers and other key players along the value chain have to overcome various challenges as the standards and certifications differ. Figure 14.4 shows the factors that affect textile companies in the adoption of certifications and standards through the fishbone diagram.

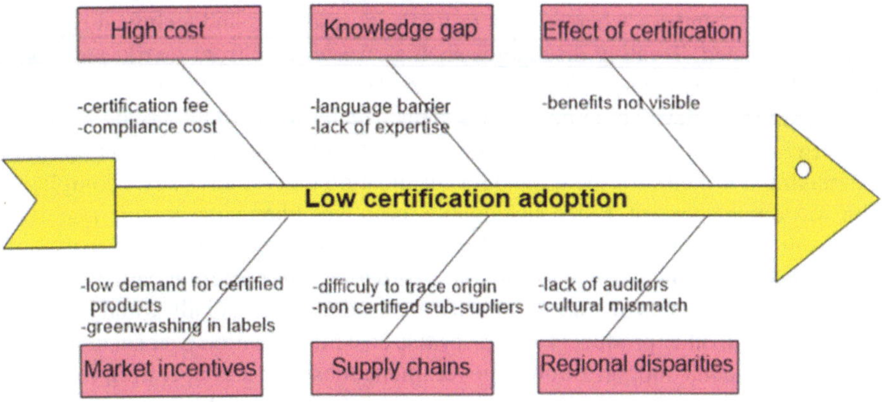

Fig. 14.4 Barriers to textile certification

14.8.1 Complex Landscape of Textile Certification

There are many sustainability certifications available in the textile industry and each has its own set of requirements, criteria, and scope [34, 45, 70]. This multiplicity of certifications and standards creates significant complexities for textile manufacturers and consumers. Some of the currently available textile certifications focus narrowly on specific aspects of sustainability, such as organic cotton farming practices [51, 92], or the use of recycled materials [58], while others adopt a broader, holistic approach encompassing all the three key pillars of sustainability [45, 47]. The absence of standardisation among various certification schemes presents a big challenge, which can result in confusion among the consumers hence undermining the overall objective of certification in textiles. This has motivated the formulation of regional standardisation bodies such as the European Committee for Standardisation, African Organisation for Standardisation (ARSO), and Sustainable Textiles of the Asian Region (STAR Network). These bodies seek to harmonise textile certification and standardisation at regional level so that challenges such as interoperability are managed.

14.8.2 Economic Barriers

The majority of economies around the world now rely heavily on SMEs for growth, and the textile sector is no exception. Synonymous with all SMEs in other sectors, SMEs in the textile industry face challenges such as access to finance and this poses a huge hurdle for them to overcome if they want to adopt sustainable certification. The costs involved in obtaining and maintaining a certification, including audits, training, and compliance adjustments, can be substantial, placing a disproportionate burden on these smaller businesses with limited resources [93]. This is especially true in developing countries where access to financial resources and technical expertise may be restricted. Treating discharges such as wastewater from textile processing requires huge investment by the manufacturer. The financial cost can be too high for manufacturers in developing economies. This usually leads to the textile manufacturer discharging untreated water into the nearby streams thus causing water pollution and making it difficult for them to attain certification that requires strict compliance with waste management. Since sustainable initiatives like using organic materials, establishing fair labour standards, and reducing waste come with higher costs, eco-friendly textile products may become less affordable. This cost implication becomes a barrier to market entry for sustainable textile producers, especially when they have to compete with other producers who prioritise cost minimisation over sustainability. Providing financial support through government subsidies and operating loans can help manufacturers overcome this obstacle and shift to sustainable practices. On the other hand, creating a successful market for certified sustainable textile products depends on manufacturers persuading consumers to pay extra for the products made sustainably.

14.8.3 Traceability and Supply Chain Management Challenges

The complicated and sometimes globalised nature of textile supply chains creates significant logistical and technological obstacles to the implementation of sustainable practices and certification. To achieve true traceability of textile products, there is a need for a robust data management system and effective collaboration across multiple players in the supply chain [83, 94]. Many supply chains lack visibility and openness, which makes it challenging to confirm the authenticity of sustainability claims and guarantee adherence to certification requirements as stated by companies on their products. This opacity creates opportunities for unethical practices, such as using non-certified materials or exploitation of labour, to go undetected. Despite possible solutions offered by investing in technologies such as blockchain to improve traceability in textile supply chains, barriers related to cost, infrastructure limitations, and the need for industry-wide standardisation pose a challenge. Overcoming these logistical and technological barriers requires a collaborative effort between businesses, technology providers, and regulatory bodies to develop and implement effective and affordable traceability solutions for the textile industry.

14.8.4 Social and Ethical Dimensions

Sustainable textile certifications and standards incorporate social and ethical criteria related to labour practices, worker rights, and fair trade [63, 93]. The elimination of child labour, forced labour, safe working conditions, and fair compensation are some of the main concerns covered under social equity of sustainability in textiles. However, the monitoring and enforcement of all these standards across the textile industry is a challenge due to its complex multi-tiered supply chains. The prevalence of unethical labour practices in certain nations, often hidden within opaque supply chains, undermines the credibility of certifications and raises concerns about the effectiveness of existing monitoring mechanisms [95]. Addressing these social and ethical issues requires a holistic approach, including stricter regulations, improved auditing practices, greater transparency, and increased consumer demand for ethically sourced products. Collaboration between fashion brands, textile manufacturers, and governments is essential for developing and implementing effective mechanisms to ensure fair labour practices throughout the textile supply chain. The development of robust and transparent traceability systems, as discussed in the previous section, is also crucial for enhancing accountability and preventing the exploitation of workers.

14.8.5 Consumer Behaviour

Several authors [27, 28, 50, 89] report that consumer behaviour is key to the success of sustainable textile certifications as it acts as a driver for implementation. While knowledge of sustainability issues is growing, customer purchase decisions are frequently impacted by considerations other than sustainability, such as price, style, and perceived quality of the textile product. The price difference between a product deemed eco-friendly and that which is not can make one purchase what they can only afford. As mentioned earlier, commonly, certified sustainable textile products have a premium price and consumers might not be willing or unable to pay that price [89]. Purchase decisions by consumers are informed by what they already know, therefore if consumer understanding of different certifications is limited it can also hinder their ability to make informed choices on the certified products. To circumvent such challenges there is a need for effective consumer educational campaigns to raise awareness of the importance of various certifications so that consumers will be able to differentiate between genuine and misleading sustainability claims. Furthermore, investigations into consumer attitudes towards recycled textile products and the prospects for upcycling in textiles can guide the creation of sustainable business models that correspond with consumer preferences and values.

14.8.6 Policy and Regulatory Frameworks

Government regulations and policies are crucial in supporting the adoption of sustainable certifications in the textile industry [3, 96]. Tax rebates and grants can mitigate the expenses incurred during certification processes, especially in developing economies. Governments and regional standards bodies should create regulations related to labour standards, environmental protection, and textile product labelling as this help create a level playing field and prevent unfair competition from businesses that do not prioritise sustainability. The effectiveness of these regulations largely relies on enforcing compliance and the ability of to police non-compliance, primarily within the multifaceted international supply chains endemic to the textile industry. There is also a need for international cooperation in establishing harmonised standards and regulations as these ensure that sustainable certifications are recognised and respected across different countries and regions. More developed economies such as the US and EU have managed to set minimum requirements for various textile products, which other regional groupings need to match so that there is harmony in standards and certifications. The current loopholes in global textile standards and certifications create a gap that can be exploited by businesses seeking to avoid compliance with sustainability requirements. Therefore, strengthening international cooperation and harmonising regulations are crucial steps toward creating a global market for sustainable textiles.

14.9 Future Directions in Standards and Certifications for Sustainable Textiles

Regulatory pressures, consumer demand, and environmental concerns are pushing the textile sector to embrace sustainable business methods. The Sustainable Development Goals (SDGs) 8, 12, 13, and 16 provide the required basis for the textile sector to produce standards and regulations to accomplish the targets established by the United Nations. With the increasing emphasis on sustainability, future certification and regulation for the textile sector are bound to change to ensure responsibility, transparency, and measurable impact. One of the most challenging issues sustainable textiles face is the fragmentation of certification schemes in different regions. To solve this problem in the future, harmonising international standards through establishing a cohesive framework that guarantees uniformity and lessens ambiguity for consumers and enterprises in the textile industry can be implemented. Governments play a key role in enforcing mandatory regulations thus in future directives they may require mandatory disclosures on environmental and social impacts, stricter penalties for non-compliance, and greater transparency in textile supply chains. A typical example of government enforcement is the EU's forthcoming ESPR which is set to impose strict sustainability criteria on textile production. Future certifications and standards can mandate the adoption of technologies that improve accountability and transparency in the textile industry. Technologies such as blockchain, AI, machine learning, IoTs and digital product passports provide real-time data on sourcing, production, and lifecycle impact, thus their integration in the textile sector is key.

Standard regulatory frameworks in the textile industry should emphasise the promotion of circularity in textiles, extended producer responsibility (EPR), take-back programs, and recycling mandates. Certifications for textiles may also expand to verify the claim by the manufacturer on the product's end-of-life sustainability as this incentivises manufacturers to design for longevity, reparability, and recyclability. Future regulatory frameworks must broaden their focus to include new alternatives including bio-based fibres, lab-grown textile fibres, and low-impact synthetic fibre alternatives as sustainable materials innovation continues to emerge. Specific certifications for such new innovative fibres are required to evaluate these materials' sustainability over their whole life cycle. One of the key challenges facing consumers is the interpretation of sustainability information on clothing labels. Standards may establish a globally acceptable label especially for the eco-label so that consumers find it easy to identify sustainable products without much facing many challenges.

14.10 Conclusion

As the world is becoming one global village, the future of sustainable textile certifications and regulatory requirements are moving toward digital integration, more harmonisation, and comprehensive sustainability evaluation. This chapter explored the various standards and certifications in the textile industry and how they have affected the sector. The chapter also explored how technology is impacting the standardisation and certifications in the textile industry. Challenges to certification, common in the textile industry were also elaborated on and possible solutions. The issue of creating standards and certifications that have a common goal is key to attaining the UN SDG goals. Governments and environmental agencies worldwide play a crucial role in the formulation of these standards hence their full commitment and participation is important. The chapter concluded by outlining the future of standardisation and certification within the industry emphasizing harmonisation of the regulatory standards in different regions.

References

1. Deshmukh, S. M., S. R. Dhokpande, A. Sankhe, and A. Khandekar. 2024. Effluent wastewater technologies for textile industry: A review. *Reviews in Inorganic Chemistry*. https://doi.org/10.1515/revic-2024-0046.
2. Naqvi, S. L. H., M. Nadeem, F. Ayub, A. Yasar, S. H. Z. Naqvi and R. Tanveer. 2024. Social and environmental impacts in textile production. In *Dye pollution from textile industry*, ed. P. Singh, 423–453. Springer. https://doi.org/10.1007/978-981-97-5341-3_19.
3. Plakantonaki, S., K. Kiskira, N. Zacharopoulos, I. Chronis, F. Coelho, A. Togiani, et al. 2023. A review of sustainability standards and ecolabeling in the textile industry. *Sustainability* 15: 11589. https://doi.org/10.3390/su151511589.
4. Botta, V., and M. Forbicini. 2023. *Voluntary certification schemes in textiles—three cases and lessons learned*. Brussels.

5. Lundy, A. 2022. Textile pollution in the United States: An analysis of the environmental impacts of textile waste and potential solutions. *The Public Purpose Journal* 20: 130–137.
6. Acquaye, R., R. Kanyire, B. Eghan, and K. Fobiri. 2023. Consumer attitude and disposal behaviour to second-hand clothing in Ghana. *Scientific African* 21. https://doi.org/10.1016/j.sciaf.2023.e01887.
7. Sumo, P. D., I. Arhin, R. Danquah, S. K. Nelson, L. O. Achaa, C. N. Nweze, et al. 2023. An assessment of Africa's second-hand clothing value chain: A systematic review and research opportunities. *Textile Research Journal* 93: 4701–4719. https://doi.org/10.1177/00405175231175057.
8. Davis Sumo, P. 2024. Textile waste management and recycling opportunities and challenges for Africa. *The Holistic Approach to Environment* 14: 57–70. https://doi.org/10.33765/thate.14.2.3.
9. Sinha, P., D. G. K. Dissanayke, R. P. Abeysooriya, and B. H. N. Bulathgama. 2022. Addressing post-consumer textile waste in developing economies. *The Journal of The Textile Institute* 113: 1887–1907. https://doi.org/10.1080/00405000.2021.1954428.
10. Lakshmanan, S. O., and G. Raghavendran. 2017. Low water-consumption technologies for textile production. *Sustainable Fibres and Textiles* 243–265. https://doi.org/10.1016/B978-0-08-102041-8.00009-3.
11. Mikucioniene, D., D. Mínguez-García, Md. R. Repon, R. Milašius, G. Priniotakis, I. Chronis, et al. 2024. Understanding and addressing the water footprint in the textile sector: A review. *AUTEX Research Journal* 24. https://doi.org/10.1515/aut-2024-0004.
12. Kumar, P. S., S. M. Prasanth, S. Harish, and M. Rishikesh. 2021. Industrial water footprint: Case study on textile industries. In: Water footprint. Environmental footprints and eco-design of products and processes, ed. S. S. Muthu, 35–60. Springer. https://doi.org/10.1007/978-981-33-4377-1_2.
13. Adebayo, G. B., G. A. Otunola, and T. A. Ajao. 2010. Assessment and biological treatment of effluent from textile industry. *African Journal of Biotechnology* 9: 8365–8368.
14. Abu Bakar, N., N. Othman, Z. M. Yunus, Z. Daud, N. Salsabila Norisman, and Hisham M. Haziq. 2020. Physico-chemical water quality parameters analysis on textile. *IOP Conference Series: Earth and Environmental Science* 498: 012077. https://doi.org/10.1088/1755-1315/498/1/012077.
15. Tüfekci, N., N. Sivri, and İ Toroz. 2007. Pollutants of textile industry wastewater and assessment of its discharge limits by water quality standards. *Turkish Journal of Fisheries and Aquatic Sciences* 7: 97–103.
16. Jahan, N., M. Tahmid, A. Z. Shoronika, A. Fariha, H. Roy, Md. N. Pervez, et al. 2022. A comprehensive review on the sustainable treatment of textile wastewater: Zero liquid discharge and resource recovery perspectives. *Sustainability* 14: 15398. https://doi.org/10.3390/su142215398.
17. Prasannamedha, G., and P. Senthilkumar. 2021. Chemical compliance and regulations in textiles and fashion. *Chemical Management in Textiles and Fashion* 135–154. https://doi.org/10.1016/B978-0-12-820494-8.00007-1.
18. Patra, A. K., and S. R. K. Pariti. 2022. Restricted substances for textiles. *Textile Progress* 54: 1–101. https://doi.org/10.1080/00405167.2022.2101302.
19. Ivy, N. 2021. Essentials of chemical management system for textiles. *Chemical Management in Textiles and Fashion* 99–112. https://doi.org/10.1016/B978-0-12-820494-8.00005-8.
20. Malinauskaite, J., and H. Jouhara. 2024. Circular economy: Policies to drive sustainability and the measures to improve energy efficiency. *Sustainable Energy Technology, Business Models, and Policies* 3–39. https://doi.org/10.1016/B978-0-443-18454-3.00002-3.
21. Schumacher, K. A., and A. L. Forster. 2022. Textiles in a circular economy: An assessment of the current landscape, challenges, and opportunities in the United States. *Frontiers in Sustainability* 3. https://doi.org/10.3389/frsus.2022.1038323.
22. Owusu-Wiredu, P. 2024. From a disposable to a sustainable fashion industry: A review of the shameful trade flows of used textiles and the need to address fast fashion. *Corporate Governance and Sustainability Review* 8: 32–44. https://doi.org/10.22495/cgsrv8i1p3.

23. Kentin, E., and G. Battaglia. 2022. *Policies and perspectives on regulating microplastic fibre pollution. polluting textiles*, 1st edn, 265–289. Routledge.
24. Shishoo, R. 2012. Introduction: Trends in the global textile industry. *The Global Textile and Clothing Industry* 1–7. https://doi.org/10.1533/9780857095626.1.
25. Maciejewski, M., and R. Bendini. 2015. The transatlantic trade and investment partnership (TTIP): Challenges and opportunities for the Internal market and consumer protection in the area of textiles and labelling.
26. Laiker, A., and P. Pande. 2024. Contribution of policy and regulations to enhance transparency and traceability in the garment industry. *The Scientific Temper* 15: 1–10. https://doi.org/10.58414/SCIENTIFICTEMPER.2024.15.spl-2.01.
27. Buchel, S., A. Hebinck, M. Lavanga, and D. Loorbach. 2022. Disrupting the status quo: a sustainability transitions analysis of the fashion system. *Sustainability: Science, Practice and Policy* 18: 231–246. https://doi.org/10.1080/15487733.2022.2040231.
28. Niinimäki, K., G. Peters, H. Dahlbo, P. Perry, T. Rissanen, and A. Gwilt. 2020. The environmental price of fast fashion. *Nat Rev Earth Environ* 1: 189–200. https://doi.org/10.1038/s43017-020-0039-9.
29. Shirvanimoghaddam, K., B. Motamed, S. Ramakrishna, and M. Naebe. 2020. Death by waste: Fashion and textile circular economy case. *Science of The Total Environment* 718: 137317. https://doi.org/10.1016/j.scitotenv.2020.137317.
30. Siliņa, L., I. Dāboliņa, and E. Lapkovska. 2024. Sustainable textile industry—wishful thinking or the new norm: A review. *Journal of Engineered Fibers and Fabrics* 19. https://doi.org/10.1177/15589250231220359.
31. Lüttin, L. 2024. Overview of U.S. sustainability regulations relevant to the fashion industry. US Textile Regulations.
32. Sugeng, R. A. N., W. Romasindah, S. Saiful. 2022. Regulatory and policy arrangement of the textile industry and national textile products for clothing resilience. *International Journal of Research and Innovation in Social Science* 6: 5–15. https://doi.org/10.47772/IJRISS.2022.6901.
33. Amutha, K. 2017. Sustainable practices in textile industry: Standards and certificates. In: *Sustainability in the textile industry. Textile science and clothing technology*, ed. S. Muthu, 79–107. Springer. https://doi.org/10.1007/978-981-10-2639-3_5.
34. Zimon, D., P. Madzik, and R. Sroufe. 2020. The influence of ISO 9001 & ISO 14001 on sustainable supply chain management in the textile industry. *Sustainability* 12: 4282. https://doi.org/10.3390/su12104282.
35. Gbolarumi, F. T., K. Y. Wong, and S. T. Olohunde. 2021. Sustainability assessment in the textile and apparel industry: A review of recent studies. *IOP Conference Series: Materials Science and Engineering* 1051: 012099. https://doi.org/10.1088/1757-899x/1051/1/012099.
36. Ulutaş, T., and Ü. Kurt. 2024. *Carbon footprint calculation methods used in the textile industry.*
37. di Poratelli, F. 2023. The role of the ISO 20121 Standard in limiting the environmental impact of fashion events. CIKIS STUDIO. https://www.cikis.studio/en/article/the-role-of-the-iso-20121-standard-in-limiting-the-environmental-impact-of-fashion-events. Accessed 31 Dec 2024.
38. Thorisdottir, T. S., and L. Johannsdottir. 2020. Corporate social responsibility influencing sustainability within the fashion industry. *A Systematic Review. Sustainability* 12: 9167. https://doi.org/10.3390/su12219167.
39. Górny, A. 2015. Occupational health and safety management in the international condition (consistent with objectives the ISO 45001 standard). *Modern Management Review*. https://doi.org/10.7862/rz.2015.mmr.55.
40. European Commission. 2024. EU strategy for sustainable and circular textiles. Energy, climate change, environment. https://environment.ec.europa.eu/strategy/textiles-strategy_en. Accessed 31 Dec 2024.
41. Crestani, M., L. Talens Peiró, S. Toboso Chavero. 2023. The environmental impact of textiles and clothing: A regional and a country approach. In: *Progress on life cycle assessment in textiles and clothing. textile science and clothing technology*, ed. S. S. Muthu, 199–230. Springer. https://doi.org/10.1007/978-981-19-9634-4_8.

42. Boschmeier, E., W. Ipsmiller, and A. Bartl. 2024. Market assessment to improve fibre recycling within the EU textile sector. *Waste Management & Research: The Journal for a Sustainable Circular Economy* 42: 135–145. https://doi.org/10.1177/0734242X231178222.
43. EPA-US. 2002. Resource conservation and recovery act. United States.
44. Zajac, G. 2023. *Toxic fashion: Evaluation of chemicals in clothing and recommendations for the amended TSCA*. Masters: University of San Francisco.
45. Rebora, M. 2024. A guide to US fashion regulations: Sustainability and compliance in 2024. https://www.traceforgood.com/blog/a-guide-to-us-fashion-regulations-sustainability-and-compliance-in-2024. Accessed 3 Jan 2025.
46. Vighini, D. 2024. Selling textiles to EU & US: A compact guide to mandatory sustainability regulations. *The Water Treatment Magazine*.
47. Xu, C., H. Cheng, and Z. Liao. 2018. Towards sustainable growth in the textile industry: A case study of environmental policy in China. *Polish Journal of Environmental Studies* 27: 2325–2236. https://doi.org/10.15244/pjoes/79720.
48. Hashim, M., M. Nazam, M. Zia-ur-Rehman, M. Abrar, S. A. Baig, M. Nazim, et al. 2022. Modeling supply chain sustainability-related risks and vulnerability: Insights from the textile sector of Pakistan. *Autex Research Journal* 22: 123–134. https://doi.org/10.2478/aut-2021-0008.
49. GIZ. 2022. *Promoting sustainability in the textile and garment industry in Asia*.
50. Peng, S.-Y., J.-Y. Liu, and Y. Geng. 2022. Assessing strategies for reducing the carbon footprint of textile products in China under the shared socioeconomic pathways framework. *Climate Change Economics (Singapore)* 13. https://doi.org/10.1142/S2010007822400048.
51. Global Standard. 2023. Global organic textile standard (GOTS). Stuttgart.
52. Rice, S. 2025. Global organic textile standard (GOTS). Organic trade association (OTA).
53. Popescu, A. 2024. Enhancing environmental standards in textiles: The role of OEKO-TEX® Standard 100. *Bulletin of the Transilvania University of Brasov Series V: Economic Sciences* 59–68. https://doi.org/10.31926/but.es.2024.17.66.1.6.
54. Fawzy Ibrahim, S. 2022. The eco-legalization and quality requirements for textiles, state of the art and current developments, vol. 2.
55. Amutha, K. 2017. Sustainable practices in textile industry: Standards and certificates. In: *Sustainability in the textile industry*, ed. S. Muthu, 79–107. Springer. https://doi.org/10.1007/978-981-10-2639-3_5.
56. IFOAM Organics. n.d. *Organic textiles protecting the credibility of the organic label*.
57. Fråne, A., C. Askham, S. Gíslason, N. Kiørboe, H. Ljungkvist, and D. McKinnon, et al. 2017. The Nordic textile reuse and recycling commitment—a certification system for used textiles and textile waste. Nordic Council of Ministers. https://doi.org/10.6027/TN2017-545.
58. Kitaychik, I. 2025. *What is the global recycled standard?* SCS Global Services.
59. Braun, G., C. Som, M. Schmutz, and R. Hischier. 2021. Environmental consequences of closing the textile loop—life cycle assessment of a circular polyester jacket. *Applied Sciences* 11: 2964. https://doi.org/10.3390/app11072964.
60. Partzsch, L., M. Zander, and H. Robinson. 2019. Cotton certification in Sub-Saharan Africa: Promotion of environmental sustainability or greenwashing? *Global Environmental Change* 57: 101924. https://doi.org/10.1016/j.gloenvcha.2019.05.008.
61. Shen, B., Q. Li, C. Dong, and P. Perry. 2017. Sustainability issues in textile and apparel supply chains. *Sustainability* 9: 1592. https://doi.org/10.3390/su9091592.
62. Shen, L., K. Govindan, and M. Shankar. 2015. Evaluation of barriers of corporate social responsibility using an analytical hierarchy process under a fuzzy environment—a textile case. *Sustainability* 7: 3493–3514. https://doi.org/10.3390/su7033493.
63. Kozar, J., and K. Connell. 2010. Socially responsible knowledge and behaviours: Comparing upper- vs lower-classmen. *Socially Responsible Knowledge and Behaviors* 1–27.
64. Asif, A. K. M. 2017. An overview of sustainability on apparel manufacturing industry in Bangladesh. *Science Journal of Energy Engineering* 5: 1. https://doi.org/10.11648/j.sjee.20170501.11.

65. Arenas, G. C., F. L. Gutierrez, M. F. Cavaa, L. M. Antonio, and J. S. Velazquez. 2022. Evaluation of fair-trade business policies and practices in garment industry: A company and consumer oriented study. *Dyna (Bilbao)* 97: 545–549. https://doi.org/10.6036/10437.
66. Rammandala, C. 2023. *What is WRAP standards and WRAP certification process*. https://www.onlineclothingstudy.com/2023/05/what-is-wrap-standards-and-wrap.html. Accessed 31 Dec 2024.
67. Tabuchi, H. 2022. How fashion giants recast plastic as good for the planet. *New York Times*.
68. Peltzer, R., and M. Brüntrup. 2023. Cotton made in Africa: A case study of sustainable production through responsible consumption.
69. Wren, B. 2022. Sustainable supply chain management in the fast fashion industry: A comparative study of current efforts and best practices to address the climate crisis. *Cleaner Logistics and Supply Chain* 4: 100032. https://doi.org/10.1016/j.clscn.2022.100032.
70. Peleg Mizrachi, M., and A. Tal. 2022. Regulation for promoting sustainable, fair and circular fashion. Sustainability 14: 502. https://doi.org/10.3390/su14010502.
71. Zimon, D., P. Madzík, S. Dellana, R. Sroufe, M. Ikram, and K. Lysenko-Ryba. 2021. Environmental effects of ISO 9001 and ISO 14001 management system implementation in SSCM. *The TQM Journal*. https://doi.org/10.1108/TQM-01-2021-0025.
72. Iftekhar Shuvo, I., J. Decaens, D. Lachapelle, I. P. Dolez. 2021. Smart textiles testing: A roadmap to standardized test methods for safety and quality-control. *Textiles for Functional Applications*. https://doi.org/10.5772/intechopen.96500.
73. Thakker, A. M., and D. Sun. 2023. Sustainable development goals for textiles and fashion. *Environmental Science and Pollution Research* 30: 101989–102009. https://doi.org/10.1007/s11356-023-29453-1.
74. H&M. 2024. H&M group annual and sustainability report 2023.
75. H&M. 2022. H&M achieves cradle to cradle Certified® gold! Inside H&M. https://www2.hm.com/en_us/life/culture/inside-h-m/hm-achieves-cradle-to-cradle-certified-gold.html. Accessed 9 Jan 2025.
76. Tarnovskaya, V. 2023. Sustainability as the source of competitive advantage. How sustainable is it? In: *Creating a sustainable competitive position: ethical challenges for international firms*, ed. P. N. Ghauri, U. Elg, and S. M. Hånell, 75–89. https://doi.org/10.1108/S1876-066X20230000037005.
77. Ponte, C. 2023. The H&M greenwashing scandal: Has business learned the lesson? Impakter. https://impakter.com/hm-greenwashing-scandal-has-business-learned-the-lesson/. Accessed 19 Jan 2025.
78. Narayanasamy, A. 2018. Better HR practices for business efficiency: A case study.
79. Ndure, I. 2025. Sri Lanka's eco-friendly apparel manufacturing surges with GOTS certification uptick. Just Style 2023. https://www.just-style.com/news/sri-lankas-eco-friendly-apparel-manufacturing-surges-with-gots-certification-uptick/. Accessed 14 Jan 2025.
80. Agrawal, T. K., V. Kumar, R. Pal, L. Wang, and Y. Chen. 2021. Blockchain-based framework for supply chain traceability: A case example of textile and clothing industry. *Computers & Industrial Engineering* 154: 107130. https://doi.org/10.1016/j.cie.2021.107130.
81. Nosirova, C. 2022. Marketing and production activities of textile companies? Blockchain technology study. In *Proceedings of the 6th international conference on future networks & distributed systems*, 152–158. ACM. https://doi.org/10.1145/3584202.3584226.
82. Shen, B. 2014. Sustainable fashion supply chain: Lessons from H&M. *Sustainability* 6: 6236–6249. https://doi.org/10.3390/su6096236.
83. Ahmed, W. A. H., and B. L. MacCarthy. 2021. Blockchain-enabled supply chain traceability in the textile and apparel supply chain: A case study of the fiber producer, lenzing. *Sustainability* 13: 10496. https://doi.org/10.3390/su131910496.
84. Sodhi, M. S., and C. S. Tang. 2019. Research opportunities in supply chain transparency. *Production and Operations Management* 28: 2946–2959. https://doi.org/10.1111/poms.13115.
85. Yang, K., S. A. McErlain-Naylor, B. Isaia, A. Callaway, and S. Beeby. 2024. E-textiles for sports and fitness sensing: current state, challenges, and future opportunities. *Sensors* 24: 1058. https://doi.org/10.3390/s24041058.

86. Sikka, M. P., A. Sarkar, and S. Garg. 2024. Artificial intelligence (AI) in textile industry operational modernization. *Research Journal of Textile and Apparel* 28: 67–83. https://doi.org/10.1108/RJTA-04-2021-0046.
87. de Oliveira Neto, G. C., L. F. R. Pinto, D. de Silva, F. L. Rodrigues, F. R. Flausino, and D. E. P. deOliveira. 2023. Industry 4.0 technologies promote micro-level circular economy but neglect strong sustainability in textile industry. *Sustainability* 15: 11076. https://doi.org/10.3390/su151411076.
88. Spangaro, A. 2024. Software solutions for regulatory compliance in fashion & textile. Renoon. https://www.renoon.com/blog/software-solutions-for-regulatory-compliance-in-fashion-textile. Accessed 13 Jan 2025.
89. Rahman, A., S. Hossen, Y. Mridha, Md. R. Amin, R. Ouboucetta. 2024. Consumer perceptions and purchasing trends of eco-friendly textile products in the U.S. Market. *Global Mainstream Journal* 1: 20–32. https://doi.org/10.62304/ijbm.v1i2.145.
90. de Abreu, M. C. S. 2015. Perspectives, drivers, and a roadmap for corporate social responsibility in the textile and clothing industry. In *Roadmap to sustainable textiles and clothing. Textile science and clothing technology*, ed. S. Muthu, 1–21. Springer. https://doi.org/10.1007/978-981-287-164-0_1.
91. Pal, R., and J. Gander. 2018. Modelling environmental value: An examination of sustainable business models within the fashion industry. *Journal of Cleaner Production* 184: 251–263. https://doi.org/10.1016/j.jclepro.2018.02.001.
92. Warasthe, R., F. Schulz, R. Enneking, and M. Brandenburg. 2020. Sustainability prerequisites and practices in textile and apparel supply chains. *Sustainability* 12: 9960. https://doi.org/10.3390/su12239960.
93. Murmura, F., and L. Bravi. 2020. Developing a corporate social responsibility strategy in India using the SA 8000 standard. *Sustainability* 12: 3481. https://doi.org/10.3390/su12083481.
94. Govindan, K., M. Shaw, and A. Majumdar. 2021. Social sustainability tensions in multi-tier supply chain: A systematic literature review towards conceptual framework development. *Journal of Cleaner Production* 279: 123075. https://doi.org/10.1016/j.jclepro.2020.123075.
95. Bick, R., E. Halsey, and C. C. Ekenga. 2018. The global environmental injustice of fast fashion. *Environmental Health* 17: 92. https://doi.org/10.1186/s12940-018-0433-7.
96. Boström, M., and M. Micheletti. 2016. Introducing the sustainability challenge of textiles and clothing. *Journal of Consumer Policy (Dordrecht)* 39: 367–375. https://doi.org/10.1007/s10603-016-9336-6.

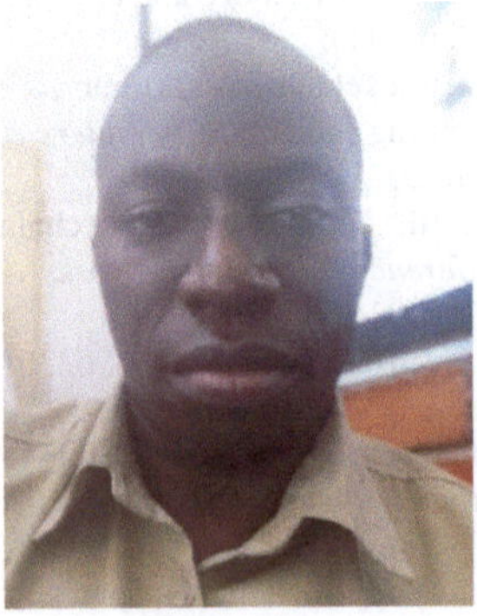

Chipambwa Walter is a senior lecturer in the department of Clothing and Textile Technology, School of Art and Design at Chinhoyi University of Technology (CUT) in Zimbabwe. Walter has extensive work in both academia and industry, as he previously worked as a production manager at Karina Textiles and Zimbabwe Grain bag for more than ten years. His research interests include sustainable design, computer aided design, digital technologies in design, textile science, design education, and he is actively involved in both academic and community development initiatives. He is a Ph.D. candidate at the University of Botswana, researching on sustainable design-driven innovation in SMEs. He has been involved in a number of projects that promote inclusivity in design at community level and human centred design.

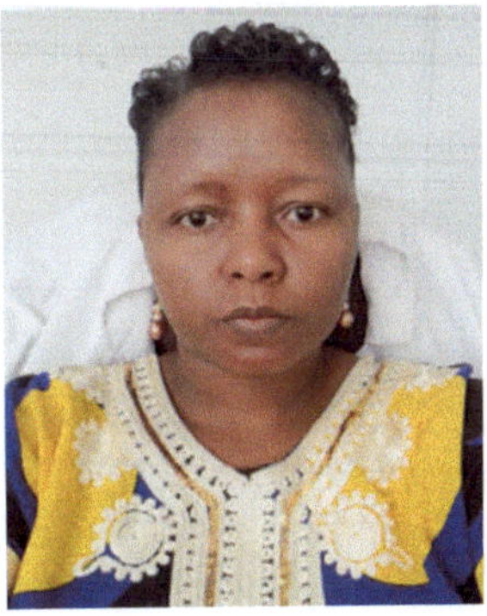

Dzingai Pethile (Nee Sibanda) is currently a senior lecturer in the department of Fibre and Polymer Materials Engineering at the National University of Science and Technology (NUST) in Bulawayo, Zimbabwe. Before she joined NUST, she worked as a clothing line production manager and as a trainee towel manufacturing production manager at Archer Clothing Manufacturers and Merlin towel manufacturers respectively. She is currently enrolled as a full-time Ph.D. student in Chemical technology at the University of Pretoria in South Africa, in the Institute of Applied Materials. Mrs. Dzingai has published several papers in peer-reviewed journals and is also a reviewer for the Zimbabwe Journal of Science and Technology. Her research interests lie at the interface of fibre and biopolymer materials engineering. She is also involved in community activities which look at the development of more efficient and eco-friendly ways of production of textiles and leather products and promoting traceability.

Chapter 15
Consumer Perception and Education in Promoting Sustainable Choices in Textiles

Sharon Nunoo, Morolake Dairo, and Tahera Kalsoom

Abstract This chapter explores consumer perception and the role of education in promoting sustainable choices within the UK textile industry. With growing concerns about the environmental and social impacts of fashion, sustainability has emerged as a critical issue for both consumers and producers. However, despite increasing awareness, a gap remains between consumers' knowledge of sustainable textiles and their actual purchasing decisions. This conceptual paper examines how consumer perceptions of sustainability can be influenced through targeted education and awareness initiatives in the UK context. Drawing on existing literature and theoretical frameworks in consumer behaviour, sustainability education, and ethical consumption, this chapter critically analyses the factors shaping consumer decision-making in the textile sector. It explores the role of government policies, NGO campaigns, and brand-led initiatives in fostering consumer awareness and engagement with sustainable textiles. By identifying key barriers, such as misconceptions, cost concerns, and limited accessibility, the chapter provides a theoretical discussion on how educational interventions can bridge the gap between awareness and action. Further, the paper evaluates potential educational strategies, including digital platforms, influencer marketing, and transparency in branding, as tools for enhancing consumer engagement. Comparative case studies of successful environmental campaigns in other sectors will be analysed to assess their applicability to the textile industry. Ultimately, this chapter offers insights into how consumer education can serve as a catalyst for driving sustainable purchasing decisions. It provides strategic recommendations for policymakers, industry stakeholders, and educators on improving consumer understanding and accessibility to sustainable textile options, fostering a more environmentally responsible and socially conscious consumer base.

S. Nunoo (✉) · M. Dairo · T. Kalsoom
Manchester Fashion Institute, Manchester Metropolitan University, Manchester, UK
e-mail: s.nunoo@mmu.ac.uk

M. Dairo
e-mail: m.dairo@mmu.ac.uk

T. Kalsoom
e-mail: t.kalsoom@mmu.ac.uk

S.-X. Jiang et al. (eds.), *Advances in Textile Materials and Processing Techniques for Sustainability*, SDGs and Textiles, https://doi.org/10.1007/978-981-95-0469-5_15

Keywords Consumer perception · Education · Sustainability · Textiles · UK · Sustainable choices

15.1 Introduction

Sustainability has become a critical concern within the textile and fashion industry, with consumers increasingly aware of its impact on their daily lives. Elkington introduced the Triple Bottom Line (TBL) framework to conceptualise sustainable development, encompassing three core dimensions: environmental, social, and economic sustainability [1]. This framework suggests that achieving sustainability requires a balanced approach to these interconnected areas. The textile and fashion industry has attempted to address environmental sustainability concerns by implementing ethical sourcing methods for material production and increasing transparency in supply chains. However, significant gaps remain in consumer understanding and industry accountability [2].

Consumers are also demanding greater transparency from brands regarding supply chain activities, particularly concerning garment workers' conditions and fair wages [3]. This consumer advocacy has led many organisations to produce Corporate Social Responsibility (CSR) reports, detailing their social interactions and ethical commitments. Conversely, brands that fail to engage in transparent reporting risk being perceived as engaging in greenwashing, where sustainability claims lack substantive action [4]. Research indicates that greenwashing leads to consumer scepticism, ultimately diminishing trust in sustainability initiatives [5].

While environmental and social sustainability have received significant attention, economic sustainability in fashion remains an underexplored area. Some organisations contribute to communities or environmental preservation as part of their economic sustainability efforts [6]. Patagonia, for instance, is widely recognised for its environmental activism, allocating all profits—approximately $100 million annually—to address climate change and preserve wilderness [7]. Similarly, Timberland engages in initiatives such as regenerative agriculture, promoting sustainable farming practices [8].

Despite these industry-led sustainability initiatives, several persistent concerns remain. Textile waste continues to be a significant global issue, with some brands disposing of excess textiles in developing countries, exacerbating environmental degradation in those regions [9]. Additionally, consumer overconsumption remains problematic, as fast fashion models encourage excessive purchasing behaviour, leading to short product lifespans [10]. The textile and fashion industry are responsible for 10% of global carbon emissions, contributing significantly to climate change [11]. Labour exploitation also persists, with many brands outsourcing production to low-cost labour markets, where workers often endure poor conditions and unfair wages [12]. The Rana Plaza disaster in Bangladesh, which resulted in over 1,100 deaths of garment workers producing for brands such as H&M and Primark, underscores the need for stricter regulations and corporate accountability [13].

As consumer perceptions of sustainability evolve, many consumers expect brands to align with social and environmental causes [14]. However, the persistence of unsustainable practices suggests a disconnect between consumer expectations and corporate actions. While organisations like the United Nations actively disseminate sustainability information, consumers often lack detailed insights into the specific challenges being addressed and the exact nature of sustainable textiles [15]. This knowledge gap highlights the need to investigate how awareness of sustainability influences consumer purchasing decisions. Generation Z, often regarded as the "woke" generation, is particularly engaged with social and environmental issues [16]. However, despite their heightened awareness, many consumers still lack a comprehensive understanding of sustainability, affecting their purchasing behaviours. The way sustainability information is communicated—whether through formal education, brand-led campaigns, or digital platforms—significantly influences consumer engagement with sustainable textile choices [17].

Currently, brands employ various communication strategies to educate consumers about their products and values. This presents an opportunity for businesses to inform consumers about existing challenges, highlight their sustainability commitments, and foster greater consumer engagement [18]. Consumers are increasingly demanding that brands take proactive sustainability measures, aligning with ethical causes while educating them on how they can contribute to sustainable consumption [17]. Brands that successfully communicate their sustainability efforts through transparency and impact-driven storytelling are more likely to gain consumer trust and loyalty [2]. Educating consumers on existing challenges and sharing insights on how brands contribute to sustainability can further strengthen this trust. Research suggests that consumers expect brands to lead the way in sustainability education, demonstrating practical ways for individuals to make informed choices [14]. Consequently, consumer perception of sustainability is influenced not only by a brand's sustainability practices but also by the clarity and credibility of sustainability messaging [5].

This chapter aims to address the following research questions, drawing on existing literature:

1. *How do consumer perceptions and misconceptions about sustainability influence purchasing decisions in the UK textile industry?*
2. *What are the most effective educational strategies for enhancing consumer engagement with sustainable textile choices?*

The subsequent sections will introduce the theories informing this chapter, including consumer perception and decision-making models, current educational interventions by brands, and barriers preventing consumers from making sustainable choices. The discussion will explore the role of consumer education in promoting sustainable choices, examining educational strategies for sustainability awareness, the use of digital media as an educational tool, and the role of brand-led initiatives and corporate responsibility in consumer education. Insights from other industries will be presented to illustrate how these educational practices could be applied to the UK context.

The final section will explore policy measures and strategic recommendations to strengthen consumer education and encourage more sustainable consumer choices.

15.2 Theoretical Foundations of Consumer Perception and Sustainable Behaviour

15.2.1 Consumer Perception and Sustainable Consumption Decisions

The Role of Knowledge, Values, and Attitudes in Shaping Consumer Choices

Consumer perception plays a crucial role in shaping purchasing behaviour, particularly in relation to sustainability. As Bhowmick [19] states, "the perception built in a consumer's mind triggers the decision-making from the consumer's end" (p. 1). Perception has been defined by Durmaz [20] as "the process by which an organism attains awareness or understanding of its environment by organising and interpreting sensory information. In general, perception is gathering information through our senses, which include seeing, hearing, touching, tasting, smelling, and sensing" (p. 218). Similarly, Madichie [21] describes perception as "the process by which people select, organise, and interpret sensations, i.e., the immediate response of sensory receptors to such basic stimuli as light, colour, odour, texture, and sound" (p. 153). These definitions highlight the role of sensory stimuli in consumer decision-making. When brands communicate with consumers, they activate sensory receptors through marketing messages and branding, creating stimuli that influence consumer responses [18]. However, consumer interpretations of these stimuli vary based on personal biases, prior experiences, and individual needs.

Figure 15.1 presents a conceptual framework outlining the interactions between external stimuli, consumer perception, internal psychological dimensions, and consumer behaviour. External stimuli—including formal educational programmes, informal campaigns by NGOs, digital platforms, advertising, and influencer marketing—initiate the consumer perception process, comprising three distinct stages: exposure (where the stimulus reaches the consumer), attention (the degree of consumer engagement with the stimulus), and interpretation (where consumers assign meaning based on their prior knowledge and experiences). As consumers progress through this perception process, they develop or reinforce their internal psychological dimensions, namely knowledge (awareness and understanding of sustainability), attitudes (evaluative feelings and beliefs about sustainable products), and values (deep-seated personal, social, and environmental principles). These dimensions interactively influence consumers' ultimate behaviour—making sustainable textile choices. This flowchart underscores the necessity of strategically designed educational and communicative interventions that effectively target and align with

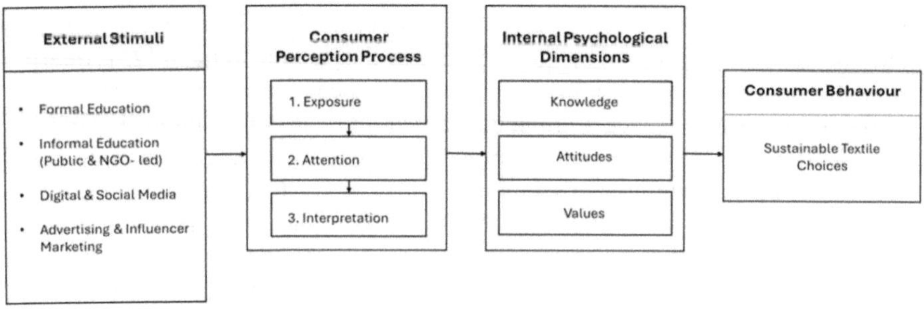

Fig. 15.1 Consumer perception and sustainable textile consumption: an integrative framework. Adapted from Ajzen [31] and [22]

consumer knowledge, attitudes, and values to foster sustained behavioural change towards sustainability.

Consumers primarily gather information from brands through advertising, marketing, and digital platforms, which include visuals, text, and, increasingly, influencer endorsements [23]. The use of influencers as intermediaries has become an essential marketing strategy, as consumers often trust peer recommendations over traditional corporate advertising [24]. This approach allows brands to communicate their values and sustainability commitments, with the hope that consumers align their personal beliefs with the messages being conveyed [25].

The process of perception consists of three key stages: exposure, attention, and interpretation, each of which influences consumer decision-making [21]. Exposure, the first stage, occurs when a stimulus enters the scope of an individual's sensory receptors, such as sight, smell, or touch. Consumers are constantly exposed to marketing messages, but they do not process every stimulus equally [23]. Instead, they selectively focus on specific stimuli while ignoring or deliberately avoiding others. To capture consumer attention, brands employ sensory marketing techniques, such as distinctive packaging, appealing visuals, and immersive brand experiences, all designed to elicit a favourable response [26]. Attention, the second stage, refers to the level of focus a consumer devotes to a particular stimulus [21]. Given the constant bombardment of advertisements, fashion brands must compete for consumer attention within highly saturated markets [27]. Some campaigns successfully engage consumers, while others fail to resonate due to lack of relevance, poor timing, or ineffective execution. The success of a brand's messaging depends on both the characteristics of the campaign and the consumer's cognitive and emotional state at the time of exposure [23]. Given the dominance of digital and social media marketing, brands increasingly rely on algorithm-driven personalisation techniques to ensure that messages reach the most receptive audiences [28]. Additionally, many fashion brands target consumers within similar demographic segments, leading to intense competition for their attention [16]. While consumers experience sensory overload, they tend to prioritise messages that align with their values or trigger an emotional response [14]. This explains why consumers increasingly expect brands to demonstrate social responsibility and align with causes that matter to them [25]. Interpretation, the final stage, is the process of assigning meaning to sensory stimuli, and it varies significantly between individuals based on their expectations, beliefs, and prior experiences [21]. In marketing, brands aim to activate the appropriate cognitive schema, influencing how consumers evaluate products and messages [23]. For instance, the McKinsey Consumer Report found that consumers prefer brands that authentically align with their personal values, reinforcing the importance of consistent and transparent sustainability messaging [16]. Misaligned messaging can result in negative brand perception, particularly when consumers perceive a discrepancy between a company's stated values and its actual practices [5].

Understanding consumer perception is critical in the context of sustainable consumption behaviour. Research suggests that economically sustainable consumption is often linked to consumer restraint, where individuals consciously decide to forgo unnecessary purchases [29]. This perspective aligns with the concept of minimalist consumption, where consumers opt for fewer, higher-quality purchases rather than frequent low-cost acquisitions [14, 30]. However, educating consumers about sustainability can significantly influence their willingness to engage in responsible consumption [14]. Effective sustainability education can help consumers overcome misinformation, enabling them to make more informed purchasing decisions that align with their values [25].

To deepen the understanding of how internal psychological factors influence consumer decision-making, this chapter will explore three theoretical models:

1. The Theory of Planned Behaviour (TPB) explains how attitudes, subjective norms, and perceived behavioural control influence purchasing decisions [31].
2. The Value-Belief-Norm (VBN) Theory highlights the link between personal values, awareness of consequences, and pro-environmental actions [22].
3. The Cognitive Dissonance (CD) Theory explores how consumers experience psychological discomfort when their behaviour contradicts their values, often prompting changes in consumption patterns [32].

Each of these theories offers valuable insights into the psychological drivers of sustainable consumption and will provide a comprehensive framework for understanding how consumer perception, values, and knowledge contribute to sustainable decision-making.

Psychological and Behavioural Theories Relevant to Sustainable Consumption

Understanding consumer behaviour in relation to sustainability requires the application of psychological and behavioural theories. These theories help explain how attitudes, values, and perceived control influence decision-making, particularly in sustainable consumption practices. Two key theoretical frameworks—the Theory of Planned Behaviour (TPB) and the Value-Belief-Norm (VBN) theory—provide insights into the cognitive processes that drive consumer choices related to sustainability.

The Theory of Planned Behaviour (TPB) was introduced by Ajzen [31] as an extension of the Theory of Reasoned Action (TRA) developed by Fishbein and Ajzen in 1975. While the TRA suggests that an individual's intention to perform a behaviour is influenced by attitude and subjective norms, the TPB adds a third component—perceived behavioural control—which accounts for an individual's assessment of their ability to engage in a particular behaviour [31]. According to the TPB, consumer behaviour stems from their individual intention, which Ajzen [31] defines as an indicator of how hard people are willing to try to perform the behaviour.

Research has demonstrated that attitude, subjective norms, and perceived behavioural control significantly influence sustainable consumption decisions [33]. Attitude refers to an individual's positive or negative evaluation of a specific behaviour, such as purchasing eco-friendly products [34]. Subjective norms reflect social pressures and expectations, influencing whether consumers adopt sustainable consumption habits [35]. Perceived behavioural control relates to an individual's perceived ease or difficulty in performing a behaviour, which may depend on external constraints such as price, accessibility, or information availability [33].

Numerous studies have applied TPB to sustainable consumption behaviours. Sheoran and Kumar [33] integrated TPB with the consumption cycle to conceptualise sustainable consumer behaviour, identifying subjective norms and perceived behavioural control as the most significant predictors of environmentally responsible purchasing. Similarly, Madichie [21] examined consumer attitudes toward sustainability and found that attitude was the strongest predictor of purchase intention, while subjective norms also played a significant role. Their study extended TPB by considering behavioural beliefs and control beliefs, revealing that consumers' perceived ease of adopting sustainable behaviour influenced their purchasing decisions only to a limited extent. Mason et al. [36] further explored the attitude-intention-behaviour gap using TPB among Italian Generation Y consumers—a demographic known for its fast fashion consumption habits. Their findings indicated that positive attitudes toward sustainability led to increased interest in responsible consumption, but this did not always translate into actual purchasing behaviour due to economic and habitual constraints. This gap highlights the limitations of attitude-driven intention models, suggesting that educational interventions and structural incentives may be required to convert sustainable attitudes into tangible consumer actions. While TPB provides valuable insights into consumer intentions, it does not fully explain how values and moral beliefs influence sustainable behaviour. To address this gap, the Value-Belief-Norm (VBN) theory offers a complementary framework by incorporating environmental responsibility and moral obligation as key drivers of pro-environmental behaviour [22].

The VBN theory, introduced by Stern [22], explores how normative values influence sustainable behaviour. Unlike TPB, which focuses on intentions and perceived control, VBN posits that personal moral beliefs and an individual's awareness of environmental consequences shape their willingness to engage in pro-environmental actions [37]. According to VBN, individuals act sustainably when they feel personally responsible for environmental harm [38].

The theory highlights two key concepts:

1. Awareness of Consequences (AC)—Individuals must first recognise the harmful environmental impact of unsustainable behaviour [38].
2. Ascription of Responsibility (AR)—Once aware, individuals must feel personally accountable for reducing environmental harm and taking corrective action [38].

However, many consumers remain unaware of the consequences of their consumption habits, resulting in low engagement with sustainable practices despite their self-perception as environmentally conscious individuals [34]. While some consumers, particularly Generation Z, are often described as the "woke" generation, their actual participation in sustainable consumption is inconsistent [16]. Studies suggest that economic factors, convenience, and limited product availability often prevent consumers from translating sustainability awareness into sustainable purchasing decisions [14].

The gap between sustainable values and purchasing behaviour can be attributed to price sensitivity and external market conditions, such as inflation and the perception that sustainable products are too expensive [33]. Additionally, many consumers fail to recognise the direct impact of environmental degradation on their personal well-being, limiting their motivation to engage in proactive sustainability efforts [35].

Fashion designer Stella McCartney [39] has been vocal about the lack of consumer awareness regarding sustainability in fashion, stating:

> Every single industry sharing this Earth needs to look at itself and be responsible, mindful, and at the same time still have a healthy business at the end of the day. What is really exciting about The New Textiles Economy Report is that it provides solutions to an industry that is incredibly harmful to the environment—these are problems that people are not even aware of [39].

This statement highlights the need for consumer education to bridge the knowledge gap regarding fashion's environmental impact. The Ellen MacArthur Foundation's New Textile Economy Report [40] underscores this point by outlining a vision for systemic change in the industry, advocating for circular economy models that could generate up to $500 billion in economic opportunities while reducing environmental damage [40]. The report presents empirical evidence of fashion's negative contributions to environmental decline, reinforcing the importance of education and industry reform.

Another psychological theory relevant to sustainable consumption is Cognitive Dissonance Theory (CDT), developed by [32]. This theory suggests that when individuals experience a discrepancy between their values, beliefs, and actions, they experience psychological discomfort, known as cognitive dissonance. To reduce this discomfort, individuals attempt to align their behaviour with their values or, conversely, adjust their attitudes to justify their existing actions [32]. In the context of sustainable consumption, cognitive dissonance arises when consumers express concern for environmental and ethical issues but continue to engage in unsustainable purchasing habits, such as fast fashion consumption [41].

Studies show that consumers resolve cognitive dissonance in different ways. Some may change their behaviour by purchasing sustainable products or reducing their consumption, while others may alter their beliefs to justify unsustainable habits [14]. For example, consumers who buy fast fashion despite being aware of its negative environmental impact might convince themselves that "one purchase won't make a difference" or that "sustainable fashion is too expensive and inaccessible" [3]. Research indicates that price, convenience, and perceived product quality are common justifications that allow consumers to maintain unsustainable consumption patterns despite their ethical concerns [42].

Cognitive dissonance is particularly relevant to greenwashing, where brands falsely claim sustainability credentials to attract ethically conscious consumers [4]. When consumers discover inconsistencies between a brand's sustainability messaging and its actual practices, dissonance is heightened, potentially leading to brand distrust and disengagement [5]. However, studies suggest that brands that actively educate consumers on sustainability issues and demonstrate transparency in their supply chains can help consumers reduce cognitive dissonance and align their purchasing decisions with their ethical beliefs [25].

Addressing cognitive dissonance in sustainable fashion consumption requires a combination of education, incentives, and systemic industry changes. Providing consumers with accessible and affordable sustainable alternatives while reinforcing positive reinforcement mechanisms, such as rewards for responsible purchasing, can reduce dissonance and encourage long-term behavioural change [41]. Given that many consumers experience dissonance between their sustainability values and actual purchasing behaviour, bridging this gap through targeted educational strategies and industry accountability is essential. The next section explores how consumer education initiatives and digital engagement tools can effectively address this challenge and promote long-term commitment to sustainability.

15.2.2 Implications for Consumer Education and Sustainable Behaviour

The Theory of Planned Behaviour (TPB) and Value-Belief-Norm (VBN) theory offer valuable perspectives on sustainable consumption, but they alone do not fully explain the disconnect between consumers' sustainability values and their actual purchasing behaviours. The Cognitive Dissonance Theory (CDT) helps bridge this gap by illustrating how consumers experience psychological discomfort when their behaviour contradicts their values [32]. Many consumers express concern for environmental and ethical issues, yet continue to purchase fast fashion due to factors such as price, convenience, and social norms [3]. To reduce this dissonance, consumers often either modify their behaviour—by choosing more sustainable options—or justify their choices by downplaying the environmental impact of their actions [41]. This is particularly relevant in the context of greenwashing, where brands mislead consumers into believing their products are sustainable, temporarily reducing cognitive dissonance but ultimately eroding consumer trust when the truth is revealed [4].

Given the persistent gap between knowledge, values, and action, educational interventions must go beyond simply raising awareness. It is not enough to inform consumers about sustainability issues; education must also provide practical tools, incentives, and behavioural nudges that reduce cognitive dissonance and encourage sustainable choices [14]. Research suggests that personalised digital engagement, such as interactive sustainability calculators, AI-driven shopping recommendations, and blockchain-enabled transparency tools, can help consumers align their values with their purchasing decisions by reinforcing their sense of agency and accountability [25]. Additionally, brands must engage in consistent, transparent communication that educates consumers about the genuine impact of their purchases, rather than employing vague sustainability claims that may lead to further dissonance and scepticism [5].

Incorporating cognitive dissonance reduction strategies into sustainability education is essential for fostering long-term consumer commitment to ethical fashion. Encouraging small, incremental behaviour changes, such as opting for second-hand clothing, participating in clothing rental schemes, or supporting brands with transparent sustainability policies, can make sustainable choices feel more accessible and manageable [42]. Furthermore, leveraging social norms and peer influence, as seen in the rise of sustainability influencers and digital advocacy campaigns, can help consumers feel socially validated in making ethical consumption choices [16].

Ultimately, addressing cognitive dissonance through targeted education and industry accountability is essential for bridging the gap between sustainability awareness and action. The following section explores educational strategies, digital engagement tools, and industry-led initiatives that can further support consumer alignment with sustainable textile consumption.

15.2.3 Educational Interventions in Shaping Consumer Attitudes

Educational interventions play a crucial role in shaping consumer attitudes toward sustainable textiles. However, their effectiveness is often mediated by factors such as the depth of engagement, cultural influences, and systemic barriers. A critical analysis of these interventions, supported by theoretical insights and empirical data, reveals both their potential and limitations in fostering genuine behavioural change.

The Impact of Formal and Informal Education on Sustainable Consumption

Integrating sustainability into formal education curricula has been a key strategy to enhance awareness of sustainable fashion consumption. Research has shown that embedding sustainability principles across various disciplines in higher education fosters a holistic understanding of responsible consumption [43]. However, the effectiveness of formal education depends not just on awareness-raising but also on experiential learning that encourages critical thinking and real-world application [44].

Despite these efforts, there remains a significant gap between knowledge acquisition and behavioural change. A Mintel report highlighted that while 62% of consumers consider sustainability important when purchasing clothing, economic constraints often override their intentions, leading them to prioritize affordability over ethical concerns. This underscores the value-action gap, where increased awareness does not necessarily translate into behavioural change. Consequently, educational institutions must employ interactive pedagogical approaches, such as problem-based learning and industry collaborations, to equip students with actionable skills in sustainable decision-making.

Beyond formal education, informal initiatives such as public campaigns, media influence, and workshops have been instrumental in driving sustainable consumption. Organizations like Fashion Revolution have utilized educational packs and digital campaigns to enhance transparency and accountability in fashion supply chains [45]. However, their latest consumer report revealed that only 19% of respondents actively seek out environmentally responsible clothing, while 14% opt for second-hand purchases instead of buying new. These statistics suggest that while public awareness efforts have increased, they are yet to effect widespread consumer behaviour shifts.

Additionally, digital and social media platforms play an increasing role in informal education. Influencers and sustainable fashion advocates on platforms such as Instagram and TikTok have leveraged digital storytelling to educate consumers on issues such as ethical production and waste reduction [46]. However, the rise of greenwashing, where brands falsely claim sustainability credentials, has led to consumer scepticism, undermining the impact of informal education [4]. Therefore, alongside increasing transparency, regulatory frameworks must be strengthened to ensure credibility in sustainability marketing.

Theories of Behavioural Change and Information Processing in Consumer Decision-Making

Understanding consumer decision-making in the context of sustainable textile consumption necessitates an exploration of behavioural change theories and information processing models. These frameworks elucidate the cognitive mechanisms and motivational factors that drive consumers toward or away from sustainable choices. The Theory of Planned Behaviour (TPB), proposed by Azjen [31], suggests that an individual's intention to engage in a behaviour is influenced by their attitude toward the behaviour, subjective norms, and perceived behavioural control. In sustainable consumption, a positive attitude towards eco-friendly products, societal approval, and the belief in one's capability to make sustainable choices collectively enhance the intention to purchase sustainable textiles. However, external barriers such as higher costs or limited availability can impede actual behaviour, highlighting the gap between intention and action. Similarly, the Value-Belief-Norm (VBN) Theory, developed by Stern et al. [38], suggests that individuals who hold strong ecological values and beliefs feel a moral obligation to engage in pro-environmental behaviours.

In the context of sustainable textiles, consumers with a deep-seated concern for environmental well-being are more likely to align their purchasing decisions with their values, opting for products that minimize ecological impact. Meanwhile, Cognitive Dissonance Theory, introduced by Festinger [32], describes the psychological discomfort experienced when one's actions contradict personal beliefs or values. Consumers aware of the environmental detriments of fast fashion may experience dissonance when purchasing such items, potentially motivating a shift toward more sustainable choices to alleviate this discomfort.

The way consumers process information significantly impacts their purchasing decisions, particularly concerning sustainable products. The Elaboration Likelihood Model (ELM), proposed by Petty and Cacioppo [47], delineates two routes of information processing: central and peripheral. Consumers engaging in the central route thoughtfully consider message content, leading to enduring attitude change, whereas those in the peripheral route rely on superficial cues, resulting in temporary change. Effective sustainability marketing should aim to engage consumers centrally by providing substantive information about product benefits and environmental impact. The Heuristic-Systematic Model (HSM), developed by Chaiken [48], identifies two processing modes: systematic (comprehensive analysis) and heuristic (mental shortcuts). Consumers often employ heuristics, such as eco-labels or brand reputation, to simplify decision-making regarding sustainable textiles. Ensuring the credibility and visibility of such cues can facilitate informed choices. However, Information Overload can negatively impact sustainable purchasing behaviour. When consumers are presented with excessive information, decision fatigue can arise, leading them to rely on familiar, unsustainable choices. Jacoby [49] found that too much information diminishes decision quality, suggesting that sustainability information should be presented in an accessible and digestible manner to enhance consumer decision-making.

Applying these theories to sustainable textile consumption highlights the need for more strategic educational and marketing interventions. Bridging the intention-behaviour gap remains a critical challenge, as many consumers intend to purchase sustainable textiles but face barriers such as high prices or limited availability. Addressing these barriers through economic incentives and increased accessibility can facilitate behaviour consistent with pro-sustainability intentions. Moreover, leveraging social norms through marketing campaigns that depict sustainable consumption as a widespread practice can enhance perceived social approval, motivating consumers to conform. Addressing cognitive dissonance by highlighting the environmental consequences of unsustainable textile choices can encourage consumers to align their purchasing behaviour with pro-environmental values. Additionally, ensuring that sustainability information is clear, concise, and credible is crucial in engaging consumers in thoughtful processing, leading to informed and lasting attitude change. By integrating these behavioural change theories with information processing models, policymakers, educators, and brands can better understand and influence consumer decision-making in sustainable textile consumption. These frameworks not only explain the psychological and cognitive barriers to sustainable fashion choices but also provide strategies to foster a shift towards more responsible consumption patterns.

Barriers to Sustainable Consumer Choices

Despite increasing awareness of environmental issues, consumers often encounter significant obstacles when attempting to make sustainable choices in the textile industry. These barriers include misconceptions and misinformation about sustainable textiles, economic constraints, accessibility issues, and the pervasive influence of greenwashing. These factors contribute to consumer scepticism and hinder the adoption of sustainable consumption habits, ultimately limiting the effectiveness of education-driven sustainability efforts [50]. A critical examination of these barriers is essential in understanding why consumer education alone is insufficient to drive widespread behavioural change in sustainable fashion.

Misconceptions and Misinformation About Sustainable Textiles

A prevalent misconception among consumers is that sustainable fashion is inherently more expensive and less accessible. This belief stems largely from the higher price points associated with some eco-friendly brands, leading to the assumption that sustainability equates to exclusivity. However, research by Lyon and Montgomery [51] indicates that sustainable textiles can be more cost-effective in the long run due to their durability and higher-quality materials. Additionally, the notion that natural fibres like organic cotton and bamboo are always sustainable is misleading, as their production still requires significant water and energy consumption [5]. Many consumers also struggle to distinguish genuinely sustainable textiles from conventional ones, as the industry lacks standardised and universally recognised labelling systems [4]. Without clear and consistent educational initiatives, misinformation continues to prevent consumers from making truly informed purchasing decisions.

Economic Constraints, Accessibility Issues, and Perceived Lack of Options

Economic barriers further complicate consumer adoption of sustainable textiles. While ethical and environmentally friendly production processes often lead to higher retail prices, research suggests that perceived affordability remains one of the biggest obstacles to sustainable fashion adoption [52]. A McKinsey & Company report [16] found that 53% of consumers cite high costs as a primary reason for not purchasing sustainable clothing, despite expressing a willingness to support ethical fashion. Moreover, accessibility remains a significant issue, as sustainable brands are not always available in mainstream retail spaces, forcing consumers to seek niche markets or online platforms that may not be convenient [4]. This lack of availability reinforces the perception that sustainable fashion is an exclusive market, catering primarily to wealthier consumers who have the financial means and access to eco-friendly brands. In contrast, fast fashion retailers continue to dominate consumer spaces with aggressive pricing strategies, making sustainable alternatives appear less competitive.

The Role of Greenwashing in Shaping Consumer Scepticism

Greenwashing, the deceptive practice of marketing products as more sustainable than they truly are, significantly contributes to consumer scepticism. Research by de Jong et al. [50] highlights that companies often exaggerate their environmental credentials or use vague terminology to mislead consumers into believing that their products are eco-friendly. This manipulation distorts consumer perceptions and makes it difficult to distinguish between genuine sustainability efforts and mere corporate branding strategies. A study by Torelli et al. [5] further found that when consumers become aware of greenwashing tactics, they develop distrust not only toward misleading brands but also toward sustainability claims in general, reducing their likelihood of engaging in eco-conscious purchasing. This widespread scepticism undermines educational initiatives aimed at promoting sustainable consumption, as consumers question whether their efforts will have any real impact. As a result, regulatory measures are needed to hold brands accountable for their claims and provide consumers with verifiable, transparent information.

15.3 The Role of Consumer Education in Promoting Sustainable Textiles

15.3.1 Educational Strategies in Sustainability Awareness

Traditional Versus Modern Approaches to Consumer Education

Early efforts to educate consumers about sustainable textiles relied on traditional, top-down methods such as public service announcements, labels, and sporadic campaigns [53]. These approaches often provided information in a one-way flow, hoping increased awareness would translate into behaviour change. However, evidence shows that while education is "prevalent in supporting an understanding of sustainable fashion practices, it does not necessarily impose changes in behaviour for most of the population" [45, 54]. Modern approaches have evolved to be more interactive and engaging, leveraging digital media and participatory learning [55]. Today's consumer education emphasizes not only imparting knowledge but also shaping values and norms through community engagement [54, 56]. This shift aligns with behavioural theories: TPB highlights that simply knowing about sustainability is not enough unless attitudes, social norms, and perceived control are also addressed [45], and VBN theory underscores that deep-seated values and personal norms must be activated to spur pro-sustainable actions [57]. Thus, contemporary strategies blend information with inspiration and empowerment, seeking to transform consumer mindsets rather than just fill knowledge gaps.

Figure 15.2 outlines the critical role that diverse educational strategies play in shaping consumer awareness and behaviours regarding sustainable textile consumption. By integrating formal educational efforts, informal campaigns, digital media,

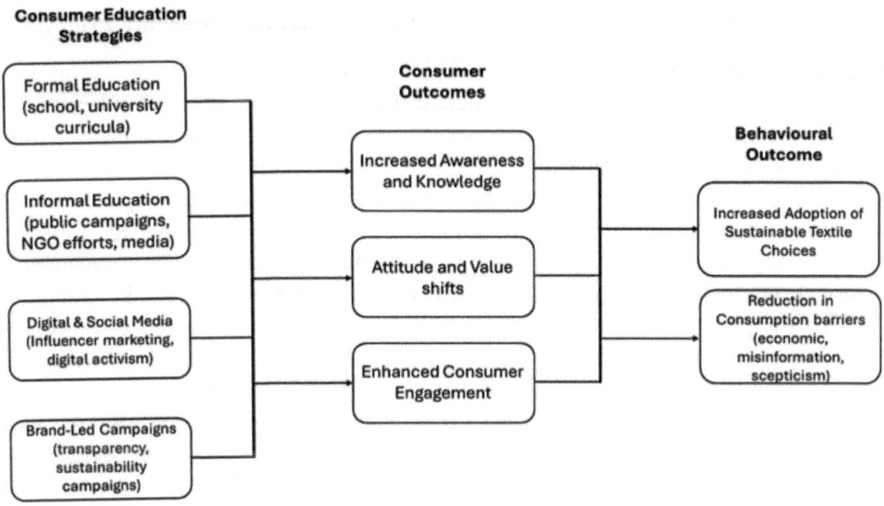

Fig. 15.2 Conceptual framework of consumer education and sustainable textile consumption

and corporate responsibility initiatives, this framework demonstrates how targeted education can effectively enhance consumer engagement and support sustainable purchasing decisions.

As illustrated in Fig. 15.2, different educational strategies collectively influence consumer attitudes and behaviours. Traditional top-down educational methods, such as labels and public announcements, initially sought behavioural change through increased awareness alone but were limited in effectiveness. Recognising these limitations, modern educational approaches have evolved into more interactive and participatory formats, embracing digital platforms and informal methods that leverage community engagement and social norms [55, 56]. Formal education embedded within school curricula and university programmes builds foundational knowledge and long-term value alignment in younger generations, thus addressing deeper attitudes and social norms identified by behavioural theories such as TPB and VBN [45, 57]. Additionally, informal education strategies—such as public campaigns, digital activism, and NGO-led initiatives—actively engage broader consumer segments, effectively increasing societal awareness and promoting behavioural change through emotional appeal and moral responsibility [53, 54]. Digital platforms further amplify educational reach, using influencer marketing and interactive content to reinforce transparency and authenticity, reducing consumer scepticism [10]. Finally, brand-led initiatives provide practical pathways for consumers, aligning product transparency and sustainability messaging directly with consumer expectations, significantly influencing their purchasing behaviours [55, 58]. Collectively, these complementary strategies reduce barriers to sustainable textile choices, empowering consumers to translate increased knowledge and favourable attitudes into concrete sustainable behaviours.

Formal Education: Sustainability in School Curricula and University Programs

Integrating sustainability into formal education is viewed as a foundational long-term strategy [45]. By embedding environmental and ethical topics in school curricula and university programs, future consumers can develop pro-sustainability values and beliefs from an early age (consistent with VBN's emphasis on values) [53, 54]. In practice, however, formal education on fashion sustainability has lagged [45]. A 2022 mixed-method study found "no formal compulsory education in place" on what sustainability means for fashion; instead, most respondents learned about sustainable fashion through informal channels like social media and documentaries. Respondents overwhelmingly believed that intervention in school curricula is needed to prepare the next generation for sustainable consumption [10], with many suggesting that learning practical skills (e.g. how to maintain clothing longevity through repair) is key to a more sustainable fashion future [57]. There are signs of progress: universities and design schools are increasingly offering courses on sustainable textiles and ethical fashion [57], and global initiatives such as the UN's Principles for Responsible Management Education encourage higher education institutions to incorporate sustainability across disciplines [45]. Formal education can shape young consumers' attitudes and perceived behavioural control [45] (per TPB) by equipping them with knowledge and skills to make sustainable choices [57]. It can also instil fundamental values of conservation and social responsibility, providing the value base that VBN theory identifies as crucial for developing personal norms [55]. For example, when students understand the environmental impact of textile production, they may internalize a sense of responsibility and moral obligation to consume more sustainably. Therefore, formal educational programs, from primary school lessons on recycling clothing to MBA coursework on sustainable supply chains, represent a vital strategy to raise sustainability awareness [45] and align consumer beliefs with pro-environmental norms [53].

Informal Education: Public Campaigns, Media Influence, and NGO Efforts

Outside the classroom, a plethora of informal educational strategies have emerged to raise public awareness about sustainable textiles [55]. Governments, non-profits, and community organizations have launched campaigns to inform and motivate consumers [57]. These range from public-information drives about textile recycling and clothing waste reduction, to workshops and clothing swap events that teach practical skills [56]. Media coverage and documentaries have been especially influential in shaping perceptions [10]. In a recent survey, 22% of consumers cited documentaries or TV as their primary source of information on fashion sustainability – a slightly higher share than those who cited social media (19%) or news outlets (15%) [46]. High-profile films (e.g. *The True Cost*) and investigative reports have exposed the environmental damage and labour abuses in the fashion industry, effectively increasing "awareness of the consequences of unsustainable fashion consumption" [53]. From a VBN perspective, such awareness of consequences (AC) is critical: when people grasp the real-world impacts of their fashion choices, it can activate their sense of moral obligation (personal norm) to change behaviour [10]. Indeed, research shows that knowledge of environmental issues has a significant influence on consumer attitudes toward sustainable apparel [53, 54, 57]. Thus, media exposés and educational content can feed into the TPB chain as well, by shifting attitudes and even perceived social pressure as more people talk about these issues [45].

NGOs and activist movements have also become powerful educators in the sustainable fashion space. One prominent example is Fashion Revolution, a global movement that since 2013 has used campaigns like #WhoMadeMyClothes to encourage consumers to question brand transparency and ethics [46]. This campaign, commemorating the Rana Plaza factory disaster, "asked people to display the labels on their clothes and ask: 'Who made my clothes?'"—a simple question that went viral on social media [53]. Through annual Fashion Revolution Week events, the movement reaches millions; in 2021 alone, 4.2 million people engaged with Fashion Revolution hashtags, dramatically raising awareness about supply chain issues [46]. Such NGO-led efforts illustrate how informal education can drive both knowledge and norms: as information spreads, consumers not only learn facts (e.g. which brands disclose their factories) but also begin to expect transparency as a new norm in fashion consumption [55]. Surveys confirm that consumers are increasingly calling on brands to be transparent. In a 2020 survey across five European countries, 69% of respondents wanted to know how their clothes were manufactured (up from 59% in 2018) and 75% agreed that brands should do more to improve the lives of garment workers. These findings suggest that public awareness campaigns have succeeded in elevating consumer expectations [57]. In line with TPB, this reflects a changing subjective norm: it is becoming more socially expected to care about ethical and sustainable clothing [45]. Moreover, many respondents in that survey believed governments should play a role (70% said governments must ensure clothing is produced sustainably) (McKinsey), indicating that informal education has also stirred public demand for policy intervention [46]. Overall, informal educational strategies, whether through compelling media content or grassroots activism, complement formal education by engaging wider audiences, leveraging emotional and moral appeals (activating values and norms per VBN) [45], and creating social momentum for sustainable textile consumption.

In the UK textile industry, consumer perceptions and sustainable choices are notably influenced by certifications, social influencers, and media narratives. Certifications such as GOTS and Fair Trade enhance consumer trust and credibility at the point of purchase, addressing concerns about greenwashing and enabling informed decisions. However, their effectiveness is closely tied to consumer awareness, necessitating continuous educational efforts. Meanwhile, social influencers significantly impact consumer attitudes by promoting sustainable practices through relatable, aspirational content on platforms like Instagram and YouTube, particularly affecting younger generations. Traditional media further amplifies these effects by highlighting sustainability issues through influential documentaries, news stories, and mainstream television programmes such as BBC's *Fashion's Dirty Secrets* and ITV's *Love Island*. Together, these elements shape a cultural environment conducive to sustainable textile choices, bridging awareness and action.

15.3.2 Digital and Social Media as Educational Tools

The Impact of Social Media on Consumer Knowledge and Attitudes

Digital and social media have revolutionized consumer education in sustainable fashion, making information more accessible and interactive than ever [55]. Platforms like Instagram, YouTube, TikTok, and Twitter facilitate the rapid dissemination of facts, stories, and opinions about sustainable textiles, reaching especially the younger demographics [59]. Social media's global reach and visual appeal enable it to "increase awareness and recognition" of sustainability issues by allowing brands and individuals to share influential content at unprecedented scale [59]. The impact on consumer knowledge is evident: complex topics (e.g. the carbon footprint of a T-shirt or the plight of garment workers) are being broken down into digestible posts, infographics, and videos that spread virally. Empirical research supports that increased knowledge can lead to attitude shifts. When consumers obtain more information about the environmental impact of fashion production, they tend to "develop a positive approach and attitude towards the purchase of sustainable clothing" [55]. One study even found that environmental knowledge influences attitudes which in turn boost intention to buy slow fashion, with attitude acting as a mediator between knowledge and intention. In the language of TPB, social media helps shape the *attitude* component by informing consumers of the downsides of fast fashion [55] and the benefits of sustainable alternatives, thus fostering more favourable attitudes toward sustainable products [59].

Social media also plays a key role in shaping perceived norms and peer influences related to sustainable fashion [55]. Through online communities, what people see others doing or endorsing can heavily influence their own behaviour [59]. Sustainable fashion conversations online often create a bandwagon effect, for instance, the popularization of thrift hauls, upcycling challenges, or outfit repeating on Instagram and TikTok has made these behaviours trendier and more socially accepted [10]. From a TPB standpoint, this corresponds to strengthening the *subjective norm*: as more influencers and peers showcase sustainable choices, individual consumers feel greater social pressure to follow suit [55]. The literature suggests that creating such social norms can significantly encourage sustainable consumption [46]. In fact, social norms are found to positively influence sustainable fashion consumption intentions [55]. One practical strategy is leveraging "dynamic norms," highlighting that an increasing number of people are adopting sustainable fashion, which serves as social proof and stimulates a desire for social approval in others. In digital spaces, dynamic norms can spread quickly – for example, a viral post stating "more than one million people have pledged to buy no new clothes this month" could motivate others to join the movement [55]. Thus, social media not only educates but also normalizes sustainable behaviours by amplifying them within society.

The Role of Influencers, Brand Transparency, and Digital Activism in Driving Sustainable Choices

Influencers and content creators on social platforms have become key educators and shapers of consumer behaviour [60]. Many environmentally conscious fashion influencers use their platforms to expose problems (e.g. doing investigative videos on wasteful practices) [10] and to model solutions (such as styling second-hand outfits or reviewing sustainable brands) [55]. These influencers effectively serve as peer educators, often more relatable to the public than traditional experts [59]. By sharing their personal journeys toward sustainable fashion, they tap into followers' values and emotions [61]. Authentic influencers can engender trust and inspire attitude change – their recommendations can carry significant weight in decision-making [61]. In TPB terms, influencers affect both *attitudes* (by providing new information and framing sustainable fashion as desirable) [60] and *subjective norms* (as followers may internalize the influencer's stance as a social expectation) [55]. At the same time, influencers who promote sustainability contribute to what VBN theory would call activating beliefs about outcomes and moral norms; for instance, an influencer highlighting the cruelty of sweatshop labour may instil a moral belief in followers that buying ethical fashion is the right thing to do [60]. Notably, companies have recognized this power and often partner with influencers or brand ambassadors to amplify their sustainability messages. Research indicates that brands leveraging social media marketing can indeed influence consumers' engagement with sustainable fashion, although care must be taken that these collaborations are genuine [10, 61].

Brand transparency has also been enhanced by digital tools, empowering consumers with information at their fingertips [55]. Many brands now use their websites and social media to share detailed sustainability reports, supply chain disclosures, or behind-the-scenes content about how products are made. Some innovators are adopting interactive approaches [59] such as digital labels or QR codes on garments that link to online databases about the product's materials, factories, and carbon footprint [60]. Such transparency initiatives function as consumer education: they inform buyers about what makes a product sustainable or not, helping to bridge the knowledge gap that often exists. For example, scholars have recommended that apparel companies provide more information on their production processes [61], even suggesting digital experiences where scanning a QR code reveals how a garment was produced and how to care for it. This level of openness can increase the value consumers attach to products and enhance their perceived behavioural control, as consumers feel better equipped to make an informed choice (a key element in TPB) [55]. Moreover, transparency can build *trust*, which is crucial since consumers must trust the information to act on it. In one study, trust in recycling and in brand claims was a significant predictor of sustainable fashion behaviour [59]. By using digital media to transparently report sustainability metrics, brands educate consumers and address the belief component of VBN (shaping beliefs about consequences and responsibility) [60].

Digital activism, often driven by grassroots movements on social media, has emerged as a powerful force to push both consumers and companies toward sustainability [55]. Hashtag campaigns and online petitions create public pressure and spread awareness rapidly [60]. A striking case was the #PayUp campaign in 2020: launched online by advocacy group Remake, this campaign mobilized citizens to demand fashion retailers honour their commitments to pay garment factories during the COVID-19 crisis. Through Instagram posts and Twitter storms, activists exposed brands that cancelled orders and urged followers to hold them accountable [10]. The result was over $22 billion unlocked for garment suppliers worldwide after massive public shaming of offending brands. This outcome demonstrates how educated consumers (in this case, consumers informed about brands not paying workers) [10] can directly influence corporate behaviour. The #PayUp movement not only educated a broad swath of the public about an obscure supply chain issue, but also reinforced a normative stance that not paying workers is unacceptable, thereby influencing purchasing intentions (many vowed to boycott brands until they paid up) [59]. Similarly, Fashion Revolution's digital activism (discussed earlier) each year floods social media with calls for transparency, effectively turning thousands of everyday consumers into campaigners for change [61]. Such digital activism operates at the intersection of education and empowerment: it provides information (often revelatory facts) and then channels that knowledge into collective action. Theoretically, this reflects both TPB and VBN in action—attitudes are influenced by revealing injustices, subjective norms are shaped as people see their peers participating in activism [61], and personal norms are triggered as individuals feel a moral duty to join in. In sum, social media and digital platforms serve as indispensable educational tools in the sustainable fashion movement, capable of shifting consumer knowledge, attitudes, norms, and intentions at a viral pace.

Case Studies of Successful Digital Sustainability Education Initiatives

Numerous real-world initiatives illustrate the efficacy of digital and social media in promoting sustainable textile choices [10]. For instance, Patagonia's Worn Wear campaign leverages social media and online content to encourage customers to repair, reuse, and resell their clothing. Through engaging videos and blog stories of people who've extended the life of their Patagonia gear, the company educates consumers on practical ways to reduce fashion waste, reinforcing both an attitude that "long-lasting clothing is cool" and a personal norm that it's right to maximize a garment's life. Patagonia's messaging often aligns with its values-driven brand identity, creating a community of consumers who feel morally connected to the cause of sustainability. Another example is Levi's "Buy Better, Wear Longer" campaign launched in 2021, a multi-platform digital ad initiative featuring youth climate activists and influencers. This campaign explicitly "raises awareness and speaks to our shared responsibility on the environmental impacts of apparel production and consumption". In the campaign, Levi's used YouTube, Instagram, and other channels to deliver the message that wearing clothes for longer and buying second-hand can significantly reduce fashion's footprint. By using popular figures like Jaden Smith and Emma Chamberlain, the campaign blended sustainability education with cultural relevance, successfully sparking conversations among young consumers. Early reports indicated the campaign not only garnered millions of views but also improved consumers' perception of Levi's as a sustainability leader. Importantly, Levi's also coupled the educational message with actionable options (highlighting its SecondHand initiative and in-store tailoring services to help extend garment life), thereby influencing perceived behavioural control and actual behaviour [10].

Outside of brand-led cases, purely informational digital initiatives have flourished as well. Websites, blogs, and forums dedicated to sustainable fashion education (such as EcoStylist or Good On You) attract large followings by reviewing brands' ethics and providing tips on sustainable shopping. These platforms often use social media to spread their content, effectively crowd-sourcing consumer education. The success of these case studies underscores a key insight of both TPB and VBN theories: making information salient and socially supported online can increase consumers' *intention* to act sustainably [10], which is a precursor to behaviour change [61]. By normalizing dialogue about sustainable fashion and providing readily available knowledge, digital initiatives reduce the barriers (information uncertainty, feeling alone in one's concerns) that previously kept consumers from translating concern into action [10].

15.3.3 Brand-Led Initiatives and Corporate Responsibility in Consumer Education

The Importance of Corporate Transparency and Sustainability Reporting

Fashion brands and retailers carry significant responsibility in educating consumers, given their position as information gatekeepers about how products are made [53]. Corporate transparency is widely regarded as a cornerstone of sustainable consumption: when companies openly report on their social and environmental performance, they not only hold themselves accountable but also inform and empower consumers [58]. Sustainability reporting (e.g. annual sustainability reports, impact dashboards, or third-party certifications) serves as an educational tool by disclosing data on supply chain conditions, resource use, carbon emissions, and more. Such disclosures can align with TPB's construct of *perceived behavioural control*—consumers feel more capable of making ethical choices when they have clear information about which brands meet sustainability criteria [53]. Moreover, transparency addresses the *awareness of consequences* element of VBN theory: by highlighting both positive impacts (fair wages, organic materials) and negative impacts (pollution, waste) of fashion production, brands help consumers understand the consequences of their purchases [58]. This knowledge can in turn foster a sense of responsibility and personal norm to support better brands.

Progress in transparency is evident, although uneven. The Fashion Transparency Index and similar assessments show a growing number of major brands publishing details about their supply chains. For example, only 12.5% of big brands published their Tier-1 supplier lists in 2016, but by the early 2020s that figure rose to 47%, reflecting increased pressure on corporate disclosure [46]. Leading companies now often have dedicated webpages for tracing a garment's journey or for explaining sustainability initiatives [10]. Such efforts respond to clear consumer demand for information, as noted, about two-thirds of surveyed consumers want to know how their clothes are produced. By meeting this demand, companies can build consumer trust and differentiate themselves as ethical leaders. However, transparency is not just about posting data [53]; it must be communicated in a consumer-friendly manner to truly educate. Best practices include using storytelling alongside data, and employing visuals or digital tools to make complex information accessible. For instance, some brands use interactive maps to show factory locations and personal stories of workers, making the abstract idea of "supply chain transparency" more tangible and resonant for the average shopper. When done effectively, corporate transparency can shift consumer attitudes and intentions [10]. If a shopper reads a brand's report and learns that the brand has drastically reduced water usage and ensured worker safety, the shopper's attitude toward buying from that brand becomes more positive (per TPB's attitude component). Simultaneously, the brand may influence the consumer's values and personal norms by signalling that caring for people and planet is a priority—effectively inviting the consumer to share those priorities (as VBN would predict) [53]. Thus, transparency and reporting are not mere compliance or PR exercises; they are key educational interventions that can realign consumer behaviour with sustainability goals.

Educational Campaigns Led by Sustainable Fashion Brands

Beyond formal reporting, many fashion companies undertake dedicated educational campaigns to promote sustainable consumption patterns [53, 58]. Sustainable fashion brands—whose business models often depend on conscious consumer choices—have pioneered creative campaigns that double as marketing and public education. These campaigns typically highlight issues (to raise awareness) and propose actions or mindset shifts (to drive behaviour change), often aligning closely with TPB constructs (influencing attitudes/norms/control) and VBN constructs (appealing to values/norms) [53, 58]. We have already mentioned Patagonia's Worn Wear and Levi's Buy Better, Wear Longer as exemplars of brand-led education. Another noteworthy example is H&M's transparency initiative: H&M, a fast-fashion giant under scrutiny, introduced a feature in its online store where customers can click "Product Sustainability" details for each item, learning about the factory, material, and even the garment's supplier. While H&M has faced criticism for greenwashing, the move represents an effort to educate shoppers at the point of purchase about what sustainability means in concrete terms (e.g. a dress containing recycled polyester, made in a factory running on renewable energy). Similarly, luxury brand Gucci launched an "Equilibrium" platform that educates the public on its sustainability and diversity projects, using articles and social media content to engage consumers in the brand's values. These initiatives illustrate how brands can use their considerable influence and resources to steer consumer understanding. They also often involve calls to action: for instance, some brands run take-back or recycling programs accompanied by campaigns urging customers to return old clothes (thus educating about circular economy principles) [53, 58].

Effective brand-led campaigns tend to share certain best practices. First, they connect sustainability to the consumer's personal values and lifestyle [58]. Brands often frame sustainability in terms of quality, longevity, or well-being—for example, promoting the idea that sustainable garments are longer-lasting and thus give more value, or that they are healthier for one's skin due to non-toxic materials [53, 58]. This positive framing is deliberate. Evidence suggests that emphasizing the benefits of sustainable fashion (positive framing) is more effective in increasing purchase intentions than harping on the negatives of unsustainable fashion. By focusing on gains (e.g. "this organic cotton shirt saves water and feels softer") rather than guilt, brand campaigns can cultivate a more welcoming attitude among consumers. Such framing can also be linked to anticipated positive outcomes, a technique known as "prefactual" messaging—for instance, a message like "*If you choose this eco-friendly jacket, you will help protect the environment*" encourages consumers to envision a favourable future scenario, which research indicates can increase willingness to behave accordingly. Second, successful campaigns *use credible voices and storytelling* [53]. Brands often involve respected activists, designers, or celebrities (as Levi's did with youth activists, or as Adidas did partnering with Parley for the Oceans on ocean-plastic sneakers) to lend credibility and inspire consumers. Storytelling—sharing real stories of artisans, or environmental heroes, or even the journey of a product—can transform abstract issues into relatable narratives, reinforcing beliefs and personal norms. Third, these campaigns make sustainability *actionable* [58]. It is not enough to tell consumers what the problems are Audrezet et al. [58]; effective education shows consumers what they can do (buy better, care for their clothes, participate in recycling programs, etc.) [58]. By providing clear action steps or services (repair workshops, clothing take-back discounts, style guides for re-wearing clothes), brands improve consumers perceived behavioural control, reducing the barriers to acting on sustainable intentions [54].

Best Practices and Challenges in Brand-Led Consumer Education

As brands ramp up their educational initiatives, a number of best practices are emerging, alongside challenges that need careful navigation. From the industry stakeholder perspective, key recommendations for effective consumer education include:

- **Be Transparent and Truthful**: Any sustainability claims or educational content must be backed by data and evidence, and shared transparently across platforms [54]. Vague or unsubstantiated claims risk being dismissed as greenwashing. Studies show that when consumers detect greenwashing, their trust in the brand and intention to purchase decline sharply [62, 63]. Conversely, providing concrete information (e.g. quantifying water saved, materials sourced, or wages paid) can enhance credibility. Brands should openly communicate both their achievements and ongoing challenges, which fosters authenticity [62, 63]. This aligns with the VBN principle of acknowledging consequences and responsibility, demonstrating that the brand itself is taking responsibility, which can, in turn, encourage consumers to do the same.

- **Leverage Positive Framing and Values**: Educational messaging should frame sustainability in a positive light—highlighting benefits, innovations, and shared values [54]. As noted, positively framed messages (how sustainable fashion benefits society and the consumer) tend to be more persuasive than fear-based or negative ones. Brands are advised to tap into values that resonate with their audience: for example, messages around quality, creativity (e.g. styling vintage clothing as a creative expression), or community (joining a movement of changemakers) can make sustainable choices personally meaningful to consumers [61]. By aligning sustainability with core consumer values (a tactic supported by VBN theory's focus on activating altruistic or biospheric values), brands can cultivate an intrinsic motivation in their customers to support sustainability as part of their identity [54].
- **Engage and Empower Through Interactive Tools**: Rather than a didactic approach, brands should use interactive education that empowers consumers. This could involve digital tools (QR codes linking to supply chain stories, mobile apps that track a product's environmental impact) [62, 63], gamification (challenges or rewards for sustainable actions), or participatory campaigns (#30Wears challenge to wear an item 30 times, etc.). Such engagement not only makes learning about sustainability more interesting, but also strengthens consumers' perceived control and personal efficacy (TPB's perceived behavioural control) by actively involving them [54]. Empowered consumers who feel their actions matter are more likely to form intentions to behave sustainably [61]. For example, when a brand provides repair tutorials or in-store tailoring services along with a campaign, it tangibly enables consumers to prolong their clothing's life, bridging the intention-behaviour gap.
- **Collaborate with Influencers and Communities**: As discussed, influencers and online communities greatly amplify educational outreach. Brands should partner with credible sustainability influencers or facilitate community-building among their customers [63]. This might mean hosting live Q&As on social media about sustainability, featuring user-generated content (customers sharing how they style thrifted clothes or care for garments) [58], or supporting independent educators and NGOs. Such collaboration helps establish a supportive subjective norm—consumers see that not just the brand [58], but many respected voices, are championing sustainable practices. It also decentralizes the education process, creating a sense of a collective movement which can reinforce individuals' commitment (personal norm) to the cause.

Despite these best practices, significant challenges persist. One major challenge is overcoming consumer scepticism [62]. With the rise of green marketing, savvy consumers (especially younger ones) have become wary of "sustainability" rhetoric. A recent survey revealed that 88% of Gen Z consumers in the U.S. do not trust brands' claims about environmental and social governance (ESG). Additionally, 42% of Gen Z said they didn't even know what makes clothing sustainable, underscoring both a trust deficit and an information gap. These statistics highlight that brands must work hard to earn credibility—it's not enough to declare a product eco-friendly; detailed education is required to clarify *why* it's sustainable and to convince a sceptical audience that the effort is genuine [45]. Cases of greenwashing in the fashion industry (from mislabelled "conscious" collections to exaggerated eco-benefits) have eroded trust and can undo the positive effects of education. If consumers suspect a brand's educational campaign is merely a marketing ploy, it may backfire, leading to cynicism instead of engagement [45]. Therefore, a key challenge is ensuring consistency and integrity in messaging. Brands must align their actions with their words—for instance, a company cannot effectively educate about reducing overconsumption if its business model aggressively encourages fast turnover of styles. Resolving this tension may require innovative business models (like resale, rental, or repair services) so that the educational message is supported by corporate practice [45].

Another challenge lies in measuring the impact of consumer education efforts on actual behaviour [45]. While it is relatively easier to track engagement metrics (views, clicks, social media interactions) for campaigns, the true test is whether consumers change their purchasing and usage habits. The intention-behaviour gap remains a concern, many consumers express intentions to buy sustainably [63] but fail to follow through due to convenience, price, or habit [62]. Education can raise intention, but without accessible sustainable options and societal support, behaviour may not shift. This implies that consumer education must be part of a broader ecosystem of change, including making sustainable products available and affordable, and perhaps nudging behaviour through policies (like eco-labelling standards or incentives). Industry stakeholders are encouraged to collaborate with regulators and NGOs to reinforce educational messages with tangible infrastructure, for example, widely available recycling drop-offs, or standard definitions for sustainability claims to avoid consumer confusion [63].

15.4 Case Studies and Comparative Insights from Other Sectors

To enhance sustainable consumer behaviour in the textile industry, it is instructive to examine strategies employed in other sectors. This section explores lessons from the food industry regarding ethical consumption and labelling, the beauty industry's approach to sustainability messaging, and waste management's adoption of circular economy models. These insights can inform the development of effective educational interventions in the textile sector.

Table 15.1 summarises the key sustainability insights drawn from the food, beauty, and waste management sectors. By examining successful educational and marketing practices from these industries, this comparison highlights how ethical labelling, transparency, influencer marketing, and circular economy principles can be effectively leveraged to promote sustainable practices in the textile industry.

Table 15.1 Comparative summary of sustainability insights from food, beauty, and waste management industries

Sector	Key strategies	Insights for textiles
Food industry	Fair trade and organic labelling	Clear ethical labelling influences consumer decisions
Beauty industry	Transparency, influencer marketing	Influencers shape consumer attitudes through credibility
Waste management	Recycling, circular economy	Education promotes consumer recycling and waste reduction

In Table 15.1, each industry provides unique but transferable strategies for enhancing consumer awareness and influencing sustainable consumer choices. The food industry demonstrates the powerful role of ethical and environmental labelling, such as Fair Trade and organic certifications, which significantly impact consumer decisions through clear, credible information [64]. The beauty industry underscores the importance of transparency and influencer-driven educational campaigns, leveraging social proof and credibility to effectively shift consumer attitudes towards sustainability [14, 65]. Waste management offers valuable insights into the implementation of circular economy models, using comprehensive education to facilitate consumer participation in recycling and waste reduction initiatives [66]. Together, these insights offer practical pathways for improving consumer engagement and behavioural outcomes within the textile sector.

15.4.1 Lessons from the Food Industry: Ethical Consumption and Labelling

The Impact of Fair Trade and Organic Labelling on Consumer Decision-Making

Ethical labelling, such as Fair Trade and organic certifications, has significantly influenced consumer purchasing decisions in the food industry. Research has shown that consumers perceive Fair Trade-labelled products more favourably and are often willing to pay a premium for them [64]. The psychological mechanism behind this behaviour is tied to the moral satisfaction associated with purchasing ethically certified goods, which is reinforced by brain imaging studies demonstrating increased neural activity in reward-processing areas when consumers view Fair Trade labels [67]. Furthermore, a field experiment in the UK found that coffee labelled as Fair Trade saw a 9% increase in sales compared to non-labelled alternatives, highlighting the tangible economic impact of ethical labelling [64].

Comparisons with Textile Sustainability Certifications (e.g., GOTS, OEKO-TEX)

The textile industry employs sustainability certifications such as the Global Organic Textile Standard (GOTS) and OEKO-TEX to indicate environmental and social responsibility. However, consumer awareness of these textile certifications is significantly lower compared to food labels like Fair Trade and organic, resulting in less impact on purchasing behaviour [68]. Studies indicate that textile consumers struggle to understand what these labels mean, with only 37% of respondents in a European survey recognizing GOTS as an eco-label [14]. This lack of awareness limits the effectiveness of textile certifications in influencing consumer choices. Increasing transparency and consumer education around these labels could help bridge this gap and enhance their impact, similar to how Fair-Trade labelling has shifted consumer behaviour in the food industry.

15.4.2 The Beauty Industry and Sustainability Messaging

Consumer Education Through Ingredient Transparency and Ethical Branding

The beauty industry has increasingly prioritized ingredient transparency and ethical branding as part of sustainability efforts. Research shows that consumers respond positively to brands that clearly communicate product ingredients, sourcing practices, and sustainability efforts [65]. Companies like The Body Shop and Lush have built their brand identities around ethical sourcing and clear labelling, which fosters consumer trust and encourages sustainable purchasing habits [66]. The rise of clean beauty—which emphasizes non-toxic, sustainable ingredients—has gained momentum, demonstrating how education on sustainability can be embedded into branding to shift consumer behaviour.

The Role of Influencer Marketing in Sustainable Beauty Choices

Influencer marketing has become a key driver of consumer engagement in the beauty industry. Research indicates that influencers who promote sustainability encourage peer-driven behaviour change, leveraging the power of social proof to influence purchasing decisions [14]. Studies suggest that 73% of Gen Z consumers trust influencers over traditional advertisements, highlighting their importance in sustainability education [67]. By working with influencers to educate consumers on ethical beauty practices and eco-friendly products, brands can create a ripple effect that extends into other industries, including textiles. The success of sustainability-focused influencers in the beauty sector suggests that similar strategies could be employed in the textile industry to enhance consumer awareness and engagement.

15.4.3 Waste Management and Circular Economy Models

Consumer Education on Recycling and Waste Reduction

Waste management has long relied on consumer education to drive participation in recycling and waste reduction initiatives. Studies show that consistent messaging, accessible information, and policy incentives significantly increase consumer engagement in recycling programs [66]. Governments and organizations have implemented large-scale educational campaigns to ensure compliance with waste separation and recycling guidelines, leading to increased participation rates and improved waste management outcomes [65]. The EU's Waste Framework Directive serves as an example of how regulatory interventions, combined with consumer education, can drive sustainable behaviour in waste management [14].

Lessons for Textile Industry Adaptation

The circular economy model, which prioritizes reduction, reuse, and recycling, has been successfully applied in waste management to create sustainable systems. Implementing circular strategies in the textile industry—such as designing for durability, facilitating garment recycling, and promoting second-hand markets—could significantly reduce the environmental footprint of clothing production [66]. One study found that textile waste could be reduced by 50% if consumers were educated on fabric longevity and proper disposal methods [65]. Brands like Patagonia and Levi's have successfully integrated circular economy principles into their business models by promoting repair services, take-back programs, and garment recycling, showcasing how these strategies can be adapted across industries [64].

15.5 Policy and Strategic Recommendations for Strengthening Consumer Education

Strengthening consumer education in the textile industry requires a holistic strategy that integrates government regulations, industry-led initiatives, and cross-sector collaborations. Policy interventions such as eco-labelling, advertising regulations, and supply chain transparency ensure that consumers receive accurate and accessible sustainability information. Simultaneously, brands and NGOs must work together to engage consumers through interactive learning experiences and digital education tools. Future research should focus on the effectiveness of emerging digital technologies, long-term consumer behaviour change, and localized sustainability education strategies. By advancing these areas, the textile industry can cultivate a more informed, responsible, and environmentally conscious consumer base.

15.5.1 The Role of Government and Public Policy in Consumer Education

Regulatory Measures for Improving Sustainable Textile Education

Governments play a pivotal role in shaping sustainable consumption through the implementation of regulatory measures. Policies that mandate transparency in the textile supply chain significantly enhance consumer awareness. For example, the EU Strategy for Sustainable and Circular Textiles mandates that textile products in the European market must be durable, recyclable, and free from hazardous substances, ensuring consumers have access to standardized sustainability information [69]. Additionally, the introduction of digital product passports (DPPs) under the Ecodesign for Sustainable Products Regulation requires brands to disclose detailed environmental impact data, fostering informed and engaged consumer behaviour [70].

National initiatives have also played a role in promoting sustainability awareness. The UK Sustainable Clothing Action Plan (SCAP) established industry-wide targets to reduce carbon, water, and waste footprints, thereby shaping consumer knowledge on textile sustainability [71]. In the United States, California's Responsible Textile Recovery Act enforces producer responsibility in textile recycling, educating consumers on sustainable disposal practices [72]. These policies highlight how government interventions can improve consumer education and sustainability engagement.

The Potential Impact of Policy Interventions (e.g., Sustainability Labelling, Advertising Regulations)

Policy interventions such as eco-labelling and advertising regulations play a critical role in ensuring consumer trust in sustainability claims. The Global Organic Textile Standard (GOTS) and OEKO-TEX certifications provide verified sustainability criteria, yet consumer understanding of these labels remains limited compared to Fair Trade or organic food labelling [68]. Government-mandated educational campaigns could increase awareness and comprehension of textile certifications, ensuring they have the same impact on purchasing decisions as their food industry counterparts [14].

Moreover, advertising regulations targeting greenwashing help prevent misleading sustainability claims, thereby reinforcing consumer confidence. Studies show that up to 40% of sustainability claims made by fashion brands are misleading, emphasizing the need for stricter guidelines [4]. Governments in the EU and UK have strengthened consumer protection laws to penalize companies engaging in greenwashing, ensuring more transparent and verifiable sustainability claims [37].

15.5.2 Industry and NGO-Led Strategies

How Brands and Non-profits Can Collaborate to Educate Consumers

The private sector and NGOs play a pivotal role in consumer education initiatives. NGOs such as Fashion Takes Action work closely with brands to provide educational resources on sustainability, including workshops, school programs, and digital learning platforms [73]. Similarly, the German Partnership for Sustainable

Textiles collaborates with brands, government bodies, and civil society organizations to improve transparency and ethical sourcing across the fashion supply chain [74].

Brands are increasingly investing in interactive consumer education platforms. For instance, Patagonia's "Worn Wear" program promotes repair, resale, and recycling, teaching consumers about product longevity and sustainability [65]. Levi's Buy Better, Wear Longer campaign similarly raises awareness about the environmental impact of fast fashion, using digital storytelling and influencer partnerships to amplify its message [14]. Such initiatives demonstrate how brands can bridge the information gap by integrating education with brand identity to cultivate sustainable consumer habits.

The Need for Cross-Sector Initiatives to Promote Sustainable Textiles

Cross-sector collaborations between governments, brands, NGOs, and academia have proven effective in promoting sustainable textile education. The UN Alliance for Sustainable Fashion unites various stakeholders to advance sustainability standards and consumer engagement [75]. Additionally, the Textile Exchange's Preferred Fiber & Materials Report provides industry-wide benchmarks for sustainable sourcing, helping consumers navigate complex textile choices [76].

Such partnerships ensure that sustainability education reaches diverse audiences, from school curricula to corporate training programs. The integration of sustainability into fashion design education further equips future industry leaders with the knowledge to drive systemic change [66].

15.5.3 Future Directions and Research Opportunities

Emerging Trends in Consumer Education for Sustainability

The rapid adoption of digital technologies has introduced new methods for consumer education. Emerging tools such as blockchain-enabled transparency platforms allow consumers to trace the lifecycle of textile products, increasing engagement with sustainability efforts [70]. Similarly, augmented reality (AR) and artificial intelligence (AI)-driven sustainability tools provide consumers with real-time product impact assessments, helping them make informed decisions at the point of purchase [65].

In addition, the use of gamification and behavioural nudging strategies has gained traction in sustainability education. Research shows that interactive and incentivized learning models significantly enhance consumer retention of sustainability knowledge, leading to more eco-conscious behaviours [14]. Future research should explore how digital interventions influence long-term consumer habits and whether AI-driven recommendations improve sustainable fashion engagement.

Gaps in Existing Literature and Suggestions for Future Research

While research on sustainable textile consumption has expanded in recent years, several key gaps remain. Firstly, longitudinal studies assessing the impact of consumer education programs on actual purchasing behaviour are limited [66]. Understanding how different educational formats—formal curricula, influencer marketing, or brand-led campaigns—affect consumer decision-making over time is essential for designing effective interventions.

Secondly, there is limited research on the intersection of cultural attitudes and sustainability education. Consumer responses to sustainability messaging vary across geographical and socio-economic contexts, and future studies should explore how tailored educational strategies can be localized to different markets [65].

Finally, research on policy effectiveness in sustainable textiles is still evolving. While policy interventions such as eco-labelling, greenwashing regulations, and circular economy initiatives have been widely studied in isolation, comparative analyses of their combined impact on consumer education and industry practices remain underexplored [69].

Further implications arising from increased consumer awareness and education include heightened demand for transparency from textile brands, greater consumer support for sustainable brands, and the strengthening of policies and regulations. Consumers increasingly expect clear, verifiable information about textile products' sustainability claims, pushing brands toward greater openness about their sourcing and production processes. This demand for transparency, coupled with consumers' willingness to support brands demonstrating genuine commitment to sustainability, incentivises industry-wide changes toward ethical practices. Additionally, informed consumer expectations catalyse stronger governmental responses, encouraging the development of robust regulations and sustainability standards, such as mandatory labelling requirements and anti-greenwashing legislation. Collectively, these outcomes further embed sustainability within the textile industry, enhancing both consumer trust and environmental outcomes.

Building on the strategic recommendations outlined in this section, Fig. 15.3 presents an integrated framework for strengthening consumer education on sustainable textiles. The model conceptualises three interconnected spheres of influence—*policy, industry*, and *education*—each contributing distinct but complementary strategies that collectively shape consumer awareness and behaviour. At the policy level, regulatory oversight, public campaigns, and funding create an enabling environment for sustainability communication. Industry actors support this through transparent practices, credible certifications, and engagement with influential communicators. Educational interventions, both formal and informal, underpin these efforts by equipping consumers with the skills and knowledge to critically evaluate sustainability claims and make informed choices. These coordinated strategies converge on a central outcome: the development of informed and empowered consumers who are more likely to engage in sustainable textile practices, such as preferring certified

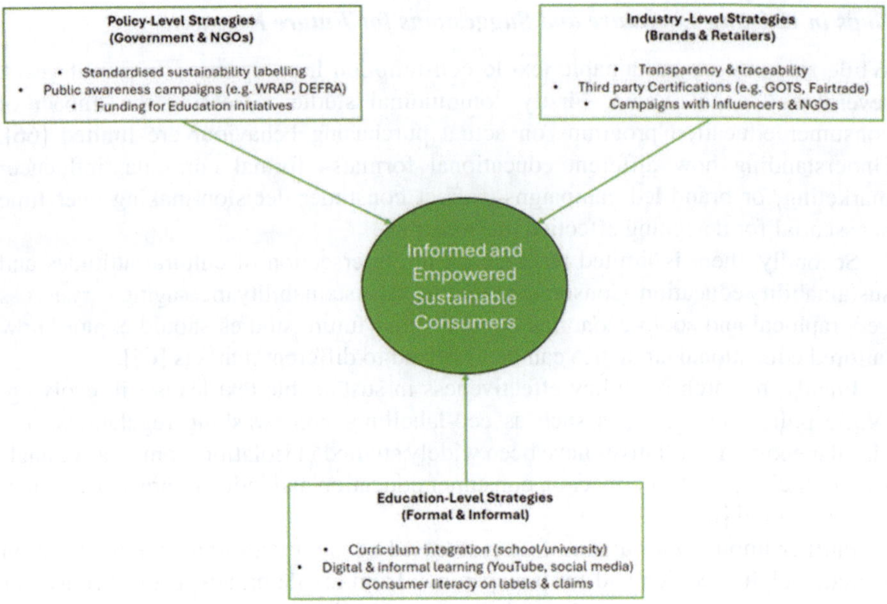

Fig. 15.3 Integrated framework for strengthening consumer education on sustainable textiles

products, supporting ethical brands, and participating in reuse and recycling initiatives. The framework highlights the importance of multi-stakeholder collaboration in achieving meaningful and lasting behavioural change in the UK textile market.

15.6 Conclusion

The findings presented in this chapter provide a comprehensive analysis of the role of consumer perception and education in promoting sustainable textile choices. By examining theoretical frameworks, consumer education strategies, and policy recommendations, this study has addressed the two central research questions: (1) *How do consumer perceptions and misconceptions about sustainability influence purchasing decisions in the UK textile industry?* and (2) *What are the most effective educational strategies for enhancing consumer engagement with sustainable textile choices?* The conclusions drawn from this research highlight the barriers, opportunities, and strategic interventions necessary to foster a more informed and responsible consumer base in the UK textile market.

15.6.1 Key Findings and Takeaways

Summary of Major Insights Discussed in the Chapter

Consumer perception significantly influences purchasing decisions, yet misconceptions and misinformation remain key barriers to adopting sustainable textile choices. Many consumers associate sustainable fashion with high costs, limited availability, and lower aesthetic appeal, deterring them from making ethical purchasing decisions [67]. Additionally, greenwashing practices by fashion brands erode consumer trust, making it difficult for individuals to distinguish genuine sustainability efforts from misleading marketing strategies [4]. The value-action gap, where consumers express concern for environmental issues but fail to translate this concern into action, continues to persist due to economic constraints, accessibility issues, and conflicting messaging [14].

Comparative insights from the food and beauty industries demonstrate that clear and standardised labelling systems, such as Fair Trade and organic food certifications, significantly influence consumer behaviour. However, the textile industry has struggled to replicate the same level of consumer trust and recognition with certifications like GOTS and OEKO-TEX due to a lack of awareness and clear messaging [64]. Furthermore, the beauty industry's success with ingredient transparency and influencer-driven marketing campaigns offers valuable lessons for sustainable textiles, particularly in enhancing consumer engagement and trust in ethical branding [65].

Consumer education emerges as a critical tool for fostering sustainable textile choices, yet its effectiveness depends on engagement depth, accessibility, and credibility. Formal education initiatives, such as integrating sustainability into school curricula and higher education programmes, provide long-term behavioural shifts, while informal educational tools, such as digital learning platforms, sustainability-driven brand campaigns, and influencer marketing, can deliver immediate impact on consumer attitudes and purchasing decisions [66]. Interactive models, such as gamification and AI-driven sustainability tools, have proven effective in engaging consumers and improving knowledge retention, suggesting promising pathways for digital learning strategies [14].

Final Thoughts on the Effectiveness of Consumer Education in Promoting Sustainable Textile Choices

Consumer education alone is insufficient to drive widespread behavioural change. While educational interventions play a vital role in shaping awareness, they must be complemented by structural and policy-level changes to ensure that sustainable choices are accessible, affordable, and credible. Governments, brands, and NGOs must collaborate to establish clear sustainability labelling, regulate misleading marketing, and integrate sustainability education into mainstream discourse [69]. Furthermore, digital innovations such as blockchain-based transparency tools and digital product passports present new opportunities for enhancing consumer engagement and trust in sustainable fashion [70].

Ultimately, educating consumers is a necessary but not sufficient condition for sustainable textile adoption. Real change requires cross-sector collaboration, regulatory oversight, and industry accountability to ensure that education translates into actionable and enduring behavioural shifts in the textile market.

15.6.2 Call to Action for Stakeholders

To achieve meaningful progress in sustainable textile consumption, multiple stakeholders must work collectively to address the key challenges identified in this study. Policymakers must implement stricter greenwashing regulations, enforce sustainability labelling requirements, and mandate corporate transparency in textile supply chains. The European Commission's strategy for sustainable textiles highlights the importance of such regulatory measures in promoting consumer awareness and informed decision-making. Additionally, integrating sustainability-focused education into national curricula can foster early-stage awareness and ethical consumption habits, ensuring that future generations develop a deeper understanding of the environmental and social impacts of textile production.

Educators also have a critical role to play in embedding sustainability education in fashion and business curricula, equipping future industry leaders with ethical decision-making skills and sustainable design principles. Universities and vocational institutions should incorporate digital literacy in sustainability-related tools, particularly in the use of AI-driven transparency platforms and blockchain verification technologies, which can enhance consumer trust in sustainability claims. Industry stakeholders, particularly fashion brands, must take greater responsibility in educating consumers about the environmental and ethical implications of textile production. Consumer engagement strategies should extend beyond traditional marketing campaigns to include interactive educational initiatives, such as transparency platforms, workshops, and sustainability-focused brand narratives that emphasise product longevity and recyclability. Some brands, such as Patagonia and Levi's, have demonstrated how transparency and repair-focused initiatives can shift consumer behaviour, reinforcing the need for companies to integrate sustainability into their core business models rather than treating it as an ancillary effort.

Collaboration between brands, non-governmental organisations (NGOs), and sustainability advocates is also crucial in amplifying educational efforts and enhancing credibility. NGOs such as Fashion Takes Action have pioneered sustainability education through partnerships with industry leaders, showcasing how collective action can bridge knowledge gaps and drive consumer engagement. At the same time, individual consumers must become more proactive in seeking reliable sources of sustainability information, supporting ethical brands, and making responsible consumption choices. Educated consumers are better positioned to challenge misleading sustainability claims, demand greater corporate responsibility, and advocate for stronger policy interventions that support a circular and ethical textile economy.

The urgency of strengthening consumer education in sustainable textiles cannot be overstated. The fashion industry is one of the largest contributors to environmental degradation, responsible for approximately 10% of global carbon emissions and 20% of global wastewater production. Without immediate interventions in education, policy, and industry practices, the sector's environmental footprint will continue to expand, undermining broader sustainability efforts. Addressing these challenges requires a multi-pronged approach, where educational interventions align with economic incentives, policy enforcement, and systemic industry change. The transition towards a circular textile economy depends on ensuring that consumers are well-informed, engaged, and empowered to make sustainable choices. By implementing the recommendations outlined in this chapter, stakeholders can bridge the gap between awareness and action, fostering a new era of responsible textile consumption. Only through collaborative and strategic action can sustainability become an industry standard rather than an exception. As we move forward, education, transparency, and policy accountability will serve as the cornerstones of a more sustainable and ethical textile industry.

References

1. Elkington, J. 1994. Towards the sustainable corporation: Win-Win-Win Business strategies for sustainable development. *California Management Review* 36: 90–100.
2. Księżak, P., and B. Fischbach. 2018. Triple bottom line: The pillars of CSR. *Journal of Corporate Responsibility and Leadership* 4 (3): 95.
3. Mcneill, L., and R. Moore. 2015. Sustainable fashion consumption and the fast fashion conundrum: Fashionable consumers and attitudes to sustainability in clothing choice. *International Journal of Consumer Studies* 39 (3): 212–222.
4. de Freitas Netto, S. V., M. F. F. Sobral, A. R. B. Ribeiro, and G. R. L. da Soares. 2020. Concepts and forms of greenwashing: a systematic review. *Environmental Science Europe* 32 (1).
5. Torelli, R., F. Balluchi, and A. Lazzini. 2020. Greenwashing and environmental communication: Effects on stakeholders' perceptions. *Business Strategy and the Environment* 29 (2): 407–421.
6. Karaosman, H., P. Perry, A. Brun, and G. Morales-Alonso. 2018. Behind the runway: Extending sustainability in luxury fashion supply chains. *Journal of Business Research* 117: 652–663.
7. Chouinard, Y. 2022. Earth is now our only shareholder. *Patagonia.*
8. Finamore, E. 2023. Step inside Timberland's award-winning regenerative projects. *Drapers.*
9. Niinimaki, K., G. Peters, H. Dahlbo, P. Perry, T. Rissanen, and A. Gwilt. 2020. The environmental price of fast fashion. *Nature Reviews Earth and Environment* 1: 189–200.
10. Zhang, B., Y. Zhang, and P. Zhou. 2021. Consumer attitude towards sustainability of fast fashion products in the UK. *Sustainability* 13 (4): 1–23.
11. Leal Filho, W., et al. 2022. An overview of the contribution of the textiles sector to climate change. *Frontiers in Environmental Science* 10: 1–5.
12. Padmanabhan, V. M., B. B. Gewehr, and S. Labowitz. 2015. The hidden price of low cost: Subcontracting in Bangladesh's garment industry. *SSRN.*
13. ILO. 2023. The Rana Plaza disaster ten years on: What has changed? *International Labour Organisation.*
14. White, K., R. Habib, and D. J. Hardisty. 2019. How to SHIFT consumer behaviors to be more sustainable: A literature review and guiding framework. *Journal of Marketing* 83 (3): 22–49.
15. U. N. E. Programme. 2022. Sustainable fashion: Communication strategy 2021–2024. *United Nations Environment Programme.*

16. Amed, I., et al. 2022. The state of fashion 2022. *McKinsey Co* 1–144.
17. Pereira, L., R. Carvalho, Á. Dias, R. Costa, and N. António. 2021. How does sustainability affect consumer choices in the fashion industry? *Resources* 10 (4): 1–30.
18. Harris, F., H. Roby, and S. Dibb. 2016. Sustainable clothing: Challenges, barriers and interventions for encouraging more sustainable consumer behaviour. *International Journal of Consumer Studies* 40 (3): 309–318.
19. Bhowmick, D. 2022. Impact of brand design on consumer perception and decision making. *International Journal of Business Forecasting and Marketing Intelligence* 7 (4): 375.
20. Durmaz, Y., and I. Diyarbakirlioglu. 2011. A theoritical approach to the role of perception on the consumer buying decision process. *Asian Journal of Business and Management Science* 1 (4): 2047–2528.
21. Madichie, N. O. 2015. *Consumer perception*, 1–22.
22. Stern, P. C. 2000. Toward a coherent theory of environmentally significant behavior. *Journal of Social Issues* 56 (3): 407–424.
23. Solomon, M. R. 2018. *Consumer behavior: Buying, having, and being*. Pearson.
24. Kapitan, S., and D. H. Silvera. 2016. From digital media influencers to celebrity endorsers: Attributions drive endorser effectiveness. *Marketing Letters* 27 (3): 553–567.
25. Minton, A. P., and R. L. Rose. 1997. The effects of environmental concern on environmentally friendly consumer behavior: An exploratory study. *Journal of Business Research* 40 (1): 37–48.
26. Krishna, A. 2012. An integrative review of sensory marketing: Engaging the senses to affect perception, judgment and behavior. *Journal of Consumer Psychology* 22 (3): 332–351.
27. Lalaounis, S. T. 2020. Brand communications and the attention economy. *Strategic Brand Management and Development Creating and Marketing* 100–134.
28. Tafesse, W., and A. Wien. 2018. Using message strategy to drive consumer behavioral engagement on social media. *Journal of Consumer Marketing* 35 (3): 241–253.
29. Hüttel, A., F. Ziesemer, M. Peyer, and I. Balderjahn. 2018. To purchase or not? Why consumers make economically (non-)sustainable consumption choices. *Journal of Cleaner Production* 174: 827–836.
30. Martin-Woodhead, A. 2022. Limited, considered and sustainable consumption: The (non)consumption practices of UK minimalists. *Journal of Consumer Culture* 22 (4): 1012–1031.
31. Azjen, I. 1991. The thoery of planned behaviour. *Organizational Behavior and Human Decision Processes* 50 (2): 179–211.
32. Festinger, L. 1957. *A theory of cognitive dissonance. A theory of cognitive dissonance*, 291–299. Stanford University Press.
33. Sheoran, M., and D. Kumar. 2022. Conceptualisation of sustainable consumer behaviour: Converging the theory of planned behaviour and consumption cycle. *Qualitative Research in Organizations and Management: An International Journal* 17 (1): 103–135.
34. Van Tonder, E., S. Fullerton, L. T. De Beer, and S. G. Saunders. 2022. Social and personal factors influencing green customer citizenship behaviours: The role of subjective norm, internal values and attitudes. *Journal of Retailing and Consumer Services* 71: 103190.
35. Hinterhuber, A., and O. Khan. 2025. What drives sustainable procurement? Insights from the theory of planned behavior. *International Journal of Operations & Production Management* 45 (13): 28–52.
36. Mason, M. C., R. Pauluzzo, and R. M. Umar. 2022. Recycling habits and environmental responses to fast-fashion consumption: Enhancing the theory of planned behaviour to predict Generation Y consumers' purchase decisions. *Waste Management* 139: 146–157.
37. Zhang, L., J. Ruiz-Menjivar, B. Luo, Z. Liang, and M. E. Swisher. 2020. Predicting climate change mitigation and adaptation behaviors in agricultural production: A comparison of the theory of planned behavior and the Value-Belief-Norm Theory. *Journal of Environmental Psychology* 68: 101408.
38. Stern, P. C., T. Dietz, T. Abel, G. A. Guagnano, and L. Kalof. 1999. A value-belief-norm theory of support for social movements: The case of environmentalism. *Human Ecology Review* 6 (2): 81–97.

39. McCartney, S. 2017. Circularity.
40. Herrmann, D. S. 2017. *A new textiles economy: Redesigning fashion's future*, 1–150. Ellen MacArthur Foundation.
41. Gleim, M. R., J. S. Smith, D. Andrews, and J. J. Cronin. 2013. Against the green: A multi-method examination of the barriers to green consumption. *Journal of Retailing* 89 (1): 44–61.
42. Joanes, T. 2019. Personal norms in a globalized world: Norm-activation processes and reduced clothing consumption. *Journal of Cleaner Production* 212: 941–949.
43. Abo-Khalil, A. G. 2024. Integrating sustainability into higher education challenges and opportunities for universities worldwide. *Heliyon* 10 (9): e29946.
44. Leal Filho, W., et al. 2018. The role of transformation in learning and education for sustainability. *Journal of Cleaner Production* 199: 286–295.
45. Bennetta, K., and J. Oeppen Hill. 2022. Educating for change?: An investigation into consumers' perception of sustainability and the educational drivers needed to support sustainable consumption. *International Journal of Fashion Design Technology and Education* 15(3), 418–429.
46. McKinsey & Company and Business of Fashion. 2023. The state of fashion industry. *Premium* 108.
47. Petty, R. E., and J. T. Cacioppo. 1986. *Communication and persuasion: Central and peripheral routes to attitude change*. Springer.
48. Chaiken, S. 1980. Heuristic versus systematic information processing and the use of source versus message cues in persuasion. *Journal of Personality and Social Psychology* 39 (5): 752–766.
49. Jacoby, J. 1984. Perspectives on information overload. *The Journal of Consumer Research* 10 (4): 432–435.
50. De Jong, M. D. T., K. M. Harkink, and S. Barth. 2018. Making green stuff? Effects of corporate greenwashing on consumers. *Journal of Business and Technical Communication* 32 (1): 77–112.
51. Lyon, T. P., and A. W. Montgomery. 2015. The means and end of greenwash. *Organization & Environment* 28 (2): 223–249.
52. Kreczmańska-Gigol, K., and T. Gigol. 2022. The impact of consumers' green skepticism on the purchase of energy-efficient and environmentally friendly products. *Energies* 15 (6): 1–12.
53. Carfora, V., G. Buscicchio, and P. Catellani. 2021. Integrating personal and pro-environmental motives to explain Italian women's purchase of sustainable clothing. *Sustainability* 13 (19).
54. Hur, E., and T. Cassidy. 2019. Perceptions and attitudes towards sustainable fashion design: Challenges and opportunities for implementing sustainability in fashion. *International Journal of Fashion Design Technology and Education* 12 (2): 208–217.
55. Vladimirova, K., et al. 2024. Exploring the influence of social media on sustainable fashion consumption: A systematic literature review and future research agenda. *Journal of Global Fashion Marketing* 15 (2): 181–202.
56. Annan-Diab, F., and C. Molinari. 2017. Interdisciplinarity: Practical approach to advancing education for sustainability and for the sustainable development goals. *International Journal of Management Educaction* 15 (2): 73–83.
57. Hong, Y., A. Al Mamun, Q. Yang, and M. Masukujjaman. 2024. Predicting sustainable fashion consumption intentions and practices. *Scientific Reports* 14 (1): 1–20.
58. Audrezet, A., G. de Kerviler, and J. Guidry Moulard. 2020. Authenticity under threat: When social media influencers need to go beyond self-presentation. *Journal of Business Research* 117: 557–569.
59. Senanu, B., T. Anning-Dorson, and N. N. Tackie. 2023. Social media insights for non-luxury fashion SMEs in emerging markets: Evidence from young consumers. *Journal of Fashion Marketing Management: An International Journal* 27 (6): 965–987.
60. Breves, P. L., N. Liebers, M. Abt, and A. Kunze. 2019. The perceived fit between Instagram influencers and the endorsed brand: How influencer–brand fit affects source credibility and persuasive effectiveness. *Journal of Advertising Research* 59 (4): 440–454.

61. Blazquez, M., C. E. Henninger, B. Alexander, and C. Franquesa. 2020. Consumers' knowledge and intentions towards sustainability: A Spanish fashion perspective. *Fashion Practice* 12 (1): 34–54.
62. Lundblad, L., and I. A. Davies. 2015. The values and motivations behind sustainable fashion consumption. *Journal of Consumer Behaviour An International Research Review* 15 (2): 149–162.
63. Martinez, J. 2019."Fashion united: Top 10 challenges facing fashion in 2020.
64. Hainmueller, J., M. J. Hiscox, and S. Sequeira. 2015. Consumer demand for fair trade: Evidence from a multistore field experiment. *Review of Economics and Statistics* 97 (2): 242–256.
65. Geissdoerfer, M., P. Savaget, N. M. P. Bocken, and E. J. Hultink. 2017. The circular economy—a new sustainability paradigm? *Journal of Cleaner Production* 143: 757–768.
66. Kirchherr, J., D. Reike, and M. Hekkert. 2017. Conceptualizing the circular economy: An analysis of 114 definitions. *Resources, Conservation and Recycling* 127: 221–232.
67. Vermeir, I., and W. Verbeke. 2006. Sustainable food consumption: Exploring the consumer 'attitude—behavioral intention' gap. *Journal of Agricultural and Environmental Ethics* 19 (2): 169–194.
68. Grankvist, G., and A. Biel. 2007. The impact of environmental information on professional purchasers' choice of products. *Business Strategy and the Environment* 16 (6): 421–429.
69. E. Commission. 2022. EU strategy for sustainable and circular textiles. *European Commission.*
70. Roberts-Islam, B. 2024. The fashion exec's guide to digital product passports.
71. Defra. 2023. Annual report and accounts 2022–23 HC1872.
72. 2024. Clothes piling up in your closet? A landmark California bill would mandate brands recycle them. *The Guardian.*
73. Colombi, C., and E. D'Itria. 2023. Fashion digital transformation: innovating business models toward circular economy and sustainability. *Sustainability* 15 (6).
74. Textilbuendnis. 2021. Partnership for sustainable textiles.
75. O. P. Network. 2018. Partnership for sustainable textiles. *One Planet Network.*
76. Jacometti, V. 2019. Circular economy and waste in the fashion industry. *Laws* 8 (4).

Dr. Sharon Nunoo is a Lecturer in Fashion Marketing at Manchester Metropolitan University teaching luxury fashion marketing and consumer behaviour. She obtained her Ph.D. at the University of Manchester in Textile Design, Fashion and Management exploring the construct of corporate strategies in the development and promotion of a textile producing community. Her research interests are in the field of fashion marketing exploring the areas of corporate heritage identity, culture, sustainability, and place branding, focusing on identity management, identity creation process, social sustainability, and second-hand fashion trade in Ghana. Link to Faculty Webpage: https://fashioninstitute.mmu.ac.uk/staff/profile/?id=5893.

Dr. Morolake Dairo is a Morolake is a lecturer at the Manchester Fashion Institute (Manchester Metropolitan University), where she teaches fashion business and marketing. She holds a Ph.D. in media and communications from the School of Media and Communication, Pan-Atlantic University, Lagos, Nigeria. Morolake's research investigates fashion cultures in emerging economies in Africa and how they interact with other elements such as identity, history, economic growth, place branding and popular culture. Prior to joining MFI, she worked as a marketing professional for 8+ years across different sectors including FMCG, Oil and gas and FINTECH. Link to Faculty Webpage: https://fashioninstitute.mmu.ac.uk/staff/profile/?id=5990.

Dr. Tahera Kalsoom is a Lecturer in Fashion Business at the Manchester Metropolitan University with a BBA in Hospitality Management from Stenden University, Qatar (2010), an M.Sc. in Financial Management from Middlesex University, London (2015), and a Ph.D. in digitalisation, IoT, and supply chains from the University of the West of Scotland (2021). She has held academic positions at the University of the West of Scotland and St. Patrick's College, London. Her research focuses on the IoT, Industry 4.0/5.0, and the digitalisation of manufacturing and supply chains. Tahera has contributed to notable EU projects like ATHIKA and Safe-RH, and her work explores the socio-technical impacts of emerging technologies such as Blockchain, AI, and smart systems on manufacturing performance. Link to Faculty Webpage: Dr. Tahera Kalsoom—Manchester Fashion Institute.

MIX
Papier aus verantwortungsvollen Quellen
Paper from responsible sources
FSC® C105338
FSC
www.fsc.org

If you have any concerns about our products,
you can contact us on
ProductSafety@springernature.com

In case Publisher is established outside the EU,
the EU authorized representative is:
Springer Nature Customer Service Center GmbH
Europaplatz 3, 69115 Heidelberg, Germany

Printed by Libri Plureos GmbH
in Hamburg, Germany